Geophysical Monograph Series

Including
Maurice Ewing Volumes
Mineral Physics

Geophysical Monograph Series

A. F. Spilhaus, Jr., Managing Editor

1 **Antarctica in the International Geophysical Year,** A. P. Crary, L. M. Gould, E. O. Hulburt, Hugh Odishaw, and Waldo E. Smith (Eds.)
2 **Geophysics and the IGY,** Hugh Odishaw and Stanley Ruttenberg (Eds.)
3 **Atmospheric Chemistry of Chlorine and Sulfur Compounds,** James P. Lodge, Jr. (Ed.)
4 **Contemporary Geodesy,** Charles A. Whitten and Kenneth H. Drummond (Eds.)
5 **Physics of Precipitation,** Helmut Weickmann (Ed.)
6 **The Crust of the Pacific Basin,** Gordon A. Macdonald and Hisashi Kuno (Eds.)
7 **Antarctic Research: The Matthew Fontaine Maury Memorial Symposium,** H. Wexler, M. J. Rubin, and J. E. Caskey, Jr. (Eds.)
8 **Terrestrial Heat Flow,** William H. K. Lee (Ed.)
9 **Gravity Anomalies: Unsurveyed Areas,** Hyman Orlin (Ed.)
10 **The Earth Beneath the Continents: A Volume of Geophysical Studies in Honor of Merle A. Tuve,** John S. Steinhart and T. Jefferson Smith (Eds.)
11 **Isotope Techniques in the Hydrologic Cycle,** Glenn E. Stout (Ed.)
12 **The Crust and Upper Mantle of the Pacific Area,** Leon Knopoff, Charles L. Drake, and Pembroke J. Hart (Eds.)
13 **The Earth's Crust and Upper Mantle,** Pembroke J. Hart (Ed.)
14 **The Structure and Physical Properties of the Earth's Crust,** John G. Heacock (Ed.)
15 **The Use of Artificial Satellites for Geodesy,** Soren W. Henriksen, Armando Mancini, and Bernard H. Chovitz (Eds.)
16 **Flow and Fracture of Rocks,** H. C. Heard, I. Y. Borg, N. L. Carter, and C. B. Raleigh (Eds.)
17 **Man-Made Lakes: Their Problems and Environmental Effects,** William C. Ackermann, Gilbert F. White, and E. B. Worthington (Eds.)
18 **The Upper Atmosphere in Motion: A Selection of Papers With Annotation,** C. O. Hines and Colleagues
19 **The Geophysics of the Pacific Ocean Basin and Its Margin: A Volume in Honor of George P. Woollard,** George H. Sutton, Murli H. Manghnani, and Ralph Moberly (Eds.)
20 **The Earth's Crust: Its Nature and Physical Properties,** John G. Heacock (Ed.)
21 **Quantitative Modeling of Magnetospheric Processes,** W. P. Olson (Ed.)
22 **Derivation, Meaning, and Use of Geomagnetic Indices,** P. N. Mayaud
23 **The Tectonic and Geologic Evolution of Southeast Asian Seas and Islands,** Dennis E. Hayes, (Ed.)
24 **Mechanical Behavior of Crustal Rocks: The Handin Volume,** N. L. Carter, M. Friedman, J. M. Logan, and D. W. Stearns (Eds.)
25 **Physics of Auroral Arc Formation,** S.-I. Akasofu and J. R. Kan (Eds.)
26 **Heterogeneous Atmospheric Chemistry,** David R. Schryer (Ed.)
27 **The Tectonic and Geologic Evolution of Southeast Asian Seas and Islands: Part 2,** Dennis E. Hayes, (Ed.)
28 **Magnetospheric Currents,** Thomas A. Potemra (Ed.)
29 **Climate Processes and Climate Sensitivity** (Maurice Ewing Volume 5), James E. Hansen and Taro Takahashi (Eds.)
30 **Magnetic Reconnection in Space and Laboratory Plasmas,** Edward W. Hones, Jr. (Ed.)

Maurice Ewing Volumes

1 **Island Arcs, Deep Sea Trenches, and Back-Arc Basins,** Manik Talwani and Walter C. Pitman III (Eds.)
2 **Deep Drilling Results in the Atlantic Ocean: Ocean Crust,** Manik Talwani, Christopher G. Harrison, and Dennis E. Hayes (Eds.)
3 **Deep Drilling Results in the Atlantic Ocean: Continental Margins and Paleoenvironment,** Manik Talwani, William Hay, and William B. F. Ryan (Eds.)
4 **Earthquake Prediction—An International Review,** David W. Simpson and Paul G. Richards (Eds.)

Geophysical Monograph 31
Mineral Physics 1

Point Defects in Minerals

Edited by
Robert N. Schock

American Geophysical Union
Washington, D.C.
1985

Point Defects in Minerals

Library of Congress Cataloging in Publication Data

Main entry under title:

Point defects in minerals.

(Geophysical monograph ; 31)
1. Mineralogical chemistry--Congresses.
2. Crystals--Defects--Congresses. I. Schock, Robert N. II. Series.

QE364.P59 1984 549'.18 84-20427
ISBN 0-87590-056-9
ISSN 0065-8448

Printed in the United States of America

CONTENTS

DEFORMATION

PHASE TRANSFORMATIONS

Preface

This volume grew from a Chapman Conference held at Fallen Leaf Lake, California, September 5-9, 1982, on the topic "Point Defects in Minerals." The aim of this conference was to bring together a variety of experts within the geosciences and those disciplines (primarily solid-state physics and chemistry and materials science) that have traditionally studied the role of point defects in solids. Defects exist in all crystals at temperatures above absolute zero and arise from the tendency of crystal structures to disorder with increasing temperature or from chemical substitution. In the absence of outside forces, the most common form of defect is one in which an atom moves from its position in the perfectly symmetric crystal to some other position not normally occupied by this atom, or perhaps any atom. Such defects are termed point defects and they are ubiquitous, although at temperatures below melting their concentration rarely exceeds several percent of the total number of atoms. Nevertheless, point defects may either control or actively participate in many physical and chemical processes in minerals, processes which in turn are important to the formation and evolution of the earth. The need to solve problems posed by recent advances in the geosciences (e.g., the development of the concept of plate tectonics) requires the extension of present concepts in point defect theory, worked out on relatively simple systems, to materials far more complex, such as the silicate minerals found in the earth.

Point defect studies have taken two paths. One is the development of detailed mathematical and physical models, and the other is the experimental determination of physical and chemical processes which are dependent on the formation and movement of point defects. The modeling approach is the subject of the first part of this volume, in which the formation, energetics, possible structures, and computer simulations of point defects are covered. Further background material to these topics can be found in work by Flynn [1972], Kröger [1974], Nowick and Burton [1975], and Schmalzried [1981]. The second half of this volume covers the application of this theoretical basis to the experimental determination of physical and chemical properties of, and processes which operate in, silicates and oxides: electrical conduction, diffusion, deformation, and phase transformation. There are, of course, many other aspects to experimental research on solids. Should the reader wish to explore these more fully, a number of general references are given in the bibliography.

It is clear from the papers included here that many physical and chemical processes in minerals (such as structural, chemical, and electronic) are controlled by the migration of point defects. An urgent need is to be able to extend the currently deficient data base. The techniques of experimental physics [Wuensch, 1983; Beniere, 1983; Barr and Lidiard, 1971; Lidiard, 1972] should be readily applicable to minerals, yet the relevance of the experiments will be dependent on the ability to control variables such as composition, pressure, temperature, microstructure, and chemical environment within conditions realistic for the earth's interior. The full realization of such an effort may eventually call for the development of a broadly based research program along the lines of the recently completed 10-year International Geodynamics Program. One of the prime requirements highlighted in this program [see Hales, 1981] is the need for more complete knowledge of the physical properties of rocks and minerals at high temperature and pressure. The papers in this volume are a partial answer to this request, setting out some of the fundamental and experimental results of point defect research on minerals and indicating directions for future research.

The organizing committee for the Chapman Conference was comprised of the editor and A. Duba, Lawrence Livermore National Laboratory; D. Kohlstedt, Cornell University; A. Lasaga, Pennsylvania State University; A. Navrotsky, Arizona State University; and T. Shankland, Los Alamos National Laboratory. The conference was convened by the American Geophysical Union and supported by the U.S. National Science Foundation and by the U.S. Department of Energy through the Office of Basic Energy Sciences. This volume indicates our enthusiasm about the possibilities for understanding the role of point defects in minerals and their role in processes in the earth.

Karen Kumpf of the Lawrence Livermore National Laboratory was responsible for coordinating the manuscripts, their editing, and for the publication schedule. Linda Hansen and Liz Garrett of the same institution spent many hours organizing the Conference itself. This volume would not have been possible without their dedicated efforts.

Dr. J. N. Boland of the State University of Utrecht, The Netherlands, made numerous suggestions about the organization of this volume. Much of his work has been incorporated. Cindy Bravo and Brenda Weaver of the American Geophysical Union provided ample organizational support.

This volume would have been impossible without the authors who enthusiastically fostered the idea of a volume and who so generously gave of their time to produce the papers contained within.

References

Barr, L. W., and A. B. Lidiard, Defects in ionic crystals. In W. Jost, Ed., Physical Chemistry, an Advanced Treatise, pp. 152-228, Academic Press, New York, 1971.

Beniere, F., Les techniques de la diffusion, in Mass Transport in Solids, NATO Adv. Study Inst. Ser., Ser. B, edited by F. Beniere and C. R. A. Catlow, vol. 97, pp. 21-41, Plenum, New York, 1983.

Flynn, C. P., Point Defects and Diffusion, Oxford University Press, New York, 1972.

Hales, A. L., Geodynamics: The unanswered questions, in Evolution of the Earth, Geodyn. Ser., vol. 5, edited by R. J. O'Connell and W. S. Fyfe, pp. 4-5, AGU, Washington, D.C., 1981.

Kröger, F. A., The Chemistry of Imperfect Crystals, 2nd ed., North-Holland, Amsterdam, 1974.

Lidiard, A. B., Atomic transport in strongly ionic crystals. In H. I. Aaronson, Ed., Diffusion, pp. 275-308. Am. Soc. Metals, 1972.

Nowick, A. S., and J. J. Burton (Eds.), Diffusion in Solids, Academic, New York, 1975.

Schmalzried, H., Solid State Reactions, Monogr. Mod. Chem., vol. 12, Verlag Chemie, Weinheim, 1981.

Wuensch, B. J., Diffusion in stoichiometric close-packed oxides, in Mass Transport in Solids, NATO Adv. Study Inst. Ser., Ser. B, edited by F. Beniere and C. R. A. Catlow, vol. 97, pp. 353-376, Plenum, New York, 1983.

Bibliography

Burnham, C. W., The importance of volatile constituents, in The Evolution of the Ioneous Rocks, edited by H. S. Yoder, pp. 439-482, Princeton University Press, Princeton, N. J., 1979.

Giletti, B. J., M. P. Semet, and R. A. Yund, Studies in diffusion III, Oxygen in feldspars, an ion microprobe determination, Geochim. Cosmochim. Acta, 42, 45-57, 1978.

Jaoul, O., and B. Houlier, Study of ^{18}O diffusion in magnesium orthosilicate by nuclear microanalysis. J. Geophys. Res., 88, 613-624, 1983.

Lasaga, A. C., and R. J. Kirkpatrick (Eds.), Kinetics of Geochemical Processes, Rev. Mineral., vol. 8, Mineralogical Society of America, Washington, D.C., 1981.

Lasaga, A. C., The atomistic basis of kinetics: Defects in minerals, in Kinetics of Geochemical Processes, Rev. Mineral., vol. 8, edited by A. C. Lasaga and R. J. Kirkpatrick, pp. 261-319, Mineralogical Society of America, Washington, D.C., 1981.

O'Connell, R. J. and W. S. Fyfe (Eds.), Evolution of the Earth, Geodyn. Ser., vol 5, AGU, Washington, D.C., 1981.

O'Keeffe, M., and A. Navrotsky (Eds.), Structure and Bonding in Crystals, vol. 11, Academic, New York, 1981.

Sammis, C. G., J. C. Smith, and G. Schubert, A critical assessment of estimation methods for activation volume, J. Geophys. Res., 86, 707-10, 718, 1981.

Sato, H., and Y. Ida, Diffusion coefficients of ions in alkali nalides determined by low-frequency impedance measurement, in High-Pressure Research in Geophysics, Adv. Earth Planet. Sci., S. Akimoto and M. H. Manghnani, pp. 159-169, D. Reidel, vol. 12, edited by Hingham, Mass., 1982.

Shankland, T. J., Electrical conductivity in mantle materials, in Evolution of the Earth, Geodyn. Ser. vol. 5, edited by R. J. O'Connell and W. S. Fyfe, pp. 256-263, AGU, Washington, D.C. 1981.

Sorensen, O. T. (Ed.), Nonstoichiometric Oxides, Academic, New York, 1981.

Stacey, F. D., M. S. Paterson, and A. Nicolas (Eds.), Anelasticity in the Earth, Geodyn. Ser., vol;. 4, AGU, Washington, D.C. 1981.

Tuller, H. L., Mixed conduction in nonstoichiometric oxides, in Nonstoichiometric Oxides, edited by T. Sorensen, pp. 271-335, Academic, New York, 1981.

Will, G., and G. Nover, Influence of oxygen partial pressure on the Mg/Fe distribution in olivines, Phys. Chem. Miner., 4, 199-208, 1979.

Robert N. Schock, Editor

POINT DEFECTS IN SOLIDS: PHYSICS, CHEMISTRY, AND THERMODYNAMICS

F. A. Kröger

Department of Materials Science, University of Southern California
University Park, Los Angeles, California 90089

Abstract. Defect formation in crystalline solids and their effects on physical properties are reviewed, paying attention to point defects as well as extended defects (shear planes). Defect chemistry describes defect formation in terms of defects as well as atoms and ions in the crystal and in adjacent phases. Differences between binary and ternary compounds are pointed out. Techniques for determining defect structures are described. Examples are given for pure and doped binary compounds (TiO_2, Al_2O_3, Fe_3O_4) and ternary compounds ($BaTiO_3$, $PbTiO_3$, $PbWO_4$, $PbMoO_4$, $LiNbO_3$, and Mg_2SiO_4).

Introduction

As recognized by Frenkel [1926, 1946], Schottky and Wagner [1931], Wagner [1931, 1933], and Schottky [1935], crystalline solids in thermodynamic equilibrium contain defects as a result of disorder, of non-stoichiometry (for compounds), and of the presence of foreign aliovalent elements [Kröger, 1974a]. The concentration of defects formed by disorder increases with increasing temperature; that of defects formed by non-stoichiometry or doping increases with the extent of non-stoichiometry and the concentration of foreign elements. These defects - usually referred to as point defects are essential for the optical, electrical, magnetic, thermal, and transport properties of the crystals [Kröger and Vink, 1956; Kröger, 1974a].

Various types of point defects have to be distinguished, viz. missing atoms (vacancies), atoms at sites which are normally not occupied (interstitials), and atoms at sites normally occupied by other atoms (misplaced atoms or antistructure defects). All these may occur as single defects or as pairs or clusters. In addition the charge of the defects may differ. Distinction has to be made between real charges and effective charges. Let us first consider the former ones. In $MgFe_2O_4$, the constituents present are Mg^{2+}, Fe^{3+}, and O^{2-} with real charges +2, +3, and -2. Removal of either of these produces vacancies $V_{Mg}2+$, $V_{Fe}3+$ and V_O2- with real charge zero. Occupation of interstitial sites by the native ions leads to charged defects Mg_i^{2+}, Fe_i^{3+} and O_i^{2-} with real charges +2, +3 and -2. Replacement of Fe^{3+} by Ti^{4+} leads to $Ti_{Fe}^{4+}3+$ with a real charge of +4. Effective charges are charges relative to the ideal unperturbed crystal. It is useful (but not universally done) to indicate these charges by different symbols: e.g., dots for positive, dashes for negative, and multiplication signs for zero charges [Kröger and Vink, 1956; Kröger, 1974a]. In this system the normal structure elements are uncharged: Mg_{Mg}^{x}, Fe_{Fe}^{x}, Fe_{Fe}^{x}, and O_O^{x}. Ion vacancies, on the other hand, are charged: $V_{Mg}2+ \equiv V_{Mg}''$, $V_{Fe}3+ \equiv V_{Fe}'''$, and $V_O2- \equiv V_O^{\cdot\cdot}$. Neutral vacancies are formed by removing atoms rather than ions: V_{Mg}^{x}, V_{Fe}^{x} and V_O^{x}. Ti^{4+} at an Fe^{3+} site, $Ti_{Fe}^{\cdot}$, has a positive effective charge of 4 - 3 = 1 unit. For interstitials, real and effective charges are the same: $Mg_i^{2+} \equiv Mg_i^{\cdot\cdot}$, $O_i^{2-} \equiv O_i''$, etc. Effective charges have the advantage of indicating how defects will interact, defects with effective charges of the same sign repelling, those of different sign attracting each other, with $E_{pot} \propto Q_1Q_2/r_{12}$, ($Q_1$ and Q_2 being the charges of, r_{12} the distance between the defects). Defects with effective charge zero may still attract each other, either due to electronic exchange or due to elastic forces. Formation of point defects usually is accompanied by shifts in the position of neighboring atoms, leading to a considerable reduction in the energy of formation. Sometimes these shifts are extensive and may move atoms from normal sites to interstitial sites, leaving vacancies at the normal sites, forming defects that are much larger and more complicated than a simple vacancy or interstitial. In such cases one speaks of microdomains [Anderson, 1974]. Examples are found in $Fe_{1-\delta}O$ [Koch and Cohen, 1969], $UO_{2+\delta}$ [Willis, 1963] and in doped fluorides [Cheetham et al., 1972]. In addition to the ionic or atomic defects introduced so far there are electronic defects. These are especially significant in crystals for which in the unperturbed ground state the valence electrons form a completely filled valence band, whereas extra electrons have to be accommodated in a normally empty conduction band, separated from the valence band by a forbidden energy gap. The basic electronic defects are electrons in the conduction band (e'), and missing electrons (holes, $h^{\cdot}$) in the valence band. Both species are mobile with a mobility decreased by thermal motion (lattice scattering) or by scattering at impurities (impurity scattering). The electronic defects may be formed by disorder (excitation across the forbidden gap) or by ionization of native or foreign atomic defects, e.g.,

$$V_O^x \rightarrow V_O^{\cdot} + e' \rightarrow V_O^{\cdot\cdot} + 2e' \tag{1}$$

$$V_{Mg}^x \rightarrow V_{Mg}' + h^{\cdot} \rightarrow V_{Mg}'' + 2h^{\cdot} \tag{2}$$

$$Ti_{Fe}^x \rightarrow Ti_{Fe}^{\cdot} + e' \tag{3}$$

In some cases it is appropriate to introduce electronic defects in the form of native ions of changed valency rather than as electrons or holes in a band. Thus in $Fe^{2+}O^{2-}$ electrons can be represented by $Fe^{+}_{Fe^{2+}}$ or Fe'_{Fe}, holes by $Fe^{3+}_{Fe^{2+}}$ or $Fe^{\cdot}_{Fe}$. Movement of these electronic defects consists of interchange of electrons between adjacent ions with normal and abnormal charges (hopping electron model):

$$Fe'_{Fe} + Fe^x_{Fe} \rightarrow Fe^x_{Fe} + Fe'_{Fe} \tag{4}$$

and

$$Fe^{\cdot}_{Fe} + Fe^x_{Fe} \rightarrow Fe^x_{Fe} + Fe^{\cdot}_{Fe} \tag{5}$$

This interchange is a thermally activated process, giving mobilities increasing with increasing temperature. The equivalence of native ions of changed charge and mobile electrons or holes is not always appreciated. Thus in a paper on $LiNbO_3$, $Nb^{4+} \equiv Nb'_{Nb}$ as well as e' were introduced as different entities and were believed to be present at different concentrations [Bollmann, 1977].

Neutral defects, ionizing with liberation of electrons, are called donors; those ionizing with liberation of holes are called acceptors. The energies of the ionization reactions define the positions of donor and acceptor levels in the forbidden gap: relative to the conduction band for donors, relative to the valence band for acceptors.

Since ionization causes a change in the charge of centers, there is a danger of ambiguity in the naming of levels. This ambiguity is removed when the level is labeled with the two charge states involved. Thus the level involved in ionization reaction (3) would be $Ti_{Fe}^{x,\cdot}$. For reactions (1) and (2) there are two levels each: $V_O^{x,\cdot}$ and $V_O^{\cdot,\cdot\cdot}$, and $V_{Mg}^{x,'}$ and $V_{Mg}^{','' }$. Alternatively one can label the level with the charge state existing when the level is occupied by an electron: Ti_{Fe}^x, V_O^x, $V_O^{\cdot}$, V_{Mg}' and V_{Mg}''. Note the asymmetry in the labeling of donors and acceptors resulting from the arbitrary choice of electrons over holes in the latter case.

Formation of atomic or ionic defects involves the removal of atoms or ions from the inside of the crystal to the surface (for vacancies) or the reverse (for interstitials). In computing energies of these processes the energies involved in the bulk processes do not offer serious difficulties, but the energies of addition to or removal from surfaces are more difficult to deal with (and may be different for different surface planes). This difficulty is avoided by calculating energies E_j to move atoms or ions from inside the crystal to a standard reservoir outside or vice versa. These, however, are not the energies of formation of the defects, but the sum of these energies and the energies of evaporation. The latter are unknown for individual components of a compound but their sum is equal to the energy of atomization (i.e., the negative of the lattice energy) for all components together. Therefore the energies of Schottky and interstitial disorder can be found by subtracting the energy of atomization per mole of compound from the sum of E's for the defects involved; for Frenkel and anti-Frenkel disorder the two atomization steps cancel one another and the disorder energy is equal to ΣE_j. The energies of formation of individual defects can only be found if an assumption is made regarding the distribution of the energy of atomization over the components. Since this assumption is usually arbitrary, the final results remain uncertain [Kröger, 1980].

In addition to the point defects considered so far, two other types of defects may occur: line defects (dislocations) and extended defects (in compounds). The former are of decisive importance for the mechanical properties of crystals (plastic deformation), but they are non-equilibrium defects. Extended defects are shear planes, i.e., planes at interfaces between sections of crystal displaced relative to each other in such a manner that what is an interstitial site in one section is a lattice site in the other [Wadsley, 1955]. When the interstitial sites in the shear plane are occupied by atoms, the shear plane represents an excess of that type of atom. When the lattice sites are unoccupied, the shear plane represents a deficiency of the type of atom normally occupying the lattice sites. Shear planes are usually equidistant, the distance depending on the amount of excess or deficiency of one component present over that characterizing the normal regions. The combination of shear planes and the regions of normal structure in between them form a shear structure [Wadsley, 1955]. Shear structures are stable configurations and may be considered new modifications. In several systems, series of shear structures occur, each characterized by a definite stoichiometry and distance between the shear planes. Examples are $Ti_nO_{(2n-1)}$ and $V_nO_{(2n-1)}$ [Gado and Magneli, 1966; Reed, 1970], the distance between the shear planes increasing with increasing values of n (being ∞ when n = ∞). Shear structures are found only in a limited number of systems, but point defects are always present (even in shear structures). In the latter case, there must be equilibrium between the point defects in the blocks with normal structure between the shear planes and those in or near the shear planes [Kröger, 1983). Such an equilibrium can only be formulated when there are also point defects in the shear planes: missing atoms in the planes representing extra atoms, or extra atoms in the planes representing missing atoms. It is to be expected that the point defects in the normal sections of the shear structures affect the physical properties in the same way as they would in the absence of shear planes. The effect of the shear planes is not known, however. If there is an effect, it could be due to the extra atoms themselves (e.g., acting as donors) and/or to the point defects in the shear plane. Non-stoichiometry in a binary compound can be discussed with the aid of G-x figures, where G is Gibbs free energy. If shear structures are absent the compound has a single hairpin shaped G-x curve with a width that is the larger, the easier it is to establish non-stoichiometry (Figure 1a). If shear structures occur, each gives rise to its own G-x curve (Figure 1b) [Anderson, 1970]. With increasing chemical potential of A, μ_A, shear structures richer in A are favored. Note, however, that each shear structure has a definite (albeit narrow) range of μ_A values at which it is stable - getting richer in A when μ_A increases. As we saw before, the change in A is the sum of the concentration change of point defects in the normal slabs and in the shear planes, the distance between the shear planes and thus the amount of excess or deficiency of A that they represent remaining constant inside a given shear structure. For any compound with an unusually wide range of non-stoichiometry, the possibility of the occurrence of shear structures should be investigated. In the following we restrict ourselves to a detailed discussion of systems containing only point defects, including methods to regulate the concentrations of defects and the physical properties dependent on these defects.

Defect Chemistry

Since many physical properties depend on the type and concentration of point defects, it is

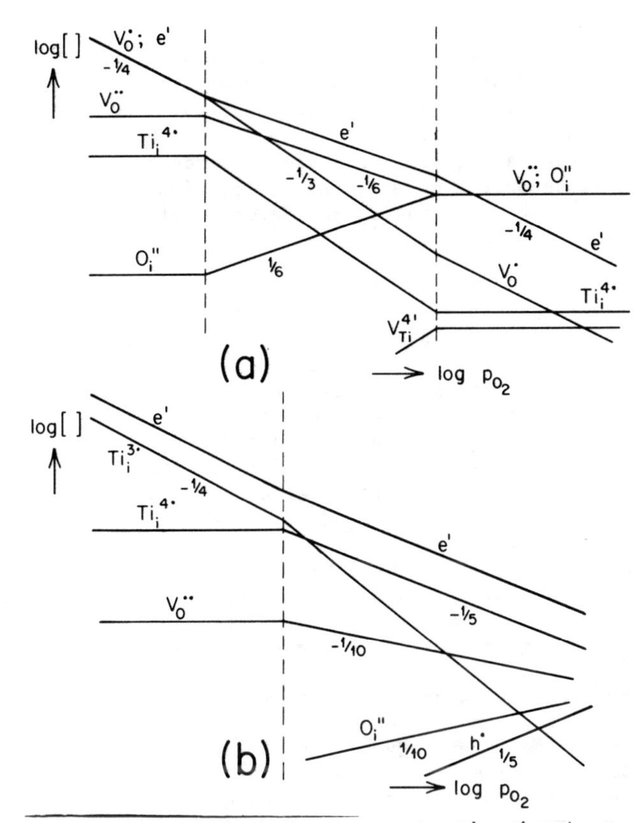

Fig. 1. Possible defect concentration isotherms for pure TiO_2 with $V_O^{\bullet}$ (a) or $Ti_i^{\bullet\bullet\bullet}$ (b) dominant.

important to understand how defect concentrations depend on experimental conditions if the system is in thermal equilibrium, or on previous history if the crystal is not in thermal equilibrium. Point defects can be considered as quasiparticles with thermodynamic properties similar to those of atoms or ions and can be discussed together with the latter in an extended chemical thermodynamics and an extended chemistry ("defect chemistry") [Kröger and Vink, 1956; Kröger, 1974a,b]. Quasichemical reactions involving atoms and point defects must maintain (1) atom balance, (2) (effective) charge balance, and (3) the site ratio as required by the crystal structure. Reactions taken into account normally involve atomic and electronic disorder processes in the crystal and processes transferring atoms between the crystal and an outer phase--either a vapor, a liquid, or another solid.

Application of the law of mass action to these reactions leads to product relations between simple powers of the activities or fugacities of the participants, valid when equilibrium is established or maintained. Often activities can be approximated by concentrations (i.e., activity coefficients f = a/c are equal to one) and fugacities by partial pressures. Inside crystals electroneutrality must be maintained, i.e., the sum of charges on positively charged defects must equal the sum of charges on negatively charged defects. Near surfaces or dislocations this is no longer true: here space charges may occur, increasing towards the surface or the core of the dislocations, with the integral of the space charge equal in magnitude (but opposite in sign) to the planar or linear charge at the surface or at the core of the dislocation [Frenkel, 1946; Lehovec, 1953; Poeppel and Blakely, 1969; Eshelby et al., 1958; Kliewer and Koehler, 1965; Kliewer, 1965; Whitworth, 1967]. If equilibrium is not complete, appropriate balance equations take the place of the equilibrium relation that is not satisfied. Thus if a fixed amount of dopant is present, a balance equation for the dopant replaces the mass action relation based on the incorporation reaction of the dopant.

Accurate solutions for the defect concentrations can be obtained as a function of basic parameters such as equilibrium constants and partial pressures (or activities) of native or foreign components (dopants) in an outer phase, using the complete neutrality condition and (if applicable) balance equations. Since generally some concentrations are far larger than other ones, a good approximation can be obtained by approximating the neutrality condition and balance equations by their dominant members (Brouwer approximation) [Brouwer, 1954]. In that case the relations between the logarithms of the concentrations and the logarithms of the variables are linear. A graphical method can be used to find the limits beyond which different approximations have to be employed. Simplest are the cases where equilibrium is complete as is often the case at high temperatures. Situations at lower temperatures often correspond to partial equilibrium, electronic equilibria being maintained, but concentrations of atomic defects being determined by the conditions pertaining to the last time the system reached complete equilibrium (previous history).

Defect chemistry is applicable to single component crystals with or without dopants, to binary compounds with or without dopants, and to pure and doped multicomponent compounds. The larger the number of components, the larger the number of thermodynamic parameters that have to be fixed to completely define the system.

Most of the work done to date refers to pure and doped single component crystals and binary compounds, but more and more work on pure and doped ternary compounds are appearing. In geophysics, multicomponent compounds will undoubtedly be the most interesting ones.

In binary compounds there is one parameter describing non-stoichiometry. Non-stoichiometry leads to formation of unsaturated bonds and to shear planes and/or point defects that act as donors or acceptors.

In ternary compounds there are two composition variables, one describing the ratio between electro-positive and electro-negative components, the other describing the ratio between the sub-compounds of the ternary (e.g., AX and B_2X_3, or AX_2 and BX, in AB_2X_4. Alternatively we can use fugacities of X and activities of the subcompounds a_{AX} or $a_{B_2X_2}$ as the variables [Schmalzried and Wagner, 1962; Schmalzried, 1965; Kröger, 1974a]. The former composition variation is the analogue of non-stoichiometry in binary compounds and leads to formation of point defects with acceptor or donor properties. The latter leads to "non-molecularity" and involves formation of ionic defects which do not act as donors or acceptors (they are in fact ionized donors and acceptors) but affect atomic transport processes such as self diffusion, ionic conductivity, and diffusion of foreign constituents. In compounds with more components the number of molecularity parameters increases, but the number of stoichiometry parameters remains equal to one.

In binary systems, phase relations are usually represented in a three-dimensional figure with total pressure P, temperature T and composition x as the coordinates, sections or projections giving simplified representations: In addition to total pressure, partial pressures of the components can be plotted along the pressure axis. As far as defect concentrations are concerned, these can be given in two-dimensional presentation with T = constant, using partial pressures as the variables (isotherms). Temperature dependence can be given at constant partial pressure of one of the components or at constant composition. Dopants make the system into a ternary one. Such systems are usually discussed on the basis of partial equilibrium, assuming a constant concentration of the dopants but component activities and/or temperature variable.

For ternary compounds a complete presentation in terms of P, T and the two composition variables would require a four-dimensional space. If the effect of P is neglected (allowed at $P \leq 1$ atm) or P is assumed to be constant at 1 atm, presentation in a three-dimensional space is possible with either T and the two compositions, or T and partial pressures and sub-compound activities as the variables [Abelard and Baumard, 1982]. Two-dimensional sections through the space figure then are sufficient to represent all states as a function of one variable if the other two are kept constant. Experimentally the situation becomes more complicated the larger the number of components.

Two approaches can be followed, one corresponding to fixation of concentrations of some of the components, the other corresponding to fixation of activities of components or sub-compounds. The former is basically one of partial equilibrium, the relative concentrations of non-volatile components being fixed by

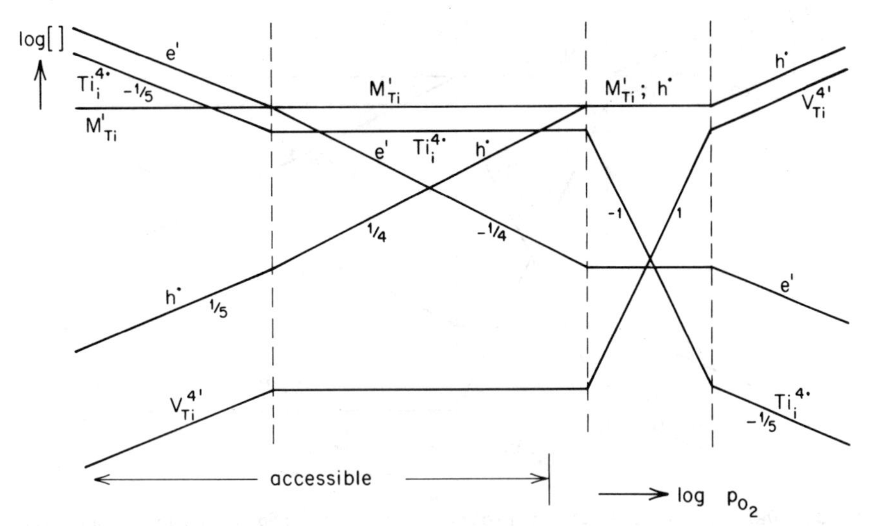

Fig. 2. Defect concentration isotherms for TiO_2:M with M = Fe or Al (acceptors) (Baumard and Tani, 1977b)

weighing (non-equilibrium), that of volatile components (usually oxygen, nitrogen or halogens) being established through equilibration with an outer phase (usually a vapor), thus regulating non-stoichiometry. Fixation of component activities involves equilibration with an outer phase (usually a vapor phase) in which both the partial pressures of the non-metals and of the metals (or the sub-compounds of these metals) are fixed.

A single vapor pressure is easily established in an open or closed two-temperature system. Establishment of several well-defined partial pressures is not possible in a closed system but requires an open system in which streams of inert gas, loaded with components at certain partial pressures by having been passed over sources of the components at regulated temperatures, are mixed and passed over the crystal. Most work published to date is of the former type.

In the following we discuss in some greater detail effects observed in a few pure and doped binary compounds (TiO_2, Al_2O_3), in a quasi-ternary compound (Fe_3O_4) and in pure and doped ternary compounds. In practice, pure compounds are usually sufficiently impure to show properties resulting from the presence of unintentional dopants. Only when the concentrations of defects caused by the impurities is smaller than those due to disorder and/or non-stoichiometry do the materials have the properties of pure crystals.

TiO_2

TiO_2 is an n-type semiconductor with electrons playing a dominant role in the defect structure. Electron conductivity and electron emission of undoped TiO_2 are proportional to $p_{O_2}^{-1/n}$ with n = 4 to 5 [Odier and Loup, 1980]. Self diffusion of oxygen is independent of oxygen pressure [Haul and Duembgen, 1965]. That of titanium is proportional to $p_{O_2}^{-1/n}$ with n = 4.86 [Akse and Whitehurst, 1978]. In donor doped TiO_2 electronic conductivity varies $\propto p_{O_2}^{-1/n}$ with n varying with increasing p_{O_2} from 5, via ∞ to 4 [Baumard and Tani, 1977a, b]. Oxygen uptake during oxidation is proportional to the donor concentration [Eror, 1981]. In acceptor doped TiO_2 the corresponding variation is from 5 via 4 to ∞ [Baumard and Tani, 1977b; Eror, 1981]. Oxygen self-diffusion is increased by doping with an acceptor (Cr). Independence of D_O^x of p_{O_2} in pure TiO_2 [Haul and Duembgen, 1965] could be explained by a model with either $V_O^{\cdot}$ or $Ti_i^{\cdot\cdot}$ as major ionic defect at low P_{O_2}, given [e'] $\propto p_{O_2}^{-1/4}$ (Figure 1). However, it is more likely that the material was impure and dominated by acceptors. Then [e'] $\propto p_{O_2}^{-1/5}$ points to $Ti_i^{\cdot\cdot\cdot\cdot}$ as the major low-p_{O_2} defect (Figure 2) [Baumard and Tani, 1977b]. This, however, is not in line with the results of calculations which show a sequence for the constants of Schottky, anti-Frenkel and Frenkel disorder $K_S > K_{F,O} > K_{F,Ti}$ with energies of formation per defect of 1.74, 4.36 and 5.98 eV [Catlow et al., 1982]. Figure 3 shows the corresponding isotherms for donor-dominated TiO_2 [Baumard and Tani, 1977b].

Al_2O_3

Al_2O_3 is an insulator at room temperature and a mixed ionic-electronic conductor at high temperature. Owing to the high lattice energy,

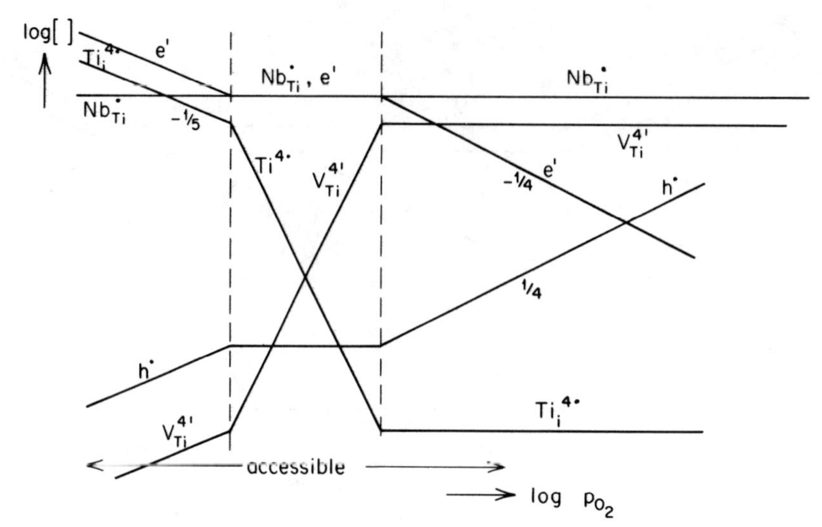

Fig. 3. Defect concentration isotherms for TiO_2:Nb (a donor) (Baumard and Tani 1977b).

defect formation energies are so large that even at impurity contents as low as 1 ppm the concentrations of defects are determined by the impurities, commonly Mg and Fe (acceptors); Ti acts as a donor. Defect isotherms are shown in Figure 4. Note the change in valency of Fe from Fe^{2+} to Fe^{3+} and of Ti from Ti^{3+} to Ti^{4+} upon oxidation. The overall oxidation process is the sum of the oxidation of Al_2O_3 and the ionization of the dopant as regulated by the position of the energy level of the dopant in the forbidden energy gap. For Al_2O_3:Fe

$$1/2\ O_2 + V_O^{\cdot\cdot} \rightarrow O_O^x + 2h^{\cdot},\ K_{ox,V(O)}$$

$$2Fe_{Al}' + 2h^{\cdot} \rightarrow 2Fe_{Al}^x,\ (K_a^{Fe})^{-2}$$

$$1/2\ O_2 + 2Fe_{Al}' + V_O^{\cdot\cdot} \rightarrow O_O^x + 2Fe_{Al}^x,\ K_{ox,V}^{Fe} = \frac{K_{ox,V(O)}}{(K_a^{Fe})^2}$$

Fe_3O_4

$Fe_3O_4 \equiv FeO{\cdot}Fe_2O_3$ crystallizing in the spinel structure is a quasi-ternary with FeO and Fe_2O_3 as the sub-compounds. It is, of course, not a true ternary, for there is a reaction transforming FeO into Fe_2O_3 or Fe^{2+} into Fe^{3+}. This transformation can be formulated in three different but equivalent ways using either real or effective charges:

$$2O_2 + 9Fe_b^{2+} \rightarrow 6Fe_b^{3+} + 3V_b + Fe_3O_4$$

$$2O_2 + 8Fe_b^{2+} \rightarrow 7Fe_b^{3+} + 3V_b + Fe_a^{3+} + 4O_O^{=}$$

$$2O_2 + 9Fe_b^{1/2'} \rightarrow 6Fe_b^{1/2\cdot} + 3V_b^{2-1/2'} + Fe_3O_4$$

Here a and b indicate cation sites with 4 and 6 coordination, respectively. For small deviations from stochiometry, application of the law of mass action with $[Fe_b^{2+}]$ and $Fe_b^{3+}] \approx$ constant leads to

$$[V_b] \equiv [V_b^{2-1/2'}] \propto p_{O_2}^{2/3}$$

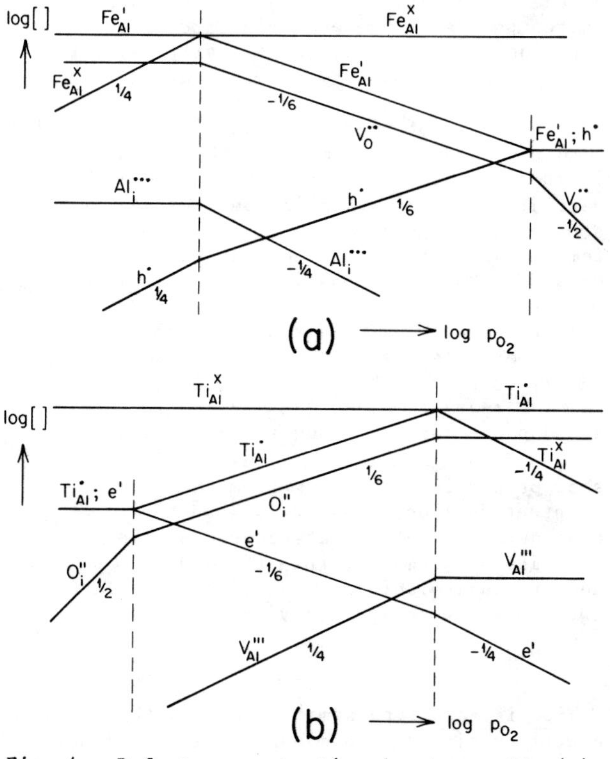

Fig. 4. Defect concentration isotherms for (a) Al_2O_3:Fe and (b) Al_2O_3:Ti.

Because of the Frenkel disorder reaction

$$Fe_b^{2+} + V_i \rightleftarrows Fe_i^{2+} + V_b; \quad K_F$$

$$[Fe_i^{2+}] = \frac{K_F}{[V_b]} \propto p_{O_2}^{-2/3}$$

(Figure 5). Studies of Fe self-diffusion show the change from diffusion by Fe^{2+} interstitials, Fe_i^{2+}, at low p_{O_2}, to diffusion by Fe vacancies at high p_{O_2} [Dieckmann and Schmalzried, 1975, 1977a, b]. A similar variation is expected and observed for diffusion of Co and Cr in Fe_3O_4 [Dieckmann et al., 1978].

The concentration of oxygen vacancies V_O is related to that of the Fe vacancies through the Schottky disorder equilibrium

$$[V_a][V_b]^2[V_O]^4 = K_S$$

and the transfer of Fe^{3+} from a to b sites

$$Fe_a^{3+} + V_b \rightleftarrows Fe_b^{3+} + V_a \; ; \; K_{dist} = \frac{[Fe_b^{3+}][V_a]}{[Fe_a^{3+}][V_b]} \approx \frac{[V_a]}{[V_b]}$$

leading to

$$[V_O] \propto [V_b]^{-3/4} \propto p_{O_2}^{-1/2}$$

The observed oxygen pressure dependence of oxygen self-diffusion $\propto (p_{H_2}/p_{H_2O})^{0.27} \propto p_{O_2}^{-0.135}$ [Castle and Surman, 1969] does not fit this model. A better (though not ideal) fit is obtained if oxygen diffuses by migration of associates of oxygen and iron interstitials

$$O_i'' + Fe_i^{\cdot\cdot} \rightarrow (O_iFe_i)^x$$

with

$$[(O_iFe_i)^x] \propto [O_i''][Fe_i^{\cdot\cdot}] \propto p_{O_2}^{-0.167}$$

Ternary Compounds

Table 1 shows a number of pure and doped ternary compounds for which attempts have been made to determine the defect structure, paying attention to oxygen-cation and cation-cation non-stoichiometry.

$BaTiO_3$ and $SrTiO_3$

A considerable amount of work has been done on $BaTiO_3$, in particular by Smyth and co-workers. Figure 6 shows defect isotherms for $BaTiO_3$ with a fixed excess of TiO_2. The TiO_2 activity is also given: note that although the concentration of TiO_2 excess is fixed, the activity is not. Figures 7 and 8 show isotherms for acceptor and donor dominated material. When transition elements such as Fe, Cr, Mn or Co dominate the neutrality condition, complications could arise because of changes of the valency with oxygen pressure [Hagemann et al., 1980; Hagemann and Ihrig, 1979; Hagemann and Hennings, 1981]. However, usually one species remains dominant. Figure 9 demonstrates this for $BaTiO_3$: Fe, with Fe^{3+} the dominant species, Fe^{2+} and Fe^{4+} the minorities. $SrTiO_3$ behaves essentially as $BaTiO_3$.

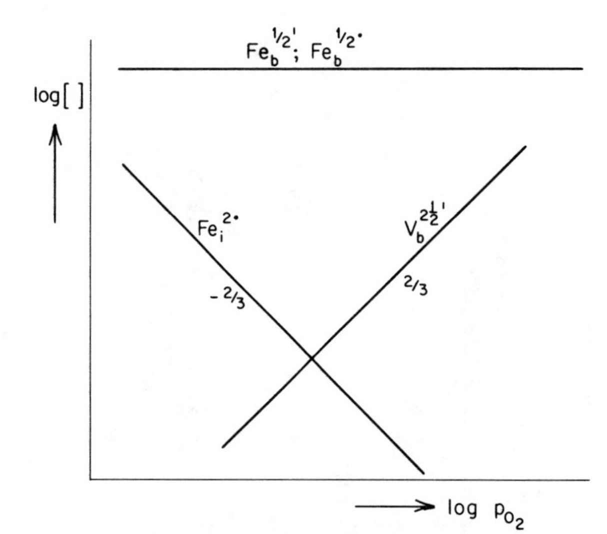

Fig. 5. Defect isotherms for Fe_3O_4.

$PbTiO_3$, $PbWO_4$, $PbMoO_4$

The lead compounds with the volatile PbO as a sub-compound were the first for which the cation-cation stoichiometry was changed and regulated via a subcompound activity. All have the tendency to lose PbO and have a defect structure dominated by $[V_{Pb}''] = [V_O^{\cdot\cdot}]$. Figure 10 shows an isotherm for PbO deficiency ($= [V_{Pb}''] = [V_O^{\cdot\cdot}]$) in $PbTiO_3$ as a function of PbO vapor pressure according to Holman and Fulrath [1973]; the isotherms for $PbZrO_3$ and $Pb(Zr,Ti)O_3$ are similar [Holman and Fulrath, 1973]. Shackelford and Holman [1975] attempted to explain these isotherms on the basis of a vapor of PbO and simple defects in the solid but were unsuccessful. They suggested that a better representation could be arrived at on the basis of a model with defect clusters in the solid. This, however, is not the case: clustering of defects increases the discrepancy between experiment and model calculations. However, there are strong indications that the major species in the vapor are Pb_2O_2 and Pb_4O_4 rather than PbO [Drowart et al., 1965]. A model based on dominance of Pb_4O_4 in the gas phase and single defects in the solid, gives a weak

TABLE 1. Defect Structure Investigations on Ternary Compounds

Compound	Pure Compounds		Doped Compounds		
	Oxygen-Cation Nonstoich.	Cation-Cation* Nonstoich.	Dopant	oxygen-cation Nonstoich.	Cation-Cation* Nonstoich.
$BaTiO_3$	Smyth (1977), Kosek & Arend (1967) Chan et al. (1981a)	Smyth (c) (1977), Seuter (c) (1974) Sharma et al. (c) (1981)	Ga, La	Daniels & Härdtl (1976)	
			Acceptor	Chan et al. (1981a) Long & Blumenthal (1971) Wernick (1976) Long & Blumenthal (c) (1971)	Chan et al. (1981a)
			Ga	Daniels (1976)	
			Al	Eror & Smyth (1978) Chan et al. (1982)	Eror & Smyth (c) (1978)
			Al, Nb	Hennings (1976) Chan & Smyth (1976)	Chan & Smyth (c) (1976)
			La	Ihrig & Hennings (1978)	Shirashi et al., (c) (1980)
			Fe	Hagemann et al. (1980)	
			Mn,Cr,Co	Hagemann et al. (1979,1981)	
$CaTiO_3$	Balachandran et al. (1982a)		Ta	Balachandran & Eror (1982b)	
			Al,Cr	Balachandran et al. (1982b) Becker & Scharmann (1974) Gupta & Weirick (1967)	
			La	Balachandran & Eror (1982c)	
$SrTiO_3$	Chan et al. (1981b) (excess TiO_2)		La	Eror & Balachandran (1981)	
			Acceptor	Balachandran & Eror (1982c) Chan et al. (1981b), Odekirk et al. (1982)	
	Balachandran & Eror (1981)	Eror & Balachandran (1982)			
	Eror & Balachandran (1982)				
$LiNbO_3$	Bergmann (1968) Bollman & Gernand (1972)	Jorgensen & Bartlett (c)(1969) Limb & Smyth (c) (1981) Krindach et al. (1976)	Fe	Belabaev et al. (1978)	
			Fe,Cu,Mn	Staebler (1975) Bollmann (1977) Dischler et al. (1974)	Staebler (1975) Phillips & Staebler (1974)
	Limb & Smyth (1981)		Ti		Burns et al. (a) (1978) Esdaile (a) 1978 Ranganath & Wang (a) (1977) Miyazawa et al. (a) (1977)

$Sr_{4/3+x}Nb_{2/3-x}O_{3-3/2x}$		Lecomte et al. (c) (1981)			
Mg_2SiO_4	Stocker & Smyth (1978)	Stocker & Smyth (a) (1978) Alelard & Baumard (a) (1982)	Fe	Stocker & Smyth (1978) Morin et al. (1977, 1979)	Stocker & Smyth (a) (1978)
Fe_2SiO_4	Sockel (1974) Ilschner et al. (1977)	Sockel (a) (1974)			
Co_2SiO_4	Greskovich & Schmalzried (1970)				
$PbTiO_3$		Holman & Fulrath (a) (1973) Shackelford & Holman (a) (1975)	Nd	Bouwma & Heilbron (1976)	Bouwma & Heilbron (c) (1976)
$Pb(Zr,Ti)O_3$	Wrobel (1979) $(Zr,Ti)O_2$ excess	Holman & Fulrath (a) (1973) Härdtl & Rau (a) (1969)			
$PbZrO_3$		Härdtl & Rau (a) (1969)			
$CaWO_4$	Rigdon & Graves (1973) Koehler & Kikuchi (1971) Nassau & Loiacono (1963)	Koehler & Kikuchi (1971)			
$CaMoO_4$	Petrov & Kofstad (1979)	Petrov & Kofstad (a) (1979)			
$PbWO_4$	van Loo (1975)		K,Bi,Y	Groenink & Binsma (1979) (excess WoO_3)	
$PbMoO_4$	van Loo (1975)		Na,Bi,Y	Groenink & Binsma (1979) (excess MoO_3)	

*(a) = activity regulation, (c) = concentration regulation

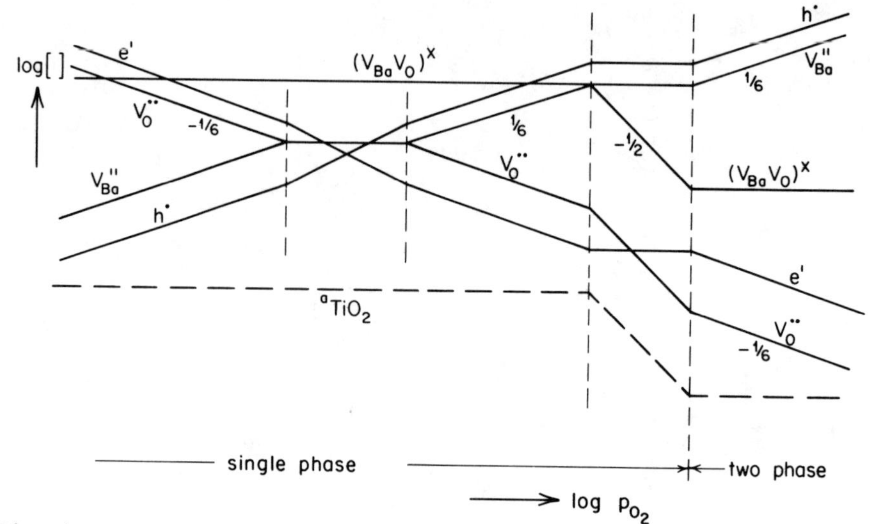

Fig. 6. Defect isotherms and TiO_2 activity, a_{TiO_2}, for $BaTi_{1+\sigma}O_{3+2\sigma}$ containing $(V_{Ba}V_O)^x$ complexes (Smyth, 1977).

partial pressure dependence of TiO_2 excess by loss of PbO:

$$Pb_{Pb}^x + O_O^x \rightarrow V_{Pb}'' + V_O^{\cdot\cdot} + 1/4\ Pb_4O_4(g)$$

$$[V_{Pb}''] = [V_O^{\cdot\cdot}]\ \text{V}\ p_{Pb_4O_4}^{1/8}$$

close to what is observed at low partial pressures. Yet the model still does not explain all the results.

Similar considerations apply to $PbWO_4$ and $PbMoO_4$. Figure 11 shows the temperature dependence of defect concentrations in $PbWO_4{:}K_{Pb}$ (an acceptor) with a fixed WO_3 excess according to data by Groenink and Binsma [1979]. The crystal is acceptor-dominated at low T but dominated by Frenkel disorder of oxygen at high T.

$LiNbO_3$

Another ternary compound for which the cation-cation stoichiometry has been regulated through the subcompound activities is $LiNbO_3$. $LiNbO_3$ is stable with [Li]/[Nb] from 1.01 to 0.89 [Byer et al., 1970]. Li_2O vapor from Li_2O [Burns et al., 1978; Ranganath and Wang, 1977], Li_2CO_3 [Miyazawa et al., 1977], or Li_3NbO_4 or $LiNb_3O_8$ [Holman et al., 1978] has been used to establish well-defined Li/Nb ratios or prevent Li_2O loss.

A change of the refractive index upon illumination (so-called optical damage) is related to the concentration of native defects which are found to be strongly affected by the Li/Nb ratio and the presence of Fe as a dopant: prevention of Li_2O loss (Li_2O excess?) gives a large photo-refraction effect.

According to Jorgensen and Bartlett [1969] conductivity is electronic, $\propto p_{O_2}^{-1/4}$ and independent of [Li]/[Nb] at low p_{O_2} but ionic, independent of p_{O_2} but increasing with Li addition at high p_{O_2}. It is not clear, however, whether the sample to which no Li_2O was added had an excess of Nb_2O_5, was stoichiometric, or had an excess of Li_2O. Attempts to find a defect model explaining the results have been made by Bollmann [1977], Bollmann and Gernand [1972] and Limb and Smyth [1981]. It seems, however, that more systematic studies have to be carried out before a satisfactory model can be found.

$(Mg,Fe)_2SiO_4$

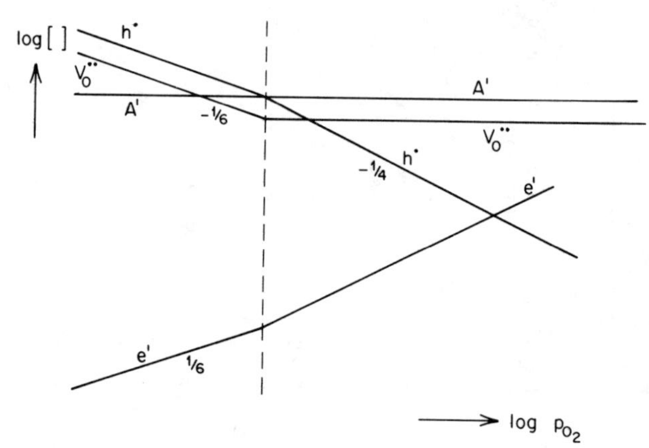

Fig. 7. Defect isotherms for $BaTi_{1+\sigma}O_{3+2\sigma}{:}Al_{Ti}$.

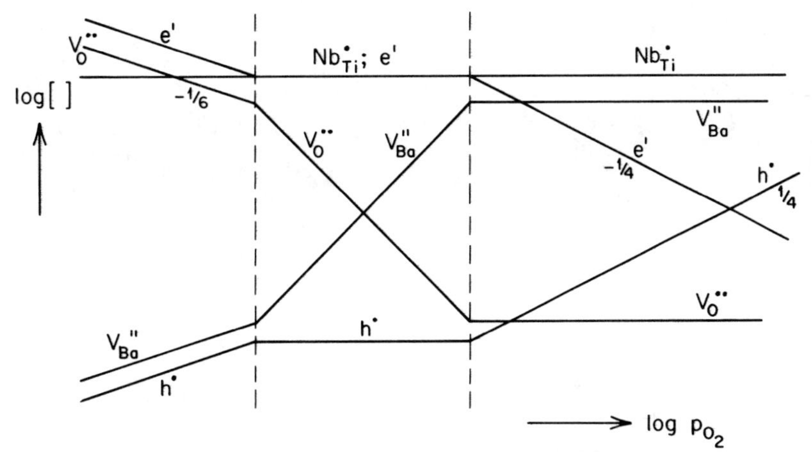

Fig. 8. Defect isotherms for $BaTi_{1+\sigma}O_{3+2\sigma}:La_{Ba}$.

Most important for the geophysicist are probably Mg_2SiO_4 (forsterite), Fe_2SiO_4 (fayalite) and their solid solutions (olivine). On the basis of crystal-chemical considerations Smyth and Stocker [1975] conclude that (1) one may expect an excess of SiO_2 but not a marked excess of MgO and (2) the major ionic disorder process is Frenkel disorder of Mg^{2+}, making $Mg_i^{\bullet\bullet}$ and V''_{Mg} the major defects in stoichiometric Mg_2SiO_4. This fits the results by Pluschkell and Engell [1968] who found ionic conduction by Mg^{2+} ions.

In crystals with an excess of SiO_2 or MgO and an excess or deficiency of oxygen, one expects different sets of major defects as indicated in Figure 12; Δx indicating SiO_2 excess, Δy oxygen excess [Abelard and Baumard, 1982]. Fields where different sets of defects dominate and isotherms as $f(p_{O_2})$ for stoichiometric Mg_2SiO_4 and Mg_2SiO_4 with excess of SiO_2 are shown in Figures 13a and 13b, respectively [Abelard and Baumard, 1982].

If iron is present at a sufficiently large concentration, it may dominate the neutrality condition at large p_{O_2} where it is oxidized to Fe^{3+} ($=Fe^{\bullet}_{Mg}$) [Smyth and Stocker, 1975; Morin et al., 1979; Sockel, 1974]. Figure 14 shows this for olivine with a ([Mg]+[Fe])/[Si] ratio close to 2 [Stocker and Smyth, 1978]. Figure 15

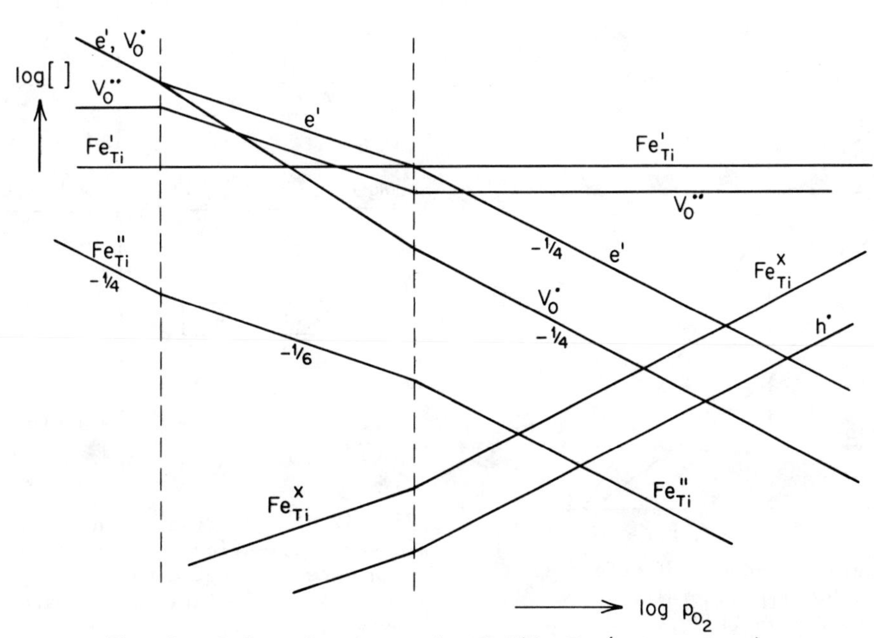

Fig. 9. Defect isotherms for $BaTiO_3$:Fe (schematical).

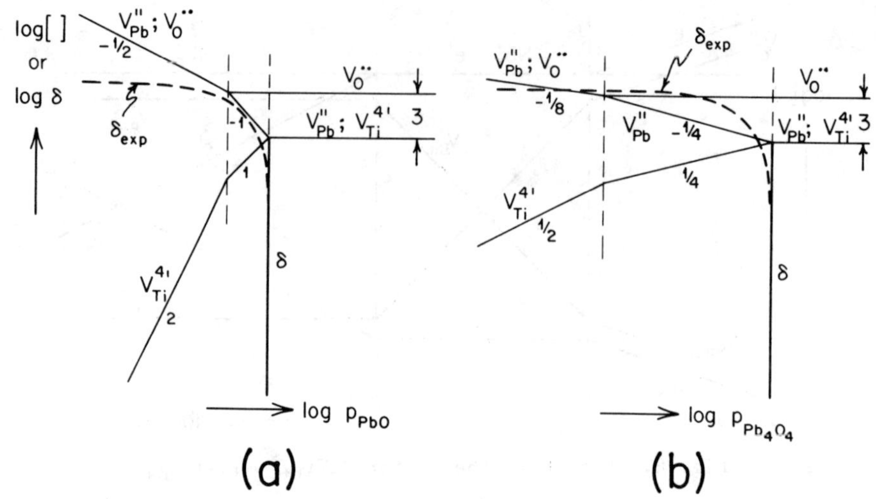

Fig. 10. Isotherms of defect concentration (and thus PbO deficiencies) as a function of (a) p_{PbO} and (b) $p_{Pb_4O_4}$ (Shackelford and Holman, 1975).

shows isotherms for Mg_2SiO_4:Fe at medium p_{O_2} (no oxygen excess or deficiency) as a function of sub-compound activity. Instead of SiO_2, here a_{En}, the activity of enstatite, $MgSiO_3$, was chosen which relates to defect concentrations through the incorporation reaction

$$4MgSiO_3 \rightleftarrows 4Mg^x_{Mg} + 2V''_{Mg} + 3Si^x_{Si} + Si^{\cdot\cdot\cdot\cdot}_i + 12O^x_O$$

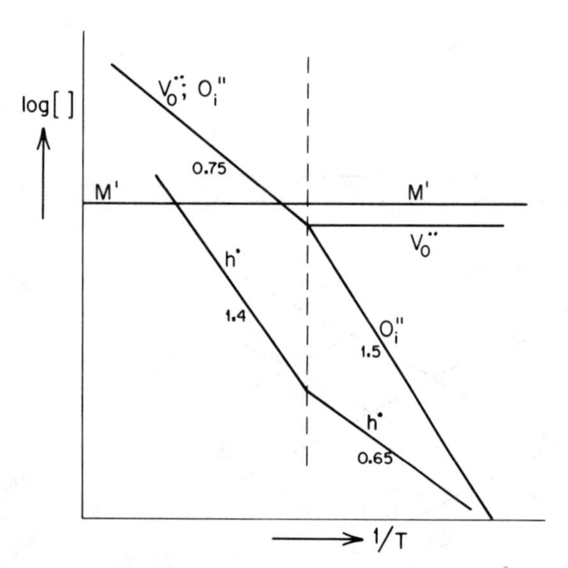

Fig. 11. Temperature dependence of defect concentrations in $PbWO_4:K_{Pb}$ according to data by Groenink and Binsma (1979).

Summary

Physical properties of crystalline solids depend on the presence of point defects and extended defects. The formation of these defects is described and it is shown how defect

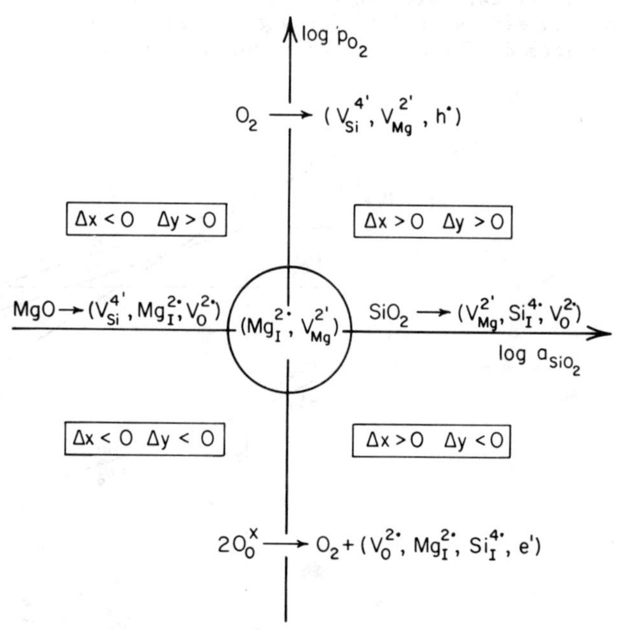

Fig. 12. Major defects in Mg_2SiO_4 expected in stoichiometric Mg_2SiO_4 and in crystals with excess or deficiency of oxygen or SiO_2 according to Abelard and Baumard (1982) and Smyth and Stocker (1975).

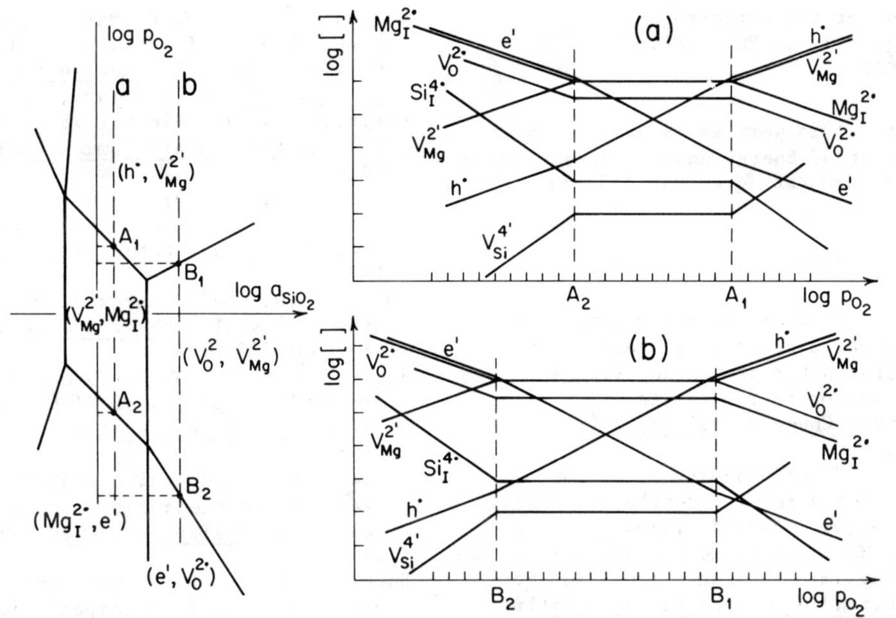

Fig. 13. Areas dominated by different sets of defects and defect concentration isotherms as $f(p_{O_2})$ for Mg_2SiO_4 with (a) [Mg]/[Si] = 2 and (b) [Mg]/[Si] < 2.

chemistry, a new branch of chemistry in which both atoms and defects play a part, can account quantitatively for variations in the concentrations of defects with variations of temperature, pressure, activities of constituents and activities or concentrations of impurities or dopants. A detailed discussion is given for the binary compounds TiO_2,

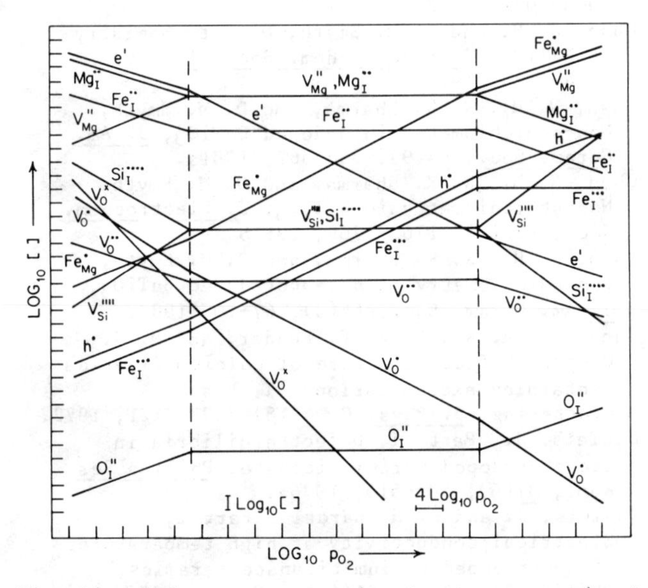

Fig. 14. Defect concentration isotherms as $f(p_{O_2})$ for Mg_2SiO_4:Fe. (Stocker and Smyth, 1978).

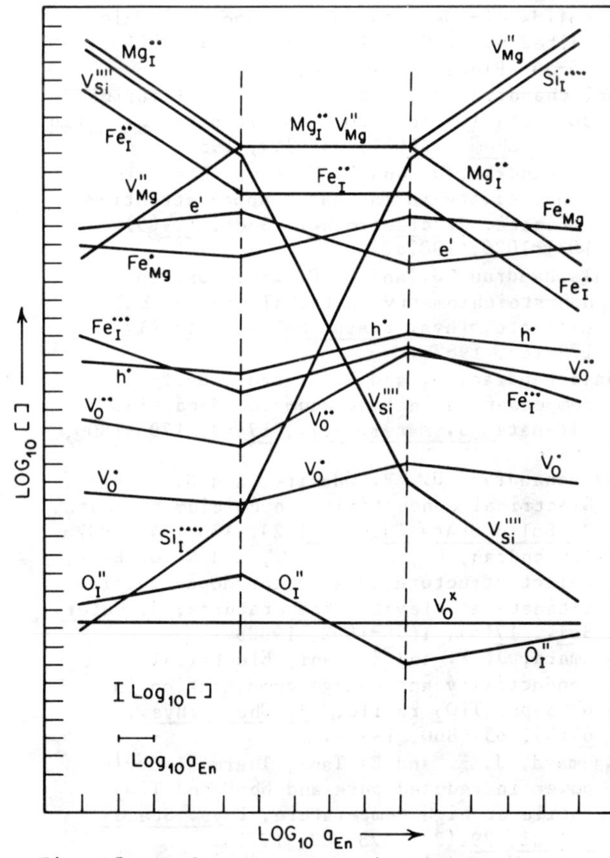

Fig. 15. Defect concentration isotherms as a function of enstatite activity a_{En} for Mg_2SiO_4:Fe without oxygen excess or deficiency (Stocker and Smyth, 1978).

Al_2O_3, Fe_3O_4 and for the ternary compounds $BaTiO_3$, $PbTiO_3$, $PbWO_4$, $PbMoO_4$, $LiNbO_3$, and Mg_2SiO_4.

Acknowledgment. This work was supported by the U.S. Department of Energy under contract AS03-76-SF001113, project agreement AT03-76 Er-71027.

References

Abelard, P. and J. F. Baumard, A new graphical representation for a systematic study of the defect structure in ternary oxides with a specific application to forsterite Mg_2SiO_4, J. Phys. Chem. of Solids,43 (7), 617-625, 1982.

Akse, J. R. and W. B. Whitehurst, Diffusion of titanium in slightly reduced rutile, J. Phys. Chem. Solids,39 (5), 457-465, 1978.

Anderson, J. S., Non-stoichiometric and ordered phases: Thermodynamic considerations, in The Chemistry of Extended Defects in Non-Metallic Solids, edited by L. Eyring and M. O'Keeffe, pp. 1-20, North-Holland, Amsterdam, 1970.

Anderson, J. S., The real structure of defect solids, in Defects and Transport in Oxides, edited by M. S. Seltzer and R. I. Jaffee, pp. 25-54, Plenum, New York, 1974.

Balachandran, U., and N. G. Eror, Electrical conductivity in strontium titanate, J. Solid State Chem., 39(3), 351-359, 1981.

Balachandran, U. and N. G. Eror, Electrical conductivity in lanthanum-doped strontium titanate, J. Electrochem. Soc., 129(5), 1021-1026, 1982a.

Balachandran, U. and N. G. Eror, Oxygen non-stoichiometry in tantalum-doped calcium titanate, Phys. Status Solidi, 71a (1), 179-184, 1982b.

Balachandran, U. and N. G. Eror, Self-compensation in lanthanum-doped calcium titanate, J. Mater. Sci., 17(6), 1795-1800, 1982c.

Balachandran, U., B. Odekirk, and N. G. Eror, Electrical conductivity in calcium titanate, J. Solid State Chem., 41(2), 185-194, 1982a.

Balachandran, U., B. Odekirk, and N. G. Eror, Defect structure of acceptor-doped calcium titanate at elevated temperatures, J. Mater. Sci., 17(6), 1656-1662, 1982b.

Baumard, J. F. and E. Tani, Electrical conductivity and charge compensation in Nb-doped TiO_2 rutile," J. Chem. Phys., 67(3), 857-860, 1977a.

Baumard, J. F. and E. Tani, Thermoelectric power in reduced pure and Nb-Doped TiO_2 rutile at high temperature, Phys. Status Solidi, 39a(2), 373-382, 1977b.

Becker, M. and A. Scharmann, Photolumineszenz und Radiothermolumineszenz in $CaWO_4$ und $CaWO_4$:Pb Einkristallen, L. Naturf. 29a (7) 1060-1064, 1974.

Belabaev, K. G., V. B. Markov and S. G. Odulov, Photovoltaic effect in reduced lithium niobate crystals, Sov. Phys. Solid State Engl. Transl., 20(8), 1458-1459, 1978.

Bergmann, G., The electrical conductivity of $LiNbO_3$, Solid State Commun. 6(2), 77-79, 1968.

Bollmann, W., The origin of photoelectrons and the concentration of point defects in $LiNbO_3$ crystals, Phys. Status Solidi 40a,(1), 83-91, 1977.

Bollmann, W. and M. Gernand, On the disorder of $LiNbO_3$ crystals, Phys. Status Solidi, 9a (1), 301-308, 1972.

Bouwma, J. and M. A. Heilbron, Non-stoichiometry and optical spectra of Nd (III) substituted $PbTiO_3$, Mater. Res. Bull., 11(6), 663-668 1976.

Brouwer, G., A general asymptotic solution of reaction equations common in solid state chemistry, Philips Res. Rep., 9(5), 366-376 1954.

Burns, W. K., C. H. Bulmer, and E. J. West, Application of Li_2O compensation techniques to Ti-diffused $LiNbO_3$ planar and channel waveguides, Appl. Phys. Lett., 33(1), 70-72, 1978.

Byer, R. L., J. E. Long and R. S. Feigelson, Growth of high-quality $LiNbO_3$ crystals from the congruent melt, J. Appl. Phys. 41(6), 2320-2325, 1970.

Castle, J. E. and P. L. Surman, The Self-diffusion of oxygen in magnetite: The effect of anion vacancy concentration and cation distribution, J. Phys. Chem., 73(3), 632-634, 1969.

Catlow, C. R. A., R. James, W. C. Mackrodt and R. T. Stewart, Defect energies in α-Al_2O_3 and rutile TiO_2, Phys. Rev. B, 25(2) 1006-1031, 1982.

Chan, N. H. and D. M. Smyth, Defect chemistry of $BaTiO_3$, J. Electrochem. Soc., 123(10) 1584-1585, 1976.

Chan, N. H., R. K. Sharma, and D. M. Smyth, Non-stoichiometry in undoped $BaTiO_3$, J. Am. Ceram. Soc., 64(9), 556-562, 1981a.

Chan, N. H., R. K. Sharma, and D. M. Smyth, Non-stoichiometry in $SrTiO_3$, J. Electrochem. Soc., 128(8), 1762-1769, 1981b.

Chan, N. H., R. K. Sharma, and D. M. Smyth, Nonstoichiometry in acceptor-doped $BaTiO_3$, J. Am. Ceram. Soc., 65(3), 67-70, 1982.

Cheetham, A. K., B. E. F. Fender, and M. J. Cooper, Defect structure of calcium fluoride containing excess cations, I, Bragg scattering, J. Phys. C, 4(18) 3107-3121, 1972.

Daniels, J., Part II, Defect equilibria in acceptor-doped barium titanate, Philips Res Rep., 31(6), 505-515, 1976a.

Daniels, J. and K. H. Härdtl, "Part I, Electrical conductivity at high temperatures of donor-doped barium titanate ceramics, Philips Res. Rep., 31(6), 489-504, 1976b.

Dieckmann, R. and H. Schmalzried, Point defects

and cation diffusion in magnetite, Z. Phys. Chem., 96(4-6), 331-333, 1975.
Dieckmann, R. and H. Schmalzried, Defects and cation diffusion in magnetite, I, Ber. Bunsenges. Phys. Chem., 81(3), 344-347, 1977a.
Dieckmann, R. and H. Schmalzried, Defects and cation diffusion in magnetite, II, Ber. Bunsenges. Phys. Chem., 81(4), 414-419, 1977b.
Dieckmann, R., T. O. Mason, J. D. Hodge and H. Schmalzried, Defects and cation diffusion in magnetite, III, Tracer diffusion of foreign tracer cations as a function of temperature and oxygen potential, Ber. Bunsenges. Phys. Chem. 82(8), 778-783, 1978.
Dischler, B., J. R. Herrington and A. Räuber, Correlation of the photorefractive sensitivity in doped $LiNbO_3$ with chemically induced changes in the optical absorption spectra, Solid State Commun., 14(11), 1233-1236, 1974.
Drowart, J., R. Colin and G. Exsteen, Mass-spectroscopic study of the vaporizaton of lead monoxide, J. Chem. Soc. Faraday Trans., 61(7), 1376-1381, 1965.
Eror, N. G., Self-compensaton in niobium doped TiO_2, J. Solid State Chem., 38(3), 281-287, 1981.
Eror, N. G. and U. Balachandran, Self-compensation in lanthanum-doped strontium titanate, J. Solid State Chem., 40(1), 85-91, 1981.
Eror, N. G., and U. Balachandran, Electrical conductivity in strontium titanate with nonideal cationic ratio, J. Solid State Chem., 42(3), 227-241, 1982.
Eror, N. G. and D. M. Smyth, Non-stoichiometric disorder in single crystalline $BaTiO_3$ at elevated temperatures, J. Solid State Chem., 24(3, 4), 235-244, 1978.
Esdaile, R. J., Closed-tube control of out-diffusion during fabrication of optical waveguides in $LiNbO_3$, Appl. Phys. Lett., 33(8), 733-734, 1978.
Eshelby, J. D., C. W. A. Newly, P. L. Pratt and A. B. Lidiard, Charged dislocations and the strength of ionic crystals, Philos. Mag., 3, 75-89, 1958.
Frenkel, J. I., Thermal agitation in solids and liquids, Z. Phys., 35, 652-669, 1926.
Frenkel, J. I., Kinetic Theory of Liquids, Oxford University Press, New York, 1946.
Gado, P., and A. Magneli, Studies on structural defects in wolfram trioxide; Influence of minor additons of tantala or niobia, Mater. Res. Bull., 1(1), 33-34, 1966.
Greskovich, C. and H. Schmalzried, Non-stoichiometry and electronic defects in Co_2SiO_4 and in $CoAl_2O_4$-Mg Al_2O_4 crystalline solutions, J. Phys. Chem. Solids, 31(4), 639-646, 1970.
Groenink, J. A. and H. Binsma, Electrical conductivity and defect chemistry of $PbMoO_4$, J. Solid State Chem., 29(2), 227-236, 1979.
Gupta, Y. P. and L. J. Weirick, Diffusion in calcium tungstate single crystals, J. Phys. Chem. Solids, 28(12), 2545-2552, 1967.
Hagemann, H. J., and D. Hennings, Reversible weight change of acceptor-doped $BaTiO_3$, J. Am. Ceram. Soc., 64(10), 590-594, 1981.
Hagemann, H. J., and H. Ihrig, Valence change and phase stability of 3d-doped $BaTiO_3$ annealed in oxygen and hydrogen, Phys. Rev. B, 20(9), 3871-3878, 1979.
Hagemann, H. J., A. Hero, and U. Gonser, The valence change of Fe in $BaTiO_3$ studied by mossbauer effect and gravimetry, Phys. Status Solidi, 61a(1), 63-72, 1980.
Härdtl, K. H., and H. Rau, PbO vapour pressure in the $Pb(Ti_{1-x}Zr_x)O_3$ system, Solid State Commun., 7(1), 41-45, 1969.
Haul, R., and G. Duembgen, Sauerstoff-Selbstdiffusion in Rutilkristallen, J. Phys. Chem. Solids, 26(1), 1-10, 1965.
Hennings, D., Part III, Thermogravimetric investigation, Philips Res. Rep., 31(6), 516-525, 1976.
Holman, R. L. and R. M. Fulrath, Intrinsic non-stoichiometry in the lead zirconate lead titanate system determined by Knudsen effect, J. Appl. Phys., 44(12), 5227-5236, 1973.
Holman, R. L., P. J. Cressman, and J. F. Revelli, Chemical control of optical damage in lithium niobate, Appl. Phys. Lett., 32(5), 280-283, 1978.
Ihrig, H. and D. Hennings, Electrical transport properties of n-type $BaTiO_3$, Phys. Rev. B, 17(12), 4593-4599, 1978.
Ilschner, B., H. G. Sockel and G. Streb, Influence of non-stoichiometry on the plasticity of oxides and silicates with special reference to Fe_2SiO_4 (in German), Z. Phys. Chem. (Neue Folge) or (Frankfurt), 108(2), 235-245, 1977.
Jorgensen, P. J. and R. W. Bartlett, High temperature transport processes in lithium niobate, J. Phys. Chem. Solids, 30(12), 2639-2648, 1969.
Kliewer, K. L., Space charge in ionic crystals, II, The electron affinity and impurity accumulation, Phys. Rev., 140a,(4), 1241-1246, 1965.
Kliewer, K. L. and J. S. Koehler, Space charge in ionic crystals, I, General approach with applications to NaCl, Phys. Rev., 140a,(4), 1226-1240, 1965.
Koch, F. and J. B. Cohen, The defect structure of $Fe_{(1-x)}O$, Acta Crystallogr. Sect. B, 25(2), 275-287, 1969.
Koehler, H. A. and C. K. Kikuchi, Identification of three trapping centers in calcium tungstate, Phys. Status Solidi, 43b(1), 423-432, 1971.
Kosek, F., and H. Arend, On high temperature conductivity of $BaTiO_3$, Phys. Status Solidi, 24(1), K69-71, 1967.
Krindach, D. P., V. S. Mriocov and L. B. Meisner, Influence of the structure of $LiNbO_3$ crystals of different stoichiometry on their refractive indices, Sov. Phys. Solid State, 18(10), 1756-1757, 1976.

Kröger, F. A., Chemistry of Imperfect Crystals, North-Holland, Amsterdam, 1974a.
Kröger, F. A., Defect thermodynamics--historical, in Defects and Transport in Oxides, edited by M. S. Seltzer and R. I. Jaffee, pp. 3-24, Plenum, New York, 1974b.
Kröger, F. A., Chemical potentials of individual structure elements of compounds, and face and dislocation specificity of thermodynamic parameters, J. Phys. Chem. Solids, 41(7), 741-746, 1980.
Kröger, F. A., Non-stoichiometry in shear structures, J. Phys. Chem. Solids, 44(4), 345-347, 1983.
Kröger, F. A. and H. J. Vink, Relations between the concentrations of imperfections in crystalline solids, in Solid State Physics vol. 3, edited by F. Seitz, and D. Turnbull, pp. 307-435, Academic, New York, 1956.
Lecomte, J., J. P. Loup, M. Hervieu, and B. Raveau, Non-stoichiometry and electrical conductivity of strontium niobates with perovskite structure, Phys. Status Solidi, 66a (2), 551-558, 1981.
Lehovec, K., Space charge layer and distribution of lattice defects at the surface of ionic crystals, J. Chem. Phys., 21(7), 1123-1128, 1953.
Limb, Y., and D. M. Smyth, Oxygen non-stoichiometry and the Li/Nb ratio in $LiNbO_3$, Am. Ceram. Soc. Bull., 60(3), 358, 1981.
Long, S. A., and R. N. Blumenthal, Ti-rich nonstoichiometric $BaTiO_3$, II, Analysis of defect structure, J. Am. Ceram. Soc., 54(II), 577-583, 1971.
Miyazawa, S., R. Gugielmi, and A. Carenco, A simple technique for suppressing Li_2O out-diffusion in Ti-$LiNbO_3$ optical waveguide, Appl. Phys. Lett., 31(11), 742-744, 1977.
Morin, F. J., J. R. Oliver, and R. M. Housley, Electrical properties of forsterite, Mg_2SiO_4, Phys. Rev. B, 16(10), 4434-4445, 1977.
Morin, F. J., J. R. Oliver, and R. M. Housley, Electrical properties of forsterite, Mg_2SiO_4, II, Phys. Rev. B, 19(6), 2886-2894, 1979.
Nassau, K. and G. M. Loiacono, Calcium tungstate, III, Trivalent rare earth substitution, J. Phys. Chem. Solids, 24(12), 1503-1510, 1963.
Odekirk, B., U. Balachandran, N. G. Eror, and J. S. Blackmore, Electronic conduction in quenched ceramic samples of highly reduced lanthanum-doped $SrTiO_3$, Mater. Res. Bull., 17(2), 199-208, 1982.
Odier, P., and J. R. Loup, An unusual technique for the study of non-stoichiometry: The thermal emission of electrons, results for Y_2O_3 and TiO_2, J. Solid State Chem., 34(1), 107-119, 1980.
Petrov, A., and P. Kofstad, Electrical conductivity of $CaMoO_4$, J. Solid State Chem., 30(1), 83-88, 1979.
Phillips, W. and D. L. Staebler, Control of the Fe^{2+} concentration in iron-doped lithium niobate, J. Electronic Mat., 3(2), 601-617, 1974.
Pluschkell, W. and H. J. Engell, Ionen und elektronen leitung in magnesium ortho silikat, Ber. Dtsch. Keram. Ges., 45(8), 388-394, 1968.
Poeppel, R. B., and J. M. Blakely, Origin of equilibrium space charge potentials in ionic crystals, Surf. Sci., 15(3), 507-523, 1969.
Ranganath, T. R., and S. Wang, Suppression of Li_2O out-diffusion from Ti-diffused $LiNbO_3$ optical waveguides, Appl. Phys. Lett., 30(8), 376-379, 1977.
Reed, T. B., The role of oxygen pressure in the control and measurement of composition in 3d metal oxides, in The Chemistry of Extended Defects in Non-Metallic Solids, edited by L. Eyring and M. O'Keeffe, pp. 21-35, North-Holland, Amsterdam, 1970.
Rigdon, M. A. and R. E. Graves, Electrical charge transport in single crystal $CaWO_4$, J. Am. Ceram. Soc., 56(9), 475-478, 1973.
Schmalzried, H., Point defects in ternary ionic crystals, in Progress in Solid State Chemistry, vol. 2, edited by H. Reiss, pp. 265-303, Pergamon, New York, 1965.
Schmalzried, H., and C. Wagner, Fehlordnung in ternären Ionenkristallen, Z. Phys. Chem. (Neue Folge) or (Frankfurt), 31(3, 4), 198-221, 1962.
Schottky, W., The mechanism of ionic motion in solid electrolytes, Z. Phys. Chem. B, 29, 335-355, 1935.
Schottky, W., and C. Wagner, Theory of ordered mixed phases, I, Z. Phys. Chem. B, 11, 163-210, 1931.
Seuter, A. M. J. H., Defect chemistry and electrical transport properties of barium titanate, Philips Res. Rep. Suppl., 3, 1-84, 1974.
Shackelford, J. F., and R. L. Holman, Non-stoichiometry in ABO_3 compounds similar to $PbTiO_3$, J. Appl. Phys., 46(4), 1429-1434, 1975.
Sharma, R. K., N. H. Chan, and D. M. Smyth, Solubility of TiO_2 in $BaTiO_3$, J. Am. Ceram. Soc., 64(8), 448-451, 1981.
Shirashi, S., H. Yamamura, H. Hanecke, K. Kakegawa, and J. Moori, Defect structure and oxygen diffusion in undoped and La-doped polycrystalline barium titanate, J. Chem. Phys., 73(9), 4640-4645, 1980.
Smyth, D. M., Thermodynamic characterization of ternary compounds, II, The case of extensive defect association, J. Solid State Chem., 20(4), 359-364, 1977.
Smyth, D. M., and R. L. Stocker, Point defects and non-stoichiometry in forsterite, Phys. Earth Planet. Inter., 10(2) 183-192, 1975.
Sockel, H. G., Defect structure and electrical conductivity of ferrous silicate, in Defects and Transport in Oxides, edited by M. S. Seltzer and R. I. Jaffee, pp 341-354, Plenum, New York, 1974.
Staebler, D. L., Oxide optical memories:

Photochromism and index change, J. Solid State Chem., 12(3, 4), 177-185, 1975.
Stocker, R. L., and D. M. Smyth, Effect of enstatite activity and oxygen partial pressure on the point defect chemistry of olivine, Phys. Earth Planet. Inter., 16(2), 145-156, 1978.
van Loo, W., Crystal growth and electrical conduction of $PbMoO_4$ and $PbWO_4$, J. Solid State Chem., 14(4), 359-365, 1975.
Wadsley, A. D., Crystal chemistry of stoichiometric compounds, Rev. Pure Appl. Chem., 5, 165-173, 1955.
Wagner, C., Theory of ordered mixed phases II, Z. Phys. Chem., Bodenstein Festband, 177-186, 1931.
Wagner, C., Theory of ordered mixed phases III, Disorder in polar compounds as a basis for ionic and electronic conduction, Z. Phys. Chem. B, 22, 181-194, 1933.
Wagner, C., and W. Schottky, Theory of ordered mixed phases I, Z. Phys. Chem. B, 11, 163-210, 1931.
Wernicke, R., Part IV, The kinetics of equilibrium restoration in barium titanate ceramics, Philips Res. Rep., 31(6), 526-543, 1976.
Whitworth, R. W., A measurement of the charge on edge dislocations in a sodium chloride crystal, Philos. Mag., 15(2), 305-319, 1967.
Willis, B. T. M., Positions of the oxygen atoms in $UO_{2.13}$, Nature London, 197(4869), 755-756, 1963.
Wrobel, Z., Electrical conductivity and Seebeck's coefficient in $Pb(Zr_xTi_{1-x})O_3$ solid solution near the morphotropic phase boundary $0.46 < x < 0.60$, Acta Phys. Pol. A, 55(3), 267-274, 1979.

POINT DEFECTS IN CRYSTALS: A QUANTUM CHEMICAL METHODOLOGY AND ITS APPLICATIONS

Alfred B. Anderson

Chemistry Department, Case Western Reserve University, Cleveland, Ohio 44106

Abstract. Theoretical methods for studying minerals and defects are briefly surveyed. The author's atom superposition and electron delocalization molecular orbital (ASED-MO) theory is introduced. Using this theory and cluster models, the author has recently calculated and discussed structural and electronic properties for wustite, hematite, magnetite, molybdite, bismite, bunsenite and α-quartz. The valence features of optical and photoemission spectra predicted by these calculations are in good agreement with experiment. This work is discussed. The conducting bronze behavior of oxygen deficient molybdite is explained. Finally, a new model is introduced for determining structures of clusters of defects in wustite, $Fe_{1-x}O$.

1. Introduction

The purpose of this section is to give a cursory overview of the status of theoretical techniques used, or of potential use, in studying minerals. The references, though broad-ranging, are not intended to be complete. It is the author's goal to provide the reader with a very recent entry into the literature of each technique. In the next section the author's theoretical method will be introduced and in the third section some of his very recent applications of his method to minerals will be discussed. The fourth and final section summarizes the prior two.

In the study of minerals several types of questions often arise. They can involve structure, stability, diffusion, phase transformations, and vibrational and electronic properties. Different theoretical methods are available for treating these properties. Some methods are relatively easy to use and others are quite difficult and some have only a tenuous relationship, from a theoretical viewpoint, to the others.

Several of the most widely used models are really empirical, and the Pauling radii and Madelung sums come to mind as examples. Advancements of this type of approach have recently been fruitful. Pairwise atom and ion interaction potential energy models have been the basis of interesting studies of structure, stability, and diffusion as applied by Catlow et al. [1982].

Simplified quantum mechanical models are being developed. The Kohn-Sham-Hohenberg [Kohn and Sham, 1965] local density functional theory provides a basis for estimating energies of many-electron systems from single particle density distribution functions. In principle, this approach is computationally simpler than conventional wave mechanical methods which deal in a higher dimensional space of multielectron wavefunctions. Several models for structural and mechanical properties of crystals have grown out of the density functional theory. The Gordon and Kim [1972] model uses rigid ionic electronic charge density functions and has been used recently by Boyer [1981] in a predictive study for a series of alkali halides. This and the Catlow approach tell us nothing about electronic properties such as optical absorption and electron photoemission spectra or conductivity.

Self-consistent valence electron delocalization using atomic pseudopotential hamiltonians has led to improvements over the Gordon-Kim rigid ion model. Using such corrections Andreoni et al. [1982] analyzed the cohesion of NaCl and obtained, presumably, band data suitable for some discussion of the electronic properties, though no discussion was made. Others have used the density functional theory in conjunction with electron delocalization to treat a broad range of properties. For example, Yin and Cohen [1982] have used the so-called ab initio pseudopotential model to predict structures, pressure-induced phase transformations, and vibrational and electronic parameters for silicon and germanium. All predicted properties except optical transition energies are in impressive agreement with experiment.

Other band theory methods are used to discuss the electronic structures of solids. The Hartree-Fock ab initio procedure of Kunz [1982] has worked well in predicting the optical properties of silver halides. However this and most other band theory techniques, excluding those using density functional theory mentioned above, are, because of cost or inaccuracy, rarely used to predict structures, phase transformations, and so on.

Since the density functional and Hartree-Fock procedures mentioned can produce good accuracy, one may wonder why they are not being used to predict and interpret properties of point defects in minerals. The answer is multifold. The methods are still new and relatively difficult to use. It may be quite difficult to treat complicated crystals with large unit cells, and, even in a simple solid, an isolated point defect must be modeled by a large unit cell. It may be mentioned that the physicists who are developing these techniques may not be aware of some of the problems in mineralogy where applications might be possible.

It is because of the difficulty of making accurate a priori theoretical predictions of properties of macroscopic mineral systems that chemical models have been so valuable. The concepts of ionic radius, valency, crystal field splitting and others have some predictive value and, importantly, interpretative value, providing theoretical frameworks for understanding. These qualitative concepts will, in all likelihood, provide the language of discussion into which results of future high accuracy calculations will need to be translated. In the interim, approximate models will continue to be used. Mineralogy cannot be expected to stand still waiting for high-accuracy calculations.

By using cluster fragment models it is possible to focus on the local chemical bonding properties of crystals while using simplified quantum mechanical methods. Models consisting of metal cations surrounded by nearest neighbor oxygen anions can be used to predict and interpret electronic properties of oxides, for example. Such properties include optical and photoemission spectra and conductivity. These studies are easily performed using readily available computer programs. The list of published examples is increasing rapidly. Recent ones include the study of chalcopyrite by Tossell et al. [1982] and the study of magnetite, hematite, and wustite by Debnath and Anderson [1982].

Most of the molecular orbital studies using clusters made thus far shy away from questions of structure preference and stability. Work of Gibbs [1982] where small hydroxyacid molecules are studied using Hartree-Fock theory and are found to bear structural similarities to various minerals is an interesting exception. Straightforward applications of bonding and electron count concepts to understanding relative bond lengths have been made, as in the work of Davies and Navrotsky [1981] in an empirical way, and in the theoretical work of Debnath and Anderson [1982]. The latter authors discuss why Fe-O distances in wustite (FeO), which contains Fe^{II}, are larger than the average Fe-O distances in hematite (Fe_2O_3), which contains Fe^{III}. The answer lies in the valence Fe d orbitals being antibonding to the neighboring oxygen anions. Such work is essentially an obvious application of well-known concepts from transition metal coordination chemistry. Tossell and Gibbs [1977] have investigated relationships between bond lengths and angles by focusing on orbital energy changes, a method of organic and inorganic chemistry.

There is a history of using the crystal field splitting parameter of coordination chemistry to understand crystal structures. This approach has recently been called into question by Burdett et al. [1982], who show the crystal field splitting parameter does not do a good job of sorting normal and inverse spinels. They argue that the metal and transition metal valence s and p orbitals may be more important. Further investigation should prove interesting.

From the studies mentioned in the previous two paragraphs, the conclusion may be drawn that the transition metal d electron energy levels may be used to rationalize some but not all features of bulk structures. It is clear, then, that in addition to the ligand-metal antibonding levels, the metal d band, the ligand-metal bonding levels, the oxygen 2p band in the case of oxides, must be considered. Furthermore, these orbital energies yield, when multiplied by their occupation, 0, 1, or 2, and summed, only a part of the total energy. To make predictions it may sometimes be necessary to include Madelung energies, when extended Hückel molecular orbital theory is being used, or the nuclear-nuclear repulsion energy minus the electron repulsion energy (which is counted twice in the orbital energy) when Hartree-Fock molecular orbital theory is being used. For further details, the reader is referred to Lowe [1978]. A pseudopotential approach such as discussed above would be handled the same way as Hartree-Fock theory in the sense that the orbital energy is only a part of the total.

In the cluster fragment studies mentioned so far, and in many other published ones, it is well established that electronic structure parameters are readily within reach using simple cluster models and that some structural features may be understood, if not predicted. What about using these models for predicting structures in crystals? The hope for doing so is far-fetched when full ab initio electronic hamiltonians are employed. For example in an FeO_6 model of hematite nine electrons must be added to the cluster to put iron in the ferric oxidation state. This is an untenable situation for structure determinations because the cluster would be predicted to fly apart due to the huge repulsive energy of all the electrons in FeO_6^{9-}. This problem is at least partially overcome in electronic structure calculations such as those of Tossell and Gibbs [1977] by placing a neutralizing spherical shell of positive charge around the cluster. Such a technique may be more appropriate for electron structure determinations than for bond length and angle predictions. For structure predictions model hamiltonians might be expected to have an edge. The extended Hückel hamiltonian omits explicit electron repulsion integrals but does include the effects of atomic orbital symmetry and overlap in determining molecular orbitals and their energies. Debnath and Anderson [1982] obtained good qualitative predictions of electronic properties using this type of hamiltonian and charged cluster models. Might such clusters be used to predict bulk structures? There is a glimmer of hope that this may at least sometimes be possible. Burdett and McLarnan [1982] have found OBe_4^{6+} clusters, when all the angles are optimized, assuming the O-Be distances are constant, produce the bulk coordination in an extended Hückel calculation. Furthermore, they find a solid linear relationship relating the orbital energies of OBe_4^{6+} clusters of various structures to extended Hückel tight-binding band calculations for the bulk. Will orbital symmetry and overlap arguments, which are the basis for our modern understanding of structures and reactions in organic and inorganic chemistry, apply to solids? It is difficult to believe they will not, at least for certain types of crystals. It is still possible that the non-orbital part of the energy may sometimes dominate, but in predicting structures and reactivity in molecules this is a rare occurrence. For interesting examples of the use of extended Hückel tight binding band calculations and molecular orbital theory to analyze structures, see the work of Burdett and Lee [1983] and Hughbanks and Hoffmann [1983].

The author has been modeling crystals with cluster fragments using a method of his own which is based on an atom superposition and electron delocalization (ASED) approach to forming chemical bonds. While this is an electronic charge density theory it is not to be confused with the density functional theory discussed above. One of the two energy components in chemical bonds in this theory is unavailable but can be well approximated using an extended Hückel one-electron molecular orbital energy. Thus the ASED-MO theory not only retains the simplicity and flexibility and interpretative value of the extended Hückel procedure but also allows full structure predictions for it includes a non-orbital part of the total energy.

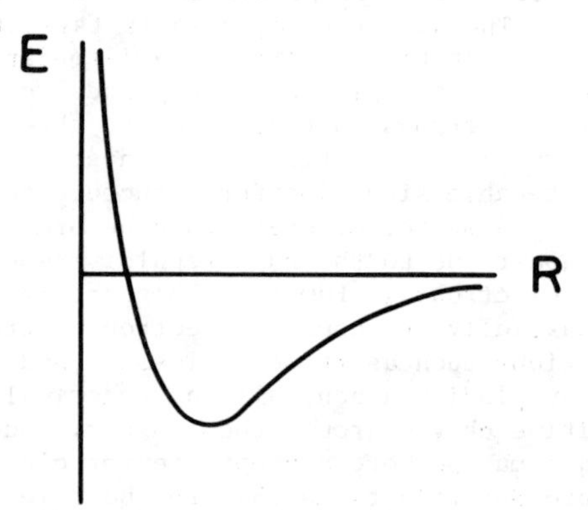

Fig. 1. Diatomic potential energy function.

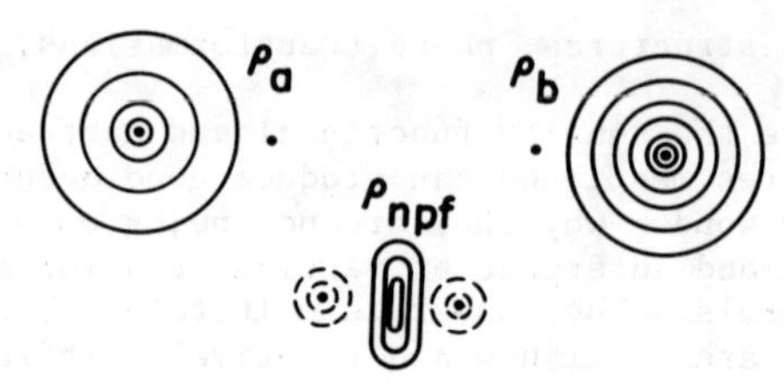

Fig. 2. Depiction of electron density components for a diatomic molecule a-b according to (1).

2. Atom Superposition and Electron Delocalization Theory

The ASED theory is physically simple and lends itself to intuitive understanding. It employs atomic densities which are used for form factors in X-ray spectroscopy. It provides a theory for force constants and atomic radii which has as yet not been fully explored. For these reasons it is appropriate to present some of the details.

The theory uses the electrostatic force theorem of Hellmann [1937] and Feynman [1939]. The theorem is easily proven and states that the force on a nucleus is equal to the electrostatic force due to the presence of the other nuclei and the electronic charge distribution in a molecule or a crystal. The integral of this force as two atoms are brought together from an infinite separation is the molecular binding energy curve as in Figure 1. This energy can be separated into parts through a partitioning of the molecular electronic charge distribution, ρ. A useful partitioning of ρ is into rigid atom components, ρ_i, that follow the nuclei and a non-rigid, non-perfectly-following part, ρ_{npf}. For a diatomic molecule, a-b, then,

$$\rho = \rho_a + \rho_b + \rho_{npf} \tag{1}$$

This is shown schematically in Figure 2. The solid contours indicate the accumulation of electron density and the dashed contours indicate depletion. Note $\int \rho_a(\underset{\sim}{r}_a)d\underset{\sim}{r}_a = Z_a$ and $\int \rho_b(\underset{\sim}{r}_b)d\underset{\sim}{r}_b = Z_b$ where Z_a and Z_b are nuclear charges of the neutral atoms. Therefore $\int \rho_{npf}(r)dr = 0$, and ρ_{npf} may be viewed as an electron delocalization or bond charge density, and is equal to the difference between ρ and the atomic parts ρ_a and ρ_b. The force on nucleus b has two non-zero components, one due to the nucleus and electronic charge distribution of atom a and the other due to ρ_{npf}. Since atomic densities are readily available, the first component is easily calculated. It is repulsive because atom a is neutral and its electrons do not completely shield its nucleus. The second component is necessarily attractive over a range in R_a but it is not

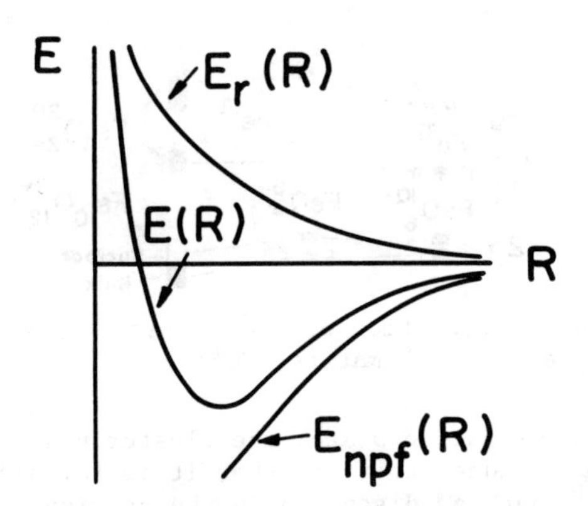

Fig. 3. Diatomic potential energy components using the densities in Figure 2 in (2).

readily available since ρ_{npf} is unknown. Chemical bonding has been examined in terms of these forces by several workers [see Deb, 1981]. To get ρ_{npf} one would have to already have solved the molecular Schrödinger equation. However, both forces, when integrated, must yield a repulsive energy, E_r and an attractive energy, E_{npf} as in (2) and Figure 3

$$E = E_r + E_{npf} \tag{2}$$

Formulas for E_r and E_{npf} are given by Anderson [1974].

Bond stretching force constants, k_e, for molecules and solids can be calculated to good accuracy using the classical electrostatic Poisson equation of Anderson and Parr [1970, 1971]:

$$k_e = 4\Pi Z_b \rho_a(b) \tag{3}$$

Anderson [1974] has shown this astonishingly simple formula is easily derived by assuming that ρ_{npf} may be replaced by point bond charges fixed in position during molecular vibrations. Then the Laplacian of the total energy, $\nabla^2_{R_b} E$ where E is obtained from (2), yields (3) because $\nabla^2_{R_b}(\underset{\sim}{R}_b - \underset{\sim}{r})^{-1} = -4\Pi\delta(0)$.

Atomic electron charge density functions decay nearly exponentially from the nucleus, so for various states of a bond the log of k_e should be linear in R. This is the case and a typical example from the work of Anderson [1975b] is shown in Figure 4.

Equation (3) has been a source of accurate empirical relationships for determining force constants from bond lengths by Anderson [1972] and Anderson and Parr [1971]. It has also provided the Anderson and Parr [1971] definition of atomic radius. These are useful and theorretically justified alternatives to the well-known force constant rules of Badger [1934, 1935] and Herschbach and Laurie [1961] and the Pauling radii. While this simple theory has interesting potential for treating elastic constants and packing problems for minerals and defects in minerals, no attempts have been made to exploit it in this context.

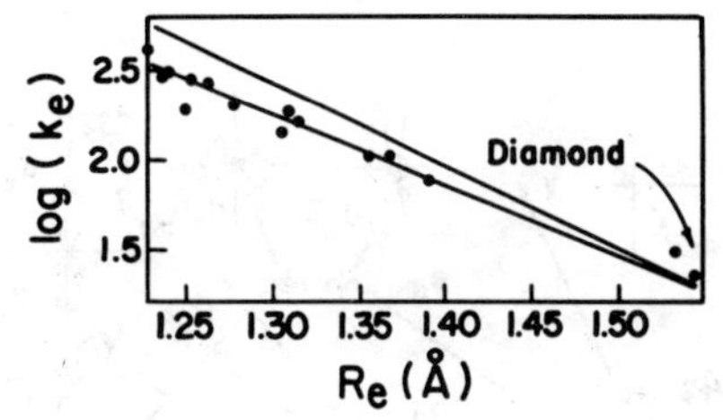

Fig. 4. The upper line is (3) plotted for C_2. Points are experimental data for various states of C_2 and for diamond.

The emphasis of the author has been directed toward finding an approximation to E_{npf} in order to make energy surface and structure predictions and to study chemical reactions using (2). To this end Anderson and Hoffmann [1974] showed the one-electron extended Hückel molecular orbital energy is a useful approximation to E_{npf}. Anderson [1975a] has provided partial theoretical justification for this intuitively reasonable discovery which, when implemented, yields the atom superposition and electron delocalization molecular orbital (ASED-MO) theory used by the author.

In summary, the ASED-MO theory of pushing together atoms, determining the energy for this, and simultaneously determining the energy associated with electron delocalization and bond formation, is an aesthetically pleasing one. The atomic densities contain, when equilibrium bond lengths are known, bond force constants to good accuracy including higher order ones, and provide a definition of atomic radius. The addition of the molecular orbital approximation allows approximate predictions of structures, stabilities, and energy surfaces as well. Furthermore, the orbital component retains all of the flexibility and interpretative attractiveness of extended Hückel theory.

3. Bonding in Representative Transition Metal Oxides

This section consists of selected case studies of minerals using cluster models and the ASED-MO theory. Optical, photoemission, magnetic, and conducting properties will be discussed. In the

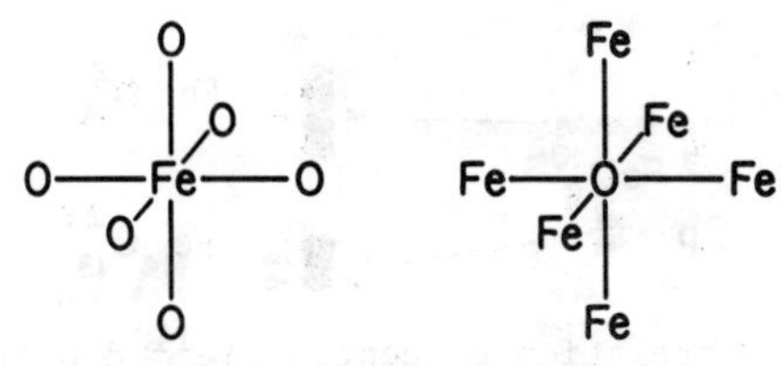

Fig. 5. Octahedral coordination of ferrous cations and oxygen anions in perfect sodium chloride structure wustite.

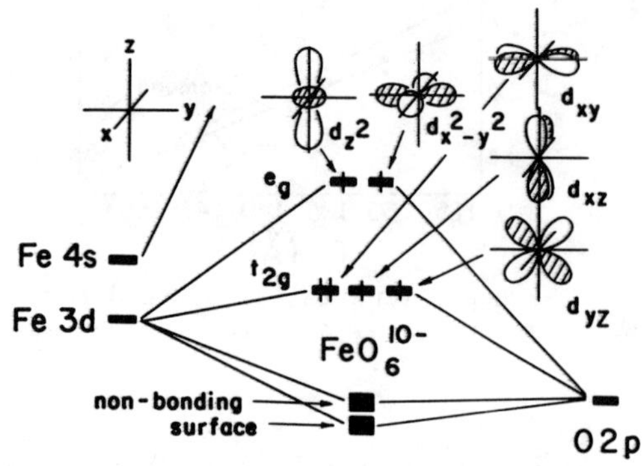

Fig. 6. Correlation diagram and Fe 3d orbitals in the FeO_6^{-10} wustite model.

case of molybdite, oxygen deficiency point defects are treated. All of this work is based on assumed structures. A recently developed approach to predicting defect cluster structures in wustite, $FeO_{1-x}O$, due to A. B. Anderson, R. W. Grimes, and A. H. Heuer (unpublished manuscript, 1983), will be discussed briefly in the last section.

Wustite, Hematite, and Magnetite

Idealized non-defective wustite, FeO, has the simple cubic sodium chloride structure. Each d^6 ferrous cation is octahedrally coordinated by six oxygen anions at a distance of 2.16 Å as in Figure 5. The anions are similarly coordinated by the cations.

Debnath and Anderson [1982] in an ASED-MO analysis using cluster models find the expected molecular orbital binding as shown in Figure 6. Quantum mechanical perturbation arguments of Salem [1968] and Hoffmann [1971] help explain why the upper t and e energy levels have large contributions from iron in their corresponding molecular orbitals and why they are antibonding between iron and the coordinating oxygen anions. Similarly, some of the lower levels are stabilized and their corresponding molecular orbitals have large oxygen contributions and are bonding between iron and oxygen. However, it is not necessary to appeal to perturbation theory since Figure 6 is the result of a molecular orbital calculation for the cluster. To achieve the

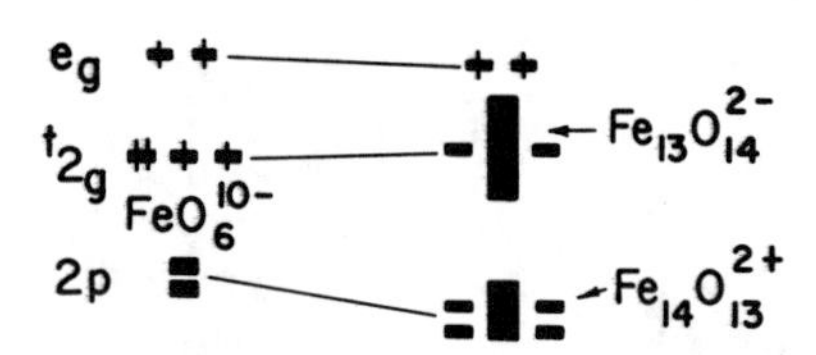

Fig. 7. Correlation of central Fe^{II} d energy levels of Figure 6 with large cluster results. Clusters are cubes with three atoms on an edge. One has iron in its center and the other oxygen.

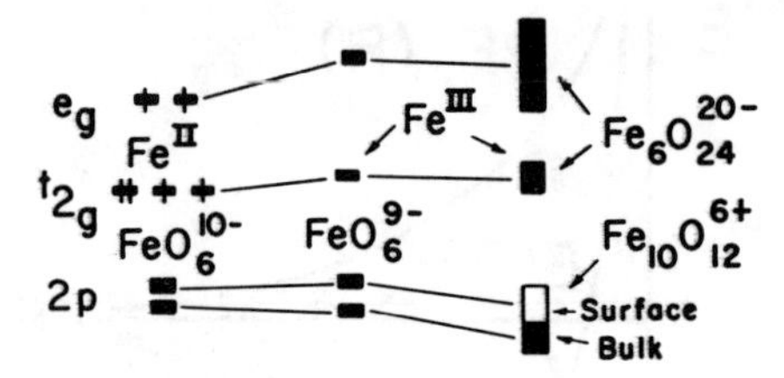

Fig. 8. Correlation of energy levels for wustite, Fe^{II}, and hematite, Fe^{III}.

ferrous oxidation state, the cluster has nine electrons added to it so that it is formally FeO_6^{10-} but, as discussed in the previous section, these electrons present no difficulty to the calculation of molecular orbitals because the hamiltonian depends only on atomic orbital symmetries and overlaps, not on occupation. It may be noted that an OFe_6^{10+} cluster would be appropriate for modeling the electronic structure of the O 2p levels in the same way.

Figure 7 illustrates how remarkably good the FeO_6^{10-} model is in determining the energy levels associated with a bulk-coordinated ferrous cation when a hamiltonian dependent on symmetries but not dependent on electron repulsions is used. The large cluster results are scarcely different. The center of gravity of the O 2p band lies lower than the singly-bonded and non-bonded O 2p state energy levels in the FeO_6^{10-} cluster.

From the electronic viewpoint, hematite, Fe_2O_3, is similar to wustite, though the octahedral symmetry is broken and the Fe-O distances range from 1.96 to 2.09 Å in this Al_2O_3 rhombohedral structure mineral. Debnath and Anderson find the d-level degeneracies are split as in Figure 8 and shifted up due to the shorter distances. Larger clusters form bands as in Figure 8.

Using the large cluster results, the center of O 2p to center of Fe 3d lower band energy differences are 3.8 eV for hematite and 3.3 eV for wustite. The difference is a consequence of structure. Debnath and Anderson argue on the basis of Fe 3s core binding energies that an additional charge state shift of 1.1 eV must be considered. If the 3.8 eV charge transfer excitation energy for hematite is taken as a base, then the corresponding energy for wustite must be increased to 4.4 eV. These results are in accord with the optical study of Tandon and Gupta [1970] producing 3.5 eV for hematite and the results of Runciman et al. [1973] producing 4.2 to 4.9 eV for Fe^{II} species in silicates.

Debnath and Anderson have gone on to show how the optical and photoemission spectra and conducting and magnetic properties of magnetite, spinel structure Fe_3O_4, may also be understood in the one-electron cluster model framework. For every pair of octahedral iron sites there is a tetrahedral site. The Fe-O distances for these sites are 2.10 and 1.82 Å, respectively. This gives rise to the electronic structure in Figure

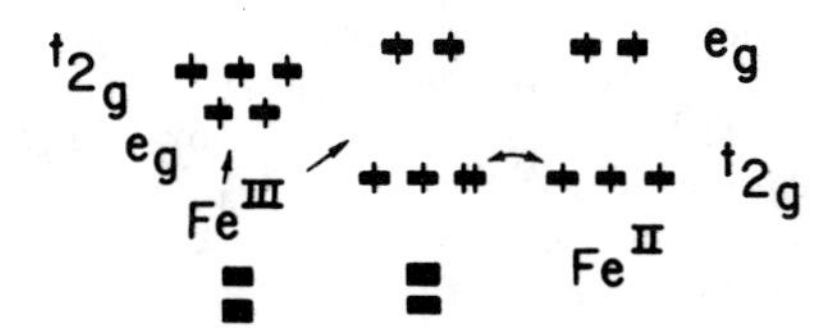

Fig. 9. Energy levels and orbital occupations for tetrahedral (left column) and octahedral sites in magnetite, Fe_3O_4. Electron delocalization between octahedral sites is suggested.

9. The average oxidation state of iron cation in the octahedral sites is 2.5 with the conducting electrons delocalized amongst them. Each tetrahedral Fe^{III} is antiferromagnetically coupled to an octahedral site resulting in a saturation magnetic moment of four Bohr magnetons due to the four unpaired electrons. Not surprisingly the Fe d ← O 2p optical charge transfer absorption spectrum is a broad band.

For further discussion the reader is referred to Debnath and Anderson (1982).

Molybdite

In molybdite, MoO_3, the d^0 Mo^{VI} cations are surrounded by six oxygen anions in a highly distorted octahedral arrangement. Further details of this orthorhombic system are given by Wyckoff [1964]. Results of ASED-MO calculations of Anderson et al. [1983] are in Figure 10. For the $Mo_6O_{25}^{14-}$ cluster all Mo^{VI} cations are bulk-coordinated. For this cluster the calculated O 2p to Mo d band gap is 3.6 eV compared to ∿ 3.2 eV measured by Ivanovskii et al. [1980].

Oxygen deficient molybdite has the interesting property of being a blue conducting bronze, acacording to Colton et al. [1978]. This may be understood by examination of the energy levels and their occupations for $Mo_6O_{25}^{14-}$ in Figure 10. Here a bulk oxygen atom is removed from the interior of the cluster, leaving two electrons in the lower d band. These electrons are seen in the photoemission work of Colton et al. [1978] and Fleisch and Mains [1982] as weak, narrow bands which strengthen and widen as oxygen deficiency increases. The electrons in the lower Mo d band are conducting and may be expected to delocalize, forming Mo^V centers. Transitions within the lower d band, which is about 2 eV wide in this model, are expected to be weak, as d ↔ d transitions are, and in the infrared to red part of the spectrum yielding a blue color. This is exactly what is observed by Colton et al. [1978].

Bismite

Bismite, the low-temperature form of bismuth oxide, α-Bi_2O_3, is pseudoorthorhombic in structure and forms monoclinic crystals. Medernach and Snyder [1978] have determined the structure of bismite and the α, γ, and δ forms of bismuth

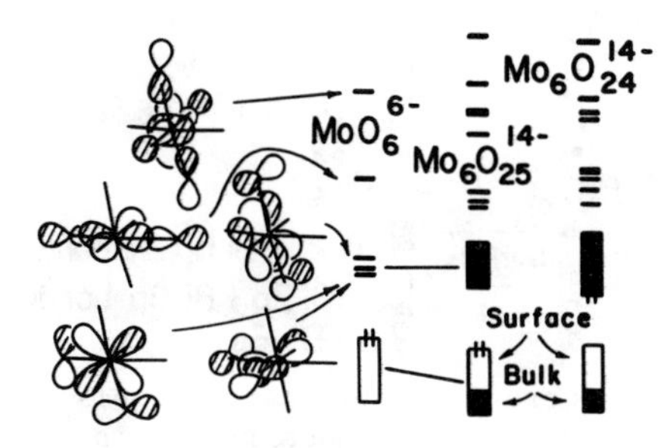

Fig. 10. Orbitals and energy levels for cluster models of molybdite, MoO_3 and for oxygen deficient molybdite. The latter has electrons in the lower d band.

oxide. Though the oxygen coordination to Bi^{III} raises considerably, Anderson et al. [1983] have found the valence electronic structure for the bismuth oxides have many similarities. The most interesting feature in common is that although in Bi_2O_3 Bi^{III} is formally $6s^26p^06d^0$, the highest filled band has considerable Bi 6p character. This is a consequence of hybridization.

ASED-MO results for the β form are in Figure 11. Here the O 2p energy levels lie between the Bi 6s and 6p levels. Perturbation theory arguments explain the resulting hybridization. The Bi 6s, 6p, 6d and O 2p atomic orbitals give rise to three bands, the top occupied one having considerable Bi 6p character and lying well above the O 2p band. Figure 12 shows the development of bands in a larger cluster calculation

Despite the structural differences, the other bismuth oxides are found to have similar electronic structures. In the case of bismite, Debies and Rabalais [1977] have seen the three filled bands using photoemission spectroscopy.

From Figure 12 it may be deduced that the optical absorption band edge in β-Bi_2O_3 is ∿ 2.6 eV. This agrees with 2.6 eV determined experimentally by Dolocan and Iova [1981].

4. Summary and Concluding Comments

Our work discussed above shows how one-electron ASED molecular orbital theory may be used

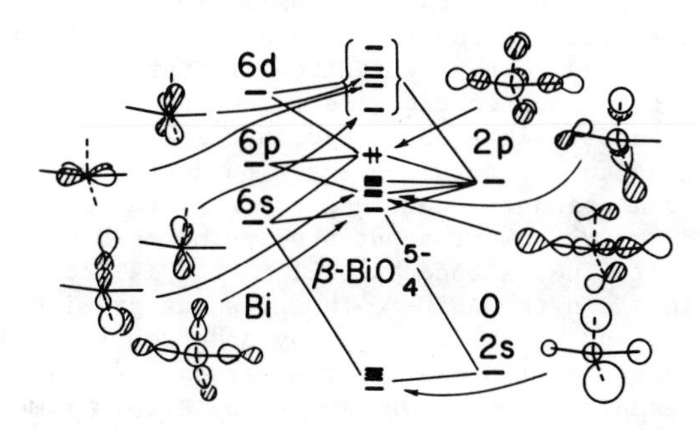

Fig. 11. Correlation diagram and bismuth orbitals in a BiO_4^{5-} model of the β form of bismuth oxide, Bi_2O_3.

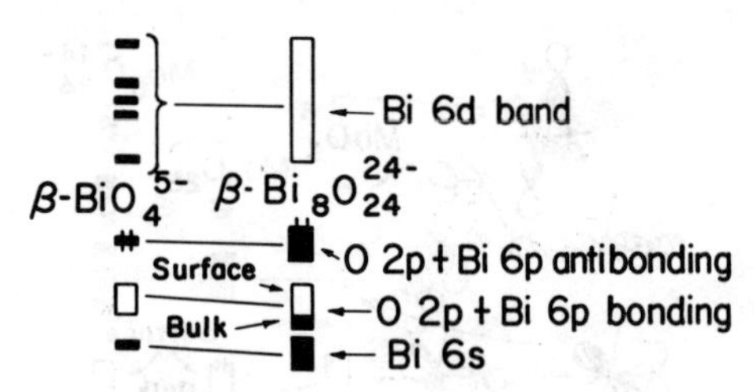

Fig. 12. Correlation of energy levels for BiO_4^{5-} model of β-Bi_2O_3 with those from a larger cluster model with eight bulk-coordinated Bi^{III}.

to predict and explain important properties of minerals using small and medium-sized cluster models. Bond length variations in ferrous and ferric oxides are readily understood in terms of the electron occupations of the antibonding iron d bands. It is shown how magnetism in magnetite may be pictured in the molecular orbital framework, though a more detailed hamiltonian and wavefunction would be needed to predict the magnetic properties of the iron oxides. The conductivity of MoO_{3-x} is shown to be readily understandable from the calculated electronic structure. It is found that in the various forms of Bi_2O_3 the highest-filled band has Bi 6p character even though Bi^{III} is formally $6s^2 6p^0$. The calculated electronic structures for all of the oxides mentioned corroborate photoemission and optical spectra when available. Virtually all minerals and point defects whose structures are known could be treated in the same manner.

It is suggested that the author's Poisson equation for force constants and definition of atomic radii might find applications to problems in mineralogy, whenever such parameters play a role. But what about predicting mineral and defect structures completely? In general a quantum mechanical technique will be necessary.

In the cluster models discussed above structures are assumed and the focus is on electronic properties. Can bulk structures be predicted using these clusters? We believe there is hope. By assuming fixed Be-O distances Burdett and McLarnan [1982] have shown the electronic energy in the extended Hückel approximation for OBe_4^{6+} is a minimum in the tetrahedral configuration. This is the correct wurtzite structure. Further, they find cluster energies for other OBe_4^{6+} structures. What about Madelung ionic contributions to the total energy? Burdett and McLarnan find the Madelung energy parallels the band energy. What about predicting Be-O distances? The extended Hückel method is incapable of this and the ASED-MO theory shows promise in a study of bunsenite, NiO, by Anderson [1980a] and a study of α-quartz by Anderson [1980b].

Recent work by A. B. Anderson, R. W. Grimes, and A. H. Heuer (unpublished manuscript, 1983) shows promise for a new technique for predicting defect structures. These workers define what they call "normalized ion energies", analogous to the OBe_4^{6+} energies of Burdett and McLarnan, for coordinated Fe^{II}, Fe^{III} and O^{2-} for all local defect structures in wustite, $Fe_{1-x}O$. The normalized ion energies, when combined with E_r of (2) allow a fair prediction of the FeO distance in wustite and yield reasonable values for vacancy nucleation energies and migration energy barriers. Certain arrangements of 4:1 clusters (four octahedral Fe^{II} vacancies surrounding a Fe^{III} interstitial) show higher stability than others and in the high vacancy limit the inverse spinel magnetite structure is most stable. Further work is planned.

Acknowledgment. This paper was prepared with the assistance of a grant from the National Science Foundation to the Materials Research Laboratory at Case Western Reserve University.

References

Anderson, A. B., On effective electronic charge densities and vibrational potential energy functions, J. Mol. Spectrosc., 44, 411, 1972.

Anderson, A. B., Derivation of and comments on Bonaccorsi-Scorocco-Tomasi potentials for electrophilic additions, J. Chem. Phys., 60, 2477, 1974.

Anderson, A. B., Derivation of the extended Hückel method with corrections: One-electron molecular orbital theory for energy level and structure determinations, J. Chem. Phys., 62, 1187, 1975a.

Anderson, A. B., Vibrational potentials and structures in molecular and solid carbon, silicon, germanium, and tin, J. Chem. Phys., 63, 4430, 1975b.

Anderson, A. B., NiO bulk properties: Initial state molecular orbital Ni_4O_4 and $Ni_{13}O_{14}$ cluster studies, Chem. Phys. Lett., 72, 514, 1980a.

Anderson, A. B., Structure and electronic properties of α-quartz from SiO_4 and Si_5O_4 models, Chem. Phys. Lett., 76, 155, 1980b.

Anderson, A. B., and R. Hoffmann, Description of diatomic molecules using one-electron configuration energies with two-body interactions, J. Chem. Phys., 60, 4270, 1974.

Anderson, A. B., and R. G. Parr, Vibrational force constants from electron densities, J. Chem. Phys., 53, 3375, 1970.

Anderson, A. B., and R. G. Parr, Universal force constant relationships and a definition of atomic radius, Chem. Phys. Lett., 10, 293, 1971.

Anderson, A. B., Y. Kim, D. W. Ewing, R. K. Grasselli, and M. Tenhover, Electronic properties of $Bi_2O_3$and MoO_3 and relationships to oxidation catalysis, Surf. Sci., in press, 1983.

Andreoni, W, K. Maschuke, and Schluter, Ab initio pseudopotential description of cohesion in NaCl, Phys. Rev. B, 26, 2314, 1982.

Badger, R. M., A relationship between internuclear distances and bond force constants, J. Chem. Phys., 2, 128, 1934.

Badger, R. M., The relationship between the internuclear distances and force constants of molecules and its application to polyatomic molecules, J. Chem. Phys., 3, 710, 1935.

Boyer, L. L., First-principles equation-of-state calculations for alkali halides, Phys. Rev. B, 23, 3673, 1981.

Burdett, J. K., and S. Lee, Peierls distortions in two and three dimensions and the structures of AB solids, J. Am. Chem. Soc., 105, 1079, 1983.

Burdett, J. K., and T. J. McLarnan, An orbital explanation of Pauling's third rule, J. Am. Chem. Soc., 104, 5229, 1982.

Burdett, J. K., G. D. Price, and S. L. Price, Role of the crystal-field theory in determining the structures of spinels, J. Am. Chem. Soc., 104, 92, 1982.

Catlow, C. R. A., J. M. Thomas, S. C. Parker, and D. A. Jefferson, Simulating silicate structures and the structural chemistry of pyroxenoids, Nature, 295, 658, 1982.

Colton, R. J., A. M. Guzman, and J. W. Rabalais, Photochromism and electrochromism in amorphous transition metal oxide films, Acc. Chem. Res., 11, 170, 1978.

Davies, P. K., and A. Navrotsky, Thermodynamics of solid solution formation in NiO-MgO and NiO-ZnO, J. Solid State Chem., 38, 264, 1981.

Deb, B. M. (Ed.), The Force Concept in Chemistry, Van Nostrand-Reinhold, New York, 1981.

Debies, T. P., and J. W. Rabalais, X-ray photoelectron spectra and electronic structure of Bi_2X_3 (X = O, S, Se, Te), Chem. Phys., 20, 277, 1977.

Debnath, N. C., and A. B. Anderson, Optical spectra of ferrous and ferric oxides and the passive film: A molecular orbital study, J. Electrochem. Soc., 129, 2170, 1982.

Dolocan, V., and F. Iova, Optical properties of Bi_2O_3 thin films, Phys. Status Solidi, 64, 755, 1981.

Feynman, R. P., Forces in molecules, Phys. Rev., 56, 340, 1939.

Fleisch, J. H., and G. J. Mains, An EXPS study of the uv reduction and photochromism of MoO_3 and WO_3, J. Chem. Phys, 76, 780, 1982.

Gibbs, G. V., Molecules as models for bonding in silicates, Am. Mineral., 67, 421, 1982.

Gordon, R. G., and Y. S. Kim, Theory for the forces between closed-shell atoms and molecules, J. Chem. Phys., 56, 3122, 1972.

Hellmann, H., Einfüring in die quantenchemie, Franz Deuticke, Leipzig, 1937.

Herschbach, D. R., and V. W. Laurie, Anharmonic potential constants and their dependence upon bond length, J. Chem. Phys., 35, 458, 1961.

Hoffmann, R., Interaction of orbitals through space and through bonds, Acc. Chem. Res., 4, 1, 1971.

Hughbanks, T., and R. Hoffmann, Molybdenum chalcogenides: Clusters, chains, and extended solids, the approach to bonding in three dimensions, J. Am. Chem. Soc., 105, 1150, 1983.

Ivanovskii, A. L., V. P. Zhukov, V. K. Slepukhin, V. A. Gubanov, and G. P. Shreikin, Influence of structural distortions on the electronic structure and optical spectra of $CaMoO_4$, $SrMoO_4$ and MoO_3, J. Struct. Chem. Engl. Transl., 21, 30, 1980.

Kohn, W., and L. J. Sham, Self-consistent equations including exchange and correlation effects, Phys. Rev. A, 140, 1133, 1965.

Kunz, A. B., Electronic structure of AgF, AgCl, and AgBr, Phys. Rev. B, 26, 2070, 1982.

Lowe, J. P., Quantum Chemistry, Academic, New York, 1978.

Medernach, J. W., and R. L. Snyder, Powder diffraction patterns and structures of the bismuth oxides, J. Am. Ceram. Soc., 61, 494, 1978.

Runciman, W. A., D. Sengupta, and J. T. Gourley, The polarized spectra of iron in silicates, II, Olivine, Am. Mineral., 58, 451, 1973.

Salem, L., Intermolecular orbital theory of the interaction between conjugated systems, I, General theory, J. Am. Chem. Soc., 90, 543, 1968.

Tandon, S. P., and J. P. Gupta, diffuse reflectance spectrum of ferrioxides, Spectrosc. Lett., 3, 297, 1970.

Tossell, J. A., and G. V. Gibbs, Molecular orbital studies of geometries and spectra of minerals and inorganic compounds, Phys. Chem. Miner., 2, 21, 1977.

Tossell, J. A., D. S. Urch, D. J. Vaughan, and G. Weich, The electronic structure of $CuFeS_2$, chalcopyrite, from X-ray emission and X-ray photoelectron spectroscopy and $X\alpha$ calculations, J. Chem. Phys., 77, 77, 1982.

Wyckoff, R. G. W., Crystal Structures, vol. 2, John Wiley, New York, 1964.

Yin, M. T., and M. L. Cohen, Theory of static structural properties, crystal stability and phase transformation: Application to Si and Ge, Phys. Rev. B, 26, 5668, 1982

COMPUTER MODELLING OF MINERALS

C. R. A. Catlow and S. C. Parker[1]

Department of Chemistry, University College London
London WC1H OAJ, United Kingdom

Abstract. We review briefly the methodology and achievements of computer simulation techniques in modelling structural and defect properties of inorganic solids. Special attention is paid to the role of interatomic potentials in such studies. We discuss the extension of the techniques to the modelling of minerals, and describe recent results on the study of structural properties of silicates. In a paper of this length, it is not possible to give a comprehensive survey of this field. We shall concentrate on the recent work of our own group. The reader should consult Tossell (1977), Gibbs (1982), and Busing (1970) for examples of other computational studies of inorganic solids. The techniques we discuss are all based on the principle of energy minimization. Simpler, "bridge-building" procedures, based on known bond-lengths, of which distance least squares (DLS) techniques are the best known are discussed, for example, in Dempsey and Strens (1974).

1. Introduction

Computer modelling is now well established as a reliable quantitative technique for investigating structural and defect properties of solids [Catlow and Mackrodt, 1982; Catlow, 1980]. Indeed for simpler materials such as the halides and oxides of the alkali and alkaline earth metals, the techniques have attained a remarkable degree of precision [Mackrodt, 1984; Catlow et al., 1979]. The present article concerns their extension to problems in mineralogy. The results we shall discuss concern primarily the prediction and rationalisation of structures, although work is presently in progress on defect calculations. The main aim of the article is, however, to demonstrate the potential that computer modelling studies could have in mineralogical studies and to draw attention to the problems that need to be overcome if the methods are to enjoy the same success in this field as in other areas of solid state science.

2. Computer Modelling: Aims and Methodology

The basis of computer modelling techniques is the specification of an interatomic potential model, from which we calculate the following properties:

1. Crystal structure. This type of prediction is classified into, first, distortions where we are concerned with deviations from models based on e.g. idealised close-packed structures or regular polyhedra, and second, structure discrimination where we attempt to discriminate between alternative structural models, for example the different silicate backbones in pyroxenoid silicates and the "normal" and "inverse" structures of spinels; examples of both types of calculation will be given in section 5.
2. Crystal properties. Presently available techniques allow us to calculate cohesive energies, elastic and dielectric constants and phonon dispersion curves. We note that accurate calculations of cohesive energies of different polymorphs of a given compound would allow us to predict phase transition energies.
3. Defect properties, included in which are the formation and activation energies of point and extended defects. Calculation of such properties may then be related to transport and thermodynamic properties of the solid.

In practice most work has been concerned with calculations of defect energies and more recently, structural properties; calculations of the crystal properties have generally been used as a way of testing the interatomic potential models. Successful calculations in all three areas could, however, be of interest in mineralogical studies.

Space permits only a brief review of the techniques involved in the calculations; these are, however, reviewed in detail elsewhere [Catlow and Mackrodt, 1982; Catlow, 1980; Mackrodt, 1984; Catlow et al., 1979]. The first feature of the methods to note is that they have, to date, been concerned only with the calculation of energy terms; the calculation of entropies

---

[1]Present affiliation is University Chemical Laboratories, Cambridge, United Kingdom

requires information on vibrational properties. Such information could, in principle, be extracted from the types of calculation we discuss; development work in this field is presently being undertaken. One general feature of the methods with which we are concerned here is they involve static lattice simulations, that is thermal motions are not included explicitly (although their effects may be at least, in part, included by use of a temperature dependent lattice parameter, thus taking into account thermal expansion). Thermal motion may be included explicitly in dynamical simulations (discussed elsewhere in this conference by Brawer). The use of such techniques increases, however, the computer time required by at least an order of magnitude and is generally unnecessary for the calculation of the properties discussed above.

The second general feature of the methods is the use of energy minimisation techniques, i.e., the calculations involve adjusting coordinates of structural parameters until a minimum energy configuration is attained.

Within the general class of calculation outlined above two types of procedure are used:

1. Perfect lattice calculations. Here we specify a unit cell (which may be very large and complex, containing several hundred atoms). Standard summations are employed to calculate the site potentials of all ions. Special attention should be paid to the use of the Ewald summation method [Tosi, 1964] for the Coulomb term; this replaces the slowly converging real space summation by rapidly converging reciprocal space sums. Given adequate site potentials, the cohesive energies may be calculated. Moreover, in addition to these potentials, we may calculate first and second derivatives of the cohesive energy with respect to coordinate displacements, and the derivatives of the energy with respect to changes in the unit cell dimensions. From these derivatives, the elastic and dielectric constants may be obtained, as well as the forces acting on the atoms in the unit cell and on the unit cell as a whole. The mathematical details of these methods are given by Catlow and Norgett [1978] and Catlow and Mackrodt [1982]. We note that the ability to calculate forces acting on atoms, allows us, by coupling the calculations with a minimisation routine to predict crystal structures, as we may use the criterion that in the equilibrium configuration the forces will be zero. Conversely, for a known crystal structure, we may assess the quality of an interatomic potential model by the magnitudes of the calculated forces, which should, of course, be zero, for a fully accurate potential. Use will be made of this latter criterion, below.

The perfect lattice simulations are coded in the PLUTO program [Catlow and Norgett, 1978] and in the METAPOCS code [Cormack et al. 1983b] which has been specifically written for crystal structure prediction. Energy minimisation codes for crystal structure prediction are also available from Busing 1972.

2. Defect lattice calculations. These calculations resemble the perfect lattice techniques as regards the type of summation techniques employed. They differ, however, in that as well as specifying a structure, a defect configuration is also specified. In most work reported to date, the defect configurations consist of a point defect or aggregate of such defects, although related techniques may be applied to planar defects such as shear planes [Cormack et al., 1982] and surfaces [Tasker, 1979]. Moreover, recent developments allow calculations to be performed on point defects at surfaces [Mackrodt and Stewart, 1977; Tasker and Duffy, 1984]. We concentrate, however, on the simple point defect calculations.

The introduction of a defect into a crystal perturbs the surrounding lattice, and it is now clear that any quantitative study of defect energetics must include a detailed treatment of the relaxation of the structure around the defect. The essential feature of the techniques involved in such calculations is the division of the crystal into two regions - an inner, and an outer; the inner region contains typically 100 ions surrounding the defect, where the forces exerted by the defect on the lattice are strong and coordinates of all ions are adjusted until they are at zero force. An explicit atomistic simulation of this type is essential in this "strong-field" region. In contrast for more distant regions (known as region II) the force exerted by the defect is weak and it is adequate to model the response of the lattice using continuum or "quasi-continuum" methods. In studies, of ionic crystals, a favoured approach is that suggested several decades ago by Mott and Littleton, 1938, in which the polarisation $\underline{P}$ at a distance r from a defect of effective charge, q, is given by the formula

$$\underline{P} = \frac{1}{4\pi}\left(1 - \frac{1}{\varepsilon_0}\right) q \frac{\underline{r}}{r^3} \qquad (1)$$

where ε_0 is the static dielectric constant of the crystal. The polarisation may then be divided into atomistic components in a manner which depends on the type of potential model that is used.

In practice, it is found necessary to introduce an interface region (region IIA) berween the inner and outer regions. Here the displacements are calculated using the Mott-Littleton procedure, but interactions with ions within region I are calculated explicitly.

Several studies [Catlow and Mackrodt, 1982; Catlow, 1977; Mackrodt and Stewart, 1979] are reported of the variation of calculated defect energy with the size of region I. For cubic crystals it has generally been found that for sizes of the inner region of greater than 80-100 ions, the energy is no longer sensitive to

TABLE 1. Calculated and Experimental Defect Formation and Activation Energies

Compound	Defect	Calculated Energy, eV	Experimental Energy, eV
	Formation Energies		
NaCl	Schottky	2.3	2.2-2.4
AgCl	Cation Frenkel	1.4	1.45
MgO	Schottky	7.5	5-7
CaF_2	Anion Frenkel	2.75	2.7
	Activation Energies		
	cation vacancy in NaCl	0.67	0.65
	cation vacancy in MgO	2.2	2.3-2.7
	anion interstitial in CaF_2	0.6	0.7

For detailed discussions including references to calculated and experimental results see Chapter 12 of Catlow and Mackrodt [1982].

further expansion; rather larger sizes (typically 150 ions) seem to be required to achieve this "convergence" in non-cubic materials. The same studies have also established that in those crystals where reliable interatomic potentials are available, the methods can yield formation and activation energies for defects that are essentially in agreement with experiment. Some typical results are present in Table 1.

Indeed it is now clear that for most applications, the limitation on the reliability of the results of the simulations arises almost entirely from the limitations on the accuracy of the potentials. The development and parameterisation of potentials is discussed below.

3. Interatomic Potentials

Simulation studies on oxides and halides, reported in recent years, have the following general features:

1. The use of the ionic model, that is the crystal is simulated by entities of integral charge. (There is no necessity to use integral values; partial charges could be specified, although in practice this has not been done.)

2. The short range forces, which come into play when the ionic charge clouds overlap, are described by two-body, central force potentials, which are generally represented by some simple analytical function, of which the Born-Mayer form with a supplementary r^{-6} term is the most commonly used; this is written as

$$V(r) = Ae^{-r/\rho} - Cr^{-6} \qquad (2)$$

The use of the two-body, central-force approximation is clearly a significant restriction which could limit applications of the potential; and indeed even for simple materials (e.g., rock salt structured halides and oxides) deviations from the predictions of such models occur. The failures of such models are most apparent in calculations of elastic and lattice dynamical properties. The simplest example of the inadequacy of the two-body, central-force potentials is provided by the prediction of such models that for crystals in which all atoms are at a centre of symmetry (e.g., NaCl structured crystals) the elastic constants C_{12} and C_{44} will be equal - a prediction which is not observed in many materials with the rock-salt structure, e.g., MgO. However, for predictions of structural and defect properties, the limitations appear to be less significant.

3. Ionic polarisation must be included in any study of lattice dynamical or defect properties, although its influence on structure seems to be far less significant. The most successful way of incorporating the effects of electronic polarisation of ions is to use the shell model originally developed for lattice dynamical studies by Dick and Overhauser [1958]. The model is based on a simple mechanical description of the ionic dipole: a core, in which the mass of the ion is concentrated is connected to a massless shell, representing the polarisable valence shell electrons. Core and shell are connected by a harmonic spring constant; and the development of a dipole moment is described in terms of the displacement of the shell relative to the core. The shell charge and spring constants are variable parameters of the model. Despite its simplicity, the model has proved to be remarkably successful in describing the dynamical and dielectric properties of solids. Moreover shell model potentials have enjoyed great success in defect studies and have proved notably superior to point dipole models [Catlow and Norgett, 1973; Catlow, 1977; Faux and Lidiard, 1971].

All potential models include variable parameters: the shell charge and spring constants referred to above, and parameters in the description of the short range interactions, e.g., A, ρ and C in the Born-Mayer potential (equation (2)). There are two ways of determining these parameters. The first is by the procedure of empirical fitting, i.e., adjustment of the potential parameters (usually via a least squares procedure) until the best agreement between calculated and experimental values for observed crystal properties is obtained. The properties included in the fitting procedure should include elastic and dielectric constants, where available, and structural properties; the latter, for cubic crystals will generally only include the lattice parameter, although for lower symmetry crystals, the structure may be characterised by a large number of parameters each of which is in principle a source of information on the interatomic potential - a feature which is exploited in our parameterisation of silicate potentials as discussed in section 5. If possible, all data should refer to 0 ^{o}K.

The empirical fitting procedure would be expected to generate reliable interatomic potentials for internuclear separations that are close to those in the perfect lattice. We have, however, no guarantee of the validity of empirically parameterised potentials for separations which are considerably different from the perfect lattice values. Such spacings would, however, be expected for certain defect configurations. For this reason considerable effort has been devoted to developing alternative non-empirical procedures for deriving interatomic potentials. The most satisfactory of these are unquestionaly ab-initio Hartree Fock methods; and notable progress has recently been made by Mackrodt and co-workers [Kendrick and Mackrodt, 1983; Mackrodt et al., 1980] in deriving such potentials for solids. At present, however, the range of application of such techniques is limited due to their excessive computational requirements. For this reason simpler procedures, based on electron-gas methods, akin to the Thomas-Fermi approach, have been widely applied to solid state potentials. Again Mackrodt and co-workers have made major contributions in this field. They have developed electron-gas potentials for a wide range of solids which they have shown work well in calculations of crystal properties and defect energies [Mackrodt, 1984, Mackrodt and Stewart, 1979].

In many cases theoretically derived and empirical potentials agree very well; examples are provided by cation-anion interactions in the alkali halides. Significant differences are found in other instances. The most important is the case of the $O^{2-} \ldots O^{2-}$ potential [Catlow, 1977; Catlow et al., 1982b] which is of critical importance in determining the magnitude of calculated oxygen interstitial energies. The calculations of Catlow [1977] were performed on the interaction of two O^{-} ions. It was argued that the additional electron in the O^{2-} ion (which is unbound in the free state) does not strongly influence the short range potential. In general, however, when applied to defect studies, there is good agreement between results obtained from empirical and non-empirical potentials (Catlow and Mackrodt, 1982, Catlow, 1980).

4. Achievements of Modelling Methods in Solid State Studies

The results obtained using the techniques and potentials outlined above have now had an impact on several areas of solid state science, most notably the following:

Calculation of Accurate Point Defect Formation Energies

In recent years accurate values have been calculated for formation and activation energies of point defects in a wide range of halide and oxide crystals; in addition binding energies of defects in complexes and substitutional energies, i.e., the energy required to replace a lattice ion by an impurity ion (or an oxidised or reduced lattice cation), have also been accurately calculated. A representative set of results were given in Table 1 which illustrates the generally excellent measure of agreement between theory and experiment. The case of MgO is of particular interest. In the past two decades there has been considerable controversy as to the role of intrinsic disorder in this material (as discussed for example in the article of Wuensch in these proceedings). Our calculations find high values for the Schottky energy in MgO, suggesting that intrinsic disorder would be negligible, except at the very highest temperatures. This prediction is in line with the latest analyses of experimental transport data for this crystal.

For further details of the present status of calculation of point defect energetics we refer to the reviews of Catlow [1980] and of Catlow and Mackrodt [1982]. For our present purposes, the importance of this work is that it clearly establishes the quantitative reliability of the techniques for those crystals where adequate interatomic potentials may be derived.

Study of Extended Defect Formation

We are concerned here with more qualitative applications of the techniques, which have been successfully applied to the elucidation of the complex modes of clustering that occur in heavily defective phases including non-stoichiometric oxides, e.g., $Fe_{1-x}O$, UO_{2+x}, TiO_{2-x} and heavily doped systems (e.g., the rare earth doped alkaline earth fluorides).

$Fe_{1-x}O$ provides a good illustration of the application of the techniques to such systems. In this rock salt structured oxide there occur large deviations from the stoichiometric composition (i.e., x may attain values of $\sim$0.15). The compositional variation is accommodated principally by formation of cation vacancies which provide charge compensation for the oxidation of Fe^{2+} to Fe^{3+}. However, diffraction studies provided clear evidence for the presence of small concentrations of cation interstitials in addition to the vacancies. This unusual observation has been rationalised by the formation of a series of vacancy-interstitial clusters whose structure is illustrated in Figure 1. The basic feature is the tetrahedral vacancy aggregate, the "4:1" cluster shown in Figure 1a which the calculations identified as a strongly bound, highly stable species; further aggregation of these clusters may then occur by edge and corner linking of these tetrahedral units to give the larger clusters shown in Figure 1. The formation of these complex types of aggregate ties in well with experimental data [Cheetham et al., 1971; Battle and Cheetham, 1979].

Similarly complex modes of aggregation have been proposed for the oxygen excess non-stoichiometric phase UO_{2+x} [Catlow, 1977] and for the analogous anion excess solid solutions CaF_2/MF_3 [Catlow et al., 1983a, b, c]. Again the predictions of the computer modelling methods agree well with experimental data.

The most recent applications in this field concern shear planes - extended planar defects observed in oxides such as TiO_{2-x} and WO_{3-x}, whose formation can be described in terms of

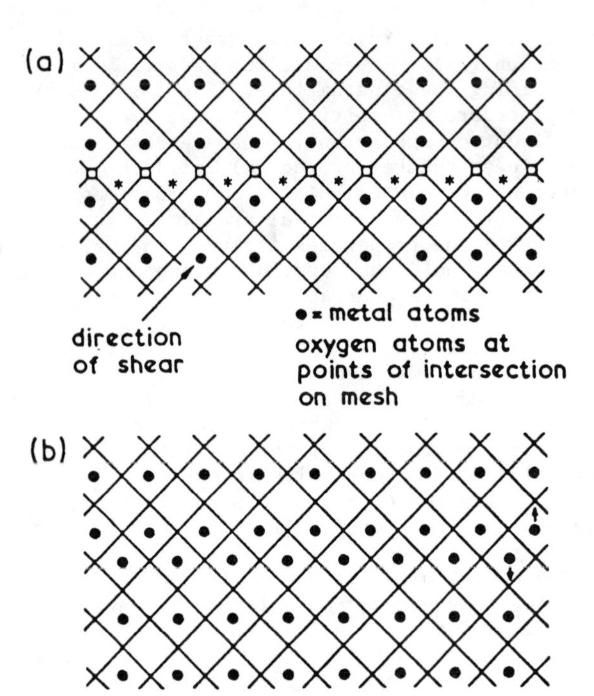

Fig. 2. Shear plane formation in oxides. Fig. (2a) shows aligned vacancies in a section through the ReO_3 structure. Shear of the lower part of the crystal superimposes the starred oxygen atoms over the vacancies with generation of the shear plane shown in Fig. (2b).

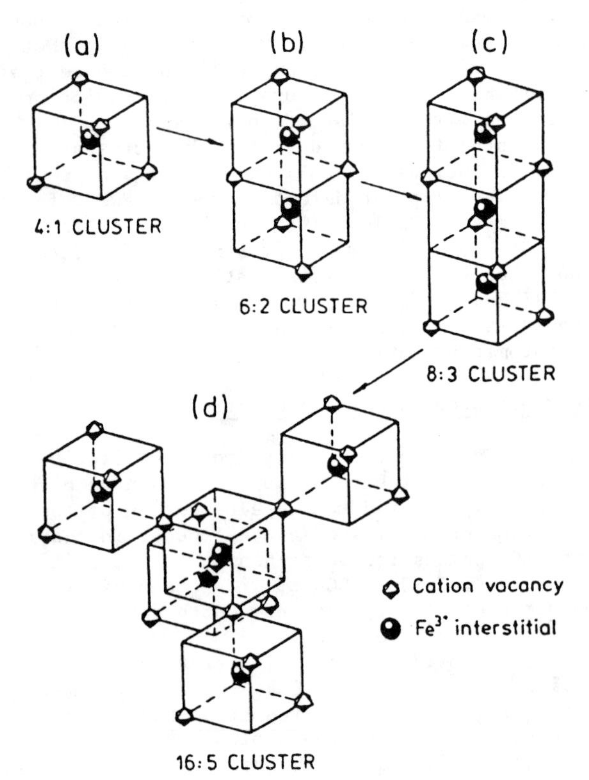

Fig. 1. Defect clusters in $Fe_{1-x}O$.

vacancy elimination as illustrated in Figure 2. Calculations [Catlow and James, 1982] have examined the stability of these species with respect to point defects, and have identified the occurrence of extensive cation relaxation in the vicinity of the planar defects as a vital factor in stabilising these species. In addition, it has clearly been shown that shear planes will exist in equilibrium with point defects. Moreover, the factors controlling the remarkable long range ordering of these shear planes that is observed [Bursill and Hyde, 1972; Tilley 1975] in both TiO_{2-x} and WO_{3-x} have recently been studied by calculations of the interaction energies of the defects as a function of their separation [Cormack et al., 1983a].

Crystal Structure Prediction

The combination of lattice energy calculations with energy minimisation techniques allows us to predict the minimum energy configuration of crystal structures. The increasing availability of reliable interatomic potentials leads to such calculations being a useful adjunct to crystallographic studies. A good illustration of the accuracy that can be achieved in such work is provided by the recently synthesised TiO_2 (B) phase. In this new phase of titanium dioxide, the linkage of the TiO_6 octahedra is as shown in

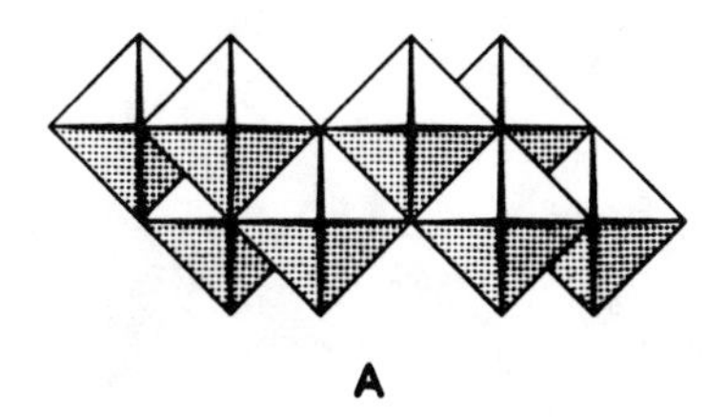

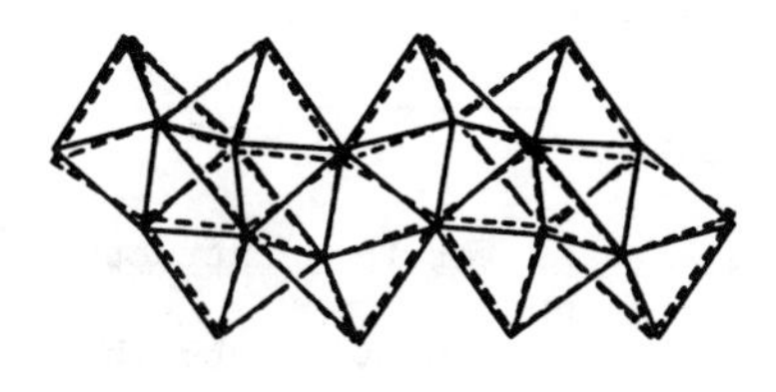

Fig. 3. TiO_2(B) structure. (a) Ideal model. (b) Observed and calculated models.

Figure 3a. The structure shown, however, is that of an idealised model containing regular octahedra. In the observed structure, the octahedra are strongly distorted as shown in Figure 3b. The same diagram also shows the predicted structure, obtained by the energy minimisation methods, using the ideal model as the starting point for the minimisation procedure. The agreement between prediction and observation is good; the discrepancies are probably within the uncertainty of the experimental coordinates. Details are given in Cormack et al., [1983b].

Structure and Transport in Superionic Solids

Superionic or fast-ion conductors are materials with ionic conductivities of the same order of magnitude as those observed for molten salts. They include the high temperature phases of simple inorganic compounds, e.g. PbF_2, $SrCl_2$, AgI, and solid solutions such as ZrO_2/CaO and CeO_2/Y_2O_3 and layer structured compounds e.g., β-Al_2O_3 and Li_3N. The materials have been the subject of extensive research in recent years owing to their potential as solid electrolytes in advanced batteries and fuel cells. Computer simulation rechniques have provided valuable information on structural and transport properties of these materials; reviews are available in the monograph of Catlow and Mackrodt [1982].

The work summarised above establishes the simulation techniques as a reliable predictive tool in a diverse range of solid state studies. The remainder of this article will concern the difficulties and successes encountered on extending the methods to silicate systems.

5. Modelling of Silicates

Extension of simulation procedures to new classes of solid devolves into the development of adequate interatomic potentials for the new systems. In modelling silicate minerals we have first investigated whether "classical" ionic model potentials of the type described in section3, are acceptable for these systems. It might be argued that such models can never be valid for silicates as it is well known that these compounds have a large measure of covalent bonding. However, care should be taken in the definition of ionicity in the present context [see e.g., Catlow and Stoneham, 1983]. In particular, we consider there to be two distinct meanings to this term. The first relates to the electron density distribution in the system, although it is notoriously difficult to derive numerical ionicity scales from measured electron density distributions. The second concerns the forces acting in a crystal; in this context, an ionic description is valid, when potential models based on such a description correctly predict properties such as the elastic and dielectric constants, the phonon dispersion curves, and, of course, the structure. It is in this latter sense that we use the term ionicity in this article. The validity of such models for silicates rests therefore on the extent to which the models reproduce the observed properties of these compounds.

In deriving potential parameters for mineral systems, a problem is encountered in that there is only a limited amount of reliable elastic and dielectric data. For these materials, however, there are abundant and accurate structural data. The structures are generally complex, often with low symmetries and commonly including a range of metal-oxygen and silicon.....oxygen bond lengths, and as such, as argued in section 3, contain a considerable amount of information on interatomic potentials. For this reason, we decided to obtain our potential parameters by "fitting to structures", that is by adjusting potential parameters until the observed structure approaches as closely as possible to the condition of being in equilibrium with the observed cell dimensions; the equilibrium conditions refer to atomic coordinates as well as the unit-cell structure. Details of the computational procedure adopted are given elsewhere [Catlow et al., 1982a; Parker, 1982].

In deriving out potential parameters, several crystal structures were considered; these are listed in Table 2. Parameters for the interactions between monovalent and divalent cations and oxygen were taken from potential models developed for the appropriate binary oxide [Catlow and Mackrodt, 1982; Stoneham, 1982]. Si...O potential parameters were then derived by the procedure described above. It was found that with two sets of Si...O potential parameters - one for structures with isolated SiO_4 groups and one for the ring and chain structures - we could adequately reproduce the observed structures of the minerals reported in Table 2. The resulting parameters are reported in Table 3. We note that the potentials are "rigid-ion" models; i.e., they

TABLE 2. Crystal Structures Used in Determining Potential Parameters

Name of Mineral	Formula
Monticellite	$CaMgSiO_4$
Silimanite	Al_2SiO_5
Titanite	$Ca(TiO)SiO_4$
Benitoite	$BaTi(SiO_3)_3$
Jadeite	$NaAl(SiO_3)_2$
β-Wollastonite	$CaSiO_3$

do not include polarisation. This is probably not a significant problem for studies of structural properties, although it could cause more serious problems in defect studies. The performance of the potentials in structure prediction is examined in the next section.

6. Structure Prediction for Silicates

We have recently undertaken structure prediction studies of minerals in both the senses discussed in the Introduction, i.e., we have investigated the extent to which our potentials can reproduce distortion from idealised models and we have applied the methods to the discrimination between alternative structural models.

In the first class, we have examined a large number of structures based on isolated SiO_4 tetrahedra, as well as structures based on $(SiO_3)_n$ rings and chains. The minerals investigated are listed in Table 4. We have shown that for all these compounds the observed structures can be reproduced by energy minimisation procedures, employing the potentials described in the previous section. The accuracy achieved is often within the experimentally reported thermal parameters; the latter are the limit of the accuracy to be expected for any method based on a static description of the lattice. We should note that good results are obtained for chain structured silicates as well as for those structures based on isolated SiO_4 groups.

Another example is provided by the case of the uranium and thorium silicates. The energy minimisation techniques find a structure that is in agreement with a more recent determination [Taylor and Ewinger, 1978] of $ThSiO_4$ and we would suggest that a re-examination of $USiO_4$ would result in a structure close to that found for $ThSiO_4$ as is predicted by the calculations.

The second set of applications concerns the structures of pyroxenoids - chain silicates of general formula $MSiO_3$ for which a number of silicate backbone structures are possible; these are illustrated in Figure 4. Calculations were performed with the aim of identifying the factors controlling the silicate backbone structure. The energies of the four structures shown in Figure 4 were calculated for four different cations - Ca^{2+}, Mg^{2+}, Fe^{2+} and Mn^{2+}. The results are displayed graphically in Figure 5; all calculations refer to structures that are fully "equilibrated", i.e., both cell dimension and atomic coordinates are adjusted until there are no residual strains.

The calculations immediately demonstrate the controlling influence of cation size on the structural type. The increase in cation size stabilises the wollastonite relative to the diopside structure. $CaSiO_3$ is thus correctly predicted to have the wollastonite structure. In

TABLE 3. Potential Parameters for Silicate Systems:

$$V(R) = A \exp(-R/\rho) - C/R^6$$

	A, eV	ρ, Å	C, eV $Å^6$
$O^{2-}-O^{2-}$	22764.3	0.1490	27.88
$Mg^{2+}-O^{2-}$	2214.39	0.2756	4.45
$Ca^{2+}-O^{2-}$	1996.35	0.3189	26.57
$Sr^{2+}-O^{2-}$	2187.61	0.3377	64.08
$Mn^{2+}-O^{2-}$	2618.35	0.3033	27.46
$Fe^{2+}-O^{2-}$	2886.3	0.2973	29.73
$Si^{4+}-O^{2-}$*	998.98	0.3455	0.0
$Si^{4+}-O^{2-}$†	473.19	0.4297	0.0

* Corresponds to an $Si^{4+}-O^{2-}$ interaction in silicates where the SiO_4 tetrahedra are linked to form rings and chains.

† Silicates where the SiO_4 tetrahedra are isolated.

TABLE 4. Minerals Investigated by Energy Minimization Techniques

	Structural Type	Formula
Titanates		
	perovskite	$CaTiO_3$
		$SrTiO_3$
		$BaTiO_3$
	ilmenite	$FeTiO_3$
		$MnTiO_3$
		$NiTiO_3$
		$CoTiO_3$
		$MgTiO_3$
Silicates		
SiO_4 tetrahedra isolated	olivine	Mg_2SiO_4
		Fe_2SiO_4
		Mn_2SiO_4
	zircon	$ZrSiO_4$
		$ThSiO_4$
		$USiO_4$
Ring	3 membered ring	$Na_2Be_2Si_3O_9$
		$\alpha Sr_3Si_3O_9$
	6 membered beryl ring	$Al_2Be_3Si_6O_8$
Chain	ortho enstatite	$MgSiO_3$
	ortho ferrosilite	$FeSiO_3$
	sodium metasilicate	Na_2SiO_3

contrast the mixed cation compound $CaMgSiO_3$ is predicted to be stable in the diopside structure–a prediction which again accords with experiment.

A remarkable feature of the calculations is the prediction of metastability for two of the structural types--the rhodonite and pyroxmangite structures--for all the cations studied. However, we find that these structures are all strongly stabilised by "mixed cation" effects; that is the energies of these structures are lowered, relative to those of the alternative structural types, by incorporating two different cations in the structure. We suggest that these structures may be thermodynamically stable only for such "mixed-cation" compounds.

Further details of these calculations are described elsewhere [Catlow et a., 1982a; Parker, 1982]. They establish, however, that the energy minimisation technique is a reliable and accurate method of predicting at least certain classes of mineral structure with the potentials described in section 5. The "validity" of the potential models in the sense discussed in section 5 is therefore clearly supported. The work on the pyroxenoids establishes that the technique may be usefully applied to the study of the factors controlling the type of structure adopted by a particular mineral.

7. Future Developments

The work we have discussed in the previous two sections established that, for certain classes of silicate structure, potential models can be developed which yield reliable predictions at least as regards structural properties for several silicate systems. We note that, to date, we have treated only systems which contain isolated SiO_4 groups, or silicate rings and chains. Extension to framework structured minerals will, we consider, require potential models of greater sophistication; in particular it is necessary to include bond-bending forces. Indeed, we have recently shown that it is possible to reproduce accurately the crystal structures of all the polymorphs of SiO_2 using a potential model which includes such terms.

However, it is clear that even with the present level of sophistication, valuable results may be obtained. Given the success of the structural studies reported in section 6, the work may be readily extended to calculation of other crystal properties and of defect energies. The scope of such calculations would be very large; and their impact on mineralogy could, we believe, be considerable.

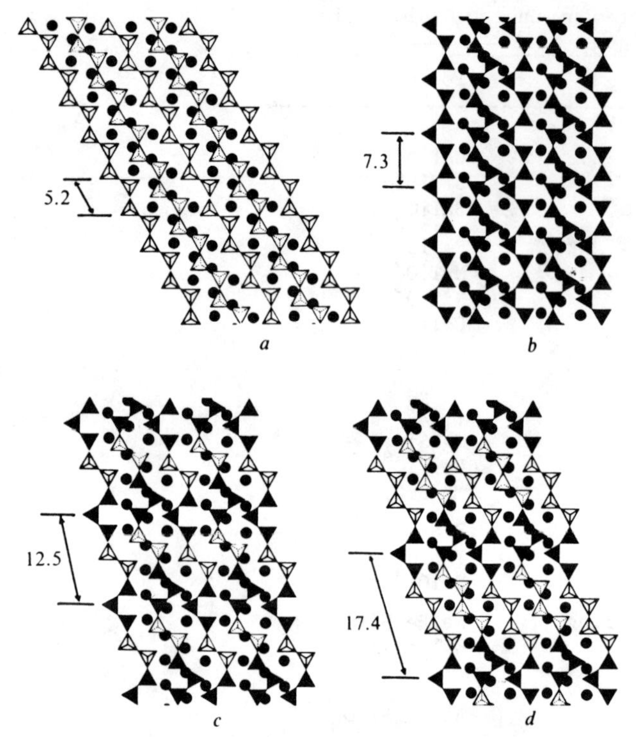

Fig. 4. Structures of silicate backbones in pyroxenoids: (a) diopside, (b) ferrosilite, (c) rhondonite, and (d) wollastonite.

Acknowledgments. We are grateful to A.N. Cormack, A.M. Stoneham, D.A. Jefferson, and J.M. Thomas for their collaboration in the work summarised in this article. We would also like to thank AERE Harwell for generous provision of computer time for the calculations we have discussed.

References

Battle, P. D., and A. K. Cheetham, The magnetic structure of non-stoichiometric FeO_2, J. Phys. C., 12, 337, 1979.

Bursill, L. A., and B. G. Hyde, CS (Crystallographic shear) families derived from rhenium (VI) oxide structure type. Electron microscopy study of reduced tungsten (VI) oxide and related pseudobinary systems, Prog. Solid State Chem., 7, 177, 1972.

Busing, W. R., A computer program to aid the understanding of interatomic forces in molecules and crystals, Acta. Cryst., A28, 252, 1970.

Catlow, C. R. A., Point defect and electronic properties of uranium dioxide, Proc. R. Soc. London Ser. A, 353, 533, 1977.

Catlow, C. R. A., Computer modelling of ionic crystals, J. Phys., 41, C6-53 (1980).

Catlow, C. R. A., Defect interaction energies in anion deficient fluorite oxides, Solid State Ionics, in press, 1984.

Catlow, C. R. A., and R. James, Disorder in

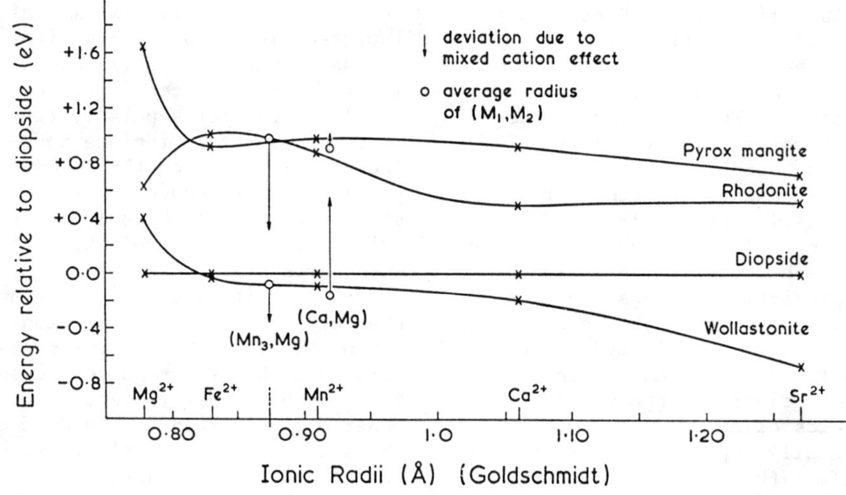

Fig. 5. The energy of the pyroxenoid silicates relative to diopside.

TiO_{2-x}, Proc. R. Soc., London Ser. A, 384, 157, 1982.
Catlow, C. R. A., and M. J. Norgett, Lattice structure and stability of ionic materials, UKAEA report, AERE-M2936, 1978.
Catlow, C. R. A., and W. C. Mackrodt (Eds.), Computer Modelling of Solids, Lect. Notes in Phys., vol.166, Springer-Verlag, Berlin, 1982.
Catlow, C. R. A., and M. J. Norgett, Defect energies in the alkaline earth fluorides, J. Phys. C., 6, 1325, 1973.
Catlow, C. R. A., J. Corish, K. M. Diller, M. J. Norgett, and P. W. M. Jacobs, The contribution of vacancy defects to mass transport in alkali halides - an assessment using theoretical calculations, J. Phys. C., 12, 451, 1979.
Catlow, C. R. A., J. M. Thomas, D. A. Jefferson, and S. C. Parker, Simulating silicate structures and the structural chemistry of pyroxenoids, Nature, 295, 658, 1982a.
Catlow, C. R. A., R. James, W. C. Mackrodt, and R. F. Stewart, Defect energies in α-Al_2O_3 and rutite, TiO_2, Phys. Rev. B, 25, 1006, 1982b.
Catlow, C. R. A., J. Corish, and A. V. Chadwick, The defect structure of anion excess CaF_2, J. Solid State Chem., 48, 65, 1983a.
Catlow, C. R. A., J. Corish, and A. V. Chadwick, Neutron diffraction studies of anion excess fluorites, Radiat. Eff., 75, 61, 1983b.
Catlow, C. R. A., A. V. Chadwick, G. N. Greaves, and L. M. Moroney, The use of EXAFS in the study of defect clusters in doped CaF_2, Radiat. Eff., 75, 159, 1983.
Catlow, C. R. A., and A. M. Stoneham, Ionicity in solids, J. Phys. C., 16, 4321, 1983.
Cheetham, A. K., B. E. F. Fender, and R. I. Taylor, High temperature neutron diffraction studies of $Fe_{1-x}O$, J. Phys. C., 4, 2160, 1971.
Cormack, A. N., R. Jones, P. W. Tasker, and C. R. A. Catlow, Extended defect formation in oxides with the ReO_3 structure, J. Solid State Chem., 44, 174, 1982.
Cormack, A. N., C. R. A. Catlow, and P. W. Tasker, Long-range ordering of extended defects in non-stoichiometric oxides, Radiat. Eff., 74, 237, 1983a.
Cormack, A. N., C. R. A. Catlow, and F. Theobald, Structure prediction of transition metal oxides by energy minimisation techniques, Acta Crystallogr., in press, 1983b.
Dempsey, M. J., and R. G. J. Strens, Physics and Chemistry of Minerals and Rocks, p.443, edited by R. G. J. Strens, John Wiley, New York, 1974.
Dick, B. G., and A. W. Overhauser, Theory of the dielectric constants of alkali halide crystals, Phys. Rev., 112, 90, 1958.
Faux, I. D., and A. B. Lidiard, Volumes of formation of Schottig defects of ionic crystals, Z. Naturforsch., 26a, 62, 1971.
Gibbs, G. V., Molecules as models for bonding in silicates, Am. Mineral., 67, 451, 1982.
Kendrick, J., and W. C. Mackrodt, Interatomic potentials for ionic materials from first principles calculations, Solid State Ionics, 8, 247, 1983.
Mackrodt, W. C., Defect energetics and their relation to non-stoichiometry in oxides, Solid State Ionics, 12, in press, 1984.
Mackrodt, W. C., and R. F. Stewart, Defect properties of ionic solids: I Point defects at the surfaces of face centred cubic crystals, J. Phys. C., 10, 1431, 1977.
Mackrodt, W. C., and R. F. Stewart, Defect properties of ionic solids: III The calculation of the point defect structure of the alkaline-earth oxides and CdO, J. Phys. C., 12, 431, 1979.
Mackcrodt, W. C., I. C. Campbell, and I. M. Hillier, Calculating defect structure of ZnO, J. Physique 1980.
Mott, N. F., and M. J. Littleton, Conduction in polar crystals I: Electrolytic conduction in solid salts, Trans. Faraday Soc., 34, 485, 1938.
Parker, S. C., Computer modelling of minerals, Ph.D. thesis, Univ. of London, London, 1982.
Stoneham, A. M., Handbook of interatomic potentials, rep. AERE-R9598, 1982.
Tasker, P. W., The surface energies, surface tensions and surface structure of the alkali halide crystals, Philos. Mag. Part A, 39, 119, 1979.
Tasker, P. W., and D. Duffy, The structure and properties of stepped surfaces of MgO and NiO, Surface Science, 137, 91, 1984.
Taylor, M., and R. C. Ewing, The crystal structures of the $ThSiO_4$ polymorphs: Huttonite and Thorite, Acta Crystallogr., B34, 1076, 1978.
Tilley, R. J. D., in Solid State Chemistry, Int. Rev. of Sci., Inorg. Chem. Ser., edited by H. F. Emeleus and L. E. G. Roberts, 1975.
Tosi, M., Cohesion of ionic solids in the Born model, in Solid State Physics, 16, edited by F. Seitz and D. Turnbull, p1, Academic, New York, 1964.
Tossell, J. A., A comparison of silicon-oxygen bonding in quartz and magnesian olivine from X-ray and molecular orbital calculations, Am. Mineral., 62, 136, 1977.

A THEORY OF THE SPECIFIC HEAT AND VISCOSITY OF LIQUID SiO_2 AND BeF_2

Steven A. Brawer

AT&T Bell Laboratories, Murray Hill, New Jersey 07974

Abstract. A theory of the transport and thermal properties of the network fluids SiO_2 and BeF_2 is presented. The theory is based on molecular dynamics simulations of BeF_2. It is observed in such simulations that a "diffusion event," requiring the separation of a Be-F neighbors, occurs only when the F ion is 3-fold coordinated, and the Be usually 5-fold coordinated. These high coordination number "defects" are thermally excited in the fluid, and their number is very small (less than 1%) below the melting point. Because application of pressure increases the average coordination number, and thus increases the number of "sites" at which diffusion occurs, it is predicted that the viscosity will decrease with increasing pressure. (This has been observed experimentally in GeO_2 and albite, but not yet in SiO_2 or BeF_2.) Further, because the number of defects is so small, their contribution to the specific heat is unmeasurably small below the melting point. Therefore, the specific heat of these fluids is vibrational, and there is no break in specific heat at the glass transition.

1. Introduction

In this article, we develop a theory for the thermodynamic and transport properties of network liquids, such as SiO_2, BeF_2 and GeO_2. (B_2O_3 and P_2O_5 are specifically excluded here.) The theory is based on molecular dynamics simulations of BeF_2, which experimentally has similar properties and structure as SiO_2. The organization of this paper is as follows: Section 2 gives a summary of properties of BeF_2 and SiO_2; In section 3, a description of the molecular dynamics technique and the limits of its applicability is presented; section 4 describes molecular dynamics simulations of BeF_2; In section 5, a theory of the properties of network liquids is developed, with predictions of new experimental results.

2. Review of Properties of Network Fluids and Glasses

1. For both SiO_2 and BeF_2, the specific heats of the supercooled fluids are nearly equal to those of the corresponding crystals all the way to the melting point. There seems to be no break in the specific heat at the glass transition temperature, in contrast to virtually all other glasses [Wong and Angell, 1976; Angell and Sichina, 1976; Angell, 1981]. For GeO_2 and $ZnCl_2$, there is a break but it is small. Further, the specific heat of SiO_2 is constant at the classical value of 9 R from 1500°K to 2600°K [Stout and Piwinskii, 1983]. Therefore, in the network liquids, there is little configurational contribution to thermodynamic properties.

2. The viscosity of network fluids appears to decrease with increasing pressure. This has been directly observed for GeO_2, as measured by Sharma and Kushiro [1979]. The viscosity of the network liquids, jadeite and albite, also decreases with increasing pressure. The viscosities of liquid SiO_2 and BeF_2 have not yet been reported as a function of pressure, but it is predicted here that they will behave in the same way as GeO_2.

3. Glassy SiO_2 can be permanently densified by applying pressure below the glass transition [see Hsich et al., 1971]. Upon annealing, also at temperatures below the glass transition, its density decreases to the value that obtained before the densification. The activation energy for density relaxation is 65 kcal/mole.

3. Molecular Dynamics Simulations of Liquids and Glasses: Methodology

The molecular dynamics (MD) technique of statistical mechanics has proved the only practical technique for studying the structure and properties of dense liquids. It is a brute-force method in which the equations of motion of a system of hundreds or thousands of particles are solved numerically with no approximation. As a consequence, most of the thermodynamic and transport properties of the system may, in principle, be evaluated exactly, with none of the approximations made in analytic theories. Several reviews are Boon and Yip [1980], Croxton [1974], Sangster and Dixon, [1976], and Woodcock [1975].

The physical system is modeled by assuming point particles of the proper mass, each particle interacting with all the others via some potential function. Most studies have assumed that the potential energy is a sum of 2-body potentials, so if $V_{ij}(r)$ is the interaction energy between particles i and j, separated by distance r, the net potential energy of particle i, say, is

$$\sum_{j(\neq i)} V_{ij}\,(r_{ij})$$

thus, the force on i, due to j, is independent of what the rest of the particles in the system are doing.

The particles are enclosed in a box of the appropriate size so that the system has the correct macroscopic density. A simple cube is usually used. Because there are so few particles, surface effects would be of overwhelming importance, and these are minimized by the use of periodic boundary conditions. The system is periodically replicated about the central cell, so in all the surrounding cells (which go to infinity) the appropriate particle has exactly the same position and velocity it has the central cell. Particles on the boundary of the central cell then interact with particles in neighboring cells, and in this way the surface of the cell is eliminated.

The temperature of the system is the average kinetic energy, so that the relation

$$\frac{3}{2}\,NkT = \sum_{i=1}^{N} \frac{1}{2} m_i v_i^2$$

defines the temperature, where m_i is the mass of particle i, v_i its speed and N the total number of particles in the central cell. To give the system a particular temperature, a Maxwell-Boltzman distribution of velocities is generated using a Gaussian random number generator.

Once these details are arranged, the motion of the system is computed by integrating numerically Newton's equations of motion

$$\bar{F}_i = m_i \frac{d^2}{dt^2} \bar{r}_i$$

where $\bar{r}_i$ is the position of ion i, m_i its mass and $\bar{F}_i$ the force on i due to all the other particles in the central cell and also in the image cells. A number of numerical schemes exist for integrating such equations, and these can be found in the references given above.

The equations of motion are integrated over the desired time interval, and so during this interval the positions and velocities of the particles are known at a discrete set of times separated by the interval used in the numerical integration scheme. The result is a sequence of snapshots of the fluid. Each snapshot is referred to as a configuration.

Any measurable physical property is then determined as a time average, which is the same as an average over the configurations. Central to the use of time averages in calculating a physical quantity is the concept of ergodicity. A system is said to be ergodic when its time average equals its ensemble average. For a simulation, ergodicity requires that the time of the simulation run be long enough [see Wood, 1968]. This idea plays a central role in simulations of glasses and arises from the following limitation of the MD technique: The longest time during which it is practical to simulate a fluid on computer is about a nanosecond.

This limitation is due to the limitations on computer time. Consider, for example, typical numbers from the author's simulations of fluorides [Brawer and Weber, 1981]. The runs were made on a mini-computer system - a PDP 11/55 and an array processor (Floating Point Systems model 120B). For 390 particles, a single time step could be done in about 6 seconds. Thus, a picosecond of fluid time required 50 minutes of CPU time, while a nanosecond of fluid time would require 35 days of CPU time! With faster computers and more careful programming, a time step could probably be done in .06 secs, and perhaps even less. This still requires 8 hours of CPU time for a nanosecond simulation.

Consider now the concept of ergodicity, as applied to computer simulations. Rigorously, in order for the time average to be used to calculate fluid properties (to equal the ensemble average), an infinite time is required. In practice, one has the intuitive feeling that the time average will give sensible results if there is enough "mixing" during the run; that is, if plenty of diffusion occurs, so each atom samples a large number of different neighbors and environments. One usually takes a practical approach to this problem by requiring three conditions to be met: (1) the property to be calculated is independent of time; (2) each particle diffuses at least a distance equal to the average interparticle separation; and (3) most particles change their nearest neighbors.

If all these conditions are not met for any run, the system is not ergodic. (For instance, if the system configuration is frozen, the first condition will be met, but the other two will not.)

A convenient measure of when the system becomes non-ergodic is the diffusion coefficient, which is

$$D = \frac{1}{6} \frac{\langle R^2 \rangle}{t}$$

where $\langle R^2 \rangle$ is the mean squared atomic displacement in time t of an atom of a particular species. If the computer run is for 10^{-11} secs of fluid time (an average run length of the typical budget) and if we set $\langle R^2 \rangle = 1A^2$, then for a time average to give a result characteristic of the equilibrium fluid we require

$$D > 0.2 \times 10^{-5}\ \text{cm}^2/\text{sec} \qquad (1)$$

In other words, the diffusion coefficient of the simulated fluid must be larger than this value in order that the time averages of the quantities be typical of a fluid in thermal equilibrium. This is the absolute minimum restriction, and will give results that are barely acceptable (for a run of .01 ns).

If the diffusion coefficient is smaller than this value, there is no diffusion on the time scale of the run. The fluid structure is essentially frozen during the run. In other words, if the fluid is not ergodic, it is a glass.

The relation (1) means that very viscous fluids cannot be directly simulated by molecular dynamics. Dense fluids have diffusion coefficients less than 10^{-16} cm^2/sec, more than 11 orders of magnitude smaller than the smallest permissible values.

The same thing can be seen by considering the viscosity, using the Maxwell relation

$$\eta = G\tau$$

where τ is the average shear relaxation time and G the shear modulus. If $G = 10^{10}$ dynes/cm^2, and the largest permissible value of τ is 10^{-9} secs, we see that only those fluids can be simulated for which the viscosity is less than 10 poise. This is a very low viscosity, considering that the glass transition generally occurs when the viscosity is about $10^{11} - 10^{13}$ poise. As a result of these considerations, if simulations are to provide any information about the behavior of dense fluids, a indirect approach must be taken. The basic idea is this: Computer simulations can be used to deduce the diffusion mechanism and trends in melt structure for high-temperature fluids. These observations must then be extrapolated in some way to lower temperatures. The numbers themselves are not extrapolated. For instance, computing diffusion coefficients at temperatures near and above the melting point and extrapolating the result to the transformation region would be quite dangerous. The diffusion coefficient versus 1/T has considerable curvature for most materials, and the greatest curvature occurs at temperatures too low for the simulations to be applicable. Rather, what is extrapolated in the mechanism for diffusion itself.

In closing, we note that details of the simulations described below are given in Brawer and Weber [1981].

4. Molecular Dynamics Simulations of BeF_2 and SiO_2

In this section we summarize some results on computer simulations of the network glasses SiO_2 and BeF_2. In the next section these results are used to develop a microscopic model which accounts for the anomalous properties of network liquids and glasses, and the results of new experiments are predicted. Computer simulations of BeF_2 have been reported by Rahman et al. [1982]. Simulations of SiO_2 can be found in Angell et al. [1981], Woodcock et al. [1976], and Soules [1979]. These studies are quite different from the one reported here, but our results agree with theirs where applicable.

The basic structure of the simulated fluid BeF_2 is approximately that of the continuous random network model, which is often used for SiO_2 [Bell and Dean, 1968a, b; Leadbetter and Wright, 1972, and references therein]. In the pure continuous random network, all the Be ions are tetrahedrally coordinated by F ions and the F ions are two-fold coordinated by Be. The tetrahedra are undistorted.

This leads to a structure of linked BeF_4 tetrahedra, the tetrahedra being joined at the vertices with randomly fluctuating Be-F-Be angles and random relative orientations of neighboring tetrahedra. However, in the simulated systems, it is always found that not all the Be and F ions have 4- and 2-fold coordination respectively. Rather, some ions have coordination numbers higher than these, and the number of such defects increases with increasing temperature. The structure is illustrated in Figure 1, where two defects are shown.

The radius of the coordination sphere for calculation of Be coordination number of F (or F by BE) is always taken as 2.4 A, approximately the first minimum of the Be-F radial distribution function (RDF). More importantly, this radius, 2.4 A, has an operational significance, so that the definition is useful. Using 2.4 Å as the radius of coordination sphere, then, the number of defects in the fluid as a function of the temperature is given in Figure 2. In this figure, f_5 is the fraction of Be ions that are five-fold coordinated by F (within 2.4 A) and f_3 is the fraction of F ions that are three-fold coordinated by Be. We note that f_5 and f_3 decreases rapidly as the temperature falls below about 1600°K. The number of defects decreases in an approximately Arrhenius manner, with activation energy of about 0.5 eV. At temperatures below 1300 K, only one or two defects are found in the entire sample. The defects exist in thermal equilibrium.

The presence of defects plays a crucial role in the kinetic properties of simulated BeF_2. The elementary event in diffusion or viscous flow in the material must be the separation of Be-F pairs which were originally nearest neighbors. Then the importance of defects is this: The separation of Be-F neighbors takes place exclusively at defect sites.

More precisely, consider the coordination number of Be and F ions which are initially within 2.4 A, and so are neighbors, and which at a later time are separated by some distance R_M, typically 3.5 A or greater. It is observed in the simulations that, just prior to the start of the separation, the coordination number of the Be and F which subsequently separate is 5 and 3 respectively. This is, diffusion occurs from high coordination sites.

We can quantify the relation between coordination number and diffusion by the following calculation. Let the criterion for a Be-F

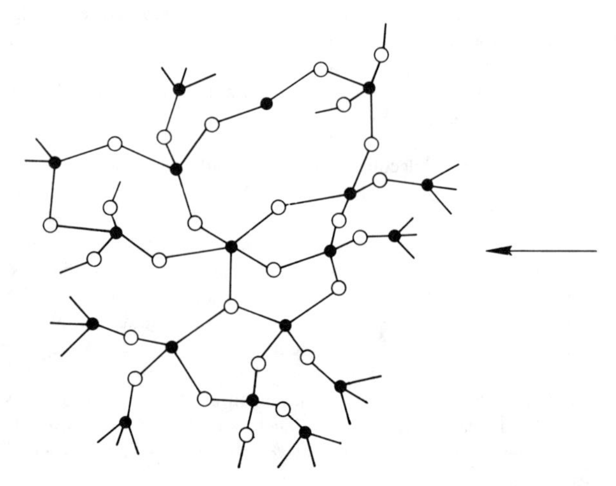

Fig. 1. Schematic diagram of the structure of BeF_2 or SiO_2 glass. Cations are filled-in circles and anions are open circles. The basic structure is the continuous random network described in the text, but computer simulations show the presence of defects in the form of 5-fold cations and 3-fold anions. An arrow points to a 5-fold cation, and right below is a 3-fold anion.

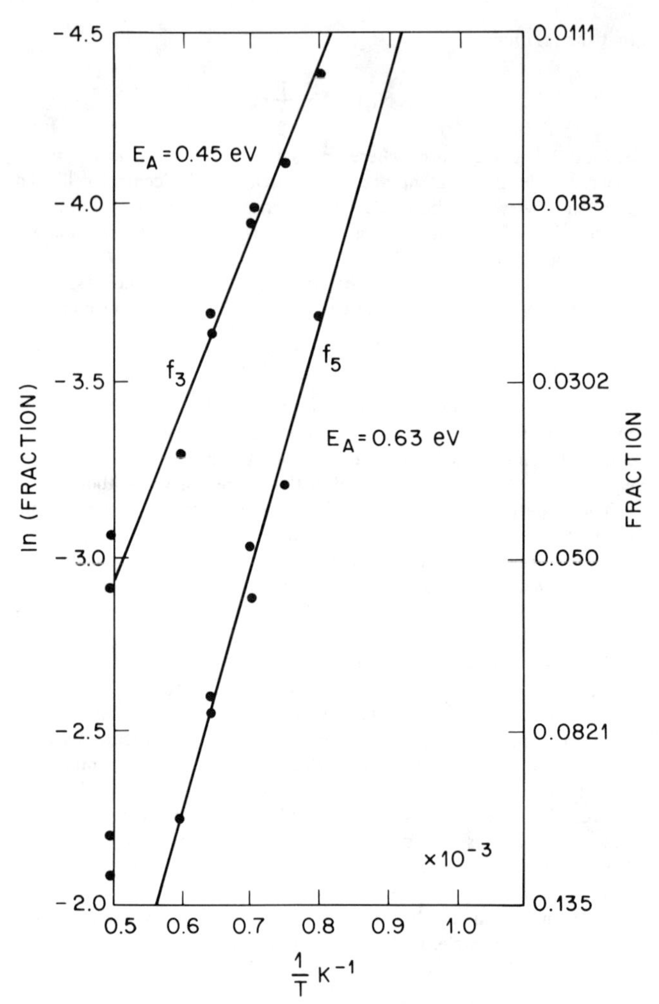

Fig. 2. The fraction f_3 of F ions that have three Be neighbor and the fraction f_5 of Be ions that have five F neighbors, as a function of temperature, for simulated BeF_2.

separation be that the pair separation goes from a distance less than 2.4 A to a separation greater than 3.5 A. All such pairs may be identified for any run. For each of these pairs, and only these pairs, the coordination numbers of the two ions involved can be calculated whenever the Be-F separation has any particular value. In other words, by examining all separating pairs, a histogram of the Be and F coordination numbers can be made, and a different histogram made for each desired Be-F distance.

The quantities that are calculated for the histograms are the fraction p(n) of Be ions that have coordination number n by F ions, and the fraction of F ions that have coordination number n by Be ions. If one were to form these histograms using all the Be and F ions in the system, one finds a large maximum at $n = 4$ and $n = 2$ respectively. These are shown as the lower histograms in Figure 3, labeled "all ions."

However, if we consider only those ions involved in separation, and average only over the time interval during which the separation is occurring, one obtains the remaining histograms of Figure 3. The first interval (0, 2.4), means that the Be-F separation is less than 2.4 A, and the histograms were formed just prior to the start of the separation. Note that the Be and F coordination numbers are largely 5 and 3 respectively, showing that separation occurs at defect

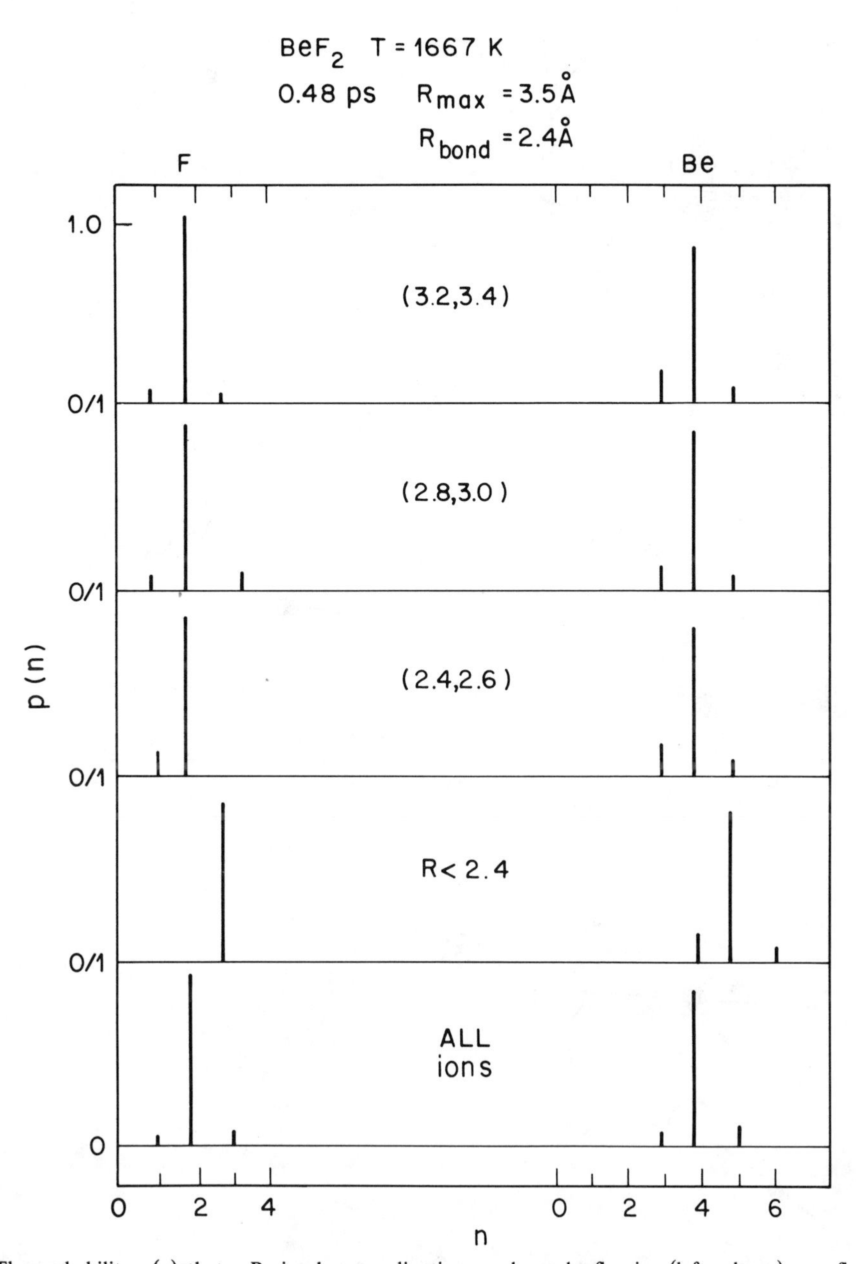

Fig. 3. The probability p(n) that a Be ion has coordination number n by fluorine (left column) or a fluorine has coordination number n by Be, for simulated BeF_2 at 1667°K. The bottom histogram is averaged over all ions. The remaining histograms are averaged only over pairs that separated (were initially within 2.4 Å and separated by more than 3.5 Å). The different histograms are averaged when the Be-F separation is within the interval noted, as explained in the text. Note that, just prior to the start of the separation, separating pairs tend to have coordination number one higher than the average, showing that diffusion occurs at defect sites.

sites. The remaining histograms give the coordination number distribution at increasing Be-F separation. Note that, after the first initial decrease in coordination, occurring because the coordination sphere radius is defined as 2.4 A, there is no further change in coordination number.

The diffusion process in computer-simulated BeF_2, then, involves the formation of defects, which are atoms with high coordination numbers, and later the separation of such defects. For diffusion and viscous flow, however, the defect must dissociate in a different way than it is formed. There must be a reshuffling of nearest neighbors, so that the topology of the network is altered.

One way this can happen is shown in Figure 4. In this case, a defect complex consisting of two 5-fold Be and two 3-fold F ions is formed. There is a bond interchange in the defect complex, so that when the defect breaks up the topology of the network is changed, and some of the ions have switched neighbors.

One can also have a mechanism like that shown in Figure 5. This can be looked upon as a transfer of a defect from one position to

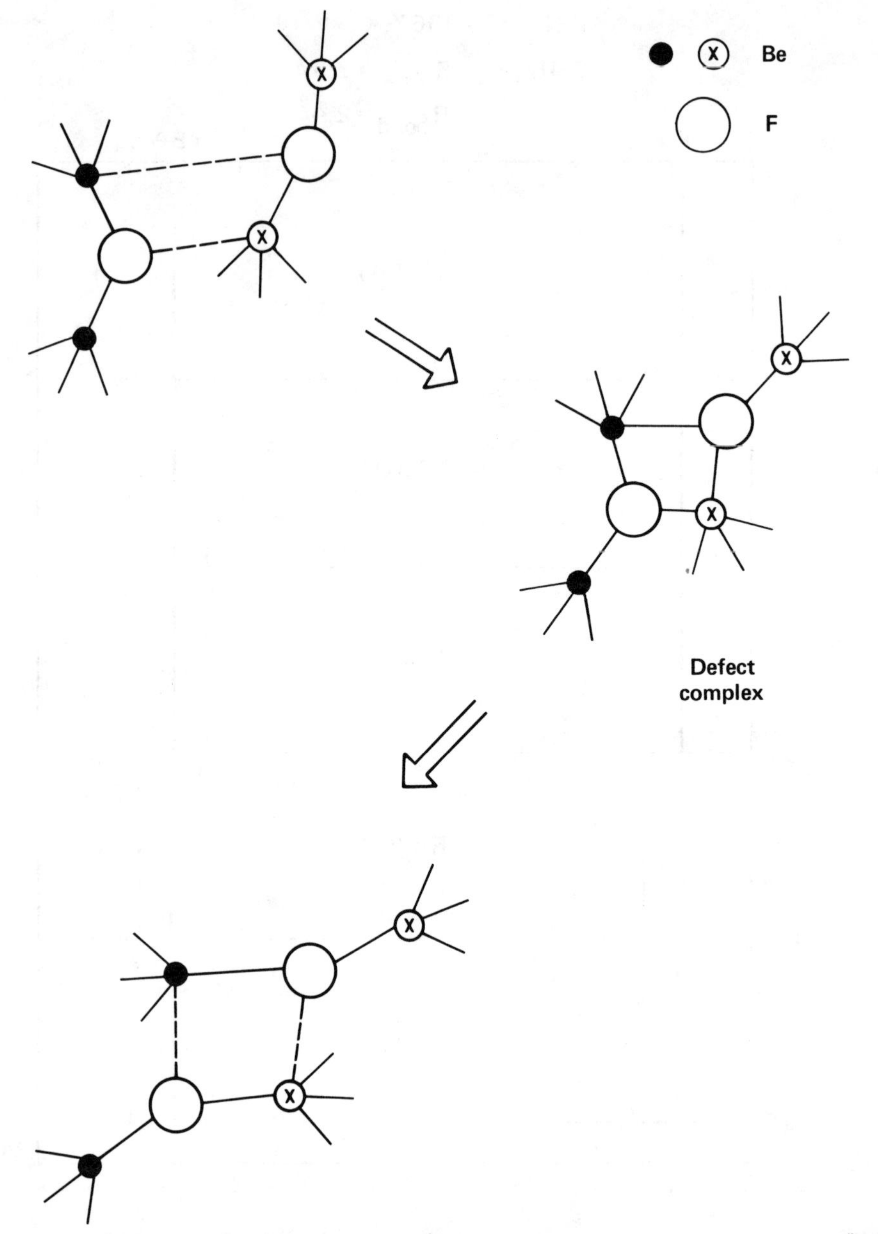

Fig. 4. A schematic representation of the observed mechanism for diffusion in molecular dynamics simulation of BeF_2. In this mechanism, a defect complex, with 3-fold F and 5-fold Be ions, is formed, and then the nearest-neighbor ions are shuffled. As a result, diffusion occurs, and the topology of the random network is altered. The mechanism can explain, at least qualitatively, the occurrence of viscous flow due to a point-defect mechanism.

another. A 5-fold Be loses an F, making another Be 5-fold. There are other possible mechanisms as well and no attempt has been made to catalog them.

In short, then, the diffusion process involves the creation of defect sites, consisting of high coordination number ions, and the subsequent breakup of the defects in such a way that the topology of the random network is altered.

The rate of structural relaxation has also been calculated for BeF_2. Structural relaxation, and viscous flow, must be related to the rate at which Be-F pairs, originally neighbors, separate. In a single relaxation time approximation, if f(t) is the fraction of Be-F neighbors which separate by time t, we have

$$f(t) = 1 - e^{-t/\tau}$$

so $1/\tau$ is an average rate for Be-F separations. If τ_s is the mean shear relaxation time, determined from the shear viscosity by

$$\tau_s = \eta/G$$

G being the infinite frequency shear modulus, then we expect

$$\tau \cong \tau_s$$

That is, the lifetime for breakup of Be-F neighbors should be

It is seen that for as much as a 50% increase of density, the F diffusion coefficient increases, and only afterwards does it decrease. This phenomenon was first observed in a simulation of SiO_2 by Woodcock, et al. [1976]. C. A. Angell, in private conversation, first connected it with the diffusion mechanism involving defects proposed by the author. The microscopic explanation of this result will be given in the next section.

Below we summarize the results of this section.

1. The fluid and glass structure is basically the continuous random network, a three-dimensionally interconnected network of BeF_4 (or SiO_4) tetrahedra, joined at the vertices with fluctuating angles and random relative orientation of tetrahedra.

2. There are defects, which are atoms of high coordination number (3-fold anions and 5-fold cations, also 3-fold cations which disappear at the lowest temperatures). These defects are thermally activated, and the number of defects decreases with decreasing

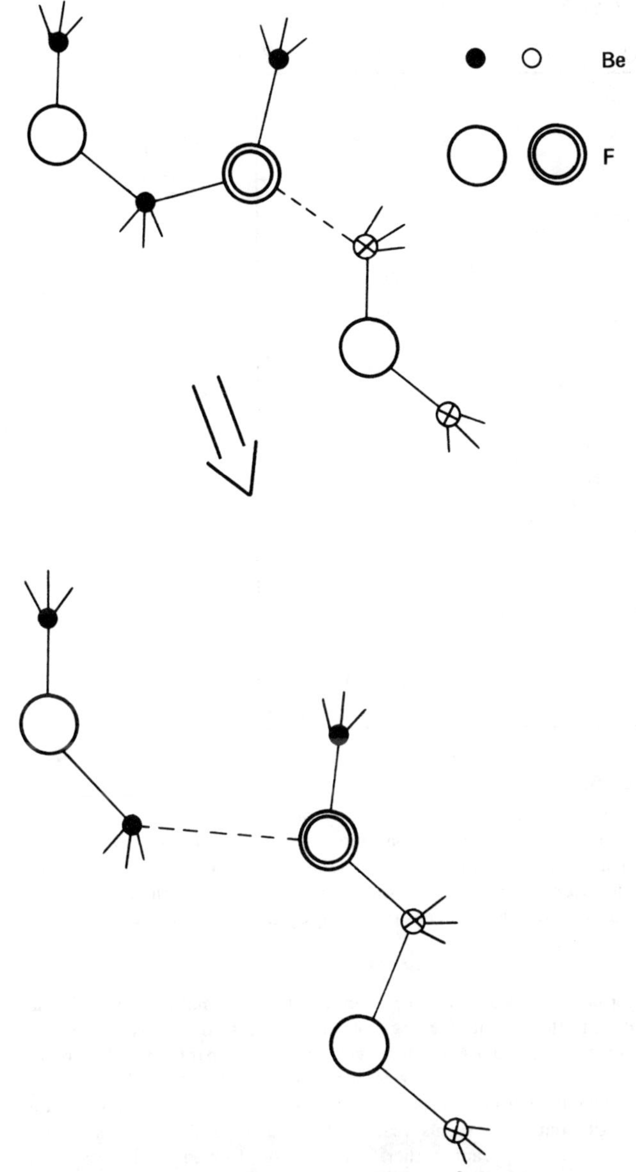

Fig. 5. Another possible mechanism for diffusion in BeF_2. This particular mechanism passes a 5-fold Be from one site to another. The transfer is catalyzed by a 3-fold F ion labeled with the interior circle.

approximately equal to the mean shear relaxation time.

In Figure 6 we show the values of τ so computed as a function of temperature. It is compared with τ_s computed from the measured viscosity [Moynihan and Cantor, 1968]. Two values of G were used, bracketing the range of G expected. The solid line in the figure corresponds to the temperature range where the viscosity was measured and the dotted line is an extrapolation.

It is seen that at the lowest temperatures there is reasonable agreement between τ and τ_s. The computed values of τ are smaller than τ_s, which is to be expected. At higher temperatures, τ rises much more slowly than the extrapolated values of τ_s.

The fluorine diffusion coefficient of the simulated fluid as a function of density at a constant temperature is shown in Figure 7.

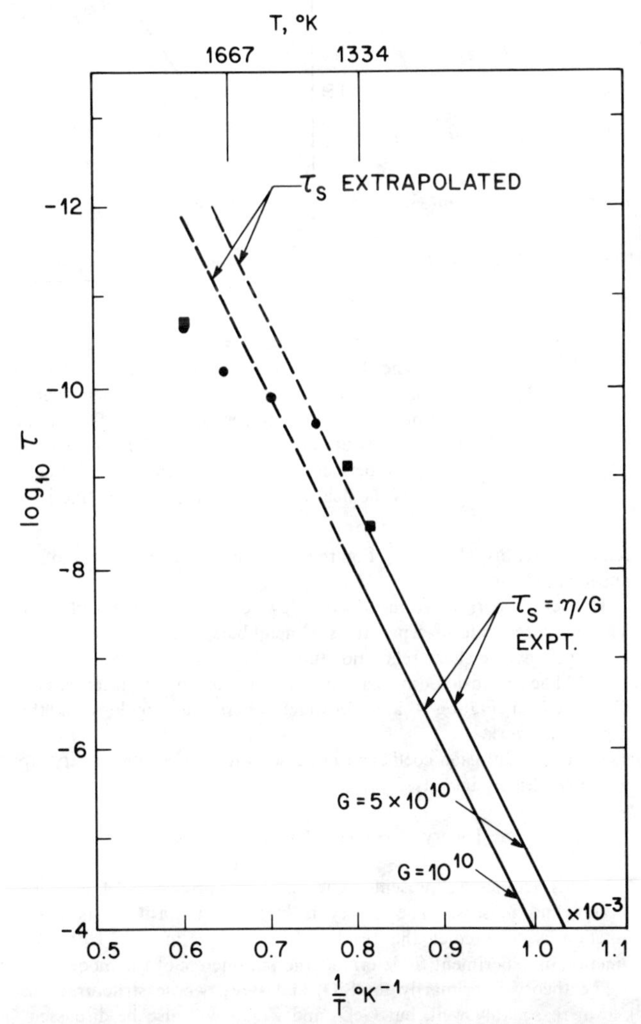

Fig. 6. The mean rate of Be-F separation, $1/\tau$, as a function of temperature T for BeF_2. τ is in seconds. The points are computed from molecular dynamics simulations. The solid line is the mean shear relaxation time τ_s as computed from the experimental shear viscosity [Moynihan and Cantor, 1968]. Two values of the infinite frequency shear modulus G were used. The dotted line is an extrapolation of the measured values.

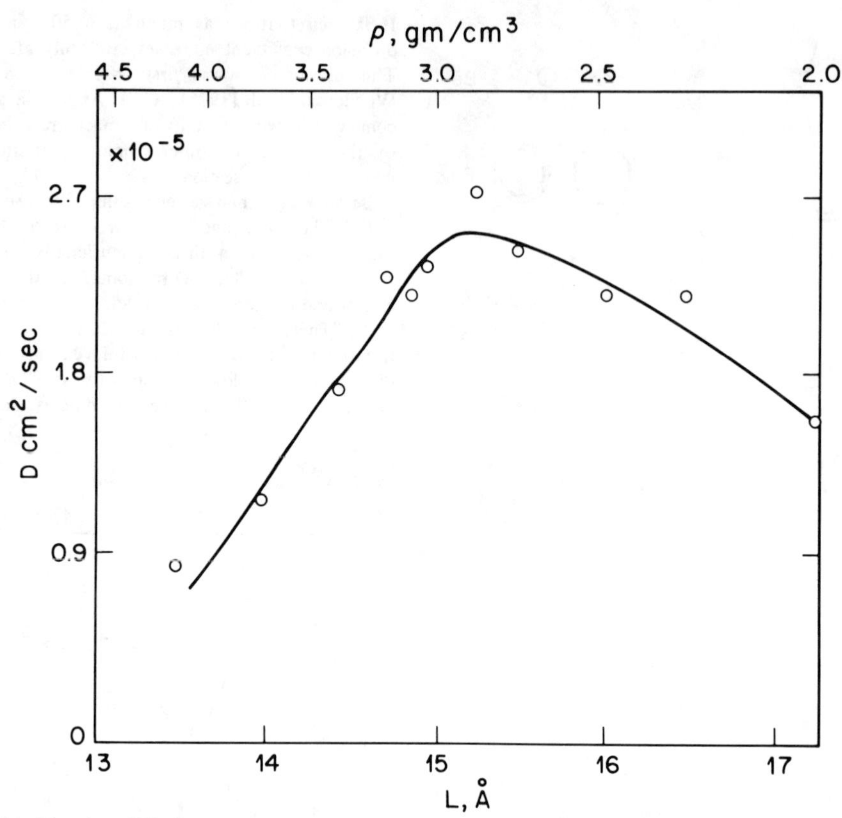

Fig. 7. The Fluorine diffusion coefficient D versus the density ρ, and box side L, for molecular dynamics simulation of BeF_2 at 1667°K. The diffusion coefficient is averaged over 3 ps. Each point in the figure is the average of three or four independent runs. The initial condition of each separate run was obtained by quenching a fluid in equilibrium at 3200°K to 1667°K and equilibrating for 10 ps. This was necessary to produce fluids with different numbers of defects. The density of the high temperature fluid and the density during equilibration were the same as the density of the final run. The line is drawn as a guide to the eye.

temperature, as shown in Figure 6 for BeF_2. The defects often appear in clusters.

3. Be-F separations, and viscous flow, occur only at defect sites. The measured rate of separations of neighbors is comparable to the measured inverse shear relaxation time.

4. The diffusion mechanism is one of bond interchange, illustrated in Figures 4 and 5, which alters the topology of the random network.

5. The F diffusion coefficient increases with increasing density, up to a 50% density increase.

5. Theory of Network Glasses and Liquids

In this section we present a theory of the properties of network liquids and glasses. The theory is largely qualitative. Its main function is to predict the results of new experiments. It is the function of experiment to determine the parameters of the model.

The theory is primarily for SiO_2 and BeF_2, whose structures are known reasonably well, but GeO_2 and $ZnCl_2$ will also be discussed. B_2O_3 is specifically excluded from this discussion. It is not a network liquid in the same sense as these other materials.

The model for a network fluid is a straightforward extrapolation of the molecular dynamics results: The fluid structure is a continuous random network in which defects are thermally activated. The model has three parameters. Two of them are illustrated in Figure 8, which shows a hypothetical potential energy surface for formation of a defect. The quantity ϵ is the formation energy for the defect, the potential energy of the defect state as opposed to the normal continuous random network. The parameter E_A is an activation energy to be surmounted when the defect breaks up.

The number of defects in the fluid below the melting point is quite small, and so it seems reasonable to assume that defects do not interact with each other. They are formed and break up independently. The assumption of non-interaction of defects, if it is valid at all, is expected to be more valid as the temperature decreases.

Before presenting the details of the model, we show how the model qualitatively rationalizes the properties described in section 2.

1. Consider first the specific heat, for which no break is observed at the glass transition temperature. The specific heat is the sum of a vibrational and configurational contribution. The vibrational contribution will be continuous through the glass transition because atomic diffusion occurs at a rate of more than 10 orders of magnitude slower than vibrations. The configurational properties of the fluid are determined by the number of defects. It is our contention that, at the glass transition temperature, the number of defects is so small that the configurational specific heat is much smaller than the vibrational contribution. In fact, it is unmeasurably small. The configurational specific heat will become large only when the number of defects is relatively large, and for BeF_2 and SiO_2 this will occur at temperature above about twice the melting point, in degs K.

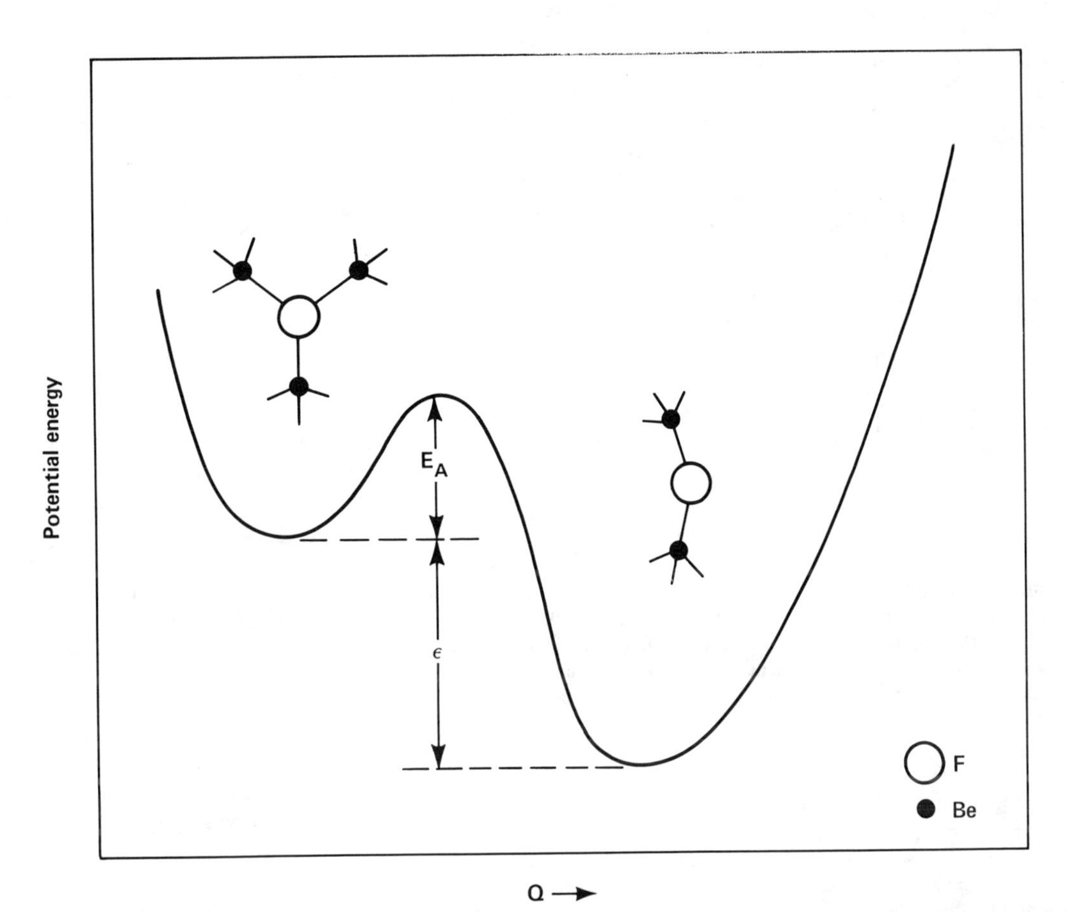

Fig. 8. Potential energy versus configurational coordinate Q for ions involves in defect formation. This potential surface shows the essence of the model of network liquids SiO_2 and BeF_2--that the defect sites have a formation energy ϵ of order 1-2 eV, and a barrier E_a that must be overcome to break up the defect.

2. The increase of viscosity with pressure, observed for GeO_2 and albite and predicted here for BeF_2 and SiO_2, is due to the fact that the number of defects increases with pressure. That is, as the pressure increases and the volume decreases, one expects a corresponding increase in the coordination number of Si and O. This corresponds to an increase in the number of defects. Since diffusion and viscous flow can occur only at defect sites, the increase in the number of defects means that there is an increase in the number of sites at which diffusion and viscous flow can take place.

3. The densification of SiO_2 which can be induced below the glass transition temperature occurs because defects are created. These defects must surmount an energy barrier to be broken up, and hence the densification can be permanent until the defects are annealed away.

If $\beta\epsilon >> 1$ ($\beta = 1/kT$, k being Boltzmann's constant), the number of defects n is (cf. Figure 8)

$$n \simeq n_o \ell^{-\beta\epsilon} \tag{2}$$

for some constant n_o. The rate at which defects are formed is

$$\omega_o \ell^{-\beta(\epsilon + E_a)} \tag{3}$$

and the rate at which they break up without any bond interchange is

$$\omega_o \ell^{-\beta E_a} \tag{4}$$

Therefore, the equation that governs the time rate of change of n is

$$\frac{dn}{dt} = -n\,\omega_o \ell^{-\beta E_a} + (n_o - n)\,\omega_o \ell^{-\beta(\epsilon + E_a)} \tag{5}$$

The third parameter of the model is an energy E_o required for bond interchange at the site of the defect. Bond interchange is required in order for actual diffusion and viscous flow to occur. Defects can be formed and break up without any bond interchange. This distinction has experimental consequences.

The idea here is that in order for a bond interchange to occur, which rearranges the local topology, a cluster of defects is required. Thus in Figures 5 and 6, 4 and 3 defects atoms respectively are present at the site. Then the rate Γ at which a bond interchange takes place at some point in the melt is

$$\Gamma = p(j)\ \Omega_o \ell^{-E_o/kT} \tag{6}$$

where p(j) is the probability of an appropriate cluster of j defects which can lead to bond interchange. The quantity Ω_o is some frequency factor for the cluster. The average relaxation time τ for diffusion is then Γ^{-1}, where Γ is given in (6).

An estimate for p(j) can be had as follows. Assuming that defects can form independently, and that one can have n_o defects per unit volume, then a defect cluster will form in a volume v. Then in such a volume, there are vn_o possible defects. Say a cluster of j defects form. The probability for this is then simply

$$p(j) = \frac{1}{Z}\,\frac{(vn_o)!}{j!\,(vn_o - j)!}\,\ell^{-\beta\epsilon j} \tag{7}$$

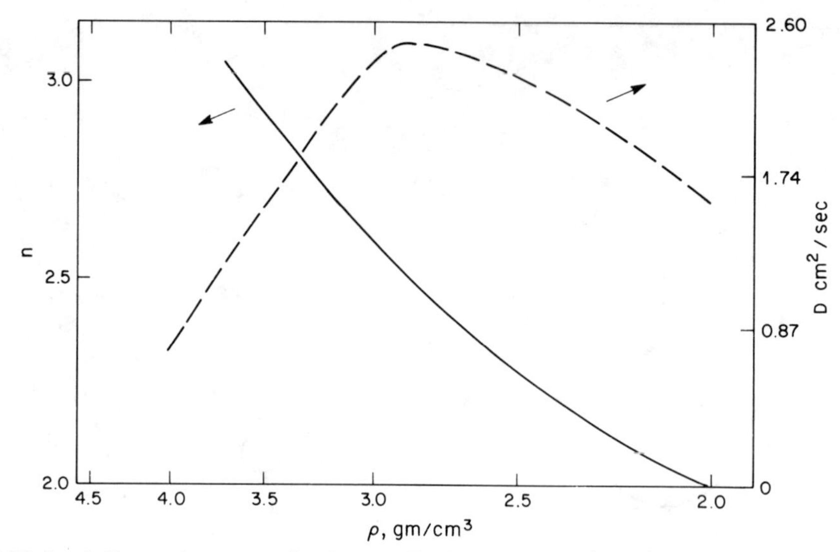

Fig. 9. Solid line indicates the average fluorine coordination number n (by Be) as a function of density for molecular dynamics simulations of BeF_2 at 1667°K. the dotted line is the F diffusion coefficient (the same as in Figure 7). The increase in fluorine coordination number at low pressure is due to formation of defects and leads to an increase in diffusion coefficient D and a decrease in viscosity.

where

$$Z = (1 + \ell^{-\beta\epsilon})^{vn_o}$$

where, as in (2), ϵ is the energy for forming a single defect, and the assumption is made that defects can form independently.

Combining (7) with (6) and using the Maxwell relation for the relaxation, $\eta = G\tau$, where G is the shear modulus and $\tau = \Gamma^{-1}$, we have (assuming also that $vn_o >> j$)

$$n \approx \frac{j!}{(vn_o)^j} \frac{G}{\Omega o} \ell^{\beta[E_o + \epsilon j]} \quad (8)$$

Now considering Figures 4 and 5, we might expect vn_o to be greater than 10. Then using $G = 5 \times 10^{10}$ dynes/cm^2, $\Omega_o = 10^{14}$ sec^{-1}, $j = 3$ and $vn_o = 10$, we find from (8) that

$$n = 3 \times 10^{-6} \ell^{\beta(Eo + \epsilon j)}$$

The actual experimental expression for liquid silica is

$$n = 10^{-10.6} \ell^{E/kT}$$

where $E = 140$ kcal/mole. Thus the pre-exponential in (8) is somewhat small, but considering the uncertainties involved, the estimate is not too bad. We could match experiment by choosing $vn_o = 100$, which means a considerable region of the glass is required to form a new defect cluster.

Actually the expression (8) is only meant to be indicative, as the true situation is quite complex. For instance, defects will not form independently, as assumed, but rather, because one defect forms, the formation of nearby ones is likely to be easier. This will cause a revision of both the expression for activation energy and for the pre-exponential.

We will see later that $\epsilon \sim 38$ kcal/mole therefore, if $j = 3$ in (8) we find for E_o

$$E_o \cong 26 \text{ kcal/mole}$$

The final feature of the model is that ϵ should decrease with increasing pressure so that the number of defects increases with pressure. In Figure 9 the average coordination number of F by Be as a function of density is shown. It is seen that the coordination number increases as density decreases. To anticipate some results, for SiO_2 the available data indicate the following numerical values for the parameters:

$$\epsilon = 38 \text{ kcal/mole}$$

$$E_a = 65 \text{ kcal/mole}$$

$$E_o = 26 \text{ kcal/mole}$$

A very important aspect of this model is that viscous flow is not required to change the number of defects because bond interchange is not involved. Thus eq. (5), describing the change in number of defects, does not contain the activation energy E_o, which is included in (8).

As a result, the rate of change of defects occurs much more rapidly than the rate of viscous flow. That is, the mean relaxation time for changing the number of defects is much shorter than the mean shear relaxation time.

To change the bonding topology does require viscous flow, however, and thus any effect associated with a changing distribution of Si-O-S-O---rings or of average cation-anion-cation angles will require viscous flow. In densification at temperatures above the glass transition, both processes are probably involved, which makes this phenomenon quite complex.

We point out one fundamental assumption of the theory: Defects can form and disassociate without a change in the topology of the random network. Viscous flow (which changes the topology of the random network) and defect formation must be completely uncoupled in order to observe the phenomena predicted here.

Consider first the specific heat. The specific heat has two contributions: (1) a vibrational one, due to the vibrations of the continuous random network; at temperatures above the glass transition this contribution is expected to be only a weak function of the number of defects; (2) a configurational contribution, which is due entirely to the presence of defects. It is assumed that the change of ring structure with temperature is insignificant, which seems to be the case experimentally.

Now below the melting point, the number of defects is small (cf, Figure 2) and therefore the configurational contribution to the

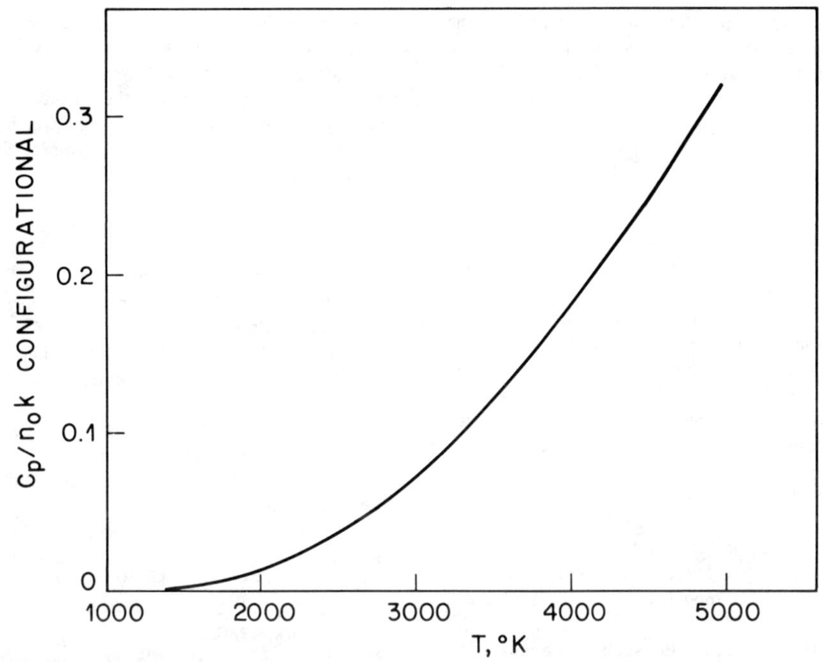

Fig. 10 Predicted configurational specific heat Cp versus temperature T for SiO_2, from eq. (9), with ϵ = 38 kcal/mole. Here, k is Boltzmann's constant and n_o is the constant in eq. (2).

specific heat is also small. For SiO_2 and BeF_2 it is hypothesized that the number of defects is so small the configurational contribution is not measurable (although it is there). As a result, even in the fluid state, the specific heat is given by the vibrational contribution alone, and therefore it is the same for the fluid as for the glass. It is not surprising, then, that there is no change in specific heat at the glass transition.

The fact that the configurational specific heat is small places a lower limit on ϵ. The internal energy is

$$E = E(vib) + \epsilon\, \ell^{-\beta\epsilon}$$

where E(vib) is the vibrational contribution, assumed independent of the number of defects, while the second term is due entirely to the defects. The specific heat at ambient pressure is the temperature derivative of the energy,

$$Cp = Cp(vib) + n_o k (\beta\epsilon)^2 \ell^{-\beta\epsilon} \qquad (9)$$

Assuming n_o = the number of fluorine (or oxygen) atoms, the fact that the configurational specific heat is small at temperature T means that

$$(\beta\epsilon)^2 \ell^{-\beta\epsilon} << 1$$

For SiO_2 the configurational specific heat is not measurable even at 2600°K. Supposing that the above expression is less than 0.05 at 2600°K gives

$$\epsilon > 38 \text{ kcal/mole}$$

Actually, a very small discontinuity in specific heat at the glass transition might be expected. The reason is that the entropy of the network--the topology of the random network--can only change by viscous flow, and is frozen in at the glass transition. If this change is thermodynamically important, it will show up as a discontinuity in specific heat. This is apparently a small effect for SiO_2 and BeF_2. It may contribute to the small discontinuities in specific heat for other network liquids like GeO_2 and $ZnCl_2$.

We can make an analogy here with crystalline systems such as alkali halides. In these systems, diffusion occurs via vacancies or interstitials which must be thermally excited. On the other hand, the number of defects is so small that the thermodynamic properties such as the specific heat of the crystal are not affected by the variation of their numbers, and thus the thermodynamic properties are due only to the lattice vibrations.

Similarly for network glasses, in the transformation range the thermodynamic properties are determined by the lattice vibrations of the random network because the number of defects is so small.

A prediction of the specific heat at the higher temperatures is shown in Figure 10. As the temperature is raised above the melting point, the number of defects increases until a significant number of defects are present. At this point, the configurational contribution to the specific heat becomes measurable and the specific heat will increase above the vibrational contribution.

For BeF_2, the computer simulations indicate that the number of defects becomes large at about 1600°K. Experimentally, the specific heat becomes greater than the classical value 9R at about 1200°K, but its slope is decreasing. The rate of increase of specific heat should start to increase again at higher temperatures.

Consider now the change of viscosity with pressure. The viscosity is determined from eq. (6), with the relation n = G/Γ,

$$\eta = \frac{G}{p(j)\,\Omega o}\, \ell^{\beta E_o}$$

As pressure is applied, the average coordination number of F ions increases (Figure 9). This means that more defects are formed, which in turn means that P(j) increases with pressure p. (Recall that p(j) is the probability of occurrence of defect cluster of j defects.) Therefore, according to the above equation, if E_o is constant, the viscosity decreases with pressure.

In other words, the viscosity decreases with pressure because the application of pressure creates more defect sites where flow can take place. We would expect a smaller effect on E_o because bond interchange involves only a local rearrangement.

This theory makes some unusual predictions concerning the rate of change viscosity following a temperature jump. One of the important aspects of the theory is that following a sudden temperature decreases, the rate of the change of the number of defects is given by (4)

$$\omega_0 \ell^{-\beta E a}$$

which has a smaller activation energy than viscous flow. On the other hand, following a sudden temperature decrease, the apparent activation energy for the viscosity itself will be E_o (cf. (6)) because flow will initially occur for constant number of defects (constant structure). However, the rate of change of viscosity will have activation energy E_a because this is the activation energy for the change of the number of the defects for SiO_2. The reason, again, is that viscous flow and structural change (change in number of defects) are uncoupled, and therefore the relaxation of the viscosity is not governed by the viscosity itself. This result is quite different from what presumably occurs in normal fluids.

Some caution must be exercised in such experiments. Since the number of defects changes so rapidly, the true $t = 0$ viscosity may be difficult to measure at constant number of defects.

Consider now densification of SiO_2 and BeF_2 below the glass transition. According to the present model, application of pressure below the glass transition creates defects. It cannot change the network topology because the time scale for such a change is too long. Permanent densification results because the breakup of the defects is a thermally-activated process, and at low temperatures the defects are frozen in. Upon annealing, also below the glass transition temperature, the defects break up, and E_a is the activation energy for break up of defects. Thus for SiO_2, it is possible to estimate from Hsich et al [1971]

$$E_a \cong 65 \text{kcal/mole} \quad (= 2.83 \text{ eV})$$

The application of pressure in the transformation range, where the fluid is in thermal equilibrium, will both create defects and will also increase the density due to a rearrangement of the random network itself. Thus two different phenomena are involved, which have quite different time scales, the defect creation being much more rapid than the change in the network. Both effects should be measurable either by determining the kinetics of densification or the annealing kinetics.

We point out that the densification due to the creation of defects will not be frozen out at the glass transition (defined as the temperature at which the viscosity if 10^{13} P) because of its lower activation energy than the viscous flow process. This also should be observable.

Another prediction is the following: If the glass is densified far below the glass transition, then if the temperature is suddenly raised to the transformation range, its viscosity should be much smaller than the equilibrium value. The reason is that many defects are created by the densification and these produce many sites where flow can take place. The viscosity should increase with time as the defects anneal away, and the activation energy for this annealing should be E_a. Experimentally this relaxation may be quite rapid, making the measurement of a changing viscosity difficult. (I thank the referee for this observation.)

Since densification below the glass transition involves only creating defects, the process can be used to determine ϵ and a volume of activation v_o per defect. Care must be taken that only a small density change is involved, so that the defects do not interact.

Acknowledgments. The author acknowledges many enlightening discussions with C. R. Kurkjian, C. A., Angell, and C. Moynihan. This work was performed while the author was at the Livermore National Laboratory, under the auspices of the Division of Materials Sciences, Office of Basic Energy Sciences, U.S. Department of Energy, and the Lawrence Livermore National Laboratory under contract W-7405-Eng-48.

References

Angell, C. A., Amorphous solids: Types, characteristics and challenges, in *Preparation and Characterization of Materials*, Academic Press, New York, 1981.

Angell, C. A., and W. Sichina, Thermodynamics of the glass transition: Empirical aspects, *Ann. N.Y. Acad. Sci., 279*, 53, 1976.

Angell, C. A., J. H. R. Clarke, and L. V. Woodcock, Interaction potentials and glass formation: A survey of computer experiments, *Adv. Chem., Phys., 48*, 397, 1981.

Bell, R. J., and P. Dean, The vibrational spectra of vitreous silica, germania and beryllium fluoride, *J. Phys. C, 1*, 299, 1968a.

Bell, R. J., and P. Dean, Properties of vitreous silica: Analysis of random network models, *Nature London, 212*, 1354, 1968b.

Boon, J. P., and S. Yip *Molecular Hydrodynamics*, McGraw-Hill, New York, 1980.

Brawer, S. A., and M. J. Weber, Molecular dynamics simulations of the structure of rare-earth-doped beryllium fluoride glasses, *J. Chem. Phys., 75*, 3522, 1981.

Croxton, C. A., *Liquid State Physics--A Statistical Mechanical Introduction*, Cambridge University, Press, New York, 1974.

Hsich, S. Y., C. J. Montrose and P. B. Macedo, The annealing dynamics of fused silica, *J. Non-Cryst. Solids, 6*, 37, 1971.

Leadbetter, A. J., and A. C. Wright, Diffraction studies of glass structure, *J. Non-Cryst. Solids, 7*, 156, 1972.

Moynihan, C. T., and S. Cantor, Viscosity and its temperature dependence in molten BeF_2, *J. Chem. Phys., 48*, 115, 1968.

Rahman, A., R. H. Fowler and A. H. Narten, Structure and motion in liquid BeF_2, $LiBeF_3$ and LiF from molecular dynamics calculations, *J. Chem. Phys., 57*, 3010, 1972.

Sangster, M. J. L., and M. Dixon, Interionic potentials in alkali halides and their use in simulations of molten salts, *Adv. Phys., 25*, 247, 1976.

Sharma, V., and J. Kushiro, Relationship between density, viscosity and structure of GeO_2 melts at low and high pressures, *Non-Cryst. Solids, 33*, 235, 1979.

Soules, T. F., A molecular dynamics calculation of the structure of sodium silicate glasses, *J. Chem Phys., 71*, 4570, 1979.

Stout, N. D., and A. J. Piwinski, Enthalpy of silicate melts from 1520 to 2600 K under ambient pressure, *High-Temp. Sci., 15*, 275, 1982.

Wong, J., and C. A. Angell, *Glass: Structure by Spectroscopy*, Marcel Dekker, New York, 1976.

Wood, W. W., Monte Carlo studies of simple liquid models, in *Physics of Simple Liquids*, edited by N. V. Temperly et al, N. Holland, Amsterdam, 1968.

Woodcock, L. V., Molecular dynamics calculations of molten ionic salts, in *Advances in Molten Salt Chemistry*, vol. 3, p. 1, Plenum, New York, 1975.

Woodcock, L. V., C. A. Angell, and P. Chesseman, Molecular dynamics study of the vitreous state: Simple ionic systems and silica, *J. Chem. Phys., 65*, 1565, 1976.

ELECTRICAL CONDUCTION IN CERAMICS: TOWARD IMPROVED DEFECT INTERPRETATION

Harry L. Tuller

Department of Materials Science and Engineering, Massachusetts Institute of Technology, Cambridge, Massachusetts 02139

Abstract. Although defects control many physical and chemical parameters of importance in solids, a detailed understanding of their generation and motion is lacking in all but a select number of materials. Methods for determining the defect properties of rather complex systems by following a systematic program of electrical measurements in conjunction with the development of defect chemical models are outlined. Specifically, methods for deconvoluting microstructure from bulk, ionic from electronic and carrier density from mobility contributions to the measured electrical conductivity are emphasized. Examples taken from the recent literature, notably MgO and ThO_2-CeO_2, are used to demonstrate the principles. Special precautions in applying this approach to rather more complex mineral systems are noted.

1. Introduction

The role of defects in controlling many physical parameters of importance in solids has been recognized for some time. Some of the more notable ones include diffusion, creep, metal oxidation and electrical conduction. Materials scientists have succeeded in fabricating denser and tougher ceramics, improving the chemical stability of solids and developing ceramics with specialized ionic and electronic functions based on an improved understanding of the generation and control of defects. In a similar fashion geophysicists would be in an improved position to model and interpret physical phenomena in the earth's mantle, e.g., mass transport, redox states, etc., as the defect structures of the major constituents of the mantle become available. In this article, I wish to demonstrate that (1) electrical transport measurements may be used to establish the defect structure of solids and that (2) these measurements also represent perhaps the most versatile means of accomplishing that objective.

It will soon become obvious from subsequent discussions (see also article by Kröger in this volume) that the defect structures of even seemingly simple systems (e.g., MgO, Al_2O_3) may be quite complex. This is a consequence of the sensitivity that such compounds exhibit simultaneously to a variety of variables including temperature, atmosphere, impurities, thermal history, and microstructure. Consequently only a systematic approach in which physical measurements are performed as a function of each variable, while maintaining as many as possible of the other constants, is likely to result in an improved understanding of the applicable defect thermodynamics and kinetics. So as not to engage in measurements on the time scale of geologic ages, one requires a measurement technique which can be applied quickly, routinely and in situ over a wide range of experimental conditions. Many conventional techniques, e.g., tracer diffusion, TGA, and ESR, are often too slow or awkward especially at elevated T or under controlled pressures. Electrical conductivity measurements, on the other hand, satisfy all of these requirements quite admirably.

As with any sensitive technique, care must be taken in its interpretation. Again, the ability to perform such measurements systematically over an extensive range of conditions enables one to make intelligent interpretation of the data in terms of applicable defect parameters.

In the following sections I illustrate how electrical measurements, in conjunction with a defect chemical approach, may be used to establish the identity of the predominant ionic and electronic defects and their transport characteristics. In doing so, I choose only a limited number of representative systems. A number of articles and books are available which attempt to review electrical properties of oxides [Adler, 1968; Kofstad, 1972; Tuller, 1981] and minerals [Shankland, 1975, 1981, and references therein] in a more comprehensive manner.

2. Electrical Conduction

Conduction by charged species is additive and so the electrical conductivity of a solid is the sum of the partial conductivities given by

$$\sigma = \sum_j \sigma_j \tag{1}$$

The partial conductivities σ_j (S/m or $1/\Omega m$) represent transport by either ionic or electronic carriers and are defined by

$$\sigma_j = c_j Z_j q \mu_j \tag{2}$$

in which c_j is the carrier density (m^{-3}), $Z_j q$ is the effective charge (coulombs), and μ_j is the mobility ($m^2/V\ s$) of the jth carrier.

Alternatively one may introduce the diffusion coefficient D_j by way of the Nernst-Einstein relation

$$D_j = (kT/Z_j q)\mu_j \tag{3}$$

and obtain the following relationship between σ_j and D_j

$$\sigma_j = c_j \frac{(Z_j q)^2}{kt} D_j \tag{4}$$

Consequently, plots of $\ln \sigma_j T$ versus reciprocal temperature are prepared when one is interested in extracting the activation energy of diffusion from ionic conductivity measurements.

The fraction of the total conductivity contribution by each type of charge carrier is defined by

$$t_j = \sigma_j/\sigma \tag{5}$$

in which t_j is known as the transference number of that species. By definition

$$\sum_j t_j = 1 \tag{6}$$

Because one is often concerned with distinguishing ionic from electronic transport, the transference numbers t_i and t_e, which reflect respectively the ionic (i.e., cations plus anions) and electronic (i.e., electrons plus holes) components of conduction, are also often utilized.

Although much of the literature on electrical properties of solids is concerned with either strictly electronic conductors ($t_e = 1$) [e.g., Sze, 1969] or ionic conductors [e.g., Hagenmuller and Van Gool, 1978] materials which exhibit mixed ionic and electronic conduction represent an equally important category [Tuller, 1981]. It is now quite clear that mixed conduction is often encountered in studies of transport in oxides. Deconvolution of electronic and ionic contributions to conduction is therefore essential in establishing the nature of the various defects and the role they play in transport.

Likewise microstructural features such as surfaces and grain boundaries can serve as alternate paths for carrier transport and may thus either enhance or depress the measured conductivity relative to the bulk value.

It has by now probably become clear to the reader, that electrical conductivity measurements by themselves cannot uniquely establish the nature of the defects for they include contributions from both bulk and microstructural features, from ionic and electronic carriers and from carrier concentration and mobility. Fortunately all of these factors can be isolated by following a systematic plan of studies as illustrated in Figure 1. In later sections I discuss briefly the basis of these techniques and illustrate how they are applied to obtain the stated objective of deconvoluting the various contributions to the overall electrical conductivity.

3. Sources of Defects

3.1. Defect Chemical Approach

Three mechanisms for defect formation in compounds need be considered simultaneously [Tuller, 1984] these are (1) thermally induced intrinsic electronic and ionic defects, (2) redox, and (3) impurity induced defects. The first two categories are predicted on the basis of statistical thermodynamics [Kroger, 1974] and the latter to satisfy electroneutrality. Examples of representative defect reactions in the three categories along with their corresponding mass action relations are given in Table 1. Mass action relations are as follows:

$$n \cdot p = K_E(T) \tag{7}$$

$$[V_M''] [V_o] = K_s(T) \tag{8}$$

$$[V_o^{\cdot\cdot}] [O_i''] = K_f(T) \tag{9}$$

$$[V_o^{\cdot\cdot}]n^2 = K_R(t) P_{O_2}^{-1/2} \tag{10}$$

$$[V_M'']p^2 = K_{ox}(T)P_{O_2}^{1/2} \tag{11}$$

$$[O_i'']p^2 = K_{ok}(T)P_{O_2}^{1/2} \tag{12}$$

$$\frac{[N_m']^2 [V_o^{\cdot\cdot}]}{a_{N_2O_3}} = K_N \tag{13}$$

$$\frac{[2\cdot_M]^2\,[V''_M]}{a_{N_2O_3}} = K'_N \qquad (14)$$

Intrinsic electronic disorder, equation (7), corresponds to excitation of electrons across the band gap between the initially filled valence and empty conduction band. The resultant creation of an electron-hole pair is schematically illustrated in Figure 2. While metals begin with a partially filled band at 0°K and thus a correspondingly high carrier density, semiconductors and insulators achieve their carriers by thermal activation. The consequence is the ten of orders of magnitude difference in intrinsic carrier concentrations at, e.g., 500°K indicated in Figure 2.

Following a quantum mechanical approach one obtains the following expression for the equilibrium constant of reaction (7) [Kittel, 1976]:

$$K_E\,(T) = N_c N_v \exp\,(-Eg/kT) \qquad (15)$$

in which Eg is the band gap energy and N_c and N_v the effective density of states in the conduction and valence bands, respectively. These are given by

$$N_{c,v} = 2\left[\frac{2\pi m^*_{e,h} kT}{h^2}\right]^{3/2} \qquad (16)$$

in which m^*_e and m^*_h are the electron and hole effective masses respectively. It follows directly from equations (7) and (15) that the intrinsic electronic carrier density is given by

$$N = p = n_i = (N_c N_v)^{1/2} \exp-\,(Eg/2kT) \qquad (17)$$

This relation is used later in the derivation of the thermal band gap of a large band gap oxide.

Two types of intrinsic ionic disorder are represented in equations (8) and (9) with the first corresponding to a pair of cations and anions leaving the interior of the crystal to form a cation-anion vacancy pair--Schottky disorder--and the second to a cation leaving a normal site to form a cation-vacancy interstitial pair--Frenkel disorder. Generally only one of these mechanisms need be considered at any one time for a given solid, the choice being determined by the crystal structure, i.e., open structures encourage Frenkel disorder while close-packed structures encourage Schottky disorder. Both mechanisms leave the stoichiometric balance intact.

The equilibrium constants for these reactions, derived from statistical thermodynamics [e.g., Kröger, 1974], are given by

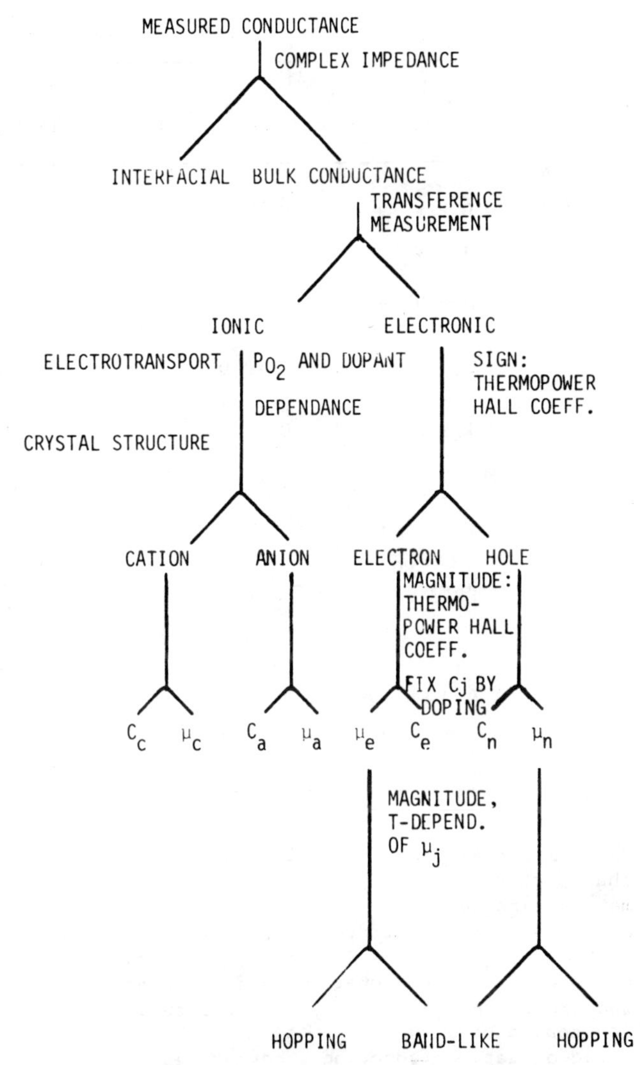

Fig. 1. A flow chart indicating the major defect and transport components contributing to the measured conductivity and methods for deconvoluting them.

$$K_S(T) = N^2 \exp\,(-E_S/kT) \qquad (18)$$

$$K_f(T) = NN^* \exp\,(-E_f/kT) \qquad (19)$$

in which N and N* are the number of normal and interstitial lattice sites and E_s and E_f are the Schottky and Frenkel formation energies, respectively.

Schottky and Frenkel defects are thermally induced as are electron-hole pairs in semiconductors. It is only over the last several decades that it has become more commonly known that non-thermally induced intrinsic ionic disorder exists in some crystals. The classic example is α-AgI in which only a fraction of the silver sites are occupied at any one time.

TABLE 1. Representative Defect Reaction

Defect Reactions	Mass Action Relations	Equation Number
$0 \rightarrow e' + h^{\bullet}$	$n \cdot p = K_E(T)$	(7)
$0 \rightarrow V_M'' + V_o^{\cdot\cdot}$	$[V_M''] [V_o] = K_s(T)$	(8)
$0 \rightarrow V_o^{\cdot\cdot} + O_i''$	$[V_o^{\cdot\cdot}] [O_i''] = K_f(T)$	(9)
$O_o \rightarrow V_o^{\cdot\cdot} + 2e' + 1/2\ O_2$	$[V_o^{\cdot\cdot}]n^2 = K_R(t)\ P_{O_2}^{-1/2}$	(10)
$1/2\ O_2 \rightarrow V_M'' + O_o + 2h^{\bullet}$	$[V_M'']P^2 = K_{ox}(T)P_{O_2}^{1/2}$	(11)
$1/2\ O_2 \rightarrow O_i'' + 2h^{\bullet}$	$[O_i'']P^2 = K_{ok}(T)P_{O_2}^{1/2}$	(12)
$N_2O_3 \xrightarrow{MO_2} 2N'_M + 3O_o + V_o^{\cdot\cdot}$	$\frac{[N'_m]^2[V_o^{\cdot\cdot}]}{a_{N_2O_3}} = K_N$	(13)
$N_2O_3 \xrightarrow{MO} 2N^{\bullet}_M + 3O_o + V_M''$	$\frac{[2^{\bullet}_M]^2[V''_M]}{a_{N_2O_3}} = K'_N$	(14)

These materials thus resemble ionic "metals" in that partial occupancy of sites (energy bands in metals) are intrinsic to the structure. This comparison is illustrated in Figure 2. Because of their high carrier densities, these materials often exhibit anomalously high ionic conduction and are therefore commonly referred to as "fast ion conductors."

The oxidation-reduction behavior, as represented by equations (10)-(12), results in an imbalance in the ideal cation to anion ratio and thus leads to "nonstoichiometry." In these cases, equilibration with the gas phase by exchange of oxygen between the crystal lattice and the gas is observed to result in the generation of electronic carriers:

Consider the two-dimensional representation of an oxide with cubic rocksalt structure illustrated in Figure 3a. On the left, oxygen enters the lattice interstitially, becomes negatively charged creating two holes to maintain charge neutrality. On the right oxygen leaves the lattice, thereby freeing two electrons while simultaneously creating a vacant site, $V_o^{\cdot\cdot}$. The oxidation and reduction reactions represented above, corresponding to equations (12) and (10), respectively, are written under the assumption that the defect states are fully ionized. For the doubly charged species considered here, two energy levels per defect are formed in the forbidden gap between conduction and valence bands. These states are illustrated in Figure 3b.

The oxidation reaction (11) can likewise be visualized by imagining that oxygen from the gas phase creates new lattice sites at the surface

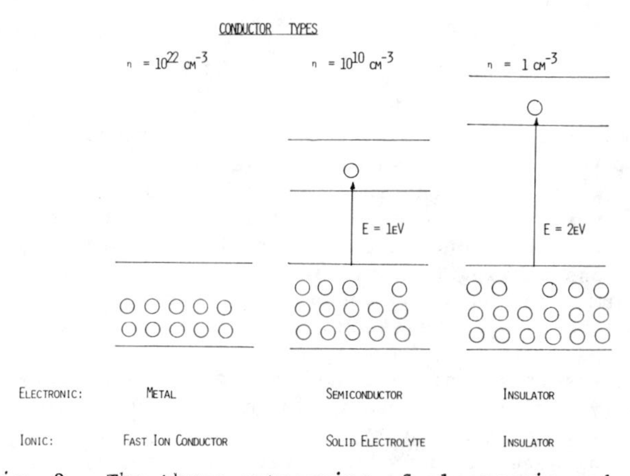

Fig. 2. The three categories of electronic and ionic conductors are distinguished by the density of free carriers available at temperature, T. For electronic carriers, this represents the relative occupancy of energy bands while for ionic conductors this represents instead relative occupancy of ionic sublattices [from Tuller, 1984].

A

M O M O M O M O
O M O M O M O M
M O M O M O M O → 1/2 O2
1/2 O2 → O M O M O M O M
M O M O M O M O

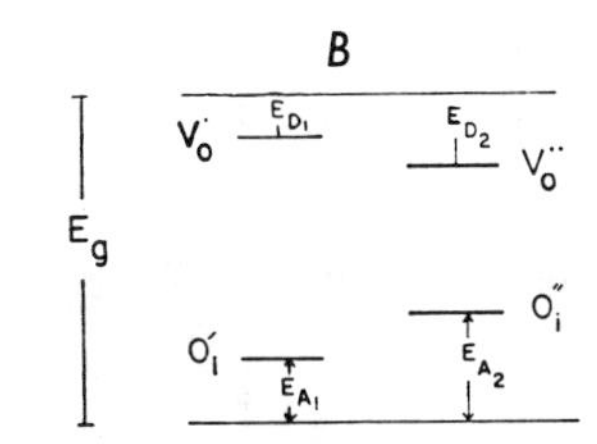

Fig. 3. (a) Schematic of two dimensional square lattice. Entry of excess oxygen into lattice at left results in formation of oxygen interstitials and holes while exit of oxygen into the gas phase on the right results in formation of oxygen vacancies and electrons. (b) Representation of the gap region between valence and conduction bands. The upper discrete levels correspond to oxygen vacancy donor states while the lower levels correspond to the oxygen interstitial acceptor states [from Tuller, 1984].

thereby resulting in the formation of acceptor-like cation vacancies to maintain site balance.

Some of the special features associated with such reactions are (1) the nature of the semiconductor can be changed from p-type to n-type under isothermal conditions by control of the oxygen partial pressure, and (2) the atomic defects which make up the acceptor and donor states can be quite mobile. This latter feature has special relevance to the kinetics of redox reactions in these systems.

Because many of the compounds of interest to the ceramist and geophysicist have relatively large band gaps and correspondingly large Schottky and Frenkel energies, control of charge or mass transport by intrinsic disorder over any extensive range of conditions is unexpected. In general the transport behavior depends on defects formed in response to impurities and deviations from stoichiometry with the former predominating at or near stoichiometry.

To characterize the electrical response of a metal oxide, for example, to temperature and atmosphere excursions, a series of simultaneous reactions of the form presented by equations (7)-(14) must be considered together with an appropriate electroneutrality equation. In the following I will treat two representative defect structures: one in which Frenkel disorder prevails on the oxygen sublattice together with an excess of acceptor impurities, and the second in which Schottky disorder prevails together with an excess of donor impurities.

3.2. Frenkel Disorder

Assuming that significant disorder occurs only on the oxygen sublattice allows us to disregard equations (8) and (11) which include metal vacancy formation. Furthermore, since the remaining equations are not mutually independent, i.e., $K_R K_{OX} = K_F K_E^2$, one may remove one of the equations, e.g., equation (12), in solving for the unknown defect densities. Only the electroneutrality equation remains to be considered, i.e.,

$$2[O_i''] + [N_m'] + n = 2[V_{\ddot{O}}] + p \qquad (20)$$

in which N_m' is, for example, a tri-valent substitutional impurity on a quadravalent M site. We further assume in the following that that value of the N_2O_3 (see equation (13)) remains within such limits that the element N remains fully soluble in the matrix over the conditions considered.

A piecewise solution to such problems is commonly attempted by sequentially choosing conditions for which only one term on each side of equation (20) need be considered. For example, under heavily reducing conditions, equation (20) simplifies to

$$n = 2\,(V_{\ddot{O}}) \qquad (21)$$

Combining with Equation [10], one obtains:

$$N = 2[V_{\ddot{O}}] = [K_R(T)/2]^{1/3}\, Po_2^{-16} \qquad (22)$$

while from equations (12) and (9):

$$p = K_E(T)[K_R(T)/2]^{-1/3}\, Po_2^{1/6} \qquad (23)$$

$$[O_i''] = K_F(T)[K_R(T)/2]^{-1/3} Po_2^{1/6} \qquad (24)$$

The three other defect regimes most likely to occur with successively increasing Po_2 include $2[V_{\ddot{O}}] = [N_m']$, $p = [N_m']$, and $p = 2[O_i'']$. A listing of solutions in these regimes may be found in the reference by Tuller [1981].

A diagram depicting the atmosphere dependencies of the defects over the four defect regimes is presented in Figure 4. Note that in drawing this diagram the transition regions existing between defect regimes have been ignored thereby resulting in straight line segments only. When necessary, more precise solutions may be obtained by solving the appropriate cubic equations for the transition regions [Kröger, 1974; Tuller and Nowick, 1979].

In examining Figure 4 we make the following observations:

1. Where they are Po_2-dependent, electrons and holes always exhibit opposite Po_2

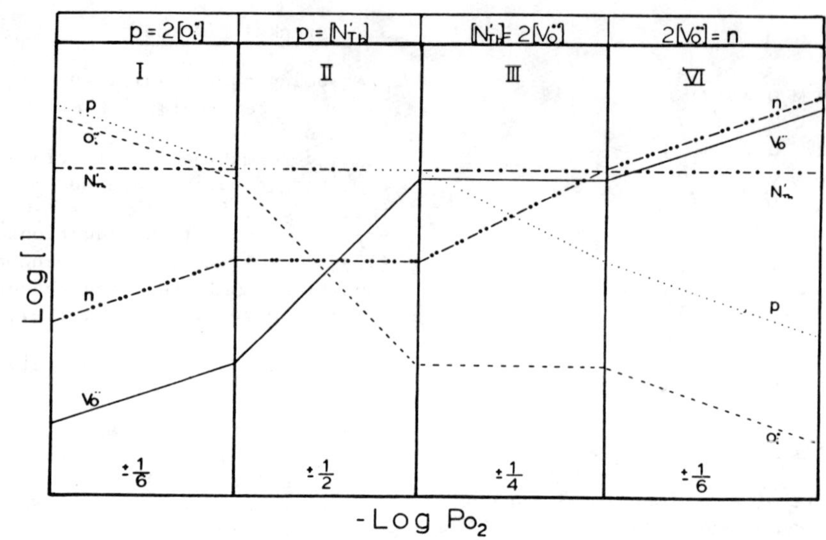

Fig. 4. Defect diagram of acceptor doped material with intrinsic frenkel equilibria [from Fujimoto, 1982].

dependencies as might be expected from equation (7). Electrons follow a $Po_2^{-1/r}$ while holes a $Po_2^{+1/r}$ dependence. Similar considerations hold for the ionic species, i.e., $[V_{\ddot{o}}] \alpha Po_2^{-1/r}$ and $[O_i''] \alpha Po_2^{+1/r}$.

2. In each defect regime, the majority carrier exhibits a unique Po_2 dependence.

3. In both extrinsic regimes the majority carrier, i.e., holes in region II and oxygen vacancies in region III, remains independent of Po_2. This is based on the assumption that the impurity N is single valent and fully ionized.

The first observation enables one to identify the type of electronic carrier (n or p) while the second and third serve to identify the defect regime. Once the defect regime is identified one may, with the assistance of the defect relations derived above, determine the controlling equilibrium constants, K(T). In a subsequent section we demonstrate how one may utilize electrical requirements to establish the nature of the predominant defects.

3.3. Schottky Disorder

The approach here is identical to the above. We again discard the defect reactions which involve defects of little consequence. In this case of Schottky disorder, equations (9) and (12), which deal with interstitial defects, may be dropped. Furthermore, since $K_R K_{OX} = K_F K_E^2$, we may likewise drop equation (10) and thereby eliminate one redundant reaction. Lastly, we choose the electroneutrality relation which includes a donor impurity as assumed above. This is given by

$$P + 2 [V_{\ddot{o}}] + [N_M^{\circ}] = n + 2[V_M''] \qquad (25)$$

Because the predictions based on this model will be compared against experimental results obtained for MgO, we assume the host cation is divalent, i.e., MO and the dopant cation trivalent.

The simplified neutrality relations will, beginning with low Po_2's, include (1) $2[V_{\ddot{o}}] = n$, (2) $[N_M^{\circ}] = n$, (3) $[N_M^{\circ}] = 2[V_M'']$, (4) $p = 2[V_M'']$. Solutions for the defects in the regime characterized by neutrality condition (3) will for example be

$$2 [V_M''] = [N_M^{\circ}] \qquad (26)$$

$$[V_o] = 2 K_S [N_M^{\circ}]^{-1} \qquad (27)$$

$$p = \{2 K_{ox} [N_M^{\circ}]^{-1}\}^{1/2} P_{o_2}^{1/4} \qquad (28)$$

$$n = K_E (2 K_{ox})^{-1/2} [N_M^{\circ}]^{1/2} P_{o_2}^{-1/4} \qquad (29)$$

A log [ ] versus log P_{O_2} dependence similar in appearance to Figure 4 is also obtained for this example.

4. Carrier Mobilities

In the preceding sections we analyzed the temperature, atmosphere, and impurity dependence of defects. In order to be in position to predict or analyze the electrical properties of ceramics, I outline the main features characteristic of ionic and electronic carrier transport as well.

TABLE 2. Selected Ionic and Electronic Carrier Probabilities [from Tuller, 1981].

Material	Ionic Defect Mobility	Electron Mobility	Hole Mobility
MgO	$\mu(V_{Mg}'') = 2\ \exp\ (-2.13\ eV/kT)$ $= 7.8 \times 10^{-7}$ (1673 K)	13 (1673 K)	4 (1673 K)
Al_2O_3	$\mu(Al_1^{3\bullet}) = 7.13 \times 10^7\ \exp\ (-4.6\ eV/kT)$ $= 1 \times 10^{-6}$ (1673 K)	3 ± 1 (100-350 K)	
CeO_2	$\mu(V_o^{\cdot\cdot}) = 1.2\ \exp\ (-0.87\ eV/kT)$ $= 4.3 \times 10^{-4}$ (1273 K) $= 2.1 \times 10^{-9}$ (500 K)	$3.9 \times 10^2/T\ \exp\ (-0.4\ eV/kT)$ 8.1×10^{-3} (1273 K) 7.3×10^{-5} (500 K)	

4.1. Ionic Mobility

A variety of ionic diffusion mechanisms tied to the nature of the mobile defect exist. Some of the common ones include (1) vacancy, (2) interstitial, and (3) interstitialcy mechanisms. The first obviously involves the jump of an ion to an adjacent and equivalent empty site; the second, the jump of an interstitial ion to an adjacent and empty interstitial site; and the last, a similar end result to the second but accomplished by an interstitial ion moving onto a normal site and thereby pushing the ion originally there to an adjacent interstitial site. All of the above cases may be described by an activated jump process for which the diffusion coefficient is given by (Goodenough, 1978]:

$$D = \gamma\ (1 - c)Za^2\ \nu_o\ \exp\ (\Delta S/k)\ \exp\ (-E_m/kT)$$

$$= D_o\ \exp\ (-\Delta G/kT) \qquad (30)$$

where a is the jump distance, ν_o the attempt frequency, $(1 - c)Z$ the number of unoccupied neighboring sites, E_m the migration energy, and γ a geometric factor.

From the preceding we find the diffusivity or, equivalently, the mobility to be a thermally activated process with activation energies typically in the electron volt range. As a consequence, ionic mobilities remain relatively small even at elevated temperatures. Some representative values for ion mobilities in oxides are shown listed in Table 2.

Exceptions do exist among the so-called fast ion conductors in which barriers to motion can be quite small, e.g. $\sim$0.15 eV for Na motion in beta alumina. Even here, ion mobilities rarely surpass 10^{-3} cm^2/V s.

To summarize, ionic mobilities tend to be strongly thermally activated with magnitudes well below 10^{-3} and more typically 10^{-7}-10^{-10} cm^2/V s even for temperatures above 1000°C.

4.2. Electronic Mobilities

In metals and conventional nonpoplar broad-band semiconductors (e.g., GaAs), electronic motion is primarily perturbed at elevated temperatures by acoustic phonon scattering. The electronic drift mobility takes the general form [Kittel, 1976]

$$\mu = \mu_o\ T^{-3/2} \qquad (31)$$

while magnitudes typically range from 500 cm^2/V s (ZnTe) to 65,000 cm^2/V s (InSb) at room temperature.

In ionic compounds, the free electronic carriers induce deformations in their surroundings, which tend to self trap the carriers in potential wells of radius r_p. The electron and the accompanying polarization cloud are generally referred to as polarons [Adler, 1968].

Where the electron-lattice interaction is weak, the well radius is greater than the interatomic distance and so the carrier remains free to move in the band but with an enhanced effective mass. The mobility of the so-called "large polaron," at temperatures above the Debye temperature, is given by [Appel, 1968]

$$\mu = \mu_o'\ T^{-1/2} \qquad (32)$$

and is expected to be $\sim$1-100 cm^2/V s at elevated temperatures. Typical results are large polaron motion in MgO and Al_2O_3 are included in Table 2.

In narrow band oxides, e.g., third transition

TABLE 3. Conduction Parameters for Electrons, Holes, and Oxygen Vacancies Corresponding to Defect Regime III in Figure 4

	$\sigma°$	E
$V_o^{\cdot\cdot}$	$[N'_M]q\mu_v^o$	E_M(=vacancy migration energy)
e	$\left(\frac{2K°_R}{[N'_M]}\right)^{1/2} q\mu_e^o$	$E_R/2 + E_H^e$
h	$\frac{K_E^o[N'_M]^{1/2}}{(2K°_R)^{1/2}} q\mu_h^o$	$E_g - E_R/2 + E_H^h$

In the above expressions, partial mobilities μ_j and equilibrium constants K_j are assumed to be of the form $\mu_j = \mu_j^o \exp - E_\mu/kT$ and $K_j(T) = K_j^o \exp - E_j/kT$. Note: if the electrons or holes are not small polarons, the hopping energies E_H^e and E_H^h may be set to zero.

metal and 4f rare earth oxides, electron-lattice interactions often become sufficiently strong to catch a carrier on a specific ion site. Under these circumstances motion to an adjacent site can only proceed via an activated hopping process similar to that discussed above for ionic diffusion. The mobility of the so-called "small polaron" is given by [Tuller and Norwick, 1977]:

$$\mu = [(1 - c)ea^2\nu_o/kT] \exp -E_H/kT \qquad (33)$$

in which E_H is the hopping energy while the other terms have the same meanings as in equation (30). Given its activated nature, the magnitude of the small polaron mobility is correspondingly reduced and generally falls within the range of 10^{-4} to 10^{-1} cm^2/V s. Values for electron small polaron mobilities in CeO_2 are listed in Table 2.

It is interesting to note that only in a system such as CeO_2, which simultaneously exhibits enhanced ionic diffusion and small polaron motion, do the ionic and electronic mobilities begin to approach each other in magnitude and only then at elevated temperatures. Normally electronic mobilities are typically > 10^6 times greater than ionic mobilities as illustrated in Table 2 for MgO and Al_2O_3.

5. Electrical Conductivity Diagrams

Carrier mobilities μ_j are commonly observed to be independent of carrier density, c_j over extensive experimental conditions. Accordingly (see equation (2)), the partial conductivities are expected to follow the same power law dependencies predicted for the defects themselves, e.g., as in Figure 4. Although the measured conductivity σ reflects the sum of the individual σ_j's, a number of features conspire to simplify the analysis. These include the following: (1) only a pair of defects dominate in any given defect regime; often, only one is mobile; (2) the partial mobilities μ_j often differ by orders of magnitude.

One may illustrate these features by reference to Figure 4. Consider first regions I and IV, in which a pair of mobile defects dominate--one electronic and one ionic. Since $\mu_{electronic} >> \mu_{ionic}$ (see preceding section), the measured conductivity is essentially equal to the hole conductivity $\sigma_h = \sigma_h^o P_{O_2}^{+1/6}$ in region I and the electron conductivity $\sigma_e = \sigma_h^o P_{O_2}^{-1/6}$ in region IV. No confusion can exist in region II, since the only mobile carriers in the neutrality relation are holes and thus $\sigma = \sigma_h$ which in this region is P_{O_2} - independent.

Region III corresponds to the only situation in which the partial ionic and electronic conductivities may be of the same order of magnitude. This follows from the fact that although $V_O^{\cdot\cdot}$ is the mobile carrier with the highest density, the minority electronic carriers typically have markedly higher mobilities. Therefore, only if the inequality

$$C_{ion}\mu_{ion} > C_{electron}\mu_{electron} \qquad (34)$$

is satisfied will ionic conductivity contribute, in a major way, to the total conductivity under any circumstances.

Even if σ_{V_O}, σ_h and σ_e all contribute to σ in a major way in region III, one may readily deconvolute their individual contributions by making use of their predicted P_{O_2} dependencies. For example, the total conductivity is given by

$$\sigma = \sigma_h + \sigma_{V_O} + \sigma_e \qquad (35)$$

in which

$$\sigma_v = \sigma_v^o \exp(-E_v/kT) \qquad (36)$$

$$\sigma_h = \sigma_h^o P_{O_2}^{+1/4} \exp(-E_p/kT) \qquad (37)$$

$$\sigma_e = \sigma_e^o P_{O_2}^{-1/4} \exp(-E_n/kT) \qquad (38)$$

Expressions for the coefficients and exponents are given in Table 3.

The partial conductivities, obtained by multiplying the charge carrier densities of

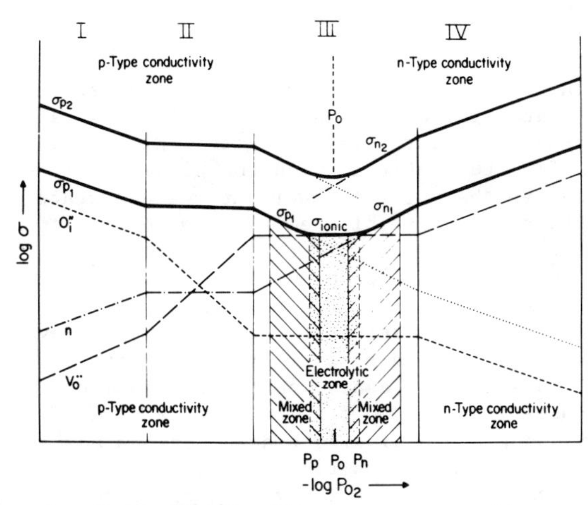

Fig. 5. Conductivity diagram corresponding to defect diagram in Figure 4. Solid curves correspond to the cases (1) μ_n, $\mu_p > \mu_v$ and (2) μ_n, $\mu_p >> \mu_v$. In case 2 only p- and n-type semiconductivity is observed. Refer to text for definition of symbols [from Tuller, 1981].

Figure 4 by their charges and mobilities, are shown plotted in Figure 5. The controlling σ_j's in each defect region are indicated by dark solid curves.

For the upper curve, I have assumed that the inequality given by equation (34) is not satisfied and so the material remains semiconducting under all conditions. The transition from p to n type semiconductivity in region III is readily distinguished by the minimum in conductivity at which σ switches from a $P_{O_2}{}^{+1/4}$ to a $P_{O_2}{}^{-1/4}$ dependence in proceeding to lower P_{O_2}'s. This minimum, designated by P_o on the P_{O_2} scale, corresponds to the intrinsic condition represented by equation (17) assuming $\mu_e = \mu_h$.

The lower curve, on the other hand, corresponds to condition (34) being satisfied. Thus, at least in the in the vicinity of P_o where the electronic conductivity is at a minimum, ionic conduction will prevail with a P_{O_2} independence as indicated in Figure 5. This segment of III is designated as the "electrolytic zone" and is bounded at high and low pressures by P_p and P_n which represent conditions where T_{ion} falls to a value of one half.

The electrolytic zone is of interest for a number of reasons. First, because it represents a situation in which conduction is fully ionic and in which the ionic carrier density is fixed by the dopant, it serves as a convenient source for obtaining the ionic mobility or equivalently the ionic diffusivity from electrical conductivity measurements. Because the conductivity measurement is highly sensitive and, when properly performed, samples the bulk properties, it does not suffer from many of the difficulties associated with tracer or exchange measurements which are often limited in their sensitivity and/or markedly influenced by damage in the near surface regions in which the tracer profiles are examined [Steele and Kilner, 1982].

Second, solids in the electrolytic zone may be used as solid electrolytes in electrochemical devices for measuring thermodynamic activities of species which are mobile in the lattice. Thus oxides such as stabilized zirconia have been utilized in concentration cells to monitor oxygen activities in equilibrium with M/MO mixtures, dissolved in liquid metals, and in the gas phase and thereby to determine standard free energies of formation of oxides, oxygen diffusion in melts, etc. [Goto and Pluschkell, 1972].

The curves drawn in Figure 5 represent an isothermal condition. As T is varied the size and position in P_{O_2} of the defect regimes shift due to the different activation energies characteristic of the defects and their mobilities. As an example, we obtain the temperature and dopant dependence of P_p and P_n, the electrolytic domain boundaries, by setting $\sigma_h = \sigma_v$ and $\sigma_e = \sigma_v$, respectively, and solving for P_{O_2} (see equations (36)-(38)). Doing so one obtains the following expressions for P_p and P_n:

$$\ln P_p = \frac{-4(E_v - E_p)}{k}\frac{1}{T} + 4\ln\left(\frac{\sigma_v^\circ}{\sigma_h^\circ}\right) \qquad (39)$$

$$\ln P_n = \frac{-4(E_n - E_v)}{k}\frac{1}{T} + 4\ln\left(\frac{\sigma_e^\circ}{\sigma_v^\circ}\right) \qquad (40)$$

When these relations are plotted as log P vs 1/T they generally give two straight line curves with opposite slopes which intersect at high T as is schematically illustrated in Figure 6 for stabilized zirconia. Thus the electrolytic domain is observed to shrink with increasing T and thereby shift from a log σ versus log P_{O_2} characteristic of the lower curve in Figure 5 at low T to one similar to the upper curve at elevated T. The above demonstrates that transitions in defect structure may and do occur as the temperature is varied even while holding the dopant concentration fixed.

Before moving on to the next section it is worth noting some of the simplifying assumptions made in the above analysis. First of all, defect-defect interactions were ignored, an assumption strictly true only for dilute concentrations. When interactions become significant, (1) activities should replace concentrations in the mass action relations and (2) defect association-dissociation reactions need be considered. At very high levels of doping one may also no longer assume composition-independent equilibrium constants. More detailed discussions of these complicating factors may be found in articles by Heyne

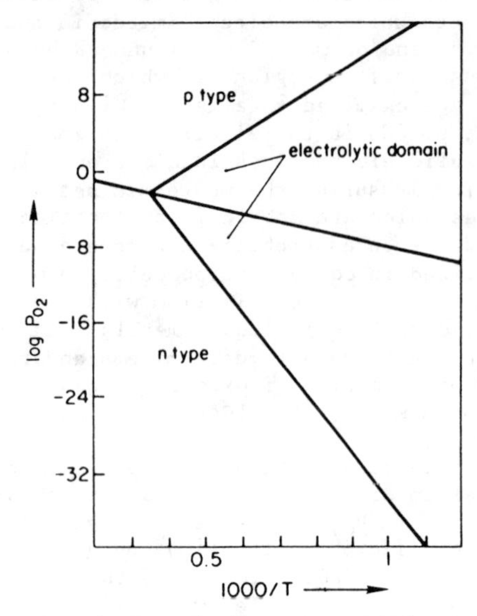

Fig. 6. Domain boundaries of stabilized zirconia as projected on the log P_{O_2}-1000/T plane [from Heyne, 1977].

[1977], Kilner and Steele [1981], and Tuller [1981].

Lastly, it is highly unlikely to observe all the defect regimes predicted in diagrams such as Figures 4 or 5 experimentally. The accessible range of P_{O_2}'s required would be enormous. Figure 7, for example, represents the predicted P_{O_2} - range for several defect regimes in CeO_2 calculated for the temperature range of ~600-2000°C. Even with 200 orders of magnitude only one full regime and parts of two others are included! Furthermore, even if one could attain such experimental conditions, many materials would decompose or change phase prior to achieving some of the predicted defect regimes. As we shall soon discover, however, each of these predicted defect regimes do occur in one material or another and so may be used in each of these cases to derive crucial thermodynamic and kinetic data.

6. Experimental Approaches

In Figure 1, I listed the more common techniques used to deconvolute (1) bulk and interfacial, (2) ionic and electronic, and (3) carrier density and mobility contributions to the measured electrical conductivity. In the following I briefly review some of these techniques and indicate when they are most likely to be used.

6.1. Impedance Measurements

Barriers to charge transport at interfaces is quite common. The most relevant interfaces for our discussion are (1) electrode-specimen and (2) grain boundary interfaces. I first briefly describe some of the blocking mechanisms and then discuss techniques to isolate them.

Consider the sequence of events which follow the contacting of two materials which initially possess different Fermi energies. Electrons flow from the material with the higher Fermi energy to that of the lower until a new state of equilibrium is established. In so doing a space charge at the interface is established which blocks further motion of majority carriers across the barrier. To generalize, any compositional or structural inhomogeneity, even within the same material, may lead to the creation of space charge barriers. Common examples include Schottky barriers at semiconductor-metal electrode junctions and semiconductor p-n junctions [Sze, 1969].

The same principles apply to ionic conductors in which one considers instead the chemical potentials of the ionic species. Additional complicating factors, however, generally arise at solid electrolyte-metal electrode interfaces. In order for a continuous current to flow through the circuit, chemical reactions must occur at the electrode-electrolyte interface. For example the following overall reaction must occur at the cathode of an oxygen ion conductor

$$1/2\ O_2(\text{gas phase}) + 2\ e'(\text{metal}) \rightarrow O^{=}(\text{electrolyte}) \qquad (41)$$

Various rate determining steps may limite the intended current flow. Some of these include [Pizzini, 1973] (1) oxygen diffusion within the pores of the electrode, (2) oxygen chemisorption and dissociation at the metal surface, and (3) transfer of electrons across the electrode/electrolyte phase boundary as well as other possibilities. Each of these processes results

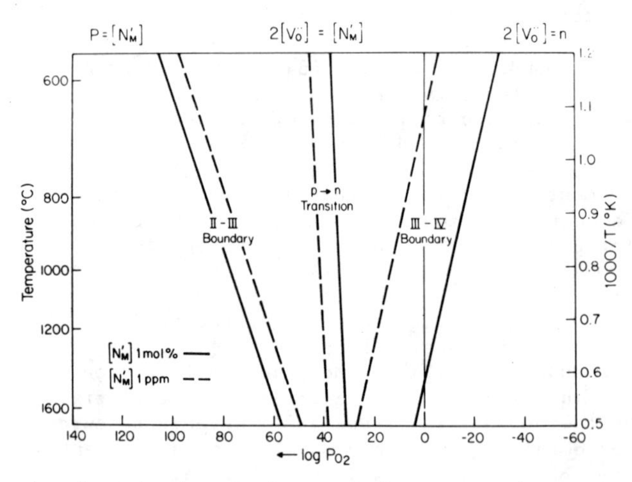

Fig. 7. The boundaries defining the region characterized by predominant ionic disorder (region III) in CeO_2 are plotted as 1000/T versus log P_{O_2} for two levels of $[N'_M]$.

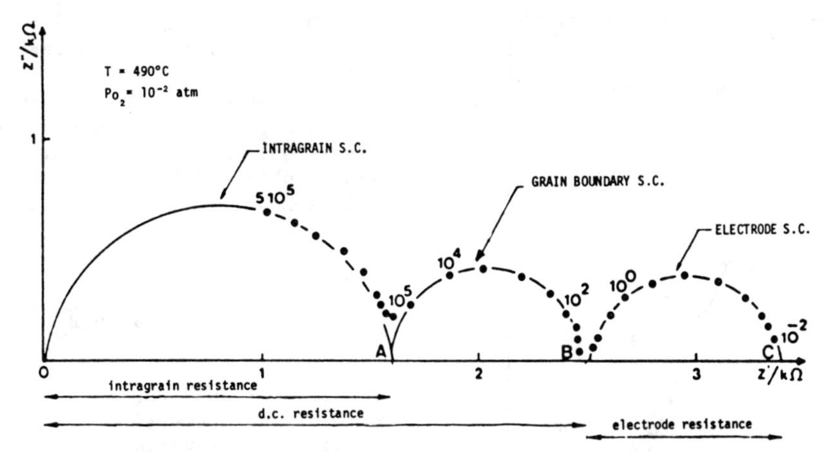

Fig. 8. Complex impedance diagram obtained on polycrystalline cubic stabilized zirconia. Semicircles corresponding to the intragrain (bulk), grain boundary and electrode contributions to the overall impedance are obtained enabling a deconvolution of all three effects [from Kleitz et al., 1981].

in electrode polarization and thus an increase in the effective impedance of the specimen.

A conventional 2 probe DC current-voltage measurement samples all of these processes and often results in conductivity values more characteristic of interfacial rather than bulk effects. A 4 probe DC measurement, in which current is injected through one set of electrodes (generally the outer ones) and the induced potential difference, measured by a high input impedance voltmeter, between the other electrodes, provides improved results since the "electrode impedance" is bypassed. This is true only if no current is drawn through the voltage electrodes. All DC measurements, however, suffer from the limitation that they cannot isolate the grain boundary contribution to the impedance.

AC measurements, performed over an extended frequency range, can however perform this deconvolution. This is based on the fact that no materials are purely resistive. They also include a reactive component which for our purposes is nearly always capacitive. The total impedance is made up of a real resistive and an imaginary capacitive term and is given by

$$Z = R + j\,(\omega C)^{-1} \qquad (42)$$

in which ω is the frequency in radians/second.

Each of the transport steps, i.e., at the electrodes, grain boundaries and within the grains involves a charge transport (resistive) and charge storage (capacitive) component. In the simplest case, these can each be modeled by a resistor R and capacitor C in parallel. In a complex impedance plot, i.e., $Z''(\omega)$ versus $Z'(\omega)$, a semicircle with one intercept at the origin (high frequency) and the other on the $Z'(w)$ axis is obtained for which (1) the x-axis intercept is the resistance R while (2) the frequency at the peak of the semicircle $\omega_p = 1/RC$.

To model the overall response of the material one assigns a parallel R-C circuit to each of the processes, i.e., grain, grain boundary, and electrode, and connects them in series. If the time constants $\tau_j = R_jC_j$ of these networks are sufficiently different, as they often are, then a series of contacting semicircles are obtained in a complex impedance plot thereby enabling individual processes to be distinguished. An actual spectrum taken on a specimen of polycrystalline stabilized zirconia is shown in Figure 8.

A recent example, in which the complex impedance technique served to assist us in establishing whether or not an observed slope change in the DC conductivity of ThO_2 corresponded to a change in bulk transport mechanism is included in Figure 9. Here we have a plot of log σ versus 1/T for ThO_2 doped with 0.1 m/o Pr. The solid curve corresponds to the 4 probe DC results. On the other hand, impedance spectra obtained on the same specimen at temperatures below the knee indicated that two components, represented by two semicircles, were contributing to σ. The conductivities derived from the intercept of the high frequency semicircle (designated as "AC bulk") are observed to fall on the dashed line extrapolated from high temperature demonstrating that the bulk transport mechanism had not in fact changed below the knee in the curve. The sum of the two semicircles, which include both bulk and grain boundary impedances, however resulted in calculated conductivities which agreed well with the DC results. Based on the above examples it becomes clear that it is well worth performing an AC impedance investigation when interfacial effects are suspected to be significant.

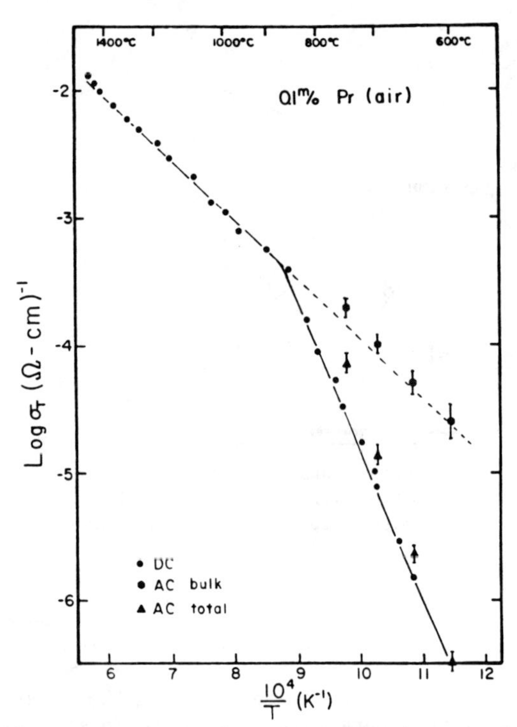

Fig. 9. Temperature dependence of electrical conductivity measured for ThO_2 + 0.1 m/o Pr by DC and AC techniques. Conductivity values derived from the "bulk" complex impedance semicircle demonstrate that kink in DC temperature dependence does not correspond to a change in the bulk transport mechanism but rather is related to grain boundary blocking [from Fujimoto, 1982].

6.2. Transference Measurements

Since mixed ionic and electronic conduction is commonly observed in ceramics, means for separating their contributions to σ are sometimes essential in deriving defect generation and transport parameters. A number of techniques are examined below.

Concentration Cells. When an ionic conductor is placed within a chemical activity gradient, an electric field is generated which for an oxide for example is given by [Tuller, 1981]

$$E = \frac{kT}{4q} \int_{I}^{II} t_{ion} \, d \ln P_{O_2} \tag{43}$$

Often one assumes t_{ion} is nearly independent of P_{O_2} and draws it out of the integral to obtain

$$E = t_{ion} \frac{kT}{4q} \ln (P_{O_2}^{II}/P_{O_2}^{I}) \tag{44}$$

Thus by fixing $P_{O_2}^{I}$, $P_{O_2}^{II}$ and T and measuring the open circuit cell voltage E, one can determine t_{ion}.

The above approximation breaks down, however, as one enters the mixed conduction regime (see Figure 5). A more appropriate expression for t ion is obtained by differentiating equation (43) with respect to the upper limit, while holding $P_{O_2}^{I}$ constant, to obtain

$$t_{ion} (P_{O_2}^{II}) = \frac{4q}{kT} \left(\frac{\partial E}{\partial \ln P_{O_2}}\right)_{P_{O_2} = P_{O_2}^{II}} \tag{45}$$

The ionic transference number may now be obtained as a function of P_{O_2} by evaluating the slope of the E versus ln P_{O_2} curve obtained at each P_{O_2}.

This approach was, for example, recently applied to MgO by Sempolinsky and Kingery [1980]. In Figure 10 log σ, log σ_{ion}, the cell emf, and t_{ion} are plotted as a function of log P_{O_2} for a high purity crystal of MgO for temperatures of 1200°-1500°C. The isotherms for σ exhibit p and n type contributions at high and low P_{O_2}'s, respectively. At intermediate P_{O_2}'s, particularly at the lower temperatures only a weak dependence, indicative of ionic conduction, is observed. This is confirmed by the maximum obtained in the σ_{ion} isotherms. Values for σ_{ion} are readily obtained by taking the production $t_{ion} \times \sigma$ and are indicated by the solid and essentially P_{O_2}-independent curves shown in Figure 10. We later examine the consequences of the ionic conductivity data in greater detail.

Electrotransport. The above technique gives the ionic transference number without specifying the individual cation and anion contributions. A DC electrotransport technique is sometimes used to determine t_{cation}. Without a source and sink for cations, mass transport by cations must result in a displacement of the material in accordance with Faraday's law.

A dilatometric method based on this principle and capable of determining very small cationic transference numbers was demonstrated by Deportes and Gauthier [1971]. In this method a highly sensitive dilatometer monitors the displacement induced by the current flow with the transference number calculated from

$$t_{cation} = \frac{\Delta x s a z F}{M I t} \tag{46}$$

in which Δx is the displacement, s the area of the interface, a the density, z the charge of the carrier, M the molar mass, I the current, and t the duration of the electrolysis. Application of this technique confirmed, for instance, that the dominant ionic carrier in MgO is the magnesium vacancy.

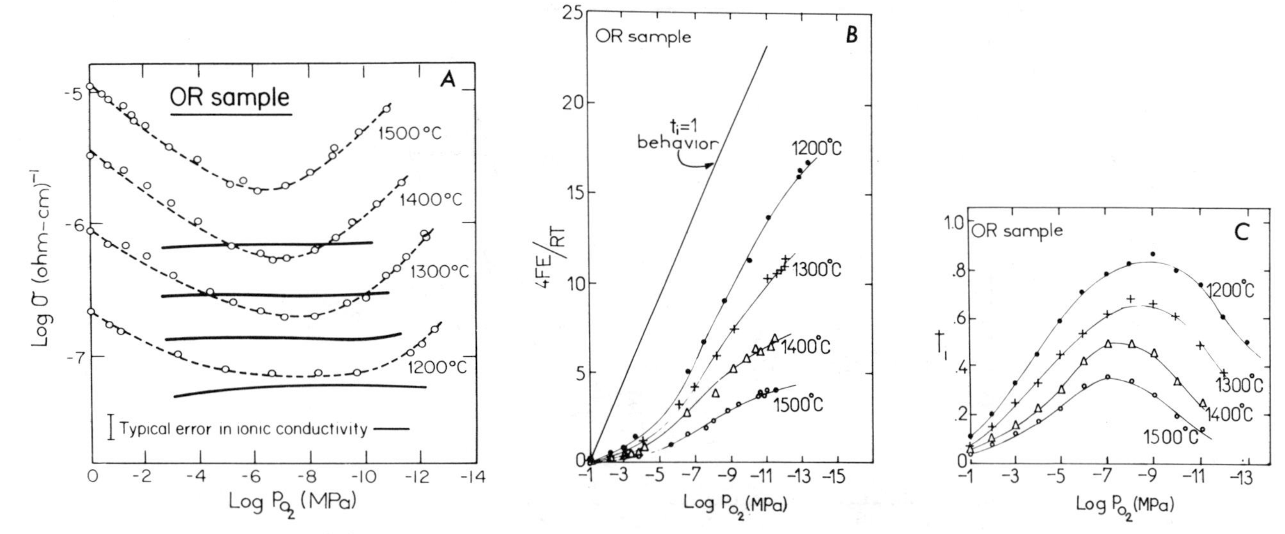

Fig. 10. Conductivity data for high purity MgO sample. Shown as functions of log P_{O_2} are (a) total (data points) and ionic (solid curves) conductivity, (b) concentration cell voltages, and (c) ionic transference number [from Sempolinski and Kingery, 1980].

6.3. Thermopower and Hall Effect

Often n and p are P_{O_2} dependent (e.g., regions I, III, and IV in Figure 4) and so may be discriminated by their opposite P_{O_2} dependencies as mentioned above, i.e., $n \propto P_{O_2}^{-1/r}$, $p \propto P_{O_2}^{+1/r}$. Often, however, a P_{O_2}-independent electronic conductivity is obtained as, e.g., recently reported for Y_2O_3 doped UO_2 [Dudney et al., 1981] and as predicted in region II, Figure 4. Under these conditions several complementary electrical techniques may be used to establish the sign of the carrier.

The thermopower, which is the ratio of the voltage induced across a specimen as a result of an imposed temperature gradient, exhibits a different polarity for electrons and holes. The thermopower Q, e.g., for electrons is given by [Wimmer and Bransky, 1974]

$$Q_e = k/e \ln \frac{N_c}{n} + A_e \tag{47}$$

in which A_e is a scattering factor (≃2-3) with the other terms already defined. The polarity of the voltage at the cold junction corresponds to the sign of the majority carrier, i.e., negative for electrons and positive for holes. Figure 11 shows recent thermopower measurements obtained in our laboratory on perovskite solid solution $Ba_{0.3}Sr_{0.7}TiO_3$ as a function of P_{O_2} at a number of temperatures [Choi, 1983]. An observed sharp transition from p-type at high P_{O_2} to n-type at low P_{O_2} corresponds nearly exactly with a corresponding transition in σ from a $P_{O_2}^{+1/4}$ to a $P_{O_2}^{-1/4}$.

Aside from furnishing the sign of the carrier, Q may also be used to determine the carrier density from relations such as (47). Combining this with σ enables the separation of the C_j and μ_j terms.

Similar information may be obtained from Hall effect measurements. Here an open circuit voltage, induced by the Lorentz force developed by crossed magnetic fields and electric currents, is measured along a direction mutually perpendicular to both current and magnetic field directions. For example, for an n-type material the Hall constant R_H is given by [Kittel, 1976]

$$R_H = \frac{E_H}{jB} = \frac{r}{nq} \tag{48}$$

in which E_H is the Hall field, j the current density, B the magnetic field and r a constant dependent on the scattering factor. Again from the sign and magnitude of R_H one may determine carrier density and type. Applications of this technique to oxide semiconductors are discussed in an article by Gvishi et al. [1968].

6.4. Doping Dependence of Conductivity

When the carrier density is fixed by a dopant impurity, e.g., $p = [N_m']$ or $2[V_O^{\cdot\cdot}] = [N_m']$ in regions II and III, respectively, in Figure 4, then one may determine the mobility of the carrier directly. Thus Dudney et al. [1981] determined $\mu_h = (200/T) \exp(-0.085 \text{ eV}/kt)$ in UO_2 doped with Y_2O_3 while Sempolinsky and Kingery [1980] found the mobility of

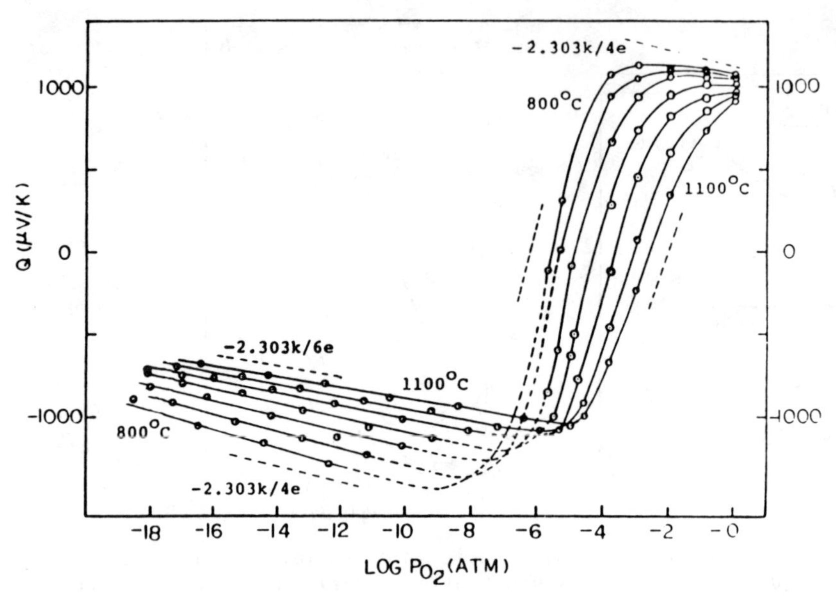

Fig. 11. Thermoelectric power (Q) versus log P_{O_2} for $(Ba_{0.3}Sr_{0.7})TiO_3$ [from Choi, 1983].

magnesium vacancies to be $\mu_{V_{Mg}}$ = (8800/T) exp (-2.28 eV/kT).

Situations arise when several carriers are predicted to exhibit the same P_{O_2} dependence, e.g., n and $[V_o]$ in Ce-doped ThO_2 [Fujimoto and Tuller, 1979]. Increasing the concentration of Ce Acceptors increased the conductivity, demonstrating that ionic conduction carried by oxygen vacancies was dominating σ rather than electrons.

6.5. Other Techniques

Other less commonly used techniques included (1) polarization cells for detecting small te's and (2) thermoelectronic emission for determining the electronic carrier density and others. These are reviewed in articles by Weppner and Huggins [1978] and Tuller [1981].

7. Defects and Transport in Oxides: An Analysis

Many of the concepts presented in preceding sections may at this juncture still remain somewhat abstract for the reader. It is therefore useful to demonstrate how such concepts have been utilized to establish defect and transport parameters in a number of representative systems. In doing so I attempt to follow the flow diagram of Figure 1 as closely as possible.

7.1. Magnesium Dioxide: MgO

Although MgO is a common ceramic with a simple cubic structure, details concerning its high temperature defect structure and carrier mobilities remained unclear until recently. Some features generally agreed upon, however, included the fact the MgO was a mixed ionic-electronic conductor at elevated temperature and that Mg was the mobile ionic species. Although activation energies for Mg diffusion, derived from high temperature tracer diffusion measurements, were in rough agreement, the magnitude of D_{Mg} differed by as much as two orders of magnitude [Harding et al., 1971; Harding and Price, 1972; Lindner and Parfitt, 1957; Wuensch et al., 1973].

The high temperature electrical studies by Sempolinski et al. [1980], data from which have already been presented in Figure 10, illustrate the approach outlined in Figure 1 quite effectively. To begin with, conductivity measurements were performed on single crystals thereby eliminating passible grain boundary contributions to σ. In order to enable concentration cell measurements and to avoid gas phase conduction, a 4 probe measurement was impractical. However, since no polarization effects were observed, the electrode impedance could be assumed to be minimal at the elevated temperature (>1200°C) of this study.

Earlier the data contained in Figure 10 was presented to illustrate how the total conductivity was deconvoluted into an ionic component (see Figure 10) and an electronic component for a crystal of high purity MgO. Similar measurements were performed on donor doped (Al, Sc, Fe) MgO in which donor levels varied from 65 to 1500 ppm. Given that σ_{ion} was found to be high and P_{O_2} independent (a small P_{O_2}-dependence in Fe-doped MgO will be discussed later), this allows for the defect reactions (26)-(29) to be applied in the interpretation.

From equation (26) we predict that (1)

$2[V_M''] = [N_M']$, i.e., carrier density is linear with and fixed by the donor density, and (2) $\mu_M = \sigma_{ion}/2[V_M'']q = \sigma_{ion}/[N_M']q$. Figure 12 in fact shows σ_{ion} is linear with dopant density over nearly two orders of magnitude at 1400°C. This, combined with the temperature dependence of σ_{ion}, enables the derivation of the mobility and diffusivity of V_{Mg}'' as illustrated in Figure 13 and already quoted above.

Donor doping requires that the ionic carrier be either V_{Mg} or O_i with the latter highly unlikely due to close packing of the lattice. This conclusion is directly supported by the electrotransport work of Deportes and Gauthier [1971] quoted above and the reasonably good agreement between magnesium tracer diffusion-derived migration energies and those obtained from ionic conduction. The spread in the magnitude of tracer diffusivities can now be accounted for on the basis of different levels of donors unintentionally added to the crystals during growth.

The electronic component of the electrical conductivity is next examined. The electronic component of σ derived from the data in Figure 10 is now presented in Figure 14 as a function of P_{O_2} for a series of isotherms. A minimum in σ is obtained at intermediate P_{O_2}'s (see Figure 15) with a p-type $P_{O}^{-1/4}$ component at high P_{O_2} and on n-type $P_{O_2}^{-1/4}$ component at low P_{O_2} as predicted in equations (28) and (29). Since N_M° is known (even in the nominally undoped crystal from the magnitude of ionic conductivity), one is now able, with the aid of (28) and (29) and the knowledge of μ_e and μ_n, to determine the equilibrium constants K_{Ox} and K_E from the electronic conductivity data. In doing so, it is often convenient to examine the temperature dependence of the electronic conductivity minimum, σ_{min}.

At the minimum, σ_n and σ_p are equal and thus [Sempolinski et al., 1980]

$$n = p(\mu_p/\mu_n) \tag{49}$$

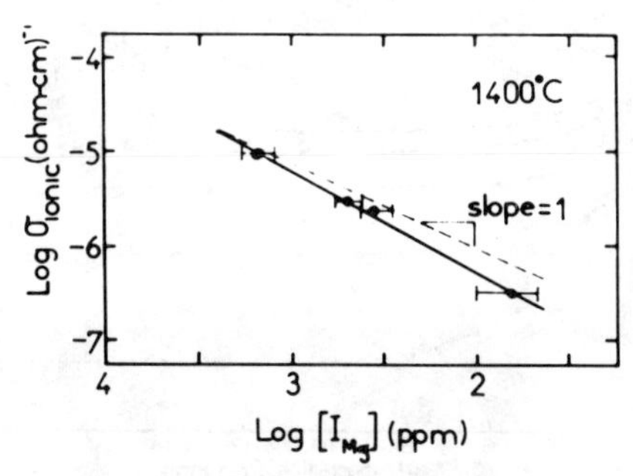

Fig. 12. Plot showing the nearly unity dependence of log σ_{ion} on log donor dopant density in MgO over nearly two orders of magnitude [from Sempolinski and Kingery, 1980].

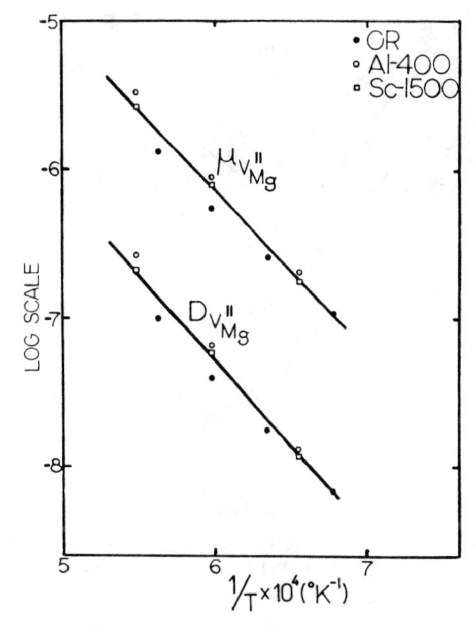

Fig. 13. Mobility (cm^2/V s) and diffusion coefficient (cm^2/s) of V_{mg}'' as a function of temperature. Different symbols, i.e., OR, AL, Sc, correspond to nominally undoped, Al-doped, and Sc-doped MgO specimens, respectively [from Sempolinski and Kingery, 1980].

and

$$\sigma_{min} = 2e\mu_p p = 2e\mu_n n \tag{50}$$

Combining these relations with equations (7) and (15) one obtains

$$p^2(\mu_p/\mu_n) = N_c N_v \exp(-E_g/kT) \tag{51}$$

Substituting the solution for p into equation (50) gives

$$\ln(\sigma_{min}) = \ln[2q(\mu_p\mu_n N_c N_v)^{1/2}] - Eg/2k(\frac{1}{T}) \tag{52}$$

Thus from a plot of ln (σ_{min}) versus 1/T one can obtain the electron-hole mobility product ($\mu_p\mu_n$) and the band gap (E_g--actually E_g^0 K is obtained) assuming only that the effective density of states terms are known. Analysis of a plot of this type for undoped MgO gives a value for $E_g = 7.0 \pm 0.5$ eV and a mobility product $\mu_p\mu_n \simeq 150$ at 1673 K. Using calculated electron and hole effective masses the individual mobilities are calculated to be $\mu_n = 21\ cm^2/V$ s and $\mu_p = 7\ cm^2/V$ s.

Both electron and hole carrier mobilities are of sufficiently high magnitude to fall within the non-activated, large polaron mobility regime and are therefore analyzed in terms of equation

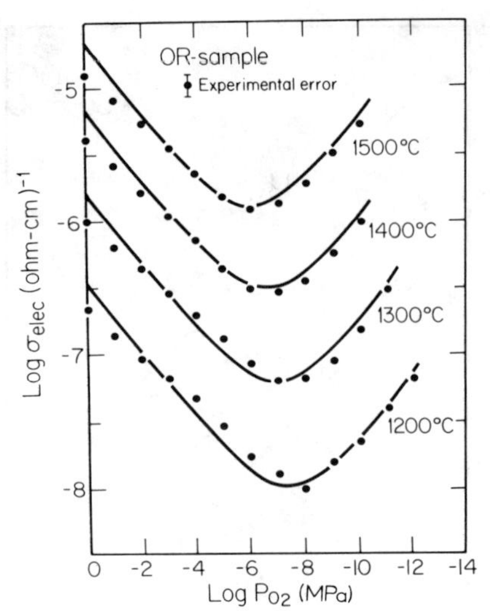

Fig. 14. Electronic conductivity as a function of P_{O_2} for undoped MgO [from Sempolinski et al., 1980].

(32) by Sempolinski et al. [1980]. This is further consistent with the broad Mg 3s conduction and O 2p valence bonds known to exist in MgO.

With the mobilities established, the following expression for the equilibrium constant $K_E(T)$ could be obtained:

$$K_E(T) = 2.2 \times 10^{33} T^3 \exp[-7.0 \pm 0.5 \text{ eV/kT}] \text{ cm}^{-6}$$

Furthermore, with a value for the hole mobility in place,it becomes a simple exercise to obtain K_{Ox} from the p-type conductivity by application of equation (28). The value for K_{Ox} obtained in this way is given by

$$K_{Ox}(T) = 7.2 \times 10^{63}$$

$$\exp(-6.3 \pm 0.8 \text{ eV/kT}) \text{ cm}^{-9} (\text{MPa})^{-1/2}$$

In concluding this section on MgO, it is worth noting the additional features introduced into the defect properties due to the variable valence of Fe in MgO. First, since Fe^{2+} is isovalent with Mg, only that fraction of the iron existing as Fe^{3+} will contribute to the formation of $[V''_{Mg}]$. Thus, in contrast to the Po_2 independence of σ_{ion} observed with dopants such as Al and Sc, s_{ion} decreases with decreasing Po_2 as $Fe^{3+}_{Mg} \rightarrow Fe^{2+}_{Mg}$. This also results in a modification of the $Po_2^{\pm 1/4}$ relations for electronic carriers towards $Po_2^{\pm 1/6}$. In the following section we examine the influence of variable valent ions in ThO_2 in some detail.

7.2. Fluorite Oxides: CeO_2, ThO_2, UO_2

In contrast to MgO, the oxides which crystallize in the cubic fluorite structure can accommodate a high degree of disorder on the oxygen sublattice. Recent studies of the electrical properties of CeO_2, ThO_2, and UO_2 doped with a variety of lower valent acceptor ions, illustrate additional features of the analytical approach outlined above.

Interfacial contributions to the electrical impedance of both ThO_2 and stabilized ZrO_2 have already been mentioned above. In both cases, frequency dependent complex impedance measurements (see Figures 8 and 9) were essential in isolating the bulk from grain boundary and electrode properties. Often at higher temperatures, however, due to higher activation energies, interfacial impedances may often be ignored, e.g., Figure 9. Complex impedance measurements should nevertheless be performed in the initial phases of measurements, to confirm that interfacial effects are in fact negligible.

Because fluorite-oxides can dissolve a high concentration of lower-valent cations into solid solution, a correspondingly high concentration of oxygen vacancies is formed as a consequence of charge compensation. The plot of log σ versus log Po_2 illustrated in Figure 15 for ThO_2 doped with 0.1 m/o Pr^{3+} shows that the inequality given in equation (34) is in fact satisfied over at least a large part of the reported experimental conditions. The existence of an electrolytic zone was further supported by oxygen concentration cell measurements [Fujimoto, 1982].

Aside from the Po_2-independent ionic conductivity, a p-type $Po_2^{+1/4}$ dependence is observed at high Po_2 and the beginnings of an n-type component at the lowest pressures of 1400° and 1500°C. All in all, the data agree well with the predictions of Figure 5 region III

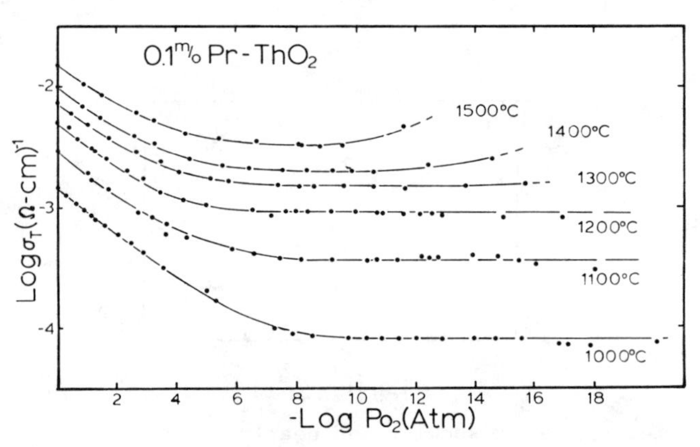

Fig. 15. Total conductivity as a function of P_{O_2} for ThO_2 + 5 0.1 m/o Pr [from Fujimoto, 1982].

and allow for the application of the defect relations applicable to that domain, e.g.,

$$\sigma_{ion} = 2[V\ddot{o}]q\mu_{Vo}[Pr'_{Th}]q\mu_{Vo}$$

$$= 1.0 \exp(-0.92 \text{ eV/kT})$$

With $[Pr'_{Th}] \cong 2.5 \times 10 \text{ cm}^{-3}$, we obtain the following value for the vacancy mobility

$$\mu_V = 0.25 \exp(-0.92 \text{ eV/kT})$$

Extensive Po_2 - independent ionic regimes bounded by electronic regimes in which $\sigma \alpha Po_2^{\pm 1/4}$ have similarly been observed experimentally in γ doped CeO_2^- [Tuller and Nowick, 1975], γ doped ThO_2 [Lasker and Rapp, 1966], and Ce doped ZrO_2 [Etsell and Flengas, 1972].

In the following we examine the electrical properties of Ce doped ThO_2 which differ substantially from the results observed above, the reason for this being the variable valent nature of Ce(+4, +3) which results in major changes in the predicted defect structure as already mentioned above for the case of Fe in Mg_O. Ce doped ThO_2 serves as a particularly good example of how electrical data may be used to confirm a particular defect model and vice versa, even under conditions where the deficit chemistry becomes relatively complex.

Let us begin by examining one of a number of ThO_2-CeO_2 solid solutions studied by Fujimoto [1982]. Figure 16 shows the conductivity data obtained for ThO_2 + 5 m/o CeO_2 as a function of Po_2 and T. At 1000°C one finds a $\sigma \alpha Po_2^{+1/4}$ component at high Po_2, $\sigma \alpha Po_2^0$ at intermediate Po_2, and the beginnings of a $Po_2^{-1/r}$ dependent term at low Po_2. This isotherm resembles the data discussed earlier for Pr doped ThO_2. Additional features, however, become evident at higher temperatures. The "n-type" component now clearly exhibits a $Po_2^{-1/6}$ dependence which is followed by a Po_2-independent component at the highest temperatures and lowest Po_2's of the experiment. This second plateau becomes even clearer in ThO_2-CeO_2 solid solutions with lower Ce levels.

The following observations agree with the predictions of Figures 4 and 5: a Po_2-independent plateau bounded at higher Po_2 by a p-type $Po_2^{-1/4}$ dependence is expected in region III as discussed above for Pr doped ThO_2. On the other hand, the following observations differ from the above expectations: (1) rather than a $Po_2^{-1/4}$ term bounding the plateau at lower Po_2, we observe a $Po_2^{-1/6}$ term; (2) an unexpected second plateau is observed at low Po_2. Furthermore, the apparent "n-type" $Po_2^{-1/6}$ component is observed

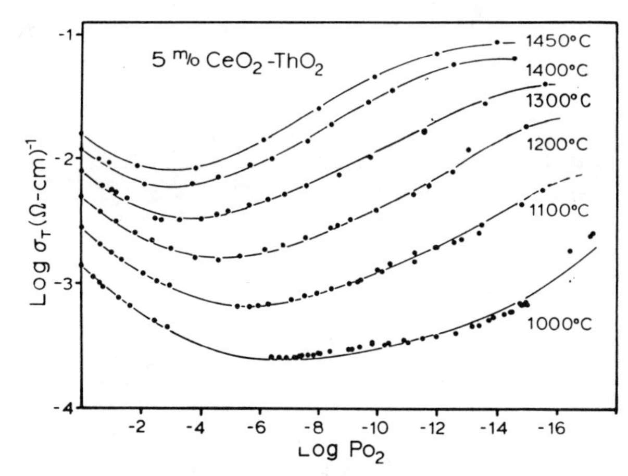

Fig. 16. Total conductivity as a function of P_{O_2} for ThO_2 + 5 m/o Ce [from Fujimoto, 1982].

to increase in magnitude at a given Po_2 and T with increasing Ce, which, given its acceptor nature (i.e., $Ce^{+4} + e' \rightarrow Ce^{+3}$), would appear to be in direct conflict with predictions.

In order to explain these observations, modifications in the defect model are in order. We note the following:

1. Ce being +4 at high Po_2, and thus isovalent with Th, should have little or no influence on the defect properties at high Po_2.
2. Low levels of lower valent background impurities are known to fix the oxygen vacancy density in undoped ThO_2.
3. Ce is known to shift to +3 at low Po_2 and should thus contribute significantly to vacancy formation at lower Po_2.

To accommodate these two extrinsic sources of oxygen vacancies, we rewrite the electroneutrality relation

$$n + 2[O_i''] + [N'_{Th}] + [Ce'_{Th}] = p + 2[V_o] \quad (53)$$

in which $[N'_{Th}]$, as before, represents the concentration of fixed valent background impurities.

At high Po_2, $[N'_{Th}] >> [Ce'_{Th}]$ and equation (53) reverts back to its old form. In particular, for the conditions studied here, we can expect that (53) can be further simplified to

$$[N'_{Th}] = 2[V_o] \quad (54)$$

which corresponds to the conditions of region III in Figure 4. As one decreases Po_2, however, one ultimately reaches the condition at which $[Ce'_{Th}] >> [N'_{Th}]$ and therefore the neutrality condition changes to

$$[Ce'_{Th}] = 2[V_o] \quad (55)$$

TABLE 4. Carrier Concentrations Corresponding to Regions IV and V of Figure 17

	Neutrality Relation $[Ce'_{Th}] = 2[2V_O^{\cdot\cdot}]$			
	$[Ce_{Th}] \simeq [Ce_{Th}]_{total}$ *	Equation Number	$[Ce'_{Th}] = [Ce_{Th}]_{total}$ *	Equation Number
n	$K_{Ce}^{-1}(2K_r[Ce_{Th}]_{tot}^{-1})^{1/3}P_{O_2}^{-1/6}$	(63)	$(2K_R[Ce_{Th}]_{tot}^{-1})^{1/2}P_{O_2}^{-1/4}$	(66)
$[V_o^{\cdot\cdot}]$	$(K_r/3)^{1/3}[Ce_{Th}]_{tot}^{2/3}P_{O_2}^{-1/6}$	(64)	$[Ce_{Th}]_{tot}/2$	(67)
$[Ce'_{Th}]$	$2K_r^{1/3}[Ce_{Th}]_{tot}^{2/3}P_{O_2}^{-1/6}$	(65)	$[Ce_{Th}]_{tot}$	(68)

*Mass balance equations.

To account for the Ce in its two valence states, a mass balance relation may be written

$$[Ce_{Th}] + [Ce'_{Th}] = [Ce_{Th}]_{total} \qquad (56)$$

in which $[Ce_{Th}]$ is the concentration of quadravalent Ce and $[Ce_{Th}]_{total}$, the total concentration of Ce added in solid solution.

It is further useful to rewrite the reduction reaction, equation (10), to account for the influence of Ce. This is given by

$$O_o + 2Ce_{Th} \rightarrow V_o + 2Ce'_{Th} + 1/2O_2(g) \qquad (57)$$

and

$$\frac{[V_o]\ [Ce'_{Th}]^2}{[Ce_{Th}]^2} = K_r P_{O_2}^{-1/2} \qquad (58)$$

This reaction differs from (10) only in that the electrons generated during reduction become trapped at the localized Ce levels in the band gap rather than entering into the conduction band. What this does mean, however, is that while reduction in an undoped oxide implies an increase in electronic conductivity, in this case only ionic carriers are markedly enhanced.

The ionization and consequent reduction of Ce_{Th} can also be described by the reaction

$$Ce_{Th} + e' \rightarrow Ce'_{Th} \qquad (59)$$

and its associated mass action relation given by

$$[Ce'_{Th}]/[Ce_{Th}]n = K_{Ce}(T) = K^{\circ}_{Ce}\ \exp\ (-E_{Ce}/kT) \qquad (60)$$

in which E_{Ce} is the impurity ionization energy.

To simplify the process of obtaining solutions of the defect densities when the electroneutrality condition (55) is controlling, it is useful to divide this region into two. The first, at higher Po_2, corresponds to conditions where

$$[Ce_{Th}] \simeq [Ce_{Th}]_{total} \qquad (61)$$

and the second lower Po_2 to

$$[Ce'_{Th}]\ [Ce_{Th}]_{total} \qquad (62)$$

Table 4 summarizes the solutions for n, $[V_o]$, and $[Ce'_{Th}]$ for these two conditions, while a defect diagram including all of the predicted defect regimes is illustrated in Figure 17. The concentrations from Table 4 are as follows:

$$K_{Ce}^{-1}(2K_r[Ce_{Th}]_{tot}^{-1})^{1/3}P_{O_2}^{-1/6} \qquad (63)$$

$$(K_r/3)^{1/3}[Ce_{Th}]_{tot}^{2/3}\ P_{O_2}^{-1/6} \qquad (64)$$

$$2K_r^{1/3}[Ce_{Th}]_{tot}^{2/3}\ P_{O_2}^{-1/6} \qquad (65)$$

$$(2K_R[Ce_{Th}]_{tot}^{-1})^{1/2}P_{O_2}^{-1/4} \qquad (66)$$

$$[Ce_{Th}]_{tot}/2 \qquad (67)$$

$$[Ce_{Th}]_{tot} \qquad (68)$$

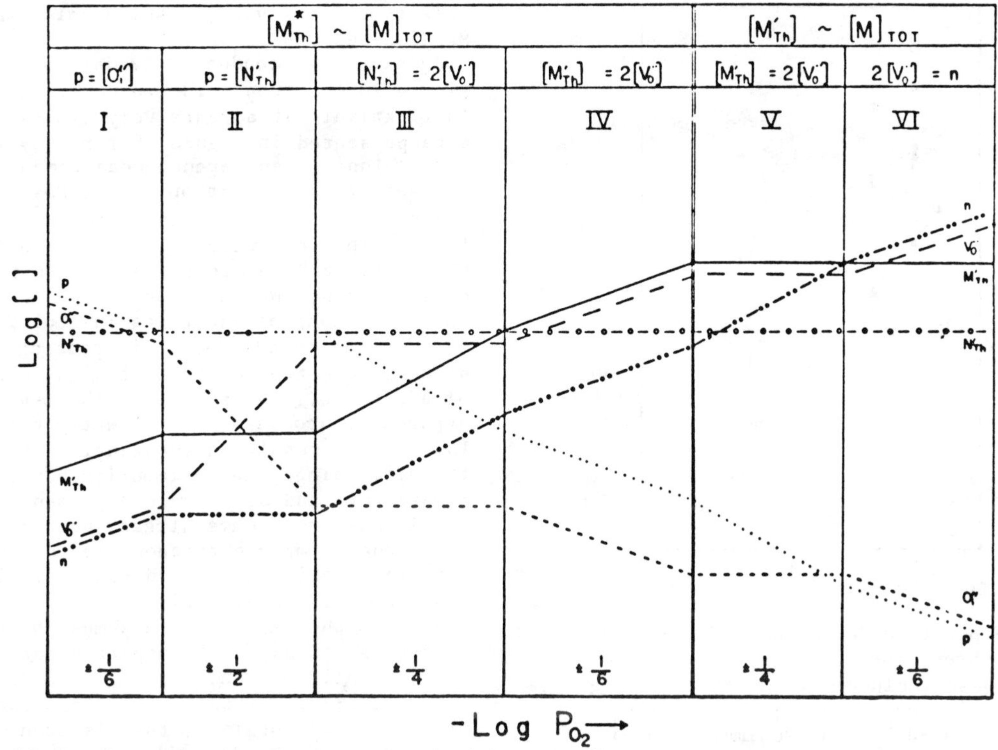

Fig. 17. Defect diagram for ThO_2 doped with both fixed valent N_{Th} and variable valent, M_{Th} acceptor impurities [from Fujimoto, 1982].

Careful evaluation of these results shows them to be in agreement with the experimental results obtained in Figure 16. First a $Po_2^{-1/6}$-dependent ionic term in region IV is indeed predicted to follow a Po_2-independent term in region III, followed in turn by a second Po_2-independent term in region V. From the predictions of Table 4, it is indeed clear that σ in all of these regions is consistent with an ionic interpretation given the experimentally observed increase in magnitude of σ with increasing $[Ce_{Th}]_{total}$. Note, on the other hand, the inverse relationships predicted between n and $[Ce_{Th}]_{total}$. These predictions are further confirmed by the ionic transference measurements shown in Figure 18 which show t_{ion} to deviate strongly from unity only at high Po_2 where $\sigma \alpha Po_2^{+1/4}$.

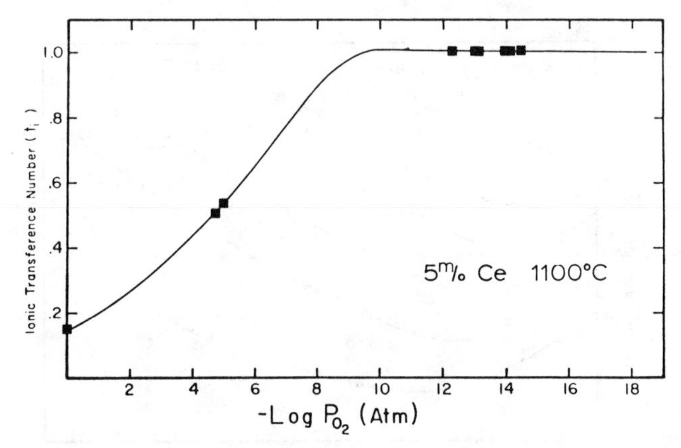

Fig. 18. Ionic transference number as a function of P_{O_2} for ThO_2 + 5 m/o Ce at 1100°C [from Fujimoto, 1982].

With the deficit model confirmed, defect and transport parameters may be deduced. Figure 19 shows the temperature dependence of σ obtained in a number of the regimes described above. For the ionic, Po_2-dependent region, an activation energy of 2.35 eV is obtained. This energy E should, according to equations (2), (30), and (64), be made up of

$$E = Er/3 + E_m \qquad (69)$$

in which E_r is the energy associated with the equilibrium constant $K_r(T)$ and E_m is the ion migration energy. On the other hand, in the "ionic background" region $[V_O]$ is fixed by $[N_{Th}]$ and so that energy must be the migration energy E_m or

$$E_m = 1.25 \text{ eV}$$

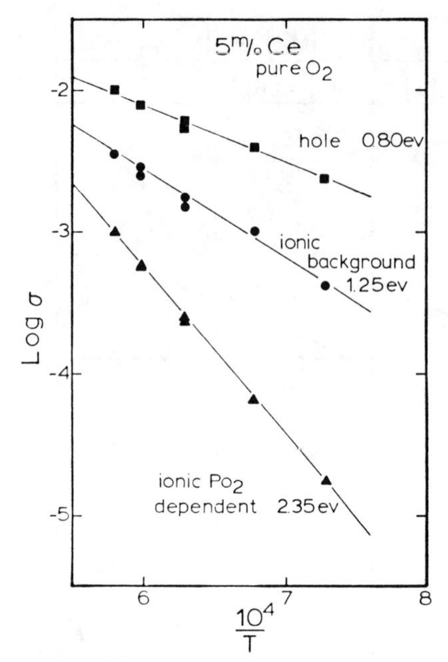

Fig. 19. Temperature dependence of conductivity obtained in three conduction regimes of ThO_2 + 5 m/o Ce. These include $\sigma \propto P_{O_2}^{+1/4}$, "hole"; $\sigma \propto P_{O_2}^{o}$, "ionic background"; $\sigma \propto P_{O_2}^{-1/6}$, "ionic P_{O_2} dependent" [from Fujimoto, 1982].

and therefore, by combination with equation (63),

$$E_r = 3.3 \text{ eV}$$

One may readily demonstrate that $K_R = K_r K_{Ce}^{-2}$ and so

$$E_{Ce} = 1/2(E_r - E_R) \qquad (70)$$

Given that a value of E_R = 10.2 eV was obtained in another study of ThO_2 [Fujimoto, 1982] one therefore obtains

$$E_{Ce} = -3.45 \text{ eV}$$

which shows that the Ce level is approximately 3.5 eV below the conduction band edge and 2 eV above the valence band edge ($E_g \sim 5.5$ eV [Strehlow and Cook, 1973]). Given that Ce is indeed a deep level impurity it is therefore not at all surprising that electrons sitting on these sites are very effectively trapped and therefore do not contribute to electronic conduction.

I conclude this section by examining the influence of exceptionally high levels of additives on the predicted defect structure. Such effects are particularly important where two components have a wide solid solubility range as exemplified by the ThO_2-CeO_2 system. An example of an important mineral of this type is olivine, a solid solution of Mg_2SiO_4 and Fe_2SiO_4.

Consider the conductivity data presented in Figure 20 for ThO_2 + 10 m/o CeO_2. Qualitatively it appears very similar to the data presented in Figure 16 for ThO_2 + 5 m/o CeO_2. Ionic transference measurements show, however, that conduction at low Po_2 is electronic rather than ionic for the 10 m/o level. This can readily be understood by noting that localized levels formed within the band gap of a solid by the addition of dilute concentrations of additives must ultimately broaden into bands due to increasing electron-electron overlap at high concentrations of additives. In ThO_2-CeO_2, the levels have apparently broadened sufficiently between 5 and 10% CeO_2 to result in substantial electron transport along the Ce impurity band. The temperature and atmosphere dependence of these mobile electrons have already been predicted since they simply correspond to the $[Ce'_{Th}]$ term in equations (65) and (68), at higher and lower Po_2's, respectively.

Because the impurity bands must be relatively narrow we can expect transport along them to be of the small polaron, hopping type described by equation (33). Without going into the details of how μ_e was obtained, the electron mobility was indeed found to be thermally activated.

8. Conclusions

I believe I have been able to demonstrate that one may, by following a systematic program of electrical measurements on appropriately doped materials, derive the defect structure and transport parameters of even relatively complex materials. Although other experimental approaches sometimes give more direct information concerning the defect properties,

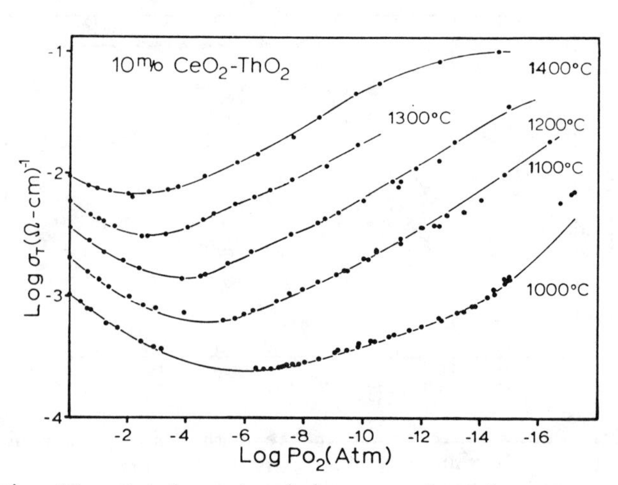

Fig. 20. Total conductivity as a function of P_{O_2} for ThO_2 + 10 m/o Ce [from Fujimoto, 1982].

e.g., tracer diffusion or electron spin resonance, they are often less sensitive or more difficult to apply. Nevertheless, confirmation by other techniques, even other limited conditions, is extremely useful in demonstrating the uniqueness of a given model.

Although the principles of applying the defect chemical approach towards characterizing the electrical properties of solids was demonstrated, many situations exist in which additional levels of complexity must be taken into account. Some of these have been noted previously, e.g., defect-defect interactions. Extensive defect interactions may ultimately lead to extended defects such as shear structures in heavily reduced TiO_2 which must be treated differently than lightly reduced TiO_2 [Catlow, 1981].

Furthermore, in extending the defect approach to ternary and higher order systems, independent thermodynamic variables in addition to T and Po_2 must be controlled. For example, in the perovskite compounds, ABO_3, one must independently control either the activity of the A or B component in addition to T and Po_2. Extension of the defect chemical approach to higher order systems is outlined in articles by Schmalzried [1965] and Smyth [1976, 1977].

Acknowledgments. Special thanks are due R. N. Schock and A. G. Duba of Lawrence Livermore Laboratories for encouraging me to complete this article and exposing me to the interesting problems existing in the minerals field. The hospitality extended to me by B. C. H. Steele of Imperial College, London (while the author was SRC visiting fellow at the Wolfson Unit of Solid State Ionics, Department of Metallurgy and Materials Science), during the writing of this article is much appreciated. Support of the work described in this article by the National Science Foundation under Grant No. 82-03697 DMR is gratefully acknowledged.

References

Adler, D., Insulating and metallic states in transition metal oxides, in Solid State Physics, vol. 21, edited by F. Seitz and D. Turnbull, pp. 1-113, Academic, New York, 1968.

Appel, J., Polarons, in Solid State Physics, vol. 21, edited by F. Seitz and D. Turnbull, pp. 193-391, Academic, New York, 1968.

Catlow, C. R. A., Defect clustering in nonstoichiometric oxides, in Nonstoichiometric Oxides, edited by O. T. Sorensen, pp. 61-98, Academic, New York, 1981.

Choi, G. M., Electrical conductivity and thermoelectric power studies in $Ba_{0.3}Sr_{0.7}TiO_3$, M. S. thesis, Dept. of Mater. Sci. and Eng., Mass. Inst. of Technol., Cambridge, June 1983.

Deportes, C., and M. Gauthier, Nouvelle méthode de détermination du nombre de transport cationique dans les oxydes solides par dilatométrie sous courant continu application à l'oxyde de magnésium, C. R. Acad. Sci. Paris, 273C, 1605-1608, 1971.

Dudney, N. J., R. L. Coble, and H. L. Tuller, Electrical conductivity of pure and yttria doped uranium dioxide, J. Am. Ceram. Soc., 64, 627-632, 1981.

Etsell, T. H., and S. N. Flengas, N-type conductivity in stabilized zirconia solid electrolytes, J. Electrochem. Soc., 119, 1-7, 1972.

Fujimoto, H. H., The electrical properties and defect structure of doped thorium dioxide, Ph.D. thesis, 235 pp., Dep. of Mater. Sci. and Eng., Mass. Inst. of Technol., Cambridge, Feb. 1982.

Fujimoto, H. H., and H. L. Tuller, Mixed ionic and electronic transport in thoria electrolytes, in Fast Ion Transport in Solids, edited by J. N. Mundy and G. K. Shenoy, pp. 649-652, North-Holland, Amsterdam, 1979.

Goodenough, J. B., Skeleton structures, in Solid Electrolytes: General Principles, Characterization, Materials, Applications, edited by P. Hagenmuller and W. Van Gool, pp. 393-415, Academic, New York, 1978.

Goto, K. S., and W. Pluschkell, Oxygen concentration cells, in Physics of Electrolytes, vol. 2, edited by J. Hladik, pp. 540-622, Academic, New York, 1972.

Gvishi, M., N. M. Tallan, and D. S. Tannhauser, The Hall mobility of electrons and holes in MnO at high temperature, Solid State Commun., 6, 135, 1968.

Hagenmuller, P., and W. Van Gool (Eds.), Solid Electrolytes: General Principles, Characterization, Materials, Applications, Academic, New York, 1978.

Harding, B. C., and D. M. Price, Cation self-diffusion in MgO up to 2350°C, Philos. Mag., 26, 253-260, 1972.

Harding, B. C., D. M. Price, and A. J. Mortlock, Cation diffusion in single crystal MgO, Philos. Mag., 23, 399-408, 1971.

Heyne, L., Electrochemistry of mixed ionic-electronic conductors, in Solid Electrolytes, edited by S. Geller, pp. 169-221, Springer-Verlag, New York, 1977.

Kilner, J. A., and B. C. H. Steele, Mass transport in anion-deficient fluorite oxides, in Nonstoichiometric Oxides, edited by O. T. Sorensen, pp. 233-269, Academic, New York, 1981.

Kittel, C. K., Introduction to Solid State Physics, 5th ed., John Wiley, New York, 1976.

Kleitz, M., H. Bernard, E. Fernandez, and E. Schouler, Impedance spectroscopy and electrical resistance measurements on stabilized zirconia, in Science and Technology of Zirconia, edited by A. H. Heuer and L. W. Hobbs, pp. 310-336, American Ceramic Society, Columbus, Ohio, 1981.

Kofstad, P., Nonstoichiometry, Diffusion and Electrical Conductivity in Binary Metal Oxides, Wiley Interscience, New York, 1972.

Kröger, F. A., The Chemistry of Imperfect Crystals, 2nd ed., North-Holland, Amsterdam, 1974.

Lasker, M. F., and R. A. Rapp, Mixed conduction in ThO_2 and ThO_2-Y_2O_3 solutions, Z. Phys. Chem., 49, 198-221, 1966.

Lindner, R., and G. D. Parfitt, Diffusion of radioactive magnesium in magnesium oxide crystals, J. Chem. Phys., 26, 182-185, 1957.

Pizzini, S., General aspects of kinetics of ion-transfer across interfaces, in Fast Ion Transport in Solids, edited by W. Van Gool, pp. 461-475, Elsevier, New York, 1973.

Schmalzried, H., Point Defects in ternary ionic crystals, Prog. Solid State Chem., 2, 265-303, 1965.

Sempolinski, D. R., and W. D. Kingery, Ionic conductivity and magnesium vacancy mobility in magnesium oxide, J. Am. Ceram. Soc., 63, 664-669, 1980.

Sempolinski, D. R., W. D. Kingery, and H. L. Tuller, Electronic conductivity of single crystal magnesium oxide, J. Am. Ceram. Soc., 63, 669-675, 1980.

Shankland, T. J., Electrical conduction in rocks and minerals: Parameters for interpretation, Phys. Earth Planet. Inter., 10, 209-219, 1975.

Shankland, T. J., Electrical conduction in mantle materials, in Evolution of the Earth, Geodynamics Ser., vol. 5, edited by R. J. O'Connell and W. S. Fyfe, pp. 256-263, AGU, Washington, D. C., 1981.

Smyth, D. M., Thermodynamic characterization of ternary compounds, I, The case of negligible defect association, J. Solid State Chem., 16, 73-81, 1976.

Smyth, D. M., Thermodynamic characterization of ternary compounds, II, The case of defect association, J. Solid State Chem., 20, 359-364, 1977.

Steele, B. C. H., and J. A. Kilner, Some characteristics of oxygen transport and surface exchange reactions in solid oxides, in Transport in Non-Stoichiometric Compounds, edited by H. Nowatry, pp. 308-324, Elsevier, New York, 1982.

Strehlow, W. H., and E. L. Cook, Compilation of energy bandgaps in elemental and binary semiconductors and insulators, J. Phys. Chem. Ref. Data, 2, 163-200, 1973.

Sze, S. M., Physics of Semiconductor Devices, Wiley Interscience, New York, 1969.

Tuller, H. L., Mixed conduction in nonstoichiometric oxides, in Nonstoichiometric Oxides, edited by O. T. Sorensen, pp. 271-335, Academic, New York, 1981.

Tuller, H. L., Highly conductive ceramics, in Glasses and Ceramics for Electronics, edited by R. Buchanan, in press, 1984.

Tuller, H. L., and A. S. Nowick, Doped ceria as a solid oxide electrolyte, J. Electrochem. Soc., 122, 255-259, 1975.

Tuller, H. L., and A. S. Nowick, Small polaron electron transport in reduced CeO_2 single crystals, J. Phys. Chem. Solids, 38, 859-867, 1977.

Tuller, H. L., and A. S. Nowick, Defect structure and electrical properties on nonstoichiometric CeO_2 single crystals, J. Electrochem. Soc., 126, 209-217, 1979.

Weppner, W., and R. A. Huggins, Electrochemical methods for determining kinetic properties of solids, Annu. Rev. Mater. Sci., 8, 269-311, 1978.

Wimmer, J. M., and I. Bransky, Electronic conduction, in Electrical Conductivity in Ceramics and Glass, part A, edited by N. M. Tallan, pp. 269-311, Marcel Dekker, New York, 1974.

Wuensch, B. J., W. C. Steele, and T. Vasilos, Cation self-diffusion in single-crystal MgO, J. Chem. Phys., 58, 5258-5266, 1973.

ELECTRICAL STUDIES OF TRANSITION METAL CATION DISTRIBUTION IN SPINELS

T. O. Mason

Department of Materials Science and Engineering and Materials Research Center
Northwestern University, Evanston, Illinois 60201

Abstract. Cation distributions in transition metal oxide spinels can be determined thermoelectrically from room temperature to their melting points. A multiprobe apparatus facilitates the collection of data. Results for magnetite, divalent ferrospinels at low temperature, Fe_3O_4-$MgFe_2O_4$ at high temperature, trivalent ferrospinels, and Fe_3O_4-Fe_2TiO_4 at high temperature are presented. Cation distributions are calculated and tested against current thermodynamic models. Extension of the technique to other mineral solid solutions is considered.

Introduction

Transition metal oxide (T.M.O.) spinels are of interest to ceramists by virtue of their useful electrical properties and applications in robust environments. In addition to their extensive use as soft magnets [Cullity, 1972], they also serve as thermistors [Sachse, 1975]. Ferrospinels have been proposed as electrodes for coal-fired magnetohydrodynamic (MHD) generators [Rossing and Bowen, 1976]. It is conceivable that T.M.O. spinels will find application as electrodes and interconnects in other energy conversion systems, e.g. batteries, fuel cells, etc., where conductivity in elevated temperature oxidizing environments is called for.

Geologists also share an interest in T.M.O. spinels. For example, titanomagnetites are the primary carriers of rock magnetism. Both aluminum ferrites [Turnock and Eugster, 1962] and titanium ferrites [Buddington and Lindsley, 1964] have been employed in the fields of geothermometry and oxygen geobarometry.

Although both cation vacancies and interstitials occur in T.M.O. spinels [Dieckmann, 1982], the principal defect type in the structure $A(B)_2O_4$, where the parentheses enclose the octahedral species, is the antisite reaction:

$$A + (B) \rightleftarrows (A) + B \tag{1}$$

Here an exchange of cations between the tetrahedral and octahedral sublattices takes place. It is well established that the law of mass action applies:

$$K_{CD} = \frac{[(A)]\,[B]}{[A]\,[(B)]} \tag{2}$$

where K_{CD} is the cation distribution constant. Three limiting cation distribution states are defined: normal, $A(B)_2O_4(K_{CD}=0)$: inverse, $B(A,B)O_4(K_{CD}=\infty)$; and random, $A_{1/3}B_{2/3}(A_{2/3}B_{4/3})O_4(K_{CD}=1)$. Of course, any intermediate distribution is possible.

Measurement of cation distribution can be accomplished in a number of ways. An excellent review is given by Navrotsky [1972]. Techniques include X-ray and neutron diffraction, Mössbauer spectroscopy, magnetometry, dilatometry, and solution calorimetry. Unfortunately, most of these techniques require a quenching step, and there is some concern as to whether high temperature distributions can be totally preserved on quenching. Furthermore, in situ studies are difficult, at best, and often require highly sophisticated equipment and/or elaborate computational methods.

We have developed a thermopower technique for cation distribution determination in T.M.O. spinels. This technique is relatively simple and can be performed in situ up to the melting points of the spinels. Furthermore, results are subject to relatively straightforward interpretation in terms of cation distributions. The multiprobe apparatus employed circumvents the need for an internal heater and greatly speeds the collection of experimental data. This is important when a large number of specimens must be quickly, but accurately, measured.

In this paper, the thermopower technique will be outlined and the multiprobe apparatus described. Results for Fe_3O_4 and various ferrospinels will be discussed according to a classification scheme based on the valence of the substituent. Ramifications of this work in the area of cation distribution thermodynamics will be stressed. Extension of our technique to other T.M.O. spinels (Co_3O_4, Mn_3O_4) and to the study of

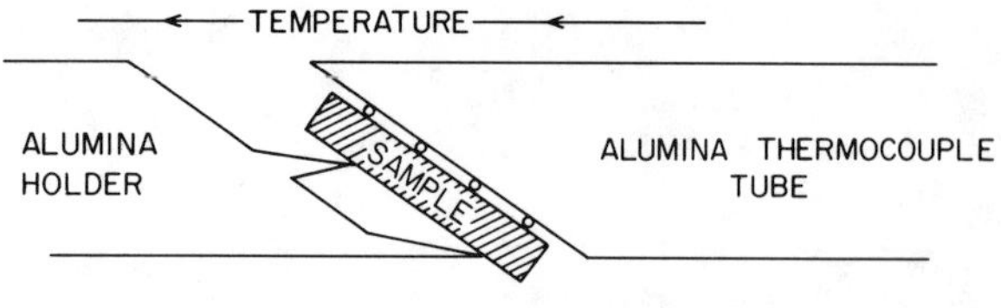

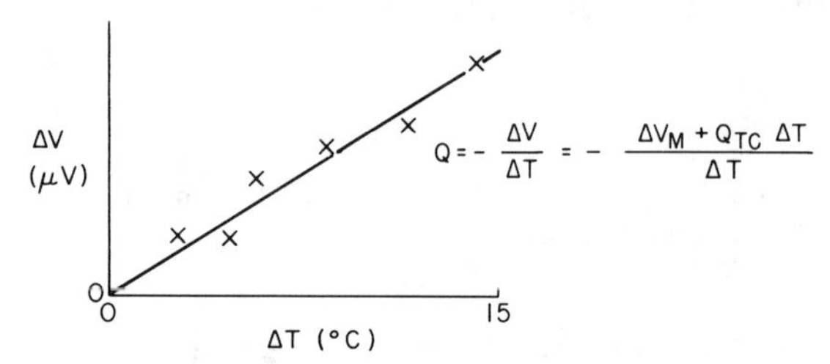

Fig. 1. Schematic of the multiprobe thermoelectric coefficient apparatus. Q_{TC} stands for the thermopower of the reference element of the thermocouple.

point defects in other T.M.O. minerals will also be discussed.

Experimental Technique and Interpretation

The basic thermopower technique and multiprobe apparatus can be explained with the aid of Figure 1. A thermoelectric voltage (ΔV) is generated across a specimen by the application of a thermal gradient (ΔT). The proportionality constant is the Seebeck or thermoelectric coefficient (Q). If the cold end is defined as positive, then the sign of Q will agree with the sign of the majority carrier. The conventional technique involves an internal heater to raise the hot end temperature with respect to the cold end. In the multiprobe technique we developed [Trestman-Matts et al., 1983a], four carefully matched thermocouples are used and the natural gradient of the experimental furnace is employed. The six ΔT's and six corresponding ΔV's, measured with the common thermocouple leads, allow the thermoelectric coefficient to be calculated. However, corrections for the thermopower of the thermocouple leads must be made. The experimental temperature is taken to be the average of the four thermocouples. Large correlation factors and/or ΔT intercepts (at $\Delta V = 0$) indicate mismatched thermocouples or contamination. The technique was tested by comparison with conventional measurements. Agreement was within two percent when Fe_3O_4 single crystals were employed.

The role of polycrystallinity must also be considered. Geological specimens will ordinarily be polycrystalline. Synthetic polycrystalline specimens are far easier to produce than single crystals. Our studies on polycrystalline Fe_3O_4 versus single crystals agree to within two percent. Except for magnetite, all of the data reported in this paper were measured on polycrystalline specimens.

The thermopower technique of cation distribution determination is based on the relationship between the thermopower (Q) and the valence ratio of conducting species $[Fe_\sigma^{3+}]/[Fe_\sigma^{2+}]$ in T.M. oxides (see references in Wu and Mason [1981]:

$$Q = \frac{-k}{e_o}\left[\ln 2\frac{[Fe_\sigma^{3+}]}{[Fe_\sigma^{2+}]} + A\right] \tag{3}$$

Here k is Boltzmann's constant, e_o is the elementary electron charge, A is a constant associated with the entropy of transport (negligibly small in the case of small polaron conduction), and the factor of 2 arises due to spin degeneracy. For equation (3) to apply, small polaron conduction must be dominant. This has been confirmed for magnetite [Wu and Mason, 1981] as well as for other ferrospinels involving aluminum [Mason and Bowen, 1981] and nickel [Griffiths et al., 1970]. It is typically observed that the activation energy of conductivity is small ($E_a \cong 0.1$-0.2 eV) and relatively constant across a solid solution until one valence state of iron is 80 to 90 percent replaced. Then E_a rises abruptly. This is consistent with small polaron conduction. Equally important is the conclusion in these studies that conduction is restricted to the octahedral sublattice.

In order to relate the thermopower to the cation distribution, a distribution model must be introduced. The lattice molecule of a ferrospinel involving a foreign cation of fixed valence, n, can be represented as:

$$Fe_a^{2+}Fe_b^{3+}Me_c^{n+}(Fe_d^{2+}Fe_e^{3+}Me_f^{n+})O_4 \tag{4}$$

where the subscripts denote the species concentrations. Site balance requires that:

$$\sum_{i=a}^{c} i + \sum_{j=d}^{f} j = 1\ (\text{tet}) + 2\ (\text{oct}) = 3 \tag{5}$$

whereas charge balance requires that:

$$2(a+d) + 3(b+e) + n(c+f) = 8 \tag{6}$$

and mass balance is given by:

$$(c+f) = 3\,\xi \tag{7}$$

where ξ is the cation mole fraction $n_{Me}/(n_{Me}+n_{Fe})$. The two cation distribution equilibria are:

$$Fe^{2+} + (Fe^{3+}) \rightleftarrows (Fe^{2+}) + Fe^{3+} \quad K_{CD}^{Fe} = \frac{bd}{ae} \tag{8}$$

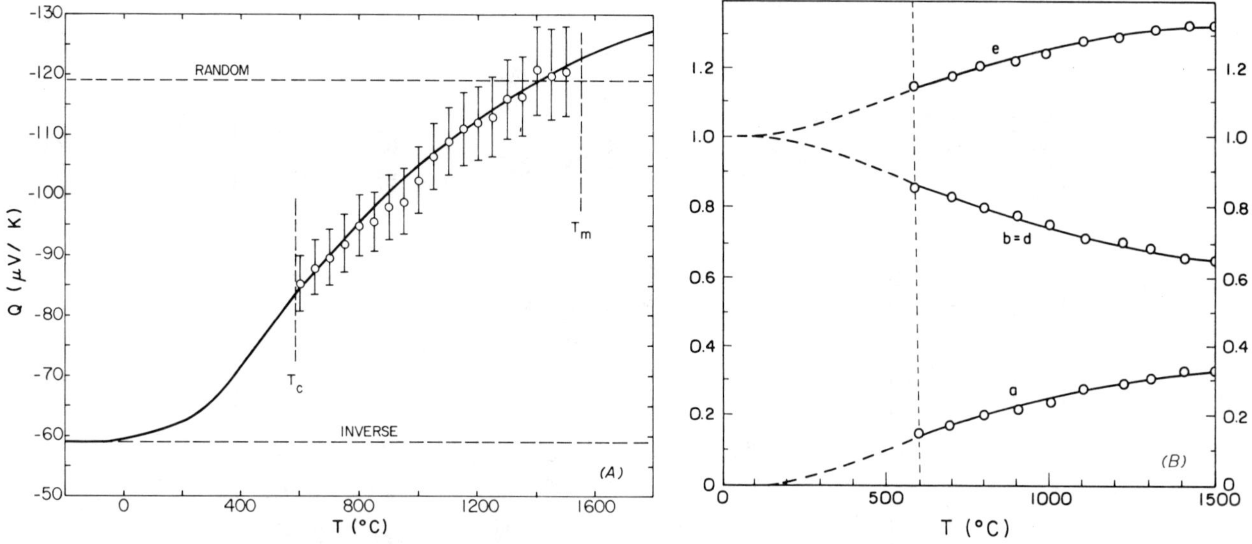

Fig. 2. (a) Experimental thermopower and (b) cation distribution, as a function of temperature in magnetite [from Wu and Mason, 1981].

$$Me^{n+} + (Fe^{3+}) \rightleftarrows (Me^{n+}) + Fe^{3+} \qquad K_{CD}^{Me} = \frac{bf}{ce} \tag{9}$$

At this point the octahedral valence ratio, $q = e/d$, is introduced. Assuming small polaron conduction via octahedral sites only, equation (3) can be rewritten as:

$$q = \frac{e}{d} \simeq \frac{1}{2} \exp\ (-Qe_o/k) \tag{10}$$

In the case that a foreign cation substitutes exclusively on the tetrahedral sublattice, it follows from site balance that:

$$q = \frac{e}{d} = \frac{2-d}{d} = \frac{e}{2-e} \tag{11}$$

whereas if the foreign cation substitutes exclusively on the octahedral sublattice:

$$q = \frac{e}{d} = \frac{2-3\xi-d}{d} = \frac{e}{2-3\xi-e} \tag{12}$$

By isolating d and e in equations (11) and (12), and substituting for d and e in equation (8) (see column 2 of Table 1) expressions result which involve only q and K_{CD}^{Fe}. Therefore it follows that a single measurement of the octahedral valence ratio, q, via thermopower (equation (10)) allows us to determine K_{CD}^{Fe} and thus the overall cation distribution. Table 1 is arranged according to the valence of the substituent and whether it occupies the tetrahedral or octahedral sublattice in Fe_3O_4. The categories chosen correspond to the most common spinel types and cover the systems to be dealt with in this study. The third and fourth columns show the octahedral valence ratio predicted in the event that $K_{CD}^{Fe} \rightarrow \infty$ or $K_{CD}^{Fe} = 1$. These will be referred to later in the paper.

Results and Discussion

Fe_3O_4

Experimental results [Wu and Mason, 1981] are displayed in Figure 2a. Also plotted in Figure 2a are the values of Q corresponding to inverse magnetite ($K_{CD}^{Fe} \rightarrow \infty$) as well as random magnetite ($K_{CD}^{Fe} = 1$), row 2, columns 3 and 4 in Table 1, respectively. It is well documented that the actual distribution is essentially inverse at low temperatures, but randomizes in the vicinity of 1300-1500°C (see references in Wu and Mason [1981]). This is precisely what the thermopower shows. Using equation (10) and the expression in row 1, column 2 of Table 1, the cation distribution as a function of temperature was calculated and is shown in Figure 2b. The solid curves (broken below 600°C) were calculated according to the following model [Wu and Mason, 1981]:

$$-\ RT\ \ln K_{CD}^{Fe} = -\ 2770R + 1.61\ RT \tag{13}$$

The first term of which corresponds to an enthalpy of cation disorder, whereas the second term would suggest a non-configurational entropy. More recently, O'Neill and Navrotsky [1983a] refitted the data to the following model:

$$-\ RT\ \ln K_{CD}^{Fe} = 2921\ R + 2\ (-2766)R(b) + 0.393\ RT \tag{14}$$

which indicates an enthalpy which varies with the

TABLE 1. Octahedral Valence Ratio and Cation Distribution Constants in Ferrospinels $Fe_a^{2+}Fe_b^{3+}Me_c^{n+}(Fe_d^{2+}Fe_e^{3+}Me_f^{n+})O_4$

(1) Substituent		(2) Cation Distribution Equilibrium	(3) Valence Ratio If	(4) Valence Ratio If
Valence	Site	$K_{CD}^{Fe} = \frac{b \cdot d}{a \cdot e}$	$K_{CD}^{Fe} \to \infty$	$K_{CD}^{Fe} = 1$
(1) none (c = f = 0)	none (c = f = 0)	$K_{CD}^{Fe} = \frac{(2-e)d}{(1-d)e} = \frac{2}{(q-1)q}$	$q = 1$	$q = 2$
(2) n = 2	tetrahedral (c = 3ξ, f = 0)	$K_{CD}^{Fe} = \frac{(2-e)d}{(1-3\xi-d)e} = \frac{2}{[q(1-3\xi)-3\xi-1]q}$	$q = \frac{1+3\xi}{1-3\xi}$	$q = \frac{2}{1-3\xi}$
(3) n = 2	octahedral (c = 0, f = 3ξ)	$K_{CD}^{Fe} = \frac{(2-e)d}{(1-3\xi-d)e} = \frac{2+3\xi q}{[q(1-3\xi)-1]q}$	$q = \frac{1}{1-3\xi}$	$q = \frac{2}{1-3\xi}$
(4) n = 2	mixed (c + f = 3ξ) (c + f = 3ξ)	$K_{CD}^{Me} = \frac{bf}{ce} = \frac{(2-qd)[2-d(q+1)]}{[3\xi-2+d(q+1)]qd}$ $K_{CD}^{Fe} = \frac{(2-e)d}{(1-3\xi-d)e} = \frac{2-qd}{(1-3\xi-d)q}$	depends on K_{CD}^{Me}	$q = \frac{2}{1-3\xi}$
(5) n = 3	octahedral (c = 0, f = 3ξ)	$K_{CD}^{Fe} = \frac{(2-3\xi-e)d}{(1-d)e} = \frac{2-3\xi}{(q-1+3\xi)q}$	$q = 1-3\xi$	$q = 2-3\xi$
(6) n = 4	octahedral (c = 0, f = 3ξ)	$K_{CD}^{Fe} = \frac{(2-6\xi-e)d}{(1+3\xi-d)e} = \frac{2-6\xi-3\xi q}{[q(1+3\xi)+6\xi-1]q}$	$q = \frac{1-6\xi}{1+3\xi}$	$q = \frac{2-6\xi}{1+3\xi}$

concentration of tetrahedral Fe^{3+}(b). Once again, however, a non-configurational entropy is required.

Divalent Substituents (Zn, Mg, Ni, Co, Mn) at Low Temperature

As indicated in Figure 2 for Fe_3O_4, K_{CD}^{Fe} should become infinitely large at low temperatures. If this behavior persists in mixed spinel systems, column 3 of Table 1 would be the predicted dependence of q, and hence Q through equation (10), on composition. In Figure 3 are plotted the functions listed in Table 1 in column 3, rows 2 and 3. These correspond to divalent substituents occupying the tetrahedral or octahedral sublattices of Fe_3O_4, respectively. Also plotted are experimental ambient temperature thermopower data from the literature for various divalent substituents. As expected [Wu et al., 1979], Zn prefers the tetrahedral sublattice whereas Ni prefers the octahedral sublattice. Pelton et al. [1979] have demonstrated that until $\xi = 1/3$ in the solid solutions Fe_3O_4-Co_3O_4 and Fe_3O_4-Mn_3O_4, Co and Mn are essentially divalent. Figure 3 shows that these ions prefer the octahedral sublattice. One surprise is the octahedral preference of Mg. Previous octahedral preference energy schemes had placed Mg^{2+} between Fe^{2+} and Fe^{3+} [see Navrotsky, 1972].

Divalent Substituent at High Temperature ($Mg_xFe_{3-x}O_4$)

Of all the substituents considered, only Mg^{2+} is known to substantially occupy both sublattices in Fe_3O_4 at elevated temperatures. This complicates analysis of the thermopower results as can be seen in Table 1, row 4, column 2. A value of K_{CD}^{Mg} is required to solve for a value of $d = Fe_{oct}^{2+}$ from the thermoelectrically derived value of q (Equation (10)). Once obtained, d and q can be used to calculate K_{CD}^{Fe} and the entire cation distribution.

Experimental results for $Mg_xFe_{3-x}O_4$ [Trestman-Matts et al., 1983b] are displayed in Figure 4a. The close resemblance of the solid solution behavior to that of Fe_3O_4 argues in favor of small polaron conduction. Fortunately, at 650°C, quench/magnetization experiments were performed by Pucher [1971]. These provide values of K_{CD}^{Mg} from which d, q, and K_{CD}^{Fe} can be determined. The resulting cation distribution is displayed in Figure 4b. Apparently, Mg^{2+} drives K_{CD}^{Fe} increasingly positive (inverse) as ξ, or x in $Mg_xFe_{3-x}O_4$, increases. Thermodynamically, K_{CD}^{Mg} obeyed the following relationship:

$$- RT \ln K_{CD}^{Mg} = 2093R + 2(- 1973)R(b) \quad (15)$$

although the tetrahedral Fe^{3+} concentration (b) varies only slightly across the solution.

The cation distribution in Figure 4b is consistent with previous electrical and thermodynamic observations in the $Mg_xFe_{3-x}O_4$ system. In quenched specimens there should be no further relaxation of the iron distribution for $x \geq 0.4$ at temperatures below 650°C. The iron distribution is totally "inverse" below this temperature. Schröder [1963] measured conductivity and thermopower for $x = 0$ (Fe_3O_4), $x = 0.5$ and $x = 0.8$ in the vicinity of room temperature. Whereas Q was temperature dependent for $x = 0$, it was not for $x = 0.5$ and $x = 0.8$. Furthermore, the activation energy of the conductivity ($d \ln \sigma T/dT^{-1}$) was 0.13 eV at these two concentrations, which is nearly identical to the small polaron hopping energy for magnetite [see Dieckmann et al., 1983]. This is what would be expected on the basis of Figure 4b. Additionally, thermodynamic activities calculated from the cation distribution show a small positive deviation from ideality. This is consistent with experimental observations [see Shishkov et al., 1980].

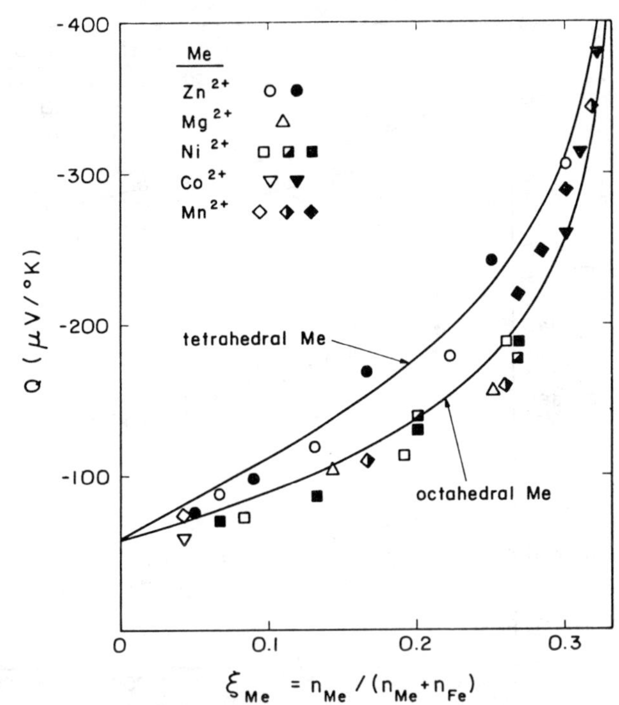

Fig. 3. Comparison of measured and calculated low temperature thermopower for divalent substituents in Fe_3O_4. References for the data shown are as follows: solid circle, Gillot [1982]; open circle and solid square, Samokhvalov and Rustamov [1965]; open square, Griffiths et al. [1970]; half-solid square, Jefferson and Baker [1968]; open triangle, Schröder [1963]; inverted open triangle and open diamond, Constantin and Rosenberg [1971]; inverted solid triangle, Jonker [1959]; half-solid diamond, Simsa [1966]; and solid diamond, Lotgering [1964].

Trivalent Substituents (Cr, Al)

Using the formulae of Table 1, row 5, columns 3 and 4, thermoelectric behavior for trivalent

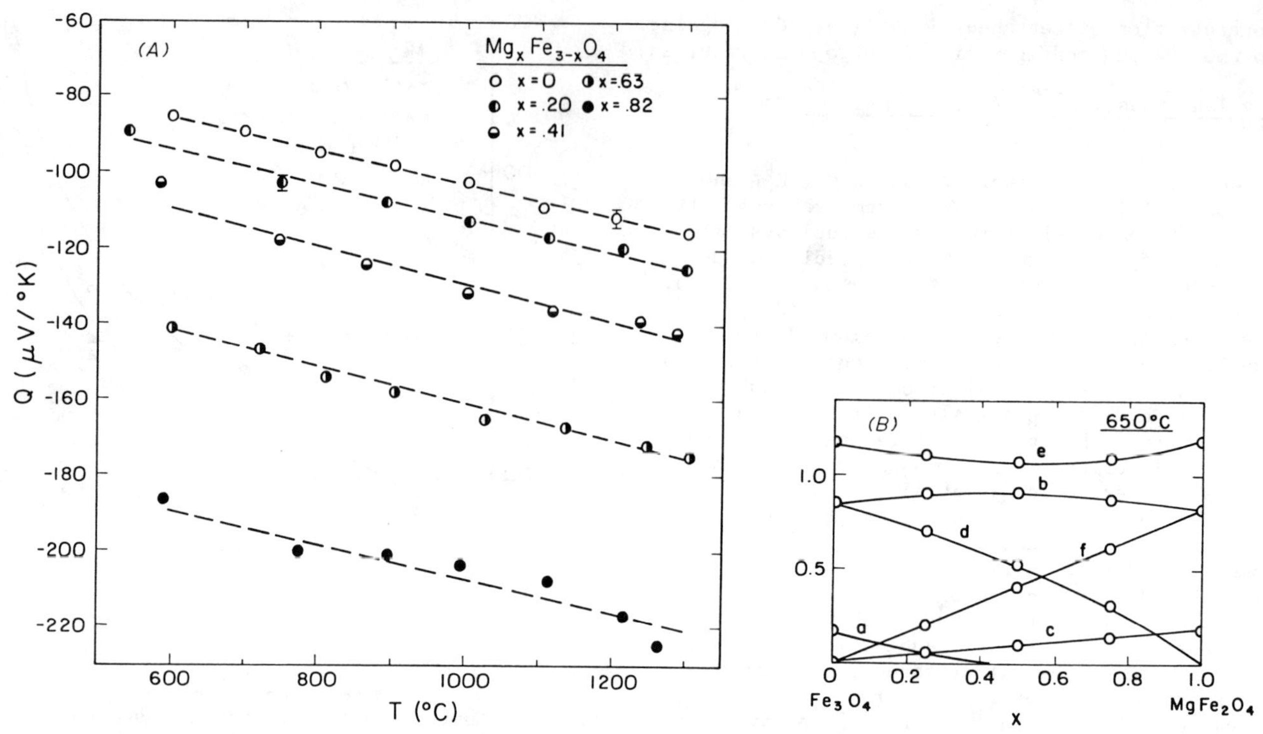

Fig. 4. (a) Experimental thermopower and (b) cation distribution (650°C) in the system $Mg_xFe_{3-x}O_4$. Cation distribution calculated at $x = 0.25$ increments from both saturation magnetization [Pucher, 1971] and interpolated thermopower data [from Trestman-Matts et al., 1983b].

substituents assumed to be in octahedral sites can be predicted via equation (10). The results are displayed in Figure 5, where "inverse iron" corresponds to $K_{CD}^{Fe} \rightarrow \infty$ and "random iron" corresponds to $K_{CD}^{Fe} = 1$. The former would be anticipated at low temperatures, whereas the latter would be anticipated at high temperatures. Both Cr^{3+} and Al^{3+} strongly prefer the octahedral sublattice [Navrotsky, 1972]. Gillot [1982] recently reported room temperature thermopower data for Fe_3O_4-$FeCr_2O_4$. These data agree well with our "inverse iron" curve. Mason and Bowen [1981] reported high temperature (1430°C) thermopower for the system Fe_3O_4-$FeAl_2O_4$, which correspond well with our "random iron" predicted behavior. This observation is somewhat difficult to reconcile with findings for divalent and tetravalent substituents, where K_{CD}^{Fe} is a strong function of disorder. Unfortunately, neither the $Fe_{3-x}Cr_xO_4$ nor $Fe_{3-x}Al_xO_4$ study were performed as a function of temperature to test the temperature and composition dependence of K_{CD}^{Fe}.

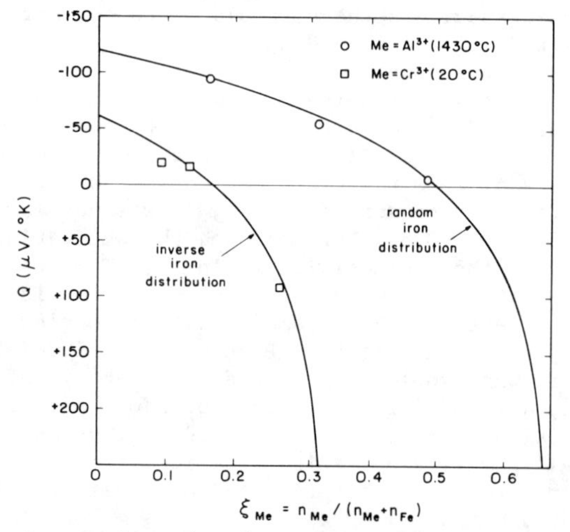

Fig. 5. Predicted and experimental low temperature and high temperature thermopower behavior in systems involving trivalent substituents in Fe_3O_4 (squares, Gillot [1982]; circles, Mason and Bowen [1981]).

Tetravalent Substituent (Fe_3O_4-Fe_2TiO_4)

The unique aspect about Ti^{4+} in Fe_3O_4 is that it resides octahedrally only. This greatly simplifies interpretation of the thermopower data shown in Figure 6 [Trestman-Matts et al., 1983c]. Once again, behavior mimics that of Fe_3O_4, albeit at increasingly more positive values. This supports our contention that the octahedral small polaron model predominates at least to $x = 0.7$ in $Fe_{3-x}Ti_xO_4$. We have converted the thermopower data into cation distributions at three temperatures in Figure 7. Thermodynamically, these distributions obey the relationship:

$$-RT \ln K_{CD}^{Fe} = -1156R + 2(-1.85T)R(b) + 3.079RT \quad (16)$$

As can be seen in Figure 7, there is a pronounced $b = [Fe^{3+}_{tet}]$ dependence on composition. Unlike $Mg_xFe_{3-x}O_4$, therefore, a good test of the O'Neill and Navrotsky [1983b] model can be made. Two novel aspects of our findings in the Fe_3O_4-Fe_2TiO_4 system are a temperature dependence of β in

$$-RT \ln K_{CD}^{Fe} = \alpha + 2\beta\,(b) \quad (17)$$

and a temperature dependence of α. This latter finding may indicate a non-configurational entropy contribution to the energy of disorder.

Using equation (16), low temperature cation distributions were generated [Trestman-Matts et al., 1983c]. Magnetometry results in the literature plus lattice parameter versus composition behavior agreed well with predictions based on our model. Furthermore, thermodynamic activities calculated from our 1300°C distribution, with the incorporation of a subregular solution term, agreed well with actual measurements. At intermediate temperatures a solvus was generated which corresponded well with experimental values of T_c and x_c in $Fe_{3-x}Ti_xO_4$.

Conclusions

The thermoelectric coefficient provides a rapid and reliable means to measure in situ the

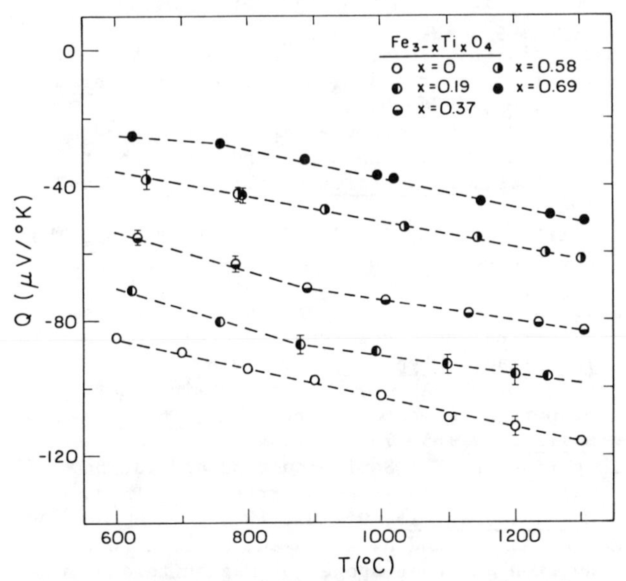

Fig. 6. Experimental thermopower as a function of temperature in the system $Fe_{3-x}Ti_xO_4$ [from Trestman-Matts et al., 1983c].

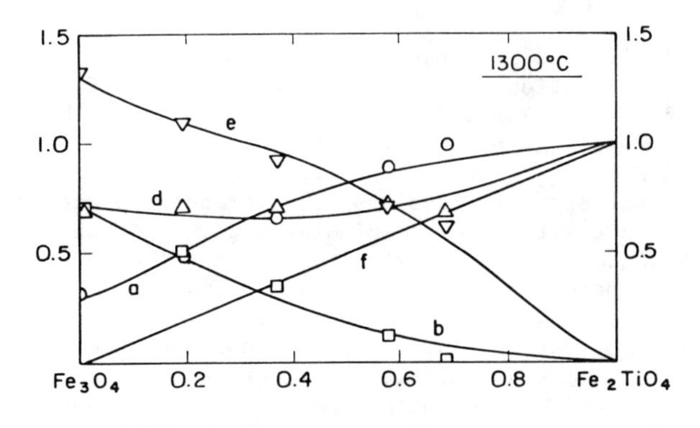

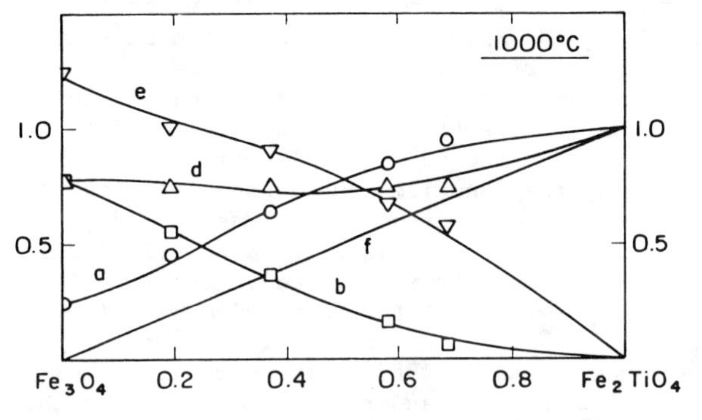

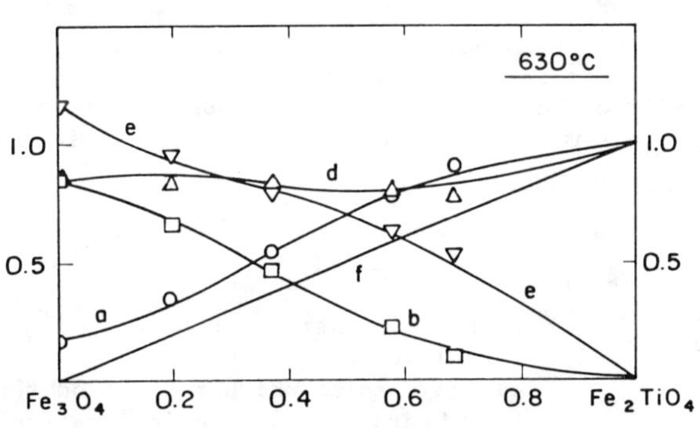

Fig. 7. High temperature cation distributions in the system $Fe_{3-x}Ti_xO_4$.

high temperature cation distributions in ferrospinels. The multiprobe apparatus greatly facilitates collection of data from a large number of samples. The major prerequisite for the technique is small polaron conduction on only one sublattice. A secondary consideration is which sublattice is occupied by the substituting cation. Where both sublattices are occupied to some degree, at least one independent measurement of K_{CD}^{Me} is necessary to determine the cation distribution thermoelectrically.

The data we have reported to date suggest that the O'Neill and Navrotsky [1983a, b] model properly addresses the disorder dependence of the cation distribution constant. Systems like Fe_3O_4-Fe_2TiO_4, where the distribution changes so dramatically with composition, are important to test such models. Novel features seen in our studies include temperature dependence of the

O'Neill and Navrotsky β factor, plus a non-configurational entropy term whenever K_{CD}^{Fe} is involved. For a discussion of this latter phenomenon, the papers by O'Neill and Navrotsky [1983a, b] are recommended.

Current work involves other transition metals which crystallize as spinels, e.g., Co_3O_4 and Mn_3O_4. Preliminary studies indicate that Co_3O_4 does not meet the requirement of small polaron conduction. Therefore, thermoelectric determination of cation distribution is not possible. Preliminary calculations indicate that Mn_3O_4 is a small polaron conductor, however via tetrahedral sites. This might mean that cation distributions in Mn_3O_4 and Mn_3O_4-based spinels can be studied thermoelectrically.

Finally, work in other T.M. oxides indicates that point defects other than cation redistribution can be studied thermoelectrically. For example, the monoxide FeO has been studied thermoelectrically [Gartstein and Mason, 1982]. The defect structure consists of vacancy-interstitial clusters in a matrix enriched in Fe^{3+}. Defects in the perovskite $LaFeO_3$ have also been investigated with thermoelectric coefficient measurements [Mizusaki et al., 1982]. The work by Sockel [1974] on Fe_2SiO_4 and results presented at this conference on the Mg_2SiO_4-Fe_2SiO_4 system [Schock and Duba, this volume] suggest that small polaron conduction is operative from fayalite out to 80-90 percent forsterite. This suggests that olivines might be amenable to thermoelectric study.

Notation

- a,b,c,d,e,f cation concentrations per spinel formula unit.
- A cation species which resides on the tetrahedral sublattice in a perfectly normal spinel.
- B cation species which resides on the octahedral sublattice in a perfectly normal spinel.
- E_a activation energy for electrical conduction.
- K_{CD} equilibrium constant for the cation distribution reaction.
- q the octahedral valence ratio, i.e., the ratio of trivalent iron to divalent iron on the octahedral sublattice.
- Q the thermoelectric power, or Seebeck coefficient.
- T_C consolute temperature.
- x composition parameter.
- x_c consolute composition.
- ΔV voltage drop developed in a sample subjected to a temperature gradient.
- ΔT temperature gradient established across a sample.
- α composition independent portion of the enthalpy for a cation distribution reaction.
- β composition dependent portion of the enthalpy for a cation distribution reaction.
- ξ cation fraction of substituent.
- σ electrical conductivity ($\Omega^{-1}cm^{-1}$).

Acknowledgements. The author acknowledges support for work on Fe_3O_4 from the Materials Research Center of Northwestern University (NSF-MRL program, grant DMR79-23573) and for the ferrospinel studies from the National Science Foundation under grant DMR 8106492. Experimental work was performed by C. C. Wu, S. Kumarakrishnan, A. Trestman-Matts, and S. E. Dorris. Thanks to T. Witt, G. Sykora, and E. Gartstein for help with the manuscript.

References

Buddington, A. F., and D. H. Lindsley, Iron-titanium oxide minerals and synthetic equivalents, J. Petrol., 5, 310-357, 1964.

Constantin, C., and M. Rosenberg, Thermoelectric power of pure and substituted magnetite above and below the Verwey transition, Solid State Commun., 9, 675-677, 1971.

Cullity, B. D., Introduction to Magnetic Materials, pp. 181-198, Addison-Wesley, Reading, Mass., 1972.

Dieckmann, R., Defect and cation diffusion in magnetite, IV, Nonstoichiometry and point defect structure of magnetite ($Fe_{3-\delta}O_4$), Ber. Bunsenges. Phys. Chem., 86, 112-118, 1982.

Dieckmann, R., C. A. Witt, and T. O. Mason, Defects and cation diffusion in magnetite, V, Electrical conduction, cation distribution and point defects in $Fe_{3-\delta}O_4$, Ber. Bunsenges. Phys. Chem., 87, 495-503, 1983.

Gartstein, E., and T. O. Mason, Reanalysis of wustite electrical properties, J. Am. Ceram. Soc., 65, C-24-26, 1982.

Gillot, B., Thermopower-composition dependence in iron-chromium, iron-zinc, and iron-cobalt ferrites, Phys. Status Solidi, A, 69, 719-725, 1982.

Griffiths, B. A., D. Elwell, and R. Parker, The thermoelectric power of the system $NiFe_2O_4$-Fe_3O_4, Philos. Mag., 22, 163-174, 1970.

Jefferson, C. F., and C. K. Baker, Mechanism of electrical conductivity in nickel-iron ferrite, IEEE Trans. Magn., 4, 460, 1968.

Jonker, G. H., Analysis of the semiconducting properties of cobalt ferrite, J. Phys. Chem. Solids, 9, 165-175, 1959.

Lotgering, F. K., Semiconduction and cation valencies in manganese ferrites, J. Phys. Chem. Solids, 25, 95-103, 1964.

Mason, T. O., and H. K. Bowen, Electronic conduction and thermopower of magnetite and iron-aluminate spinels, J. Am. Ceram. Soc., 64, 237-242, 1981.

Mizusaki, J., T. Sasamoto, W. R. Cannon, and H. K. Bowen, Electronic conductivity, Seebeck coefficient, and defect structure of $LaFeO_3$, J. Am. Ceram. Soc., 65, 363-368, 1982.

Navrotsky, A., Thermodynamics of binary and ternary transition metal oxides in the solid state, in Transition Metals, edited by D. W. A. Sharp, University Park Press, Baltimore, Md., 1972.

O'Neill, H. S. C., and A. Navrotsky, Simple spinels: Crystallographic parameters, cation radii, lattice energies, and cation distributions, Am. Mineral., 68, 181-194, 1983a.

O'Neill, H. S. C., and A. Navrotsky, Cation distributions and thermodynamic properties of binary spinel solid solutions, Am. Mineral., in press, 1983b.

Pelton, A. D., H. Schmalzried, and J. Sticher, Thermodynamics of Mn_3O_4-Co_3O_4, Fe_3O_4-Co_3O_4, Fe_3O_4-Mn_3O_4 spinels by phase diagram analysis, Ber. Bunsenges. Phys. Chem., 83, 241-252, 1979.

Pucher, R., Magnetic and X-ray diffraction measurements of the synthetic spinel system $FeFe_2O_4$-$MgFe_2O_4$-$NiFe_2O_4$, Z. Geophys., 37, 344-356, 1971.

Rossing, B. R., and H. K. Bowen, Materials for open cycle MHD channels, in Critical Materials Problems in Energy Production, edited by C. Stein, Academic, New York, 1976.

Sachse, H. B., Semiconducting Temperature Sensors and Their Application, pp. 45-52, John Wiley, New York, 1975.

Samokhvalov, A. A., and A. G. Rustamov, Electrical properties of ferrite spinels with a variable content of divalent iron ions, Sov. Phys. Solid State Engl. Transl., 7, 961-966, 1965.

Schock, R. N., and A. G. Duba, Point defects and the mechanisms of electrical conduction in olivine, in Point Defects in Minerals, Geophys. Monogr. Ser., AGU, Washington, D. C., this volume.

Schröder, H., Electrical properties and redox characteristics in magnesium-iron ferrite solution, Ph.D. thesis, Univ. of Jena, Jena, German Democratic Republic, 1963.

Shishkov, V. I., A. A. Lykasov, and A. F. Il'ina, Activity of the components of iron-magnesium spinel, Russ. J. Phys. Chem. Engl. Transl., 54, 440-441, 1980.

Simsa, Z., The thermoelectrical properties of manganese ferrites, J. Phys. B, 16, 919-928, 1966.

Sockel, H. G., Defect structure and electrical conductivity of crystalline ferrous silicate, in Defects and Transport in Oxides, edited by M. S. Seltzer and R. I. Jaffe, pp. 341-356, Plenum, New York, 1974.

Trestman-Matts, A., S. E. Dorris, and T. O. Mason, Measurement and interpretation of thermopower in oxides, J. Am. Ceram. Soc., 66, 589-592, 1983a.

Trestman-Matts, A., S. E. Dorris, and T. O. Mason, Thermoelectric determination of cation distributions in Fe_3O_4-$MgFe_2O_4$, J. Am. Ceram. Soc., in press, 1983b.

Trestman-Matts, A., S. E. Dorris, S. Kumarakrishnan, and T. O. Mason, Thermoelectric determination of cation distributions in Fe_3O_4-Fe_2TiO_4, J. Am. Ceram. Soc., in press, 1983c.

Turnock, A. C., and H. P. Eugster, Fe-Al oxides: Phase relationships below 1000°C, J. Petrol., 3, 533-565, 1962.

Wu, C. C., and T. O. Mason, Thermopower measurement of cation distribution in magnetite, J. Am. Ceram. Soc., 64, 520-522, 1981.

Wu, C. C., S. Kumarakrishnan, and T. O. Mason, Thermopower composition dependence in ferrospinels, J. Solid State Chem., 37, 144-150, 1979.

HIGH PRESSURE ELECTRICAL CONDUCTIVITY IN NATURALLY OCCURRING SILICATE LIQUIDS

James A. Tyburczy and Harve S. Waff

Department of Geology, University of Oregon
Eugene, Oregon 97403

Abstract. Electrical conductivities of molten Hawaiian rhyodacite and Yellowstone rhyolite obsidian were measured between 1200° C and 1400° C and at pressures up to 25 kilobars. The two melts exhibit similar trends. Arrhenius behavior is observed at all pressures studied. Isobaric activation enthalpies increase from about 0.5 eV at atmospheric pressure to about 0.9 eV at 25 kbars, and the magnitude of the conductivity decreases by about a factor of 4 between 0 and 25 kbar. At pressures between about 10 and 15 kbar an abrupt decrease in the slopes of isothermal log σ versus pressure plots is observed. In each pressure range an equation of the form $\sigma = \sigma_0' \exp[-(E_\sigma' + P\Delta V_\sigma')/kT]$, where σ_0', E_σ', and $\Delta V_\sigma'$ are constants, describes the polybaric, polythermal data. Comparison of these data with high pressure electrical conductivities of molten basalt and andesite reveals that relatively silica-rich melts, from andesitic to rhyolitic in composition, display similar trends, while the basaltic melt has analogous, but quantitatively different trends. Comparison of zero-pressure electrical conductivity and sodium diffusivity by means of the Nernst-Einstein relation indicates that sodium ion transport is the dominant mechanism of charge transport in the obsidian melt at zero pressure. The tholeiitic melt, on the other hand, displays only order of magnitude agreement between the electrical conductivity and sodium diffusivity, indicating that either ions other than sodium play a significant role in electrical transport or that the motions of the sodium ions are strongly correlated, or both. Comparison of the isobaric and isochoric activation enthalpies indicates that electrical conduction is energy restrained, as opposed to volume restrained. Conductivities in the andesitic, rhyodacitic, and rhyolitic melts conform to a single compensation law line, with no indication of the change in activation volume. The tholeiitic melt has a slightly different compensation line. In light of Jambon's (1982) semiempirical activation energy theory, this fact indicates that under all conditions studied ionic conduction is due to transport of cations of similar size, and that these conducting ions are small relative to the size of an oxygen anion. The changes in activation volume occur at roughly the same pressures that changes in liquidus mineral assemblage occur. Changes in melt polymerization may or may not be reflected in changes in liquidus phase degree of polymerization.

Introduction

Measurements of the electrical conductivity of molten natural silicates can yield information on ionic transport in these liquids. At low pressures the conductivity of molten basalts and andesites appears to be strongly influenced by sodium ion transport [Oppenheim, 1968, 1970; Waff and Weill, 1975]. Our previous work indicates that this is probably true at pressures up to 25 kbar, with the added conclusion that melt structure and melt structural changes with pressure exert an influence on the mobilities of the modifying cations [Tyburczy and Waff, 1983]. This conclusion contrasts with the conclusion of Jambon [1982] that zero-pressure ionic diffusivities in obsidian and other silica-rich glasses and melts are independent of the matrix.

Relatively few measurements of ionic diffusivities in natural or complex silicates at high pressures exist. Watson [1979, 1981] has measured tracer diffusion coefficients of Ca and Cs in a sodium-calcium-aluminosilicate melt at high pressure. The measured diffusivities decrease monotonically with pressure up to about 30 kbar. Dunn [1983] found that the oxygen chemical diffusion coefficient in several basaltic melts shows sharp discontinuities as a function of pressure. He correlated the jumps in oxygen diffusivity with changes in liquidus phase, even though the measurements were made above the liquidus temperature.

We have measured the electrical conductivities of molten Hawaiian rhyodacite and Yellowstone rhyolite obsidian between 1200° C and 1400° C and at pressures up to 25 kbar. These measurements coupled with previous measurements to high pressures on a basaltic and an andesitic melt

TABLE 1. Chemical Compositions of the Rock Melts Studied in Oxide Weight Per Cent

Oxide	YRO Yellowstone Obsidian	HR-1 Hawaiian Rhyodacite
SiO_2	78.07	68.00
TiO_2	0.13	0.58
Al_2O_3	12.04	15.29
FeO*	1.41	3.00
MnO	0.04	0.25
MgO	0.00	2.00
CaO	0.43	2.76
Na_2O	3.80	4.59
K_2O	5.08	3.53
Total	101.00	100.00

[Tyburczy and Waff, 1983] allow us to compare conductivity trends with pressure and temperature for a broad compositional range of naturally occurring silicate liquids.

Experimental Procedures

The experimental techniques used in this work have been described in detail by Tyburczy and Waff [1983]. Zero-pressure electrical conductivity measurements were made using the loop technique [Waff, 1976] in a CO_2 atmosphere. The high-pressure measurements were made in an internally heated, solid medium, piston-cylinder apparatus [Haygarth et al., 1969]. A cylindrical core of fused sample approximately 3 mm in length and 3 mm in diameter was inserted into a fused quartz ring and placed between Pt 70%/Rh 30% end electrodes. The remainder of the sample chamber assembly, both inside and outside the graphite furnace, was composed of soft-fired pyrophyllite (fired at 930° C for 3 to 6 hours). The sample impedance was measured with a frequency selective voltmeter. No frequency dependence was observed between 50 Hz and 10 kHz. All reported measurements were made at 2000 Hz. Temperature was controlled to within ±5° C with a constant-power controller and measured with a Pt 70-Rh 30/Pt 94-Rh 6 thermocouple. The pressure values reported reflect a -15% friction correction. In each experimental run the sample was brought up to a reference temperature and pressure (commonly 1400° C and 8.5 kbar) and held there until the resistance reading stabilized. Measurements were then taken by cycling T and P, with frequent returns to the reference conditions. Measurements were made at 50° C intervals between 1200° C and 1400° C and at 4.3 kbar intervals between 0 and 25.5 kbar. Resistance readings were converted into electrical conductivity values by correcting for parallel resistance and minor geometrical distortions in the manner described previously.

Electron microprobe analyses of the rock melts studied in this work are listed in Table 1. The analyses were done using an electron microprobe. The melts studied were a Hawaiian rhyodacite, HR-1, and a Yellowstone rhyolite obsidian, YRO. Post-run compositional analyses demonstrate that little chemical contamination from the silica ring occurred. The pressure and temperature conditions employed during the runs extended above and below the liquidus of the rock melts. No evidence for crystallization was detected during the runs, nor do the data show any kinks or inflections at the liquidus.

Duplicate runs were performed on each melt, and the results agree to within 0.10 units in $\log_{10} \sigma$. The absolute accuracy of the measurements, taking into account resistance measurement, leakage, and geometrical factors, is estimated to be about ±15%.

Results

Figure 1 is a plot of log σ versus 1/T at 8.5 kbar for the molten Hawaiian rhyodacite, HR-1. The data for HR-1 at other pressures, and the high pressure data for the Yellowstone obsidian melt, YRO, show similar reproducibility. At each pressure Arrhenius behavior is observed, i.e.,

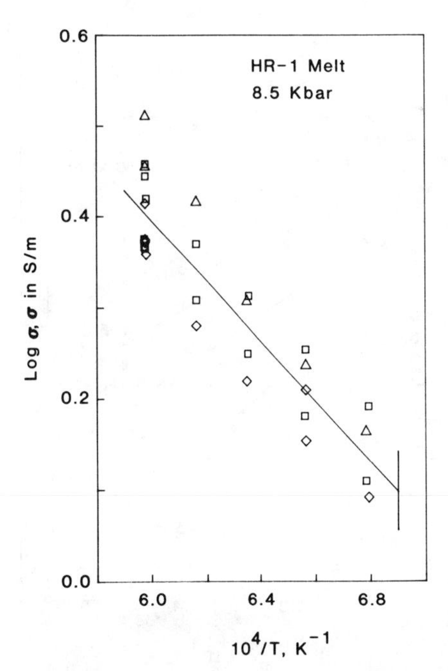

Fig. 1. Log σ versus 1/T at 8.5 kbar for the molten Hawaiian rhyodacite, HR-1. Different symbols represent different experimental runs. The line is the least squares fit to the Arrhenius equation, equation (1). The vertical bar is plus or minus one standard error of regression.

$$\log \sigma = \log \sigma_0 - \frac{H_{\sigma,p}}{2.3kT} \qquad (1)$$

where σ_0 is a preexponential constant, $H_{\sigma,p}$ is the isobaric activation enthalpy, and k is the Boltzmann constant. Figures 2 and 3 show the isobaric Arrhenius fits for the two rock melts, and the parameters for these fits are given in Table 2. The zero-pressure results for YRO agree well with the results of Murase and McBirney [1973] for a melt of similar composition. The conductivity decreases with increasing pressure, dropping to about 1/4 of its zero-pressure value at 25 kbar. Activation enthalpies and preexponential constants increase with increasing pressure. In each rock melt the pressure dependence is smaller at higher pressures. This is shown more clearly in Figure 4, which shows isothermal log σ versus pressure data at 1400° C for molten HR-1. A distinct change in slope occurs between about 9 and 12 kbar. This change in slope occurs at all temperatures investigated (Figure 5), and analogous changes occur for molten YRO at about 13 to 18 kbar (Figure 6). Molten tholeiite and andesite also show a discontinuity in the pressure dependence of the electrical conductivity [Tyburczy and Waff, 1983]. Accordingly, each data set was broken into a low-pressure subset and a high-pressure subset. For each polythermal, polybaric subset the data has been fit to an equation of the form

$$\log \sigma = \log \sigma_0' - (E_\sigma' + P\Delta V_\sigma')/kT \qquad (2)$$

by means of multiple linear regression [Bevington, 1969]. In equation (2) σ_0' is a preexponential constant, E_σ' is the activation volume, and $\Delta V_\sigma'$ is the activation volume. Primed symbols are used in equation (2) to distinguish the parameters from those in equation (1). The fits are shown in Figures 5 and 6 and the parameters are listed in Table 3. F-test statistics indicate that the reduction in variance brought about by breaking the data into two subsets is significant at the 99% level [Green and Margerison, 1978]. For YRO and HR-1, the activation energy E_σ' increases by about 50% at the transition pressure, while the activation volume $\Delta V_\sigma'$ decreases by about the same amount. The parameters derived by fitting equation (1) to an isobaric data set differ from those derived by fitting equation (2) to a polybaric data set. If $H_{\sigma,p}$ varied linearly with pressure and σ_0 were independent of pressure, then the parameters obtained using either equation would be the same ($H_{\sigma,p}$ would be equal to $E_\sigma' + P\Delta V_\sigma'$). However, σ_0 does vary slightly with pressure (Table 2); thus the parameters derived using each of the two forms differ slightly. Equation (2) is a useful descriptive representation of the data over a particular pressure and temperature interval. For analysis

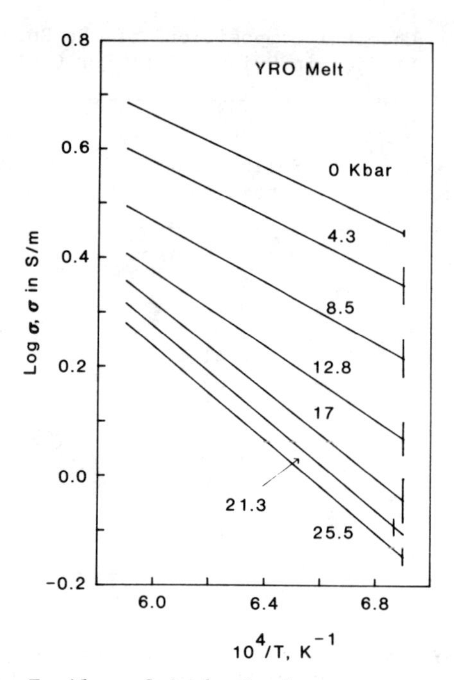

Fig. 3. Family of isobaric least squares fits to the Arrhenius equation, equation (1), for the molten Yellowstone rhyolite obsidian, YRO. Parameters for the fits are given in Table 2. Vertical bars represent plus or minus one standard error of regression.

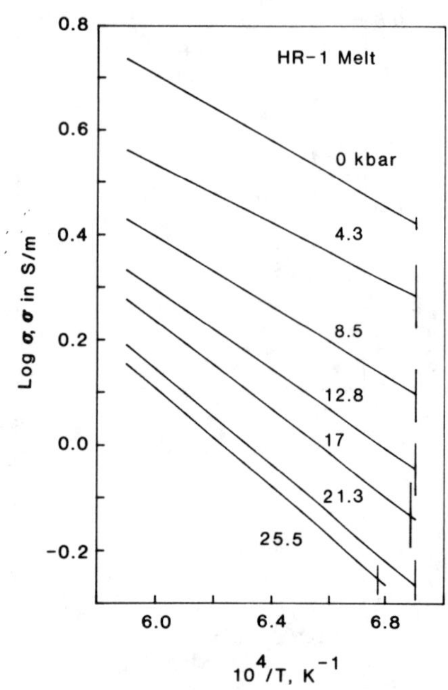

Fig. 2. Family of isobaric least squares fits to the Arrhenius equation, equation (1), for the molten Hawaiian rhyodacite, HR-1. Parameters for the fits are given in Table 2. Vertical bars represent plus or minus one standard error of regression.

TABLE 2. Isobaric Activation Enthalpies $H_{\sigma,p}$ and Preexponential Terms Log σ_o for Electrical Conductivity Data Fit to the Equation $\log \sigma = \log \sigma_o - H_{\sigma,p}/2.3kT$

Pressure, kbar	log σ_o, σ_o in s/m	95% Confidence Interval for log σ_o	$H_{\sigma,p}$, eV	95% Confidence Interval for $H_{\sigma,p}$	Standard Error Estimate for log σ
		YRO			
0.0	2.09	0.04	0.473	0.01	0.003
4.3	2.07	0.29	0.495	0.09	0.034
8.5	2.13	0.32	0.552	0.10	0.033
12.8	2.40	0.26	0.672	0.08	0.030
17.0	2.73	0.39	0.796	0.12	0.041
21.3	2.80	0.16	0.836	0.05	0.015
25.5	2.80	0.17	0.848	0.05	0.015
		HR-1			
0.0	2.58	0.05	0.621	0.02	0.005
4.3	2.19	0.44	0.548	0.14	0.058
8.5	2.38	0.29	0.657	0.09	0.045
12.8	2.56	0.50	0.749	0.16	0.048
17.0	2.74	0.53	0.829	0.17	0.061
21.3	2.87	0.54	0.902	0.17	0.047
25.5	2.90	0.23	0.922	0.07	0.021

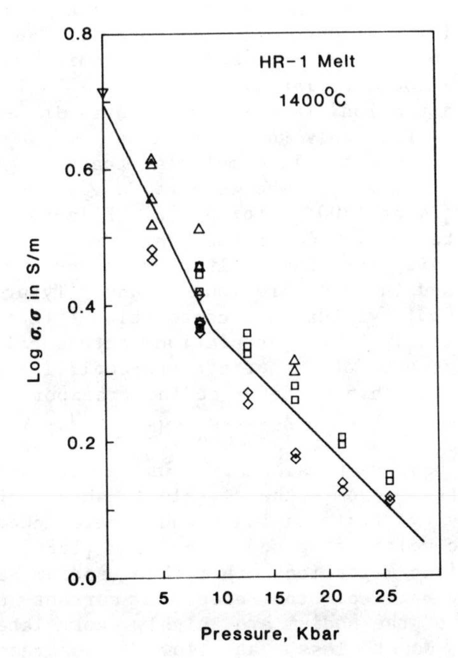

Fig. 4. Isothermal log σ versus pressure at 1400° C for the molten Hawaiian rhyodacite, HR-1. Different symbols represent different experimental runs. The solid line is the best fit to equation (2) in the low- and high-pressure regions, respectively.

of the data in terms of transport processes and melt structure, most theoretical and quasi-thermodynamic treatments have been done in terms of equation (1) and the corresponding isochoric and isothermal equations, equations (7) and (9), respectively [e.g., Nachtrieb, 1980]. Watson [1979, 1981] employed an equation similar to equation (2) to describe high-pressure self diffusion in silicate melts, but allowed the activation volume to be a linear function of temperature. For the data reported here such temperature coefficients are small in magnitude and negative in sign, and do not yield much improvement in the quality of the fit to the data. Therefore we employ the temperature-independent form of the activation volume in this work.

Discussion

If the electrical conductivity in a given melt is largely through transport of a given ion, the self-diffusion coefficient of the ion and the conductivity are related through the Nernst-Einstein equation,

$$\sigma = Dn \mid z \mid e^2/kT \tag{3}$$

where D is the self-diffusion coefficient, n is the number of mobile cations per cm^3, z is the charge of the ion, e is the charge on the electron, k is the Boltzmann constant, and T is the absolute temperature. The degree to which the Nernst-Einstein equation holds is given by the correlation factor f,

$$f = D_{meas} n \mid z \mid e^2 / \sigma_{meas} kT \qquad (4)$$

where D_{meas} and σ_{meas} are experimentally measured values. For simple silicate melts which conduct via sodium ion transport, correlation factors on the order of 0.3 to 0.5 have been determined experimentally. Correlation factors of this magnitude have been shown to be due to a correlation of ionic jumps in either a vacancy mechanism or an indirect interstitial mechanism [Haven and Verkerk, 1965].

Carron [1968] found good agreement (f = 0.76 to 1.21) between electrical conductivity and the sodium tracer diffusion coefficient in obsidian glasses between 645° C and 830° C. Extrapolating the sodium tracer diffusion coefficient values of Jambon [1982], determined in the range 140° C to 850° C, up to the 1200° C to 1400° C range, and using the conductivity values for the molten obsidian reported here, yields values of f between about 0.9 and 1.1. Extrapolating over this large temperature interval is reasonable because in granitic systems Arrhenius plots are linear over broad temperature ranges (in the case of Na a range of at least 700° C) and because in this system no discontinuities in diffusivity were observed in the glass transition region

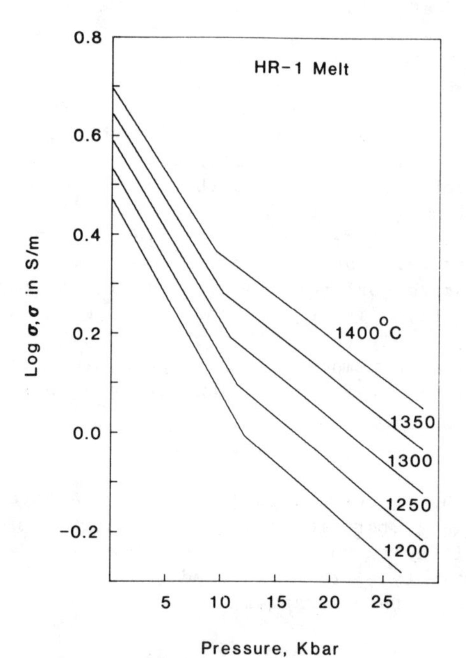

Fig. 5. Family of log σ versus pressure lines for the molten Hawaiian rhyodacite, HR-1. The lines represent the best fit to equation (2) in the low- and high-pressure regions, respectively. Parameters for the lines are given in Table 3.

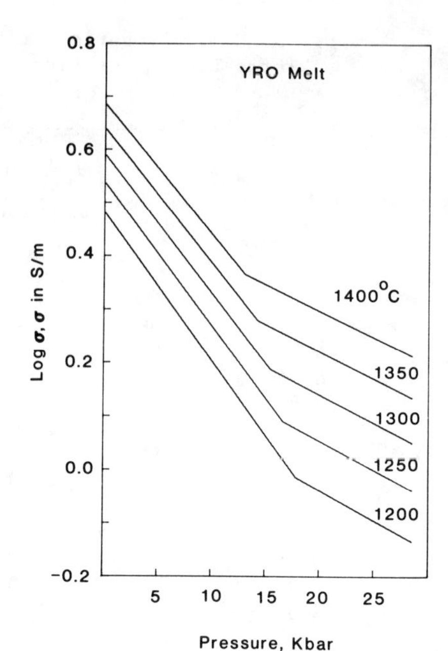

Fig. 6. Family of log σ versus pressure lines for the molten Yellowstone obsidian, YRO. The lines represent the best fit to equation (2) in the low- and high-pressure regions, respectively. Parameters for the lines are given in Table 3.

(approximately 650 to 750° C) [Jambon, 1982]. Thus, in obsidian melts at zero pressure conductivity appears to be totally caused by sodium ion transport, and the ionic jumps are probably not correlated.

The situation is not so clear in basaltic melts. The only determination of Na tracer diffusion in a basaltic melt is that of Hofmann and Brown [1976], who report a value of 2.45 x 10^{-6} cm^2/s at 1300° C for a 1921 Kilauea olivine tholeiite melt. Combining this value with the conductivity for the molten Hawaiian tholeiite determined in our previous study [Tyburczy and Waff, 1983] yields a correlation factor of approximately 0.1. Correlation factors calculated for transport of other network-modifying cations are smaller than that for sodium transport, but of the same order of magnitude ($f_{Mg} \sim 0.09$, $f_{Ca} \sim 0.06$, $f_{Fe} \sim 0.07$, $f_K \sim 0.01$). These correlation factors have been calculated using diffusivities calculated from the relationship between diffusivity, ionic radius, and ionic charge in basaltic melts proposed by Hofmann [1980]. Thus, in basaltic melts ions other than sodium may be carrying part of the electric current or the motions of the sodium are highly correlated, or both. Nonetheless, at low pressures the electrical conductivity and sodium diffusivity agree to within about an order of magnitude. Thus, low-pressure electrical conductivity values for basaltic melts [Waff and Weill, 1975; Rai and Manghnani, 1977; Tyburczy and Waff, 1983] may be used to estimate sodium diffusivity values in

TABLE 3. Parameters for the Fits to the Equation $\log \sigma = \log \sigma_0' - (E_\sigma' + P\Delta V_\sigma')/2.3kT$

	YRO		HR-1	
Pressure range, kb	0-12.8	17-25.5	0-8.5	12.8-25.5
$\log \sigma_0'$, σ_0' in s/m	2.19	2.77	2.36	2.75
E_σ', eV	0.50	0.76	0.55	0.74
$\Delta V_\sigma'$, cm^3/mol	7.9	3.2	11.1	5.3
	VC-4W		**HR-1**	
Pressure range, kb	0-8.5	12.8-25.5	0-4.3	8.5-25.5
$\text{Log } \sigma_0'$, σ_0' in s/m	3.00	3.82	5.05	5.33
E_σ', eV	0.78	1.17	1.30	1.53
$\Delta V_\sigma'$, cm^3/mol	17.9	3.3	4.6	-0.1

these melts to within about one order of magnitude.

Figure 7 shows the zero-pressure log σ versus 1/T plots for the four rock melts for which high pressure conductivity data are available. The

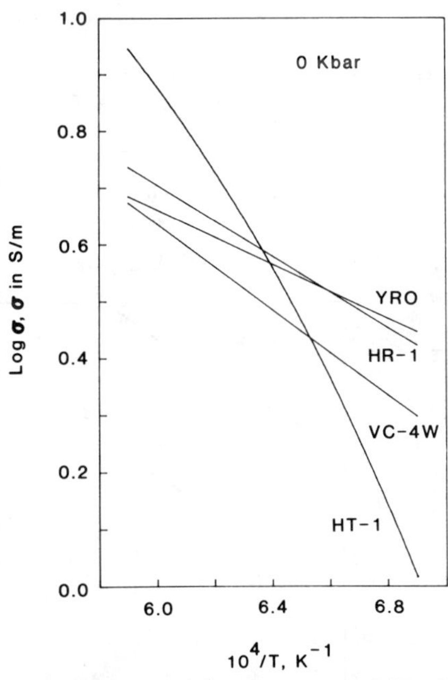

Fig. 7. Zero-pressure log σ versus 1/T plots for the rock melts for which high-pressure electrical conductivity data have been obtained. For the molten andesite (VC-4W), rhyodacite (HR-1), and rhyolite (YRO), the lines represent the best fit to the Arrhenius equation, equation (1). Parameters for HR-1 and YRO are given in Table 2. Parameters for VC-4W are given in Tyburczy and Waff [1983]. The data for HT-1, the Hawaiian tholeiite, are fit to an equation of the form $\log \sigma = A + B/(T - T_0)$ where A, B, and T_0 are constants. Parameters for HT-1 are given in Tyburczy and Waff [1983].

tholeiitic melt, HT-1, shows distinctly different behavior than that of the other three melts. The andesitic, rhyodacitic, and rhyolitic melts exhibit subparallel straight-line trends (see the discussion of the compensation law below). The tholeiitic melt has a curved trend which crosscuts the trends for the other melts. Curved trends have been observed previously for basaltic melts and seem to be related to the iron content of these melts [Waff and Weill, 1975; Rai and Manghnani, 1977]. Jambon [1982] noted that in relatively silica-rich compositions diffusivities are comparatively insensitive to composition, but that in basaltic glasses and melts diffusivities were distinctly different from those in rhyolitic melts. For the silica-rich compositions the electrical conductivity increases with increasing silica content. There is no correlation with sodium or total alkali content. Data on log σ versus 1/T for 17 kbar data are shown in Figure 8. Owing to the different pressure dependences for each melt, the lines are more spread out than at zero pressure. At high pressures the tholeiitic melt has a distinctly different activation enthalpy than the other melts. Figure 9 is a plot of log σ versus P at 1400° C for the four rock melts. The molten tholeiite displays a much smaller activation volume than the other melts. It decreases from about 4.6 cm^3/mole to near 0 cm^3/mole at about 9 kbar (Table 3). The acidic rocks display subparallel conductivity trends with pressure. The activation volume change occurs at about 8, 11, and 14 kbar for the molten andesite, rhyodacite, and rhyolite, respectively. The activation volume changes are from about 18 to about 3.3, 11 to about 5.3, and 7.9 to 3.2 cm^3/mole, respectively. The molar volume of a sodium ion in six-fold coordination with oxygen is about 3.4 cm^3 and that of a sodium ion in eight-fold coordination is about 4.8 cm^3 [Whittaker and Muntus, 1970]. The activation volumes bear no clear relation to the volume of the mobile species.

Figure 10 is a plot of log σ_0 versus $H_{\sigma,p}$ for the rock melts studied, where σ_0 and $H_{\sigma,p}$ are

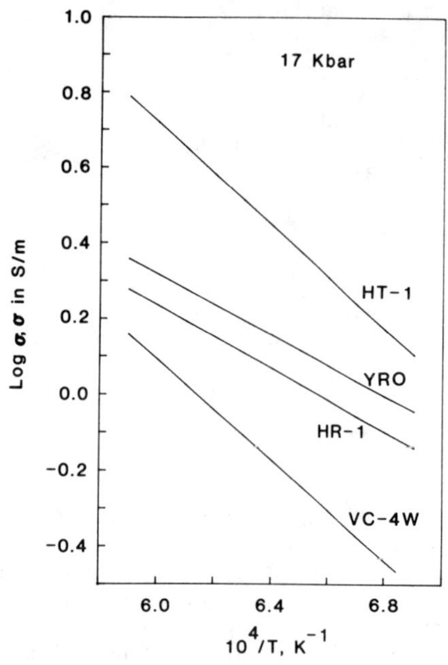

Fig. 8. Log σ versus at 17 kbar for the four rock melts. The lines represent the best fit to the Arrhenius equation, equation (1). Parameters for HR-1 and YRO are given in Table 2. Parameters for HT-1 and VC-4W are given in Tyburczy and Waff [1983].

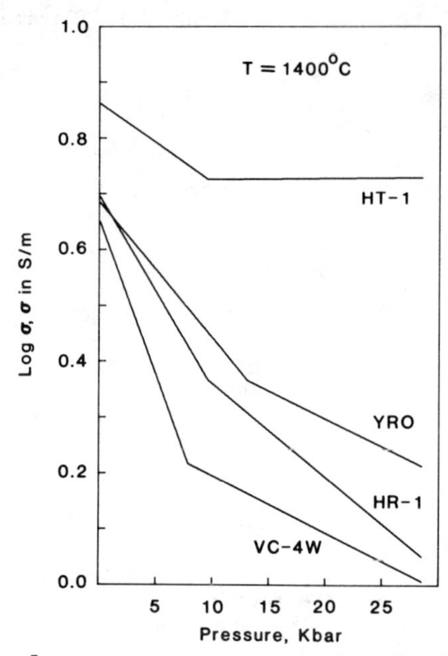

Fig. 9a. Log σ versus pressure for the four rock melts studied. Lines represent best fits to equation (2) in the low- and high-pressure regions, respectively. Parameters for HR-1 and YRO are given in Table 3. Parameters for HR-1 and VC-4W are given in Tyburczy and Waff [1983].

obtained from the isobaric Arrhenius equation, equation (1). The observed linear relationship between σ_0 and $H_{\sigma,p}$ is termed a `compensation law.´ Compensation has been observed for diffusion in a variety of silicate crystals, glasses, and melts [Winchell, 1969; Watson, 1979; Hofmann, 1980; Hart, 1981; Jambon, 1982] and for electrical conductivity in ionically-conducting silicate glasses and melts [Hughes and Isard, 1972; Tyburczy and Waff, 1983]. The correlation is an empirical one, although some theoretical justification has been developed (see discussions in Lasaga [1981], Hart [1981], and Jambon [1982]). For each melt, the compensation law is followed, and no difference between the behavior at low and high pressures is observed. The data for the three silica-rich rocks overlap and may be described by a single line. The equation of the line for the siliceous melts is given by

$$\log \sigma_0 = 0.75 + 0.11\, H_{\sigma,p} \quad (5)$$

where σ_0 is in S/m and $H_{\sigma,p}$ is in kcal/mole. For 21 data points the standard error in $H_{\sigma,p}$ is 1.1 kcal/mole ($r^2 = 0.979$). Jambon [1982] points out that for the compensation law to be significant the linear correlation must be very high. He has developed a semiempirical model of ionic transport through an elastic medium which accounts for trends in diffusivity observed in natural obsidian glasses and melts. In this model the activation energy for tracer diffusion is composed of the

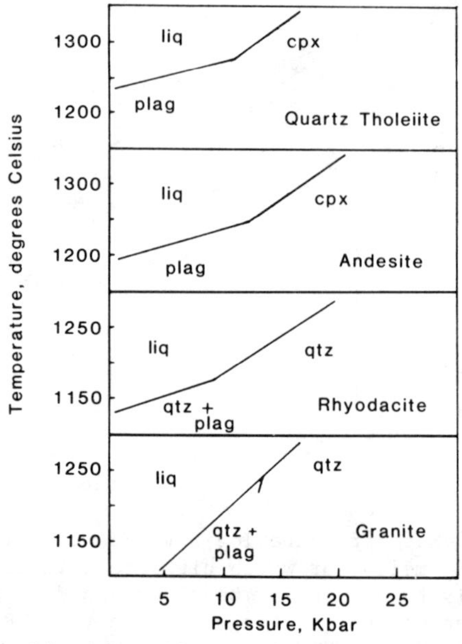

Fig. 9b. Liquidus phase relations for rock melts of composition similar to the ones under study in this work. Quartz tholeiite liquidus phase relations are from Green et al. [1967] and Green and Ringwood [1968]. Andesite liquidus relations are from Green [1972]. Rhyodacite phase relations are from Green and Ringwood [1968]. Granite (rhyolite) phase relations are from Stern and Wyllie [1981].

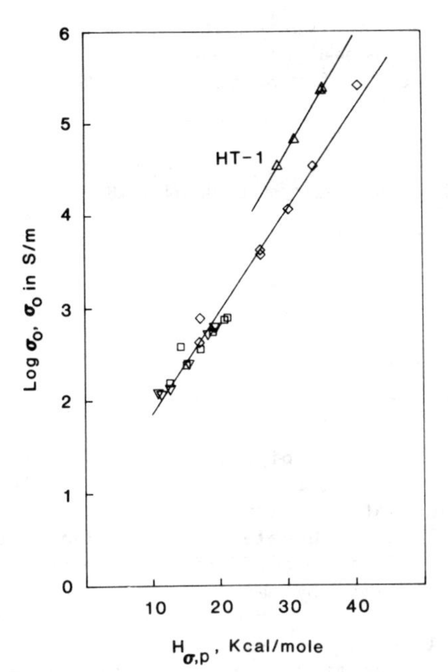

Fig. 10. Compensation plot, log σ_0 versus $H_{\sigma,p}$ (from equation (1)), for the rock melts studied. Symbols are as follows: triangles, molten HT-1 [Tyburyzy and Waff, 1983]; diamonds, molten VC-4W [Tyburczy and Waff, 1983]; squares, molten HR-1 (this study); inverted triangles, molten YRO (this study). Lines are linear least squares fits to the data for HT-1 only, and to the data for the remaining three rock melts together, respectively.

coulombic energy associated with removal of an ion from its site E_C and the elastic shear energy required to make a jump E_S. For diffusion in obsidian glass and melt the semi-empirical equation for the activation energy is

$$E = 8 + 128(r-1.34)^2 + 33Z^2/(r+1.34) \quad (6)$$

where r is the eight-fold coordinated ionic radius in A, Z is the cationic charge, and E is in kcal/mole. This model is successful in reproducing the activation energies for tracer diffusion for inert, alkali, and alkaline earth elements in obsidian glasses and melts. If the cationic radius is less than the radius of the oxygen anion, the elastic energy term is small, and the activation energy is composed largely of the coulombic term. Under these conditions the compensation law is shown to hold for diffusion of ions of the same size in an obsidian matrix, but not for ions of different size. The high degree of correlation obtained for the compensation plot in this study suggests that for electrical conductivity at low and high pressures these conditions are met. Thus, the ions contributing to the conductivity at high pressures have a radius similar to those contributing to the conductivity at low pressure.

The ratio of the isochoric activation enthalpy to the isobaric activation enthalpy is a measure of the relative magnitudes of purely energetic versus dilational terms in the conduction process. The isochoric activation enthalpy $H_{\sigma,v}$ is defined as

$$H_{\sigma,v} = -R \left[\frac{\partial \ln\sigma}{\partial(1/T)} \right]_V \quad (7)$$

Note the distinction between the isobaric activation enthalpy $H_{\sigma,p}$ as defined by equation (1) and the isochoric activation enthalpy. The two activation enthalpies are related by

$$H_{\sigma,p} - H_{\sigma,v} = T\alpha\Delta V_\sigma/\beta \quad (8)$$

where α is the isobaric thermal expansion coefficient, β is the isothermal compressibility and ΔV_σ is the activation volume, defined by

$$\Delta V_\sigma = -RT \left[\partial \ln \sigma/\partial p\right]_T \quad (9)$$

[Brummer and Hills, 1961; Nachtrieb, 1980]. $H_{\sigma,v}$ can be considered to be the energy required for an ion to jump into an existing hole or equilibrium site, while $H_{\sigma,p}$ is the energy required to jump plus the energy required to create the hole [Brummer and Hills, 1961]. Thus, depending on the magnitude of $H_{\sigma,v}/H_{\sigma,p}$ the conduction process may be termed energy restrained or volume restrained [Barton, 1971]. For example, Barton et al. [1968] determined that $H_{\sigma,v}/H_{\sigma,p}$ at one atmosphere pressure is 0.97 for molten $LiNO_3$ and 0.28 for molten $CsNO_3$. The activation volume for conduction is approximately equal to the ionic volume of the Li^+ ion in the case of $LiNO_3$ and slightly smaller than the volume of the Cs^+ ion in the case of $CsNO_3$. Thus transport of the larger Cs^+ ion is limited by the volume required for the ion to move, while for the smaller Li^+ ion the volume requirements are small and transport is limited by the difficulty of removing the ion from its initial "site".

For naturally occurring silicate melts, data on the pressure dependence of density are meager, so it is not possible to do a rigorous analysis of $H_{\sigma,v}$ as a function of T and V. However, examination of equation (8) shows that with measured values of $H_{\sigma,p}$ from atmospheric pressure studies, ΔV_σ from high pressure studies, and zero-pressure values of α and β, the value of $H_{\sigma,v}$ at atmospheric pressure can be determined. The use of values of α extracted from Bottinga and Weill [1970], values of β from Murase and McBirney [1973], and values of $H_{\sigma,p}$ and ΔV_σ from this work, yields the $H_{\sigma,v}/H_{\sigma,p}$ values shown in Table 4. The values of the $H_{\sigma,v}/H_{\sigma,p}$ range from a minimum of 0.73 for the molten andesite to a maximum of 0.90 for the molten tholeiite. Thus in these melts the conduction process at one atmosphere pressure is energy restrained. $H_{\sigma,v}$ cannot be equated directly with the coulombic term in the formulation of the total activation energy. Nonetheless, the high values of $H_{\sigma,v}/H_{\sigma,p}$ determined in this study indicate that the elastic energy terms for conduction in these melts are small, and that the type of compensation law derived by Jambon [1982] for the diffusion of small cations in obsidian glass and melts will also be valid for conduction in the melts studied in this work.

In our previous work [Tyburczy and Waff, 1983] we suggested that the change in the pressure dependence of the conductivity of the melts was related to a density increase caused by the depolymerization of the melt as pressure increased. This conclusion was based on a comparison of density and viscosity versus pressure trends with the observed electrical transport trends. Dunn [1983] found that large discontinuities in the oxygen chemical diffusion coefficient in basalts occur at the pressures at which the liquidus phase changed from olivine to pyroxene, even though the measurements were made at temperatures above the liquidus. He proposed that continuous changes in the relative proportions of the anionic species occur over broad pressure ranges, and that anionic disproportionation reactions occur within the narrow pressure ranges corresponding to the liquidus phase changes. Because phase equilibria have not been determined for the specific melt compositions we have studied, the correspondence between changes in activation volume and changes in liquidus phase cannot be made with complete confidence. Nonetheless comparison with phase equilibria of similar rock types allows us to make an approximate correlation. Figure 9b shows the representative liquidus phase relations for several melts with compositions similar to the ones we have studied. For the rhyodacitic and rhyolitic melts the agreement is good, for the tholeiitic melt it is only approximate, while for the andesitic melt it is poor. It is interesting to note that although we have proposed that a progressive depolymerization of the melt occurs with increased pressure, based on melt viscosity and density data, the liquidus mineral assemblage does not necessarily reflect this trend. In the tholeiitic and andesitic melts a change from liquidus plagioclase at low pressures to pyroxene at about 9 kbar represents decreased polymerization in the solid. In granitic melts the change from quartz to quartz plus plagioclase on the liquidus represents no change in the polymerization of the solid. The manner in which melt structure influences mineral crystallization is very poorly understood.

Conclusions

The magnitude of the electrical conductivity and its variations with temperature and pressure are similar for andesitic, rhyodacitic, and rhyolitic melts. Activation energies range from about 0.5 to 1 eV while activation volumes vary from about 3 to 18 cm^3/mole, depending on the pressure range. In this compositional range activation energies and activation volumes tend to decrease with increasing silica content. Tholeiitic melt has slightly different trends, tending to have higher absolute values of conductivity, larger activation energies, and smaller activation volumes.

TABLE 4. Activation Volume and Activation Enthalpies at Atmospheric Pressure and 1400° C for Natural Silicate Melts

Melt	$H_{\sigma,p}$ kcal/mol	ΔV_σ cm^3/mol	$H_{\sigma,v}$ kcal/mol	$H_{\sigma,v}/H_{\sigma,p}$
HT-1	31.9	9.2	28.7	0.90
VC-4W	17.2	16.5	12.6	0.73
HR-1	14.3	10.2	11.9	0.83
YRO	10.9	7.3	9.3	0.85

Comparison of the electrical conductivity with sodium diffusion coefficients for granitic (rhyolitic) melts by means of the Nernst-Einstein relation indicates that at zero pressure the electrical conductivity can be explained largely by sodium ion transport. In the case of the tholeiitic melt, however, the correlation factor is about 0.1, indicating that the transport of ions other than sodium contribute to the conductivity, or that the sodium ion jumps are highly correlated. The conductivity differences between the two groups may therefore be related to differences in ionic transport mechanism.

The ratio of the atmospheric pressure values of the isochoric activation enthalpy to the isobaric activation enthalpy indicates that for all the melts studied the conduction process is energy restrained at low pressures. This suggests that the elastic energy terms are small. If the semiempirical activation energy theory of Jambon [1982] is correct, the compensation law should hold in this case, as is observed.

In each of these melts there is a decrease in activation volume at low to intermediate pressure. The pressures at which the decreases occur correspond roughly to the pressures at which changes in liquidus mineral assemblage occur, even though the measurements were made at temperatures above the liquidus. It is difficult to explain the change in slope on the basis of changes in the degree of polymerization of the liquidus phase. Compensation law plots show no reflection of the change in activation volume. The silica-rich melts show a very high degree of correlation on the compensation law plot. Following the model of Jambon [1982], this may indicate that the dominant charge-carrying cations are the same at low and high pressure. This is also true for the tholeiitic melt, although the compensation law line for this melt differs from that of the silica-rich melts. Thus the change in activation volume probably reflects a structural rearrangement in the melt.

Acknowledgments. We would like to thank J.S. Huebner for a thoughtful review. This work was supported by NSF grants EAR 80-09529 and 82-13664.

References

Barton, A. F. M., The significance of the constant volume principle in liquid transport properties, Rev. Pure Appl. Chem., 21, 49-66, 1971.

Barton, A. F. M., B. Cleaver, and G. J. Hills, High pressure studies on fused salt systems, Trans. Faraday Soc., 64, 208-220, 1968.

Bevington, P. R., Data Reduction and Error Analysis for the Physical Sciences, 336 pp., McGraw-Hill, New York, 1969.

Bottinga, Y., and D. F. Weill, Densities of liquid silicate systems calculated from partial molar volumes of oxide components, Am. J. Sci., 269, 169-182, 1970.

Brummer, S. B., and G. J. Hills, Kinetics of ionic conductance, 1, Energies of activation and the constant volume principle, Trans. Faraday Soc., 57, 1816-1822, 1961.

Carron, J. P., Auto diffusion du sodium et conductivite electrique dans les obsidiennes granitiques, C. R. Acad. Sci. Paris, 266, 854-856, 1968.

Dunn, T., Oxygen chemical diffusion in three basaltic liquids at elevated temperatures and pressures, Geochim. Cosmochim. Acta, in press, 1984.

Green, J. R., and D. Margerison, Statistical Treatment of Experimental Data, Physical Sciences Data 2, 382 pp., Elsevier, New York, 1978.

Green, T. H., Crystallization of calc-alkaline andesite controlled high pressure hydrous conditions, Contrib. Mineral. Petrol., 34, 150-166, 1972.

Green, T. H., and A. E. Ringwood, Genesis of the calc-alkaline igneous rock suite Contrib. Mineral. Petrol., 18, 105-162, 1968.

Green, T. H., D. H. Green, and A. E. Ringwood, The origin of high-alumina basalts and their relationships to quartz tholeiites and alkali basalts, Earth Planet. Sci. Lett., 2, 41-51, 1967.

Hart, S. R., Diffusion compensation in natural silicates, Geochim. Cosmochim. Acta, 45, 279-291, 1981.

Haven, Y., and B. Verkerk, Diffusion and conductivity of sodium in sodium silicate glasses, Phys. Chem. Glasses, 6, 38-45, 1965.

Haygarth, J. C., H. D. Luedmann, I. C. Getting, and G. C. Kennedy, Determination of portions of the bismuth III-V and IV-V equilibrium boundaries in single-stage piston cylinder apparatus, J. Phys. Chem. Solids, 30, 1417-1424, 1969.

Hofmann, A. W., Diffusion in natural silicate melts: A critical review, in Physics of Magmatic Processes, edited by R.B. Hargraves, pp. 385-417, Princeton University Press, Princeton, N. J., 1980.

Hofmann, A. W., and L. Brown, Diffusion measurements using fast deuterons for in situ production of radioactive tracers, Year Book Carnegie Inst. Washington, 75, 259-262, 1976.

Hughes, K., and J. O. Isard, Ionic transport in glasses, in Physics of Electrolytes , vol. 1, edited by J. Hladik, Academic Press, New York, 1972.

Jambon, A., Tracer diffusion in granitic melts: Experimental results for Na, K, Rb, Cs, Ca, Sr, Ba, Ce, Eu, to 1300° C and a model of calculation, J. Geophys. Res., 87, 10,797-10,810, 1982.

Lasaga, A. C., Transition state theory, in Reviews in Mineralogy, vol. 8, Kinetics of Geochemical Processes, edited by A. C. Lasaga and R. J. Kirkpatrick, pp. 135-170, Mineralogical Society of America, Washington, D. C., 1981.

Moynihan, C. T., and A. V. Lesikar, Weak-electrolyte models for the mixed-alkali effect in glass, J. Am. Ceram. Soc., 64, 40-46, 1981.

Murase, T., and A. R. McBirney, Properties of some common igneous rocks and their melts at high temperatures, Geol. Soc. Am. Bull., 84, 3563-3592, 1973.

Nachtrieb, N. H., Conduction in fused salts and salt-metal solutions, Ann. Rev. Phys. Chem., 31, 131-156, 1980.

Oppenheim, M. J., On the electrolysis of molten basalt, Mineral. Mag., 36, 1104-1122, 1968.

Oppenheim, M. J., On the electrolysis of basalt, II, Experiments in an inert atmosphere, Mineral. Mag., 37, 568-577, 1970.

Rai, C. S., and M. H. Manghnani, Electrical conductivity of basalts to 1550° C, in Magma Genesis, Bull. 96, edited by H. J. B. Dick, pp. 219-237, Oregon Department of Geology and Mineral Industries, Portland, 1977.

Stern, C. R., and P. J. Wyllie, Phase relationships of I-type granite with H_2O to 35 kilobars: The Dinkey Lakes biotite-granite from the Sierra Nevada batholith, J. Geophys. Res., 86, 10,412-10,422, 1981.

Tyburczy, J. A., and H. S. Waff, Electrical conductivity of molten basalt and andesite to 25 kilobars pressure: Geophysical significance and implications for charge transport, J. Geophys. Res., 88, 2413-2430, 1983.

Waff, H. S., Electrical conductivity measurements on silicate melts using the loop technique, Rev. Sci. Instrum., 47, 877-879, 1976.

Waff, H. S., and D. F. Weill, Electrical conductivity of magmatic liquids: Effects of temperature, oxygen fugacity, and composition, Earth Planet. Sci. Lett., 28, 254-260, 1975.

Watson, E. B., Calcium diffusion in a simple silicate melt to 30 kbar, Geochim. Cosmochim. Acta, 43, 313-322, 1979.

Watson, E. B., Diffusion in magmas at depth in the earth: The effects of pressure and dissolved H_2O, Earth Planet. Sci. Lett., 52, 291-301, 1981.

Whittaker, E. J. W., and R. Muntus, Ionic radii for use in geochemistry, Geochim. Cosmochim. Acta, 34, 945-956, 1970.

Winchell, P., The compensation law for diffusion in silicates, High Temp. Sci., 1, 200-215, 1969.

POINT DEFECTS AND THE MECHANISMS OF ELECTRICAL CONDUCTION IN OLIVINE

R. N. Schock and A. G. Duba

University of California, Lawrence Livermore National Laboratory
Livermore, California 94550

Abstract. Measurements of electrical conductivity on single crystals of naturally occurring olivine and synthetic forsterite as a function of oxygen fugacity indicate that point defects play a crucial role in this process. Conductivity data (at 1200°C) reveal n-type conduction in forsterite and p-type conduction in natural peridot. Since the activation energies for conduction (2-3 eV) are well below the measured band gap in these olivines, conduction must be either an extrinsic electronic or an ionic process. Expected energy levels of impurities such as iron are such that either ionic or electronic charge carriers are energetically favorable. The electrical conductivity of forsterite decreases with increasing oxygen fugacity, which is consistent with conduction by electrons. Electrons are made available for conduction by the loss of oxygen from the crystal, which produces point defects and free electrons. Favorable defects include oxygen vacancies and magnesium and silicon interstitials. However, in the presence of iron, as in olivine, conductivity increases with increasing oxygen fugacity. Ferrous iron in the olivine can be oxidized to the ferric state as oxygen fugacity increases. In this case it is not necessary to produce electrons to balance the charge, since the creation of magnesium vacancies accomplishes the same result. Trivalent iron now carries the charge as if it were an electron hole. It is likely that the magnesium vacancies created to balance the charge also affect other physical properties, including creep, diffusivity, and possibly acoustic attenuation.

Introduction

In order to predict conditions in the earth's mantle, we must understand the mechanisms which control the properties of the minerals which make up the mantle. Point defects play a significant role in electrical conduction of supposed mantle minerals [Pluschkell and Engell, 1968; Parkin, 1972; Duba et al., 1973; Cemic et al., 1978; Schock et al., 1980], and recent evidence demonstrates that they play a role in solid-state deformation as well [Kohlstedt and Hornack, 1981; Jaoul et al., 1980; Ricoult and Kohlstedt, this publication]. These same point defects may also affect other processes, such as seismic wave velocity and attenuation.

We present electrical conductivity data on the mineral olivine ($Mg_{1.8}Fe_{0.2}SiO_4$) and on synthetic samples of the pure magnesium form, forsterite (Mg_2SiO_4). We then compare these data with available literature data on the creep of olivine and discuss these results in terms of the possible point defect structure of olivine. We use the effects of composition, either through naturally occurring differences or by laboratory-induced changes, to gain some insight into the operable processes. By controlling the intensive thermodynamic variables temperature and pressure, and having knowledge of the effect of composition, we can limit the degrees of thermodynamic freedom. If the degrees of freedom could be reduced to zero, we would eliminate any ambiguity in determining the possible point defect mechanism which controls the physical process.

If we ignore minor components, olivine is a four-component system. By controlling pressure, temperature, and oxygen fugacity, we eliminate all but two degrees of thermodynamic freedom. For forsterite (which contains no iron) under these same pressure and temperature conditions, the degrees of freedom are reduced to one. To reduce the degrees of freedom even further in a laboratory experiment, it is necessary to control more variables. This is extremely difficult, although Will and Nover [1979] have made an effort to do so using solid-state buffers. In the experimental configuration described here and in most laboratory experiments, there are other constraints. Most notable is that the sample, because it is surrounded by a gas, can be considered to be a closed system with respect to the metal and silicon components. That is, neither of these components can be gained or lost during the

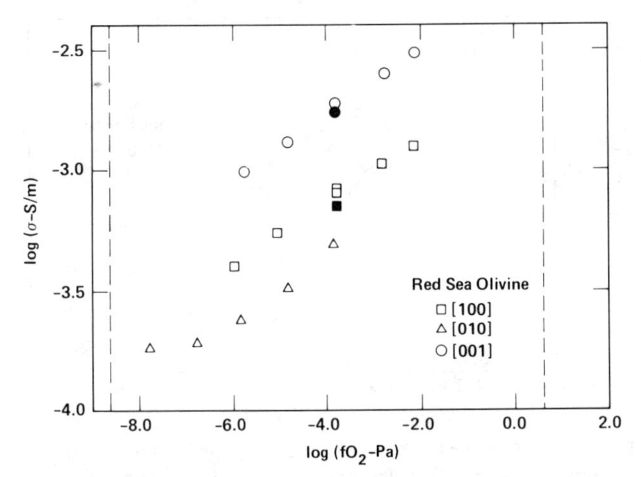

Fig. 1. Electrical conductivity as a function of oxygen fugacity in three directions for an olivine crystal from St. John's Island at 1200°C and 0.1 MPa total pressure. The filled symbols are data taken after the full excursions in oxygen fugacity (see text). The vertical dashed lines represent the stability field of olivine of this composition at this temperature [Nitsan, 1974].

course of the experiments, and so their ratios are fixed, though their specific concentrations may be unknown. This fixes another degree of freedom.

Experimental Data

Electrical conductivity is calculated from measurements made on flat rectangular parallelepipeds that had been cut from a single crystal so that a crystallographic axis lies along the shortest dimension. The conductance apparatus is described in detail by Netherton and Duba [1978]. The sample, 0.3 to 0.5 mm in its smallest dimension, is held between platinum electrodes welded to platinum-platinum 10% rhodium thermocouples. The thermocouple leads thereby serve both to measure temperature and conductance. Temperature is accurate to within 3°C. Conductance was measured with an impedance measuring assembly at a frequency of 1 kHz, with a stated accuracy of 1%, and verified with precision resistors. Conductivity was then calculated from this conductance and the dimensions of the sample. Oxygen fugacity (f_{O_2}) is controlled through mixtures of CO and CO_2 and determined using the data of Deines et al. [1974]. The actual fugacity values were checked with a calcia-doped zirconia cell and found to be accurate within 0.1 of an order of magnitude for $f_{O_2} \geq 10^{-7}$ Pa (1 Pa = 10^{-5} bar). After changing to a given gas mixture, conductance varied with time for 4-6 hours, before settling to a new equilibrium value. At least 24 hours elapsed between successive data points in Figures 1-3.

Figure 1 shows the electrical conductivity of an olivine (peridot) from St. John's Island in the Red Sea (RSP) at 1200°C as a function of oxygen fugacity (f_{O_2}). This sample has a composition of Fo92 (i.e., 92% forsterite). Data for all three crystallographic directions are shown. Similar data for a sample of olivine (chrysolite) from the San Carlos Indian Reservation (SCC) in Arizona (Fo91) are shown in Figure 2. The logarithm of the conductivity for RSP olivine varies by about the 1/7 power of the oxygen fugacity in the [001] direction. The variability of the conductivity in the [100] direction is, for all intents and purposes, the same. The average slope between the three data points at $f_{O_2} > 10^{-6}$ Pa, in the [010] direction, is about 1/6.5 and increases to 1/6 between the two data points at highest fugacity. In the case of the SCC olivine, the logarithm of the conductivity varies by about the 1/7 power of the oxygen fugacity at the highest fugacities. At f_{O_2} < about 10^{-3} to 10^{-5} Pa, the slopes of the SCC data in Figure 2 decrease gradually.

The solid symbols in Figures 1 and 2 show reproducibility of data measured at a given gas mix after measurements at other gas mixtures. Reproducibility at a gas mix in the range 10^{-2} Pa > f_{O_2} > 10^{-6} Pa for iron-bearing olivines is ±0.03 of a log unit in conductivity. However, if the gas mixture is more reducing than (less than) 10^{-6} Pa, reproducibility of

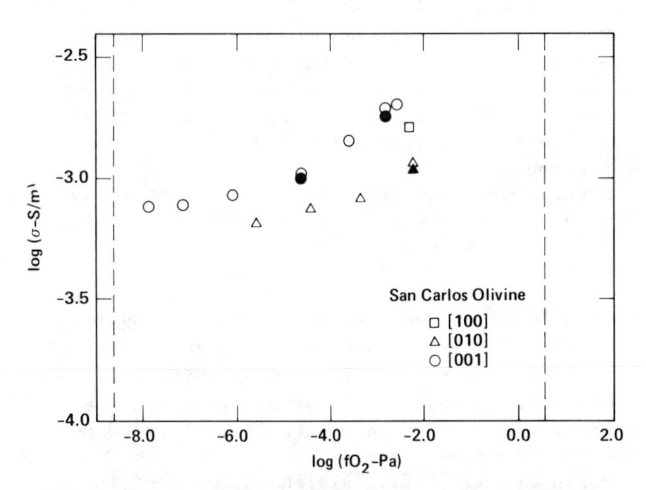

Fig. 2. Electrical conductivity as a function of oxygen fugacity in three directions for an olivine crystal from the San Carlos Indian reservation, Arizona, at 1200°C and 0.1 MPa total pressure. The filled symbols are explained in the text. The vertical dashed lines represent the stability field of olivine of this composition at this temperature [Nitsan, 1974].

conductivity is much worse and the later conductivity is always lower; the magnitude of the decrease is related to the time spent at the low f_{O_2}. Although olivine of this composition is stable at $f_{O_2} \leq 10^{-8.3}$ Pa at 1200°C, electron microscopic analysis shows that loss of iron from our sample to the platinum foil (J. N. Boland, private communication, 1982) is the cause of this decrease. While loss of iron to platinum is a commonly observed phenomenon in petrological investigations [Green and Ringwood, 1970; Merrill and Wyllie, 1973], our results indicate that keeping the f_{O_2} of the experimental system above 10^{-8} Pa reduces that loss.

The conductivity data shown in Figure 3 were obtained on a sample of synthetic forsterite (Fo100) grown by the Linde division of the Union Carbide Corporation. In contrast to the data on the olivines in Figures 1 and 2, the logarithm of these conductivities, also at 1200°C, varies as the oxygen fugacity to the -1/6 power. Data on other synthetic forsterites obtained by Parkin [1972], although at higher temperatures (1400-1500°C), show similar behavior, as do data on a polycrystalline forsterite by Pluschkell and Engell [1968] at 1400°C. Parkin's data show a decrease in slope at oxygen fugacities above about 10^{-3} Pa (10^{-8} atm). At Fe/Mg ratios above 0.2 percent in synthetic olivine, Parkin observed a reversal in slope such that a positive dependence was observed at high f_{O_2} ($>10^{-3}$ Pa).

Creep data for forsterite and olivine have been obtained by Jaoul et al. [1980] and by Kohlstedt and Hornack [1981]. At 1477°C the exponent of the creep rate of SCC olivine varies as f_{O_2} to the +1/6 power. In contrast the creep rate of synthetic forsterite has no dependence on f_{O_2}. Although the data of Kohlstedt and Hornack on SCC show more scatter than those of Jaoul et al. [1980], it is still possible to construct a positive slope of about 1/6 through the data.

Several other measurements are also possibly relevant. Buening and Buseck [1973] measured the diffusivity for iron-magnesium interdiffusion in San Carlos single crystals at 1100°C for four different compositions between 10 and 40 weight percent iron. The diffusivity they report has a +1/6 power dependence on P_{O_2}, in agreement with both the conductivity and the creep data. Jaoul et al. [1980] measured the diffusion of oxygen in forsterite and found no dependence on the oxygen fugacity.

Morin et al. [1977, 1979] have measured the electrical conductivity of a synthetic forsterite single crystal, although not as a function of oxygen pressure. Morin et al. [1977] developed a model based on the difference between the directional dependence of their conductivity in forsterite and the directional dependence of diffusion in olivine as measured by Buening and Buseck [1973]. Our conductivity data on RSP and SCC olivines are in agreement with the Fe-Mg diffusivities on olivine of the same composition [Buening and Buseck, 1973], with respect to crystallographic orientation. There is therefore no a priori reason to appeal to different mechanisms for diffusion and conductivity in forsterite, and so the model of Morin et al. [1977] is not necessary.

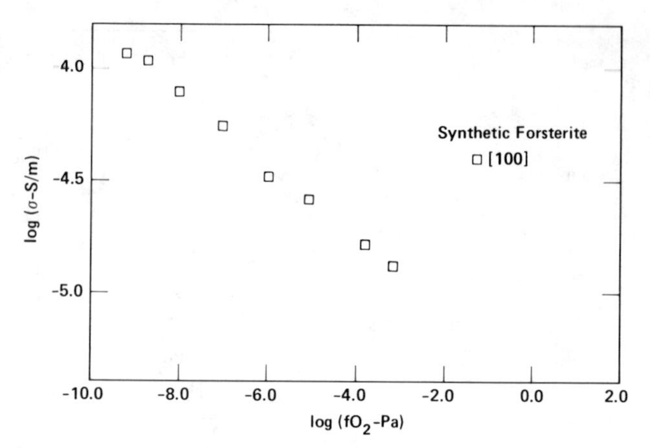

Fig. 3. Electrical conductivity as a function of oxygen fugacity in the [100] direction of a synthetic forsterite, at 1200°C and 0.1 MPa total pressure.

Defect Reactions

Since the activation energies for electrical conduction in both olivine and forsterite are in the range of 1.5-3 eV [Duba, 1972; Duba et al., 1974; Schock et al., 1977] and are well below the measured band gaps (6-7 eV [Shankland, 1968; Nitsan and Shankland, 1976]), conduction must be either an extrinsic electronic or an ionic process. We rule out mixed conduction because of regular slopes at these temperatures when conductivity is plotted against reciprocal temperature. Expected energy levels of impurities such as iron are such that either ionic or electronic charge carriers are energetically favorable [Morin et al., 1977].

There are a number of possible reactions involving defects which, if the concentration of these defects were directly responsible for the resultant data on forsterite, would produce slopes with the magnitudes observed. Various possible defects are involved. The defect reactions themselves have been written by Smyth and Stocker [1975] for olivine and forsterite under varying conditions. Oxygen might simply be removed from the crystal leaving vacancies, as in

$$O_O^x \rightleftarrows 1/2\ O_2 + V_O^{\cdot\cdot} + 2e' \qquad (1)$$

where the notation of Kröger and Vink (1956) is used, O_O^x indicating an oxygen atom occupying a normal oxygen lattice position (with its electrons), $V_O^{\cdot\cdot}$ a doubly charged (+) vacant oxygen lattice position, and e' an electron. The law of mass action allows the concentration of oxygen vacancies $[V_O^{\cdot\cdot}]$ in (1) to be calculated as a function of the oxygen partial pressure P_{O_2}, e.g.,

$$K_1 \, P_{O_2}^{-1/2} = [V_o^{\cdot\cdot}] \, [e']^2$$

$$2[V_o^{\cdot\cdot}] = [e'] \propto P_{O_2}^{-1/6}$$

where K_1 is the equilibrium constant for reaction (1). (Henceforth, we will use P_{O_2} and f_{O_2} interchangeably, aware that experimentally we are controlling f_{O_2}, and that f = P only for an ideal gas.) In order for reaction (1) to control charge neutrality and the concentrations of the two defects produced, other possible reactions must either not be operable in this P_{O_2} range, or must not produce these defects in any significant concentration. In the following analysis, we shall make a number of simplifying assumptions about the possible types of defects. These allow an analysis of the data in terms of the types of defects most likely to be present in the largest quantities. The assumptions are that impurities other than iron either play a minor role or behave like iron (i.e., they have the same valence states), that electrons (e') and electron holes (h•) produced by defect reactions but not trapped are more mobile than ionic defects, and that we are not dealing with associated defects. Other assumptions which deal with levels of ionization, donor and acceptor energy levels, and unfavorable defects have been stated by Smyth and Stocker [1975] and will not be repeated here.

Another possible reaction which yields P_{O_2} dependence similar to those observed in forsterite involves the loss of oxygen from the crystal with the production of magnesium and silicon interstitials ($Mg_I^{\cdot\cdot}$ and $Si_I^{\cdot\cdot\cdot\cdot}$) and free electrons (e'),

$$2Mg_{Mg}^x + Si_{Si}^x + 4O_o^x \rightleftarrows 2O_2 + 2Mg_I^{\cdot\cdot} + Si_I^{\cdot\cdot\cdot\cdot} + 8e' \quad (2)$$

If charge neutrality were controlled by this reaction, the concentration of electrons [e'] would vary as the -2/11 power of P_{O_2}, i.e.,

$$K_3 P_{O_2}^{-2} = [Mg_I^{\cdot\cdot}]^2 \, [Si_I^{\cdot\cdot\cdot\cdot}] \, [e']^8$$

$$8[Si_I^{\cdot\cdot\cdot\cdot}] = 4[Mg_I^{\cdot\cdot}] = [e'] \propto P_{O_2}^{-2/11}$$

The present data alone do not allow a specific reaction to be chosen, and experiments to deduce the non-charge-carrying defects are necessary to distinguish these and other possible reactions.

It is clear from the conductivity data that the presence of iron changes the mechanism of conduction. There are many reactions that yield the logarithm of concentrations of Fe defects proportional to a power of P_{O_2} between +1/4 and +1/7. If, for example, the oxidation state of Fe on Mg sites were controlled by the following reaction,

$$Fe_{Mg}^x \rightleftarrows Fe_{Mg}^{\cdot} + e' \quad (3)$$

and the concentration of electrons [e'] were controlled by (1), then the concentration of trivalent iron, $[Fe_{Mg}^{\cdot}]$, would vary as the 1/6 power of the oxygen fugacity, since

$$K_2 = [Fe_{Mg}^{\cdot}] \, [e'] = [Fe_{Mg}^{\cdot}] \, P_{O_2}^{-1/6}$$

and

$$[Fe_{Mg}^{\cdot}] = K_2 P_{O_2}^{1/6}$$

If the concentration of electrons in the iron oxidation reaction (3) were controlled by reaction (2), then $[Fe_{Mg}^{\cdot}]$ in the system would vary as the +2/11 power of the oxygen fugacity. These reactions are shown to illustrate that one cannot make any simple conclusions based on the slope of conductivity or creep data. Other reactions, and combinations, are given in Smyth and Stocker [1975] and Stocker [1978].

Discussion

The simplest explanation for the observed behavior of conductivity in forsterite at low P_{O_2} is that the charge carriers are electrons and that their mobility is sufficiently high that the conductivity depends only on the formation of electrons. The reasons for this are straightforward. First, there are numerous defect reactions that are dependent on P_{O_2} and that produce electron concentrations which vary negatively with P_{O_2} at slopes of -1/5 to -1/10 (see, for example Stocker [1978]). Second, the mobility of any free electrons present could be significantly higher than any possible ionic charge carriers, based simply on their size. Electrical conductivity, and any other process involving defects, depends on the product of the concentration and the mobility of the relevant charge carrier. In order for measured slopes to correspond to calculated P_{O_2} dependence, mobilities of the relevant species must be such that the variability in concentration can be reflected in the data. Two prerequisites must be met: first that the mobility be high enough so that the effect of concentration can be observed, and second that the mobility of a defect species not be dependent on P_{O_2}. If electrons were carrying the charge, these assumptions would be reasonable. At high P_{O_2} the decrease in slope of the forsterite data shown by Parkin may be due to either a change in the charge neutrality condition, perhaps to the onset of an electron trapping mechanism, or to mixed conduction due to the presence of small amounts of iron.

Unfortunately, the conductivity-vs-P_{O_2} data alone are not sufficient to indicate electrons as the charge-carrying species in forsterite under these conditions. Examination of Brouwer [1954] diagrams of defect concentrations as a function of the logarithm of P_{O_2} [e.g., Stocker, 1978] indicates that other species, such as $Mg_I^{\cdot\cdot}$ or $Si_I^{\cdot\cdot\cdot\cdot}$, or even oxygen vacancies ($V_O^{\cdot\cdot}$) could, under the appropriate charge neutrality conditions, be responsible for the observed negative P_{O_2} dependence of conductivity in forsterite. Recent measurements of the Seebeck coefficient [Shankland and Duba, unpublished manuscript; Schock et al., 1984] indicate that the sign of the charge carrier in forsterite is negative. Therefore, of the possible charge carriers based on P_{O_2} dependence of concentration, only electrons have the measured sign.

As with the data on forsterite, there are many possible defect reactions [Stocker, 1978] that produce defects in olivine whose concentrations vary with oxygen pressure like the data in Figures 1 and 2. The specific defects in addition to $Fe_{Mg}^{\cdot}$ are electron holes and magnesium and silicon vacancies (V_{Mg}'' and V_{Si}''''). In olivine the conductivity data could be explained by the oxidation of iron, creating free electrons in the process as in reaction (3). Since the observed slopes are positive with oxygen fugacity, the charge carriers would be the oxidized iron acting as an electron hole. This mechanism is supported by recent measurements of the Seebeck coefficient in SCC olivine that indicate that the dominant charge carrier is positive [Shankland et al., 1982], thereby precluding the two vacancy types. An electron hole, strongly localized on an iron ion, and the resulting lattice distortion, can be referred to as a small polaron [Adler, 1973]. The concentration of small polarons in iron-bearing olivine is increased over that in pure forsterite, simply by the presence of small amounts of oxidized iron. In forsterite with very low iron concentrations (i.e., <0.2 percent [Parkin, 1972]), the concentration of polarons is obviously too small, and electrons still carry the charge.

However, there are reactions involving $Fe_{Mg}^{\cdot}$ other than (3) which are more likely to operate in olivine. As mentioned previously, the presence of iron changes the creep behavior between forsterite and olivine [Jaoul et al., 1980; Kohlstedt and Hornack, 1981], and ionic defects are likely to be involved in the creep process, if not in the conduction process. These reactions involve the oxidation of iron and the concurrent production of magnesium vacancies to maintain charge neutrality. One possible reaction, outlined by Smyth and Stocker [1975], involves the incorporation of excess oxygen into the olivine lattice, with the production of magnesium and silicon vacancies,

$$8Fe_{Mg}^{x} + 2O_2 \rightleftarrows 4O_o^{x} + 2V_{Mg}'' + V_{Si}'''' + 8Fe_{Mg}^{\cdot} \quad (4)$$

If excess silicon were present as silicon interstitials, then iron could be oxidized without the production of a supposedly energetically unfavorable defect, the silicon vacancy, by

$$8Fe_{Mg}^{x} + Si_I^{\cdot\cdot\cdot\cdot} + 2O_2 \rightleftarrows 8Fe_{Mg}^{\cdot} + Si_{Si}^{x} + 4O_o^{x} + 2V_{Mg}'' \quad (5)$$

The $[Fe_{Mg}^{\cdot}]$ would vary as f_{O_2} to the 1/5.5 power if reaction (4) were operable and to the 1/4.5 power if reaction (5) were involved. The lower slopes in Figures 1 and 2 would seem to favor reaction (4).

The cumulative evidence thus points to electrons, produced by as yet unknown defect reactions, as responsible for conduction in forsterite. However, when iron is present, as in olivine, electron holes localize on the iron ion as part of the oxidation process and carry the charge. If we make the assumption that the mechanical and electrical data are the result of a consistent defect reaction or reactions which occur in these olivines during the mechanical and electrical experiments, even though the observed effects may be different, we may draw some conclusions.

It has often been assumed that oxygen because of its relatively large atomic volume will exhibit the slowest self-diffusivity of any of the atoms present in olivine and therefore would define the limit to the rate of mechanical deformation. As the slowest species it would not be involved in electrical conduction. Considering the observations [Jaoul et al., 1980] on the creep of forsterite as a function of oxygen pressure in terms of oxygen vacancies, we find that only under one set of conditions are there simple reactions [see Stocker, 1978, Table 1] that produce these defects independent of P_{O_2} under the conditions of the experiments (i.e., a closed chemical system in which the ratios of the remaining ions are fixed). The one condition involves a non-stoichiometric excess of silicon, sufficient to control charge neutrality [see Stocker, 1978, reaction 12]. There are none at all that produce concentrations which vary as the positive power of the oxygen fugacity. The zero slopes in the creep data for forsterite argue therefore either against oxygen vacancies controlling creep directly, or they argue for the presence of excess silicon within the olivine phase. The recent work of Jaoul et al. [1980, 1981], which showed no P_{O_2} dependence of oxygen or silicon diffusivity in forsterite, would tend to support the conclusion that oxygen movement is not directly responsible for the observations of creep in olivine. Also, the values of oxygen diffusivity found by Jaoul et al. are too low by several orders of magnitude to account for the equilibration times noted in our work. Obviously some other mechanism is also operating.

It is difficult to imagine a situation where enough excess silica would be present to control charge neutrality without the appearance of a second phase, and to our knowledge no one has reported second phases within single crystals of these minerals. Natural crystals generally have enough time to equilibrate and grow recognizable phases. However, there are a number of reactions involving magnesium defects which, if the concentrations of these defects controlled the property being measured, could result in the observed data [see Stocker, 1978]. They are derivatives of the basic Frenkel defect reaction, $Mg_{Mg}^{x} \rightleftarrows V_{Mg}'' + Mg_{I}^{\cdot\cdot}$, and $[V_{Mg}'']$ is independent of P_{O_2}

Obviously iron plays a major role in creep as well as in conductivity. However, it is difficult to see immediately why the mere presence of iron would affect the creep rate in the olivine structure. Either the site occupancy of iron or the production of specific defects in association with oxidation could conceivably affect creep. Several laboratories have studied the site occupancy of iron in the olivine structure. If iron were found to have a preference for one of the two non-equivalent octahedral sites normally occupied by magnesium, then one might look for an explanation for the creep data in terms of the iron atom and specific site symmetries. However, the site occupancy studies yield ambiguous results. Finger [1970] and Finger and Virgo [1971] found a preference for iron on the M1 site in a lunar and a terrestrial olivine using x-ray techniques. In contrast, Wenk and Raymond [1973] show a preference for iron on the M2 site in a metamorphic olivine. Chatelain and Weeks [1973], using electron spin resonance (ESR) techniques on a synthetic forsterite with a small amount of trivalent iron, assigned this iron to the M2 site. However, Zeira and Hafner [1974] dispute this assignment and reassign the iron to M1. Their own ESR work on another synthetic crystal doped with a small amount of trivalent iron indicated that this iron was equally distributed between M1 and M2.

Will and Nover [1979], in an x-ray structural refinement on a natural iron-bearing olivine, attempt to clarify this situation and claim not only that iron does show a preference, but that its preference for the M1 and M2 sites changes as a function of P_{O_2} in crystals equilibrated at 750°C. Based on these studies, iron would seem to prefer the M1 site at low P_{O_2} and the M2 site at high P_{O_2}. While it is tempting to conclude that this purported site occupancy change may be related to the observed creep behavior in olivine, there is reason to question Will and Nover's use of solid-state buffers, in particular enstatite, $MgSiO_3$. Olivine and enstatite ($MgSiO_3$) have different Fe/Mg partitioning coefficients [Mori and Green, 1978]. This use raises the possibility that either the olivine system was not closed, and Fe and Mg were free to migrate to or from the olivine phase, or that the system was closed and therefore the crystal could not have attained equilibrium with the buffer because of the temperatures and the times chosen. These and earlier x-ray results are also suspect because the reduction of the data involves assumptions about the phases present in the sample and their compositions. Two recent studies circumvent this problem through the use of an electron microscope beam, and therefore a small sample area, together with energy dispersive analysis [Taftø and Spence, 1982; Smyth and Taftø, 1982]. These studies show equal distribution of iron between M1 and M2 sites. We conclude that there is no site preference for iron yet demonstrated.

If $Fe_{Mg}^{\cdot}$ were produced by oxidation as in either reaction (4) or (5), then $[Fe_{Mg}^{\cdot}]$ would vary as P_{O_2} to the power 1/5.5 and 1/4.5, respectively. Electrical conduction would in either case still be by $Fe_{Mg}^{\cdot}$. The oxidation of iron in either reaction (4) or (5) creates V_{Mg}'' and, if the concentration of V_{Mg}'' were controlling creep, could thereby control the creep process. However, it is not immediately obvious how the presence of a magnesium vacancy would control creep.

The olivine structure is shown in Figure 4 in the (100) plane. If a magnesium atom in an M1 or M2 position were to be removed, then the diffusion of another atom along the [100] axis would be facilitated. It is important to realize that, if the diffusion were related to creep, it would take place in response to stress, not in response to a change in P_{O_2}. When iron is present in sufficient amounts, as in olivine, V_{Mg}'' could be created as oxidation takes place, as in reactions (4) and (5). Whether creep is in fact diffusion-controlled is arguable. The activation energy for diffusion of oxygen, at least in forsterite, is much smaller (3.5-3.9 eV [Jaoul et al., 1980; Reddy et al., 1980]) than that for creep (6-7 eV [Jaoul et al., 1980]). This observation indicates a more complicated process, but does not rule out oxygen diffusion as one step in that more complicated process. Hobbs [1981, 1983] has in fact argued that positively charged kinks [Hirsch, 1979, 1981] that migrate along dislocation lines control creep in both forsterite and olivine, rather than diffusion. For olivine, Hobbs assumes that the neutrality condition is $2[V''_{Mg}] = [Fe^{\bullet}_{Mg}]$, a condition that arises from the reaction often written as

$$1/2\ O_2 + 3Fe^x_{Mg} \rightleftarrows 2Fe^{\bullet}_{Mg} + V''_{Mg} + FeO \quad (6)$$

whose equilibrium constant is

$$K_6 P_{O_2}^{1/2} = [Fe^{\bullet}_{Mg}]^2\ [V''_{Mg}]$$

An equally plausible reaction, which yields the same equilibrium constant expression, is

$$O_2 + 4Fe^x_{Mg} + SiO_2 \rightleftarrows 4Fe^{\bullet}_{Mg} + 2V''_{Mg} + Si^x_{Si} + 4O^x_o \quad (7)$$

If (6) or (7) were controlling charge neutrality, $[Fe^{\bullet}_{Mg}]$ would vary as P_{O_2} to the 1/6 power, since $2[V''_{Mg}] = [Fe^{\bullet}_{Mg}]$. Reaction (6) is often cited in the literature [Buening and Buseck, 1973; Will and Nover, 1979; Hobbs, 1983] without the specific acknowledgment that it demands excess SiO_2 or FeO be present if it is to take place in olivine. We know of no evidence which argues for excess silica or FeO in any of the relevant experiments, with the exception of those of Buening and Buseck where both Fe_3O_4 and SiO_2 were present in the experimental material in amounts totalling less than 3 percent. However, at the experimental conditions cited by Buening and Buseck, the Fe_3O_4 would be reduced to FeO, which would in turn react with any SiO_2 present to form Fe_2SiO_4 [Williams, 1971]. The final result would be a mixture of fayalite with an excess of either SiO_2 or FeO, possibly neither.

Finally, in all of the foregoing we have made the assumption that defect association was not taking place and affecting any process. We have only circumstantial evidence that this assumption is valid, and thus some discussion is warranted. Defects are known to associate in solids [Lidiard, 1957], although they tend to become dominant only at lower temperatures (<1/2 the melting temperature). Gourdin et al. [1979] and Gourdin and Kingery [1979] have calculated the energies of various defect associations in MgO and concluded that when $Fe^{\bullet}_{Mg}$ is present, aggregates with magnesium vacancies are energetically stable and their concentration is intimately related to the oxidation state of iron. In effect, $Fe^{\bullet}_{Mg}$ and V''_{Mg} combine in various polymer-like combinations, in some cases with oxygen. The presence of aluminum as an impurity is calculated to also affect defect association in the same manner. Gourdin et al. [1979] used such a model to explain data on the amount of trivalent iron in MgO as a function of thermodynamic conditions at 1678 K (melting

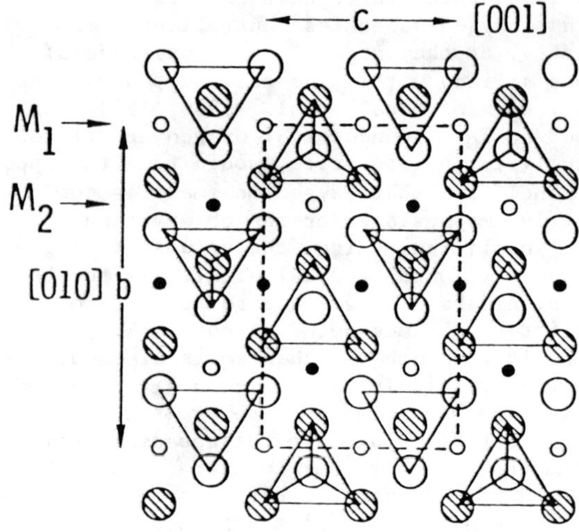

Fig. 4. Projection of the olivine structure looking in the [100] direction [after Morin et al., 1977]. The M1 and the M2 cation sites are indicated and Si atoms are in centers of tetrahedra. Open and closed cation symbols are cations at the unit cell edge and 1/2 distance into the cell, respectively.

temperature 3178 K). Whether such effects operate in olivine is speculative, but it would seem highly probable owing to the similarities in defect reactions and band-gap energies in MgO and olivine. However, the relatively high temperatures (1500 K) for the experimental data in olivine discussed here (melting temperature ∿2050 K) and the ability to explain the data based on simple unassociated defects would argue that, at least at these temperatures, associations do not need to be taken into account. At lower temperatures they would become very suspect to dominate the electrical and physical properties.

Conclusions

Data presented in this paper on electrical conductivity and literature data on creep as a function of the fugacity of oxygen indicate that defects play a crucial role in these processes. Electrons are indicated as the dominant charge carriers in forsterite, because of their inherent high mobility. In iron-bearing olivine, the iron obviously plays a major role and alters both the conduction mechanism and the creep process. Iron can act as an electron donor and allows conduction to take place through small polarons, a process suggested by Shankland [1975], whereby oxidized iron in normal lattice positions acts as an electron hole. It seems likely that production of magnesium vacancies in conjunction with oxidation of iron is related to the deformation properties of olivine. If magnesium vacancies are the defect which controls transport properties in olivine, as we postulate here, and since oxygen diffusion in forsterite shows no dependence on oxygen pressure [Jaoul et al., 1980], we conclude that the diffusivity of oxygen in olivine should be positively dependent on P_{O_2}.

In any measurements on olivine, or any other mineral, care must be taken to control thermodynamic conditions during the experiment and to be certain either that equilibrium has been attained or that the state of disequilibrium is known. The stability field of olivine in particular is very narrow and puts severe constraints on the range of experimental conditions. In addition, as we have shown in this paper, chemical effects such as the uptake of iron from silicates in contact with platinum, at $f_{O_2} \leq 10^{-8}$ Pa, must also be taken into account.

Acknowledgments. It is impossible to mention by name all those who provided the stimulation for this work and the comments to improve it. We thank them all. Discussions with Rick Stocker, Tom Shankland, and David Kohlstedt were particularly valuable and the manuscript benefitted measurably from their comments. Don Anderson at California Institute of Technology, Chuck Sonett at University of Arizona, and Ted Ringwood at Australian National University provided space and, in some cases, financial support to facilitate these studies. This paper is dedicated to the memory of John C. Jamieson, without whose interest and insight neither of us would likely have set out on this endeavor. This work was supported by the Office of Basic Energy Sciences and performed under the auspices of the U.S. Department of Energy by the Lawrence Livermore National Laboratory under contract W-7405-ENG-48.

References

Adler, D., Electronic configuration and electrical conductivity in ceramics, Am. Ceram. Soc. Bull., 52, 154-159, 1973.

Brouwer, G., A general asymptotic solution of reaction equations common in solid-state chemistry, Philips Res. Rep., 9, 366-376, 1954.

Buening, D. K., and P. R. Buseck, Fe-Mg lattice diffusion in olivine, J. Geophys. Res., 78, 6852-6862, 1973.

Cemic, L., E. Hinze, and G. Will, Messungen der elektrischen Leitfahigkeit bei kontrolliert, Sauerstoffaktivitaten in Druckapparaturen mit festen Druckubertragungsmedien, High Temp.-High Pressures, 10, 469-472, 1978.

Chatelain, A., and R. A. Weeks, Electron paramagnetic resonance of Fe3+ in forsterite, J. Chem. Phys., 58, 3722-3726, 1973.

Deines, P., R. H. Nafziger, G. C. Ulmer, E. Voermann, "Temperature-Oxygen Fugacity Tables for Selected Gas Mixtures in the System C-H-O at 1 Atmosphere Total Pressure," Bull. Earth Miner. Sci. Exp. Stn. Pa. State Univ., 88, 129, 1974.

Duba, A., Electrical conductivity of olivine, J. Geophys. Res., 77, 2483-2495, 1972.

Duba, A., J. N. Boland, and A. E. Ringwood, Electrical conductivity of pyroxene, J. Geol., 81, 727-735, 1973.

Duba, A., H. C. Heard and R. N. Schock, Electrical conductivity of olivine at high pressure and under controlled oxygen fugacity, J. Geophys. Res., 79, 1667-1673, 1974.

Finger, L., Fe/Mg ordering in olivines, Year Book Carnegie Inst. Washington, 69, 302-305, 1970.

Finger, L., and D. Virgo, Confirmation of Fe/Mg ordering in olivines, Year Book Carnegie Inst. Washington, 70, 221-225, 1971.

Gourdin, W. H., and W. D. Kingery, The defect structure of MgO containing trivalent cation solutes: Shell model calculations, J. Mater. Sci., 14, 2053-2073, 1979.

Gourdin, W. H., W. D. Kingery, and J. Driear, The defect structure of MgO containing trivalent cation solutes: The oxidation-reduction behavior of iron, J. Mater. Sci., 14, 2074-2082, 1979.

Green, D. H., and A. E. Ringwood, Mineralogy of peridotitics compositions under upper mantle conditions, Phys. Earth Planet. Inter., 3, 359-371, 1970.

Hirsch, P. B., A mechanism for the effect of doping on dislocation mobility, J. Phys. Colloq. Orsay Fr., C3-40, 117-121, 1979.

Hirsch, P. B., Plastic deformation and electronic mechanisms in semiconductors and insulators, J. Phys. Colloq. Orsay Fr., C3-42, 149-160, 1981.

Hobbs, B. E., The influence of metamorphic environment upon the deformation of minerals, Tectonophysics, 78, 335-383, 1981.

Hobbs, B. E., Constraints on the mechanism of deformation of olivine imposed by defect chemistry, Tectonophysics, 92, 35-69, 1983.

Jaoul, O., C. Froidevaux, W. B. Durham, and M. Michaut, Oxygen self-diffusion in forsterite: Implications for the high temperature creep mechanism, Earth Planet. Sci. Lett., 47, 391-397, 1980.

Jaoul, O., M. Poumellec, C. Froidevaux, and A. Havette, Silicon diffusion in forsterite: A new constraint for understanding mantle deformation, in Anelasticity in the Earth, Geodynamics Ser., vol. 4, edited by F. D. Stacey, M. S. Paterson, and A. Nicholas, p. 95, AGU, Washington, D.C., 1981.

Kohlstedt, D., and P. Hornack, Effect of Oxygen Partial Pressure on the Creep of Olivine, in Anelasticity in the Earth, Geodynamics Ser., vol. 4, edited by F. D. Stacey, M. S. Paterson, and A. Nicholas, p. 101, AGU, Washington, D.C., 1981.

Kröger, F. A., and H. J. Vink, in Solid State Physics, Adv. in Res. and Appl., edited by F.

Seitz and D. Turnbull, pp.307-435, Academic, New York, 1956.
Lidiard, A. B., Ionic conductivity, Handb. Phys., 20, 246-349, 1957.
Merrill, R. B., and P. J. Wyllie, Absorption of iron by platinum capsules in high pressure rock melting experiments, Am. Mineral., 58, 16-20, 1973.
Mori, T., and D. H. Green, Laboratory duplication of phase equilibria observed in natural garnet lherzolites, J. Geol., 86, 83-97, 1978.
Morin, F. J., J. R. Oliver, and R. M. Housley, Electrical properties of forsterite, Mg SiO, Phys. Rev. B, 16, 4434-4445, 1977.
Morin, F. J., J. R. Oliver, and R. M. Housley, Electrical properties of forsterite, Mg SiO, II, Phys. Rev. B, 19, 2886-2894, 1979.
Netherton, R., and A. Duba, An apparatus for simultaneously measuring electrical conductivity and oxygen fugacity, Rep. UCRL-52394, Univ. of Calif., Lawrence Livermore Natl. Lab., Livermore, 1978.
Nitsan, U., Stability field of olivine with respect to oxidation and reduction, J. Geophys. Res., 79, 706-711, 1974.
Nitsan, U., and T. J. Shankland, Optical properties and electronic structure of mantle silicates, Geophys. J. R. Astron. Soc., 45, 59-87, 1976.
Parkin, T., The electrical conductivity of synthetic forsterite and periclase, Ph.D. thesis, Univ. of Newcastle upon Tyne, Newcastle upon Tyne, England, 1972.
Pluschkell, W., and H. J. Engell, Ionen und Elektronenleitung im Magnesiumorthosilikat, Ber. Dtsch. Keram. Ges., 45, 388-394, 1968.
Reddy, K. P. R., S. M. Oh, L. D. Major Jr., and A. R. Cooper, Oxygen diffusion in forsterite, J. Geophys. Res., 85, 322-326, 1980.
Ricoult, D., and D. L. Kohlstedt, Experimental evidence for the effect of chemical environment upon the creep rate of olivine, this publication.
Schock, R. N., A. G. Duba, H. C. Heard, and H. D. Stromberg, The Electrical conductivity of polycrystalline olivine and pyroxene under pressure, in High Pressure Research: Applications in Geophysics, edited by M. Manghnani and S. Akimoto, Academic, New York, 1977.
Schock, R. N., A. G. Duba, and R. L. Stocker, Defect production and electrical conductivity in olivine (abstract), Lunar Planet, Sci., 11, 710, 1980.
Schock, R. N., A. G. Duba, and T. J. Shankland, Mechanisms of electrical conductivity in olivine, Proc. Int. Geol. Congr., in press, 1984.
Shankland, T. J., Band gap of forsterite, Science, 161, 51-53, 1968.
Shankland, T. J., Electrical conduction in rocks and minerals: Parameters for interpretation, Phys. Earth Planet. Inter., 10, 209-219, 1975.
Shankland, T. J., R. N. Schock, and A. G. Duba, Thermoelectric effect in olivine (abstract), Eos Trans. AGU., 45, 1090, 1982.
Smyth, D., and R. L. Stocker, Point defects and non-stoichiometry in forsterite, Phys. Earth Planet. Inter., 10, 183-192, 1975.
Smyth, J. R., and J. Taftø, Major and minor element site occupancies in heated natural forsterite, Geophys. Res. Lett., 9, 1113-1116, 1982.
Stocker, R. L., Influence of oxygen pressure on defect concentrations in olivine with a fixed cationic ratio, Phys. Earth Planet. Inter., 17, 118-129, 1978.
Taftø, J., and J. C. H. Spence, Crystal site location of iron and trace elements in a magnesium-iron olivine by a new crystallographic technique, Science, 218, 49-51, 1982.
Wenk, H. R., and K. N. Raymond, Four new structure refinements of olivine, Z. Kristallogr., 137, 86-105, 1973.
Will, G., and G. Nover, Influence of oxygen partial pressure on the Mg/Fe distribution in olivines, Phys. Chem. Miner., 4, 199-208, 1979.
Williams, R. J., Reaction constants in the system Fe-MgO-SiO_2-O_2 at 1 atm between 900° and 1300°C: Experimental results, Am. J. Sci., 270, 334, 1971.
Zeira, S., and S. S. Hafner, The location of Fe3+ ions in forsterite, Earth Planet. Sci. Lett., 21, 210-208, 1974.

A TECHNIQUE FOR OBSERVING OXYGEN DIFFUSION ALONG GRAIN BOUNDARY REGIONS IN SYNTHETIC FORSTERITE

R. H. Condit, H. C. Weed, and A. J. Piwinskii

Lawrence Livermore National Laboratory
Livermore, California 94550

Abstract. We describe a technique for observing oxygen diffusion into synthetic forsterite, Mg_2SiO_4. An oxygen-18 isotope tracer was used in a gas-solid interchange anneal. Tracer penetration into Mg_2SiO_4 was examined by the technique of proton bombardment activation to convert the oxygen-18 into radioactive fluorine-18, followed by autoradiography. This is the first application of this technique to silicate diffusion studies and is concerned with grain boundary regions which contain other phases than the matrix material. High diffusion rates measured at 1374°C are related to the presence of molten material, while at 1107°C there is evidence for penetration of oxygen gas along cracks in the grain boundary region. Our best diffusion coefficient obtained at 1325°C is in general accord with earlier predictions for grain boundary diffusion, albeit higher. This suggests that the grain boundary material is enriched in dislocations and interphase contacts. This technique for observing oxygen-18 tracer should be applicable to studies of grain boundary and dislocation diffusion and to studies of solid-solid and solid-gas reactions.

Introduction

There are several techniques using nuclear reactions to observe diffusion [Calvert et al., 1974]. This paper reports a way to observe the oxygen-18 isotope as if it were a radiotracer. The oxygen-18 is converted to radioactive fluorine-18 by 3 MeV proton bombardment and autoradiography then registers the location of this radioactivity. The method has been previously used to study oxygen grain boundary diffusion in MgO [McKenzie et al., 1971], but has been improved since then as described below and will be described in greater detail in a future paper. In this study we have elected to measure oxygen diffusion along grain boundaries in forsterite, Mg_2SiO_4, containing second phase materials because: (1) oxygen is a pervasive constituent of minerals and ceramics and its atomic transport probably dominates the kinetics of many solid state processes; (2) data are available on the bulk diffusion for most of the ions in the forsterite structure; (3) forsterite is representative of geologically important materials constituting a major part of the earth's upper mantle; and (4) commercially synthesized forsterite is available having conveniently oriented grain boundaries containing second phase materials.

We illustrate the oxygen tracer technique by displaying material transport details of the gas-solid exchange process. This exchange type of experiment is so commonly used for obtaining diffusion data that it has merit to know where the oxygen tracer actually goes. Such information is easily obtained using radiotracers or their equivalent as we have done.

Materials

The forsterite employed in our experiments was obtained from Muscle Shoals Electrochemical Division, Glasrock Products, Inc., Tuscumbia, Alabama. It is made by the fusion of a mixture of MgO and SiO_2 in an electric arc furnace. The crystals in a few portions of the resultant ingot grow as thin platelets stacked with their b-axes parallel with their thin dimension. The a- and c-axes are randomly oriented. Impurities were determined by arc spectrum analysis as parts per million by weight: Al 2000, Ca 2000, Fe 2000, Mn 200, B 30, Zr 30, Cr 10, Ti 10, Sr 7, and Cu 1. The following were not detected and the limits of detection are indicated: <300 Na; <100 Ba; <30 Cd, Nb, Sb, Sn; <10 Pb, V; <3 Mo, Ni, Sn; <1 Ag, Be, Bi, Co. In addition to impurities, the platelets initially contained scattered globules of silicon mostly smaller than 20 μm diameter, a volume fraction of less than 0.1% of the total volume or 2% of the grain boundary material. The platelets varied in thickness from a few tens to hundreds of micrometers. The boundary regions contained phases in addition to the forsterite and their widths ranged from less than one up to about 50 μm in a few cases. Many were 3 to 10 μm. In

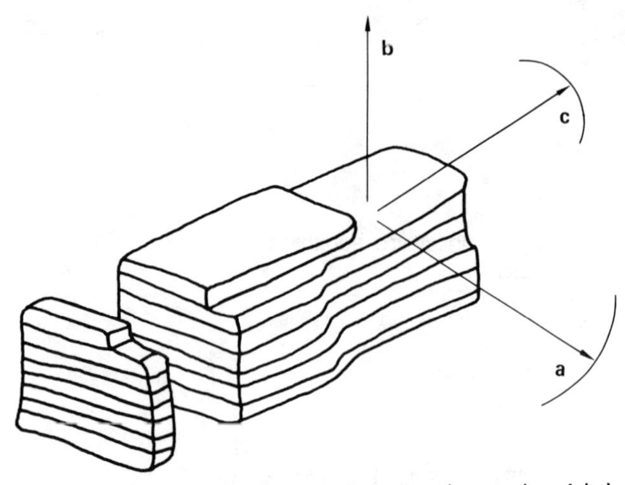

Fig. 1. The sample is cut from an ingot in which the forsterite has crystallized as platelets. These have their c-axes normal to their thin direction and the other axes randomly oriented.

specimens as a whole, the boundary regions comprised about 5% of the total volume. There often were variations in width of individual boundaries from one point to another.

We prepared samples from this material as rectangular blocks about a centimeter square by a few millimeters thick so as to expose the platelets edge on, see Figure 1. The grain boundaries were mostly aligned in parallel strips across the face of the sample. These specimens were preannealed in air at 1300°C for four to ten hours to equilibrate them with 0.2 atmosphere of oxygen, to relieve residual stresses, and to remove traces of free silicon. During the air anneal the color changed from a slight gray tinge to milky white. This may be due to oxidation of free silicon to finely-dispersed silica. They were then ground flat with front and back surfaces parallel and were polished on 0.03 μm alumina grit. Under a stereoscopic microscope, the grains were clear and the boundaries showed the refractive index discontinuities of a two phase system.

Electron microprobe examination has revealed a depletion of magnesim in the boundary region but an enhancement of silicon, calcium, and aluminum (Figure 2). Figure 2a is a target current image; Figures 2b-2e are Kd, x-ray images of the concentrations of various elements. This suggests that the phases might include pyroxene, (Mg, Ca)SiO_3, and augite, a pyroxene having aluminum in the structure. Thin section examination indicated

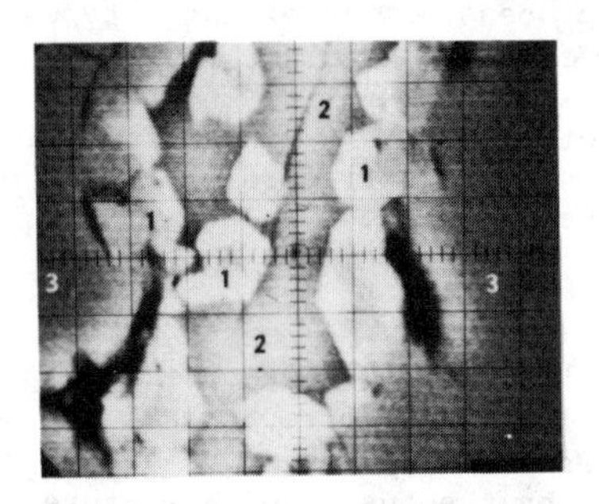

Fig. 2a

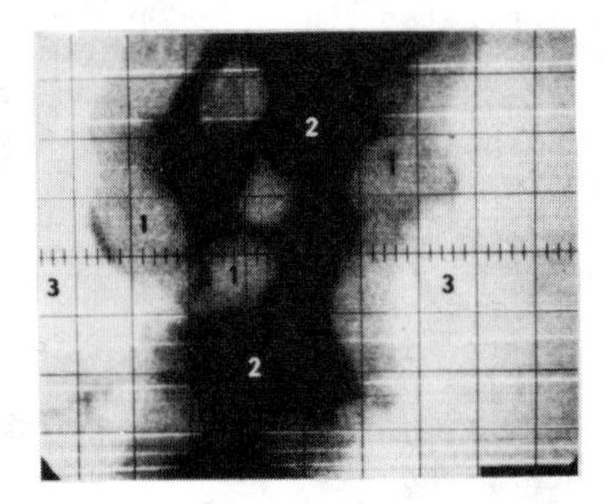

Fig. 2b

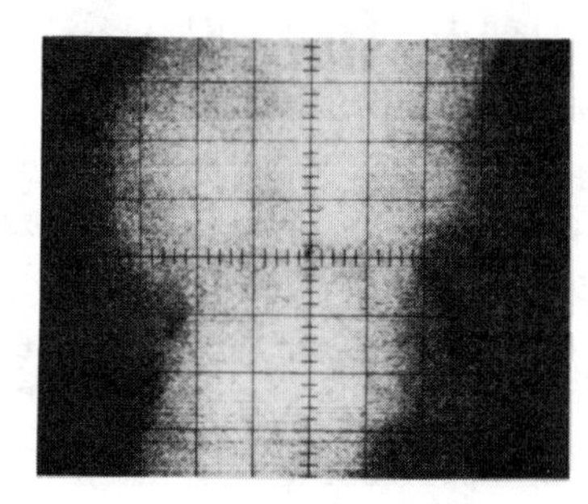

Fig. 2c

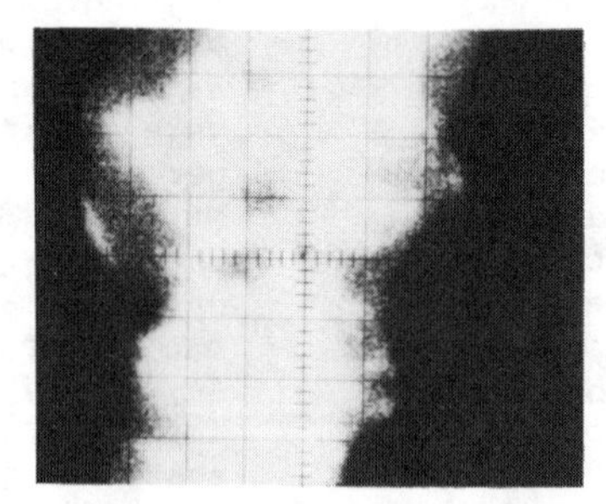

Fig. 2d

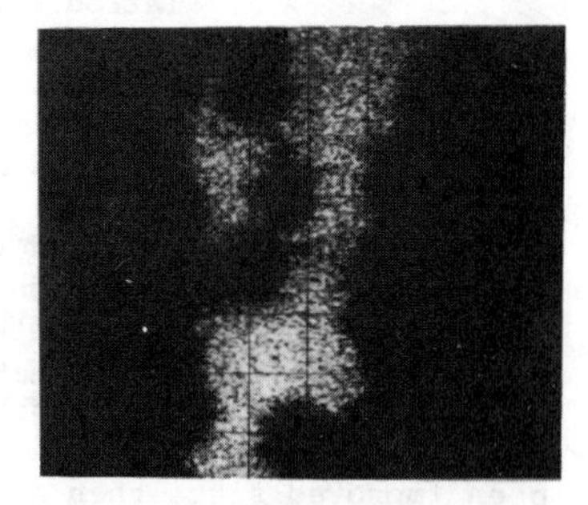

Fig. 2e

Fig. 2. (a) Electron microprobe target current scan of a boundary region shows three phases, (1) the forsterite matrix, (2) a second crystalline phase, and (3) an amorphous phase. The cross hatching on the grid marks are 1 μm intervals; electron microprobe x-ray image scans show (b) a depletion of magnesium, (c) an enhancement of silicon, (d) an enhancement of calcium, and (e) a slight concentration of aluminum.

that about half of the grain boundary material was amorphous.

Experimental Procedures

The specimens were annealed in oxygen-18 in a fairly conventional apparatus. They were loaded onto an alumina boat used in previous oxygen-18 studies, the only other ceramic in the hot zone, and this rested inside a platinum tube closed at one end and covered with a removable platinum door at the other. The tube was 1.5 cm in diameter and 10 cm long. It served as the susceptor for an induction heater. The door had a sight hole which allowed pyrometric measurement of temperatures of the specimens within the tube. The platinum tube rested inside a water cooled quartz tube envelope and the induction heating coil spiraled outside it.

We used an optical pyrometer to measure temperature in our experiments. It was calibrated with respect to an NBS standard lamp, window corrections were made, and the temperature gradients within the platinum tube were determined. The temperature wandered during a run within a span of 14°C for our longest run at 1107°C, and otherwise within 5°C.

There could be a surface barrier to oxygen exchange between the solid and the gas phase, but from Reddy et al., [1980] it appears that this should not be a problem. On the basis of the foregoing we believe that the diffusion anneal parameters were satisfactory for obtaining meaningful data.

Bevel

Following the diffusion anneal, the specimens were beveled so that we could follow each grain boundary across the original surface and down the bevel, see Figure 3. The bevel surface was polished flat with 0.03 μm alumina. The bevel typically extended from the middle of one face to the edge and was measured with a micrometer mounted above the specimen which was held on a traveling microscope stage. This allowed us to translate the traverse along the bevel (ℓ) into a measure of distance (x) below the original solid-gas diffusion interface. This is recorded as the slope ratio, $d\ell/dx$, (bevel) distance to corresponding depth, in Table 1.

Activation

The oxygen-18 tracer is activated using a Van de Graaff accelerator. The reaction, $^{18}O(p,n)^{18}F$ gives fluorine-18. This is a positron-emitting isotope with a half-life of 109 minutes which decays back to oxygen-18. We have verified the reported yield for this reaction as a function of proton energy [Mark and Goodman, 1956]. The fraction of the original oxygen atoms which are activated is about 10^{-8} and so chemical contamination induced by the

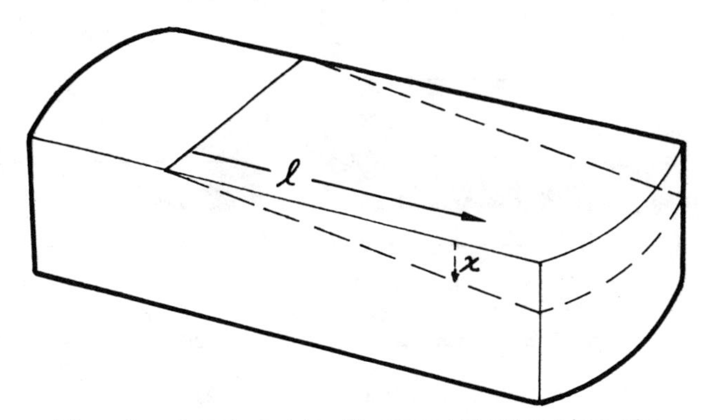

Fig. 3. A schematic diagram illustrating the beveling geometry.

transmutation is insignificant. The recoil distance of the fluorine-18 is about 0.2 μm, a distance small compared with the resolving power of the emulsion. An interesting feature of this technique is that one can bombard with proton energies only slightly greater than needed for the reaction, 3.1 Mev being optimum. Since the protons lose energy as they pass into the sample, only the oxygen-18 within a mean depth of 2 μm of the surface is activated. Thus, surface distribution features can be easily resolved using autoradiography without having to contend with a high background from subsurface activation. This distinguishes ion bombardment from neutron bombardment which produces a more homogeneous activation.

The beam, normally 3 μamp current, was expanded to about a five millimeter spot diameter and was swept back and forth across the specimen surface several times per second to achieve homogeneous activation. In order to monitor the uniformity of the activation, we used the $^{50}V(p,n)^{50}Cr$ reaction on a vanadium foil followed by autoradiography of the foil.

Care was taken to assure that the specimen was not overheated by the proton beam; it was mounted on a cold finger filled with liquid nitrogen using a special thermally conducting silicone grease as an adhesive. The specimen's surface temperature was conservatively estimated to be less than 500°C during the 60 to 90-minute irradiations. This was determined from calculations using thermal conductivity data on forsterite and from scaling up the beam flux and observing when melting in test specimens occurred. Thus, there should have been negligible migration of the tracer during the irradiation.

Autoradiography

Although we are using a positron emitter, fluorine-18, the positrons may be treated as a beta-rays for the purposes of interpreting our data and we can consult the large literature on the behavior of beta rays [Paul and Steinwedel, 1955; Rogers, 1973]: their absorption on passage

TABLE 1. Diffusion Anneals

Anneal	Specimen Number D36	D27	D28	D46
Temperature °C	1107	1221	1325	1374
Duration, s	35000	7200	7200	7200
Bevel: $d\ell/dx$	22.7	215	41.5	2.87
Profile d log C/dt, mm	0.28	0.14	0.50	0.042
d ln C/dx, mm	14.6	70.6	52.3	2.75
Apparent mean penetration, μm	68	14	10	364
G.B. analysis bulk m^2/s	2.1×10^{-21}	2.0×10^{-20}	1.4×10^{-19}	4.0×10^{-19}
Width, μm	10	3	3	10
Boundary diffusion D_b, m^2/s	1.3×10^{-16}	1.1×10^{-16}	6×10^{-16}	1.1×10^{-13}
Slab diffusion D, m^2/s	1.4×10^{-13}	1.3×10^{-14}	1.8×10^{-14}	9.1×10^{-12}

through matter, their reflections, and their effect on photographic film. Thus, we know how radiation flux falls off with distance from its source due to absorption and to an inverse squared distance law. Beta rays suffer reflections at material density discontinuities, but since we preserve the same geometry for all measurements, this has not concerned us.

Fluorine-18 activity was induced at those points where the original oxygen-18 tracer was distributed and photographic film was used to record its distribution and concentration. We used a strippable emulsion, Eastman Kodak AR-10. This is supplied on a glass plate from which it is stripped and floated on water. The specimen is brought up under it so that the emulsion drapes over the specimen to make intimate contact across both the unbeveled and beveled surfaces. The emulsion is initially 5 μm thick and is supported on a 10 μm gelatin backing but after becoming wet, the film distends a little. Thus, as it dries on the specimen the effective thicknesses are 3.7 (the emulsion in direct contact with the specimen) and 7.4 um (the backing over it). In our procedures we draped a second emulsion over the first so that two autoradiographs were obtained simultaneously, one from direct contact with the specimen and the second elevated above at a distance, 11.1 μm. After exposure together the two were separated again. They were developed as separate film pieces but in the same developer chemicals. The first emulsion provides good spatial resolution, less than 5 μm, while the second more accurately integrates activity from a broader surface area and is used to measure tracer activity along the grain boundaries, see Figures 4a and 4b.

In general, we took exposures for several different times to calibrate the response characteristics of our film using the same spot on our specimen, and correcting for the decay of the fluorine-18. With the activations and film we used, the most convenient optical densities were obtained after an hour exposure, starting with a freshly irradiated specimen.

Grain Boundary Profile Analyses

Analysis of the autoradiographic data requires careful attention, because: (1) the relationship between activity in a narrow region such as a grain boundary and the optical density induced in a film is not necessarily simple and (2) our grain boundaries have widths which can vary along their length and the autoradiographic trace must be

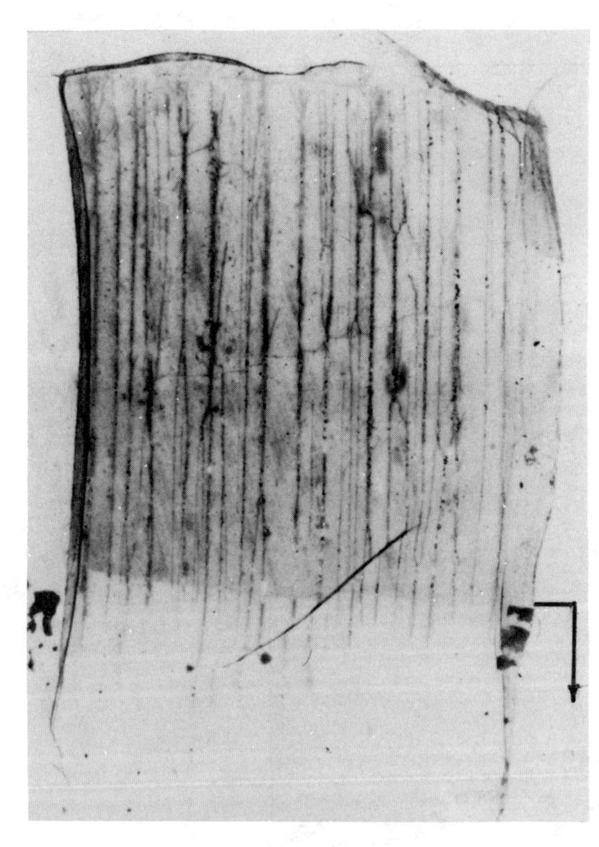

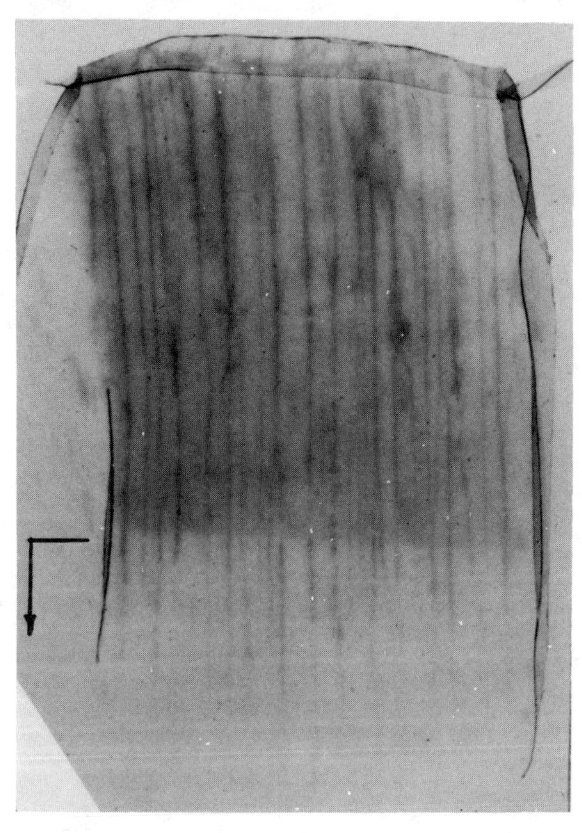

Fig. 4a

Fig. 4b

Fig. 4. Autoradiographs of specimen D46. The emulsion in direct contact (a) gives good resolution while the second emulsion, above it (b) integrates radiation exposure from a larger area. The bevel starts about two thirds of the way along the specimen surface.

related to specific features on each specimen to properly evaluate the data.

We scanned our film using a Perkin-Elmer Micro-10 Microdensitometer-50 which automatically traverses the whole autoradiograph and records the measurements of optical density for each point on computer tape. Using a 5 μm square spot size gives four million values per square centimeter. We can process the data in various ways using a program called SOCKITTOME [Derby et al., 1980]. We can excerpt portions of the data so as to draw isodensity contours for large or small areas with any desired detail or to automatically track along grain boundaries and plot an integrated value for optical density straddling the boundary.

We need to relate our measurements of film optical density to exposure which is defined as the product of duration times radiation flux. Fortunately, a "reciprocal" relationship can be assumed to apply [Rogers, 1973]; the optical density of a film is only dependent on the exposure, not on the time or flux separately. We have been able to use a range of exposures where the optical density is approximately proportional to exposure corresponding to the straight portion of the exposure-density curve for photographic films.

The flux through a given point in the film does not depend only on the activity at the corresponding spot on the specimen; more distant points contribute to the beta-ray flux as seen at the point. This is not a serious issue when a homogeneous source of activity is used. However, when there are steep concentration gradients in activity as there are from a grain boundary region to the neighboring crystal matrix, the film density at a given point will be a sum of contributions from near and distant points. The effect is known in autoradiography as "cross-fire" [Passioura, 1972] and has been a matter of continual concern to users of autoradiography.

No completely satisfactory method for handling cross-fire has yet been developed, but for our purposes we have been able to cope: (1) by using a small film scanning spot, the 5x5 μm square, for our optical density measurements, and (2) by making our optical density measurements on the second emulsion which was elevated 11.1 μm above the specimen surface in the double emulsion technique mentioned above. Our computer modelling then indicates that a satisfactory way to

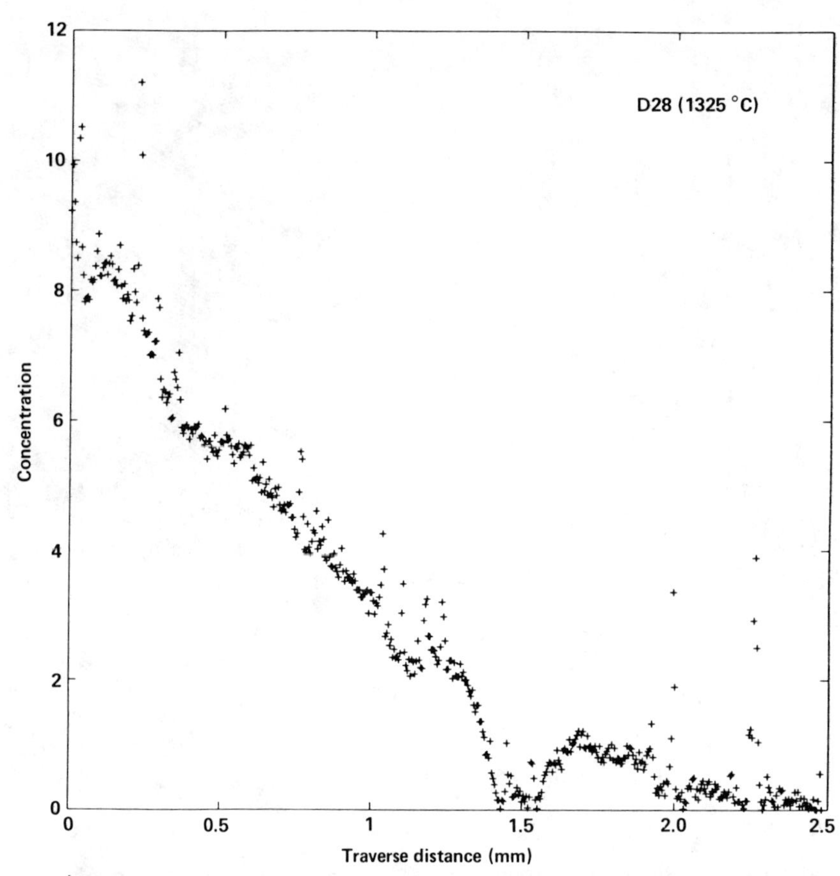

Fig. 5. Densitometer traces for the specimen annealed at 1325°C. This is plotted as a measured function of traverse distance across the autoradiograph. The bevel starts at the left edge of the plot. Concentration is measured in arbitrary units.

calculate the activity at a point along a grain boundary region is to integrate the optical density across the region width plus whatever distance to either side exhibits measurable darkening from the boundary. In practice this straddle extends a few tens of micrometers to either side of the region. Integration could be achieved using a rectangular densitometer spot scan which spanned the width of the region or, as we did, numerically summing from data arrays.

The varying width of the grain boundary regions means that there is more tracer registered at one point than another, but we are only interested in the concentration of tracer per unit volume of grain boundary region material. Analyzing the regions in full detail would require that we trace out the width of each region over its full length and depth below the original gas-solid exchange surface. This could be accomplished by beveling in small slices and keeping a photomicrographic record of the surface after each slice. Thus far, we have been content to select regions which appear fairly constant in width.

The densitometer plots are printed from the computer as a function of traverse distance (ℓ) along the autoradiograph or along the sample bevel. Our results are shown in Figure 5. It is a simple matter to translate these into depths below the initial surface using measurements of the surface contours of the beveled specimen and noting how they relate to positions on the autoradiograph.

Deriving Grain Boundary Diffusion Coefficients

After the foregoing analysis, one can feel justified in taking the oxygen concentration to be directly proportional to the film optical density trace along a grain boundary through the series of linear relationships, (1) optical density proportional to exposure, (2) exposure proportional to F-18 concentration, and (3) F-18 concentration proportional to O-18 concentration. In recording values of optical density, we have eliminated those portions where visual inspection of the autoradiograph reveals dust or other artifacts in the film or where there has been pull-out of grain boundary material during polishing so that no O-18 enriched material was present. Also, we have ignored regions where there were obvious fractures in the grain boundaries. The best fit of log of concentration, log C, versus traverse distance ℓ, is given as d log C/dℓ in Table 1. It is obtained by least-squares analysis of log C vs ℓ, assuming that log $C = A_o + A_1 \ell$. Other data and results are also summarized in Table 1.

For a general discussion of diffusion and the treatment of penetration profiles the reader should see Shewmon [1963] who includes a derivation of the Fisher equation [Fisher, 1951] for grain boundary diffusion with references to some subsequent analyses.

We have plotted the profiles in two ways in Figure 6, (1) as log (C) versus x, the Fisher equation applicable to grain boundary diffusion; and (2) as error function complement, erfc (C/Co) versus x, applicable to slab diffusion. The formula for grain boundary diffusion is

$$D_b = \frac{2}{d} \left(\frac{D_v}{\pi t}\right)^{1/2} \left(\frac{d \ln C}{d x}\right)^{-2} \quad (1)$$

and for slab diffusion is

$$C/C_o = \text{erfc } (n) \quad (2)$$

where n is related to D by

$$D = x^2/4 \ t \ n^2 \quad (3)$$

where x is penetration distance, d, is grain boundary width, Dv is the bulk diffusion coefficient, t is the duration of the high

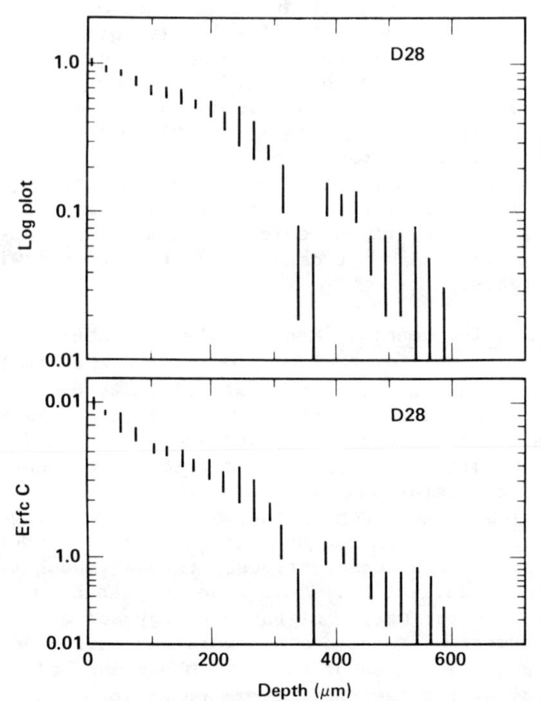

Fig. 6. Results from equations (1) and (2) are plotted as log C and erf (C/Co) versus depth, χ.

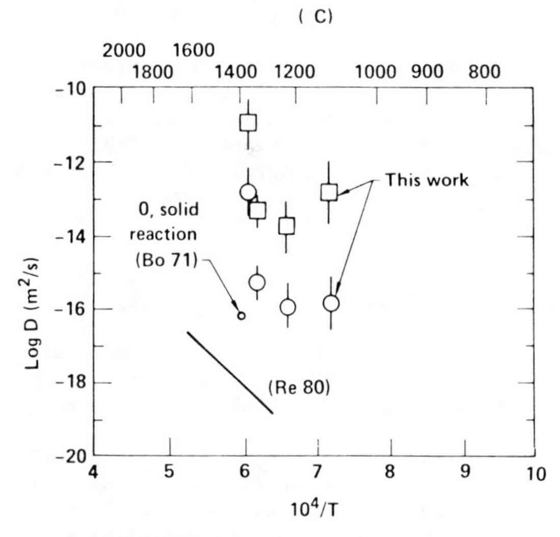

Fig. 7. Data for oxygen diffusion coefficients in forsterite as a function of reciprocal temperature kelvins include our data (as computed by the grain boundary and the slab models, "log plot" and "erfc plot," respectively), self-diffusion data as determined by Reddy et al. [1980] and as derived from the $MgO-SiO_2$ reaction by Borchardt and Schmalzried [1971].

temperature anneal, and C/Co is the ratio of concentration at x to that at the surface. The Fisher equation calls for values of the bulk oxygen diffusion coefficient which are available for forsterite single crystals [Reddy et al., 1980]. Experimental results summarized in Table 1 are plotted in Figure 7.

Discussion of the Data

It is clear from the experimental data that oxygen has preferentially penetrated the forsterite specimens along grain boundary regions. The data do not allow us to determine whether a better fit is provided by an error function plot or a log plot, to clearly differentiate between slab or a grain boundary diffusion. The marginal quality of the data is due in part to the intrinsic scatter in grain boundary transport rates. This is always likely when the distribution of foreign material along a grain boundary is inhomogeneous. It also results from the initial difficulties we have had in working with a Van de Graaff proton accelerator. We required that the machine be adapted to a use for which it was not originally intended; we called for a broad, uniform beam, tight control over beam voltage, and limitation of beam current to avoid sample burn-out. We have spent a considerable amount of time in the development of the techniques, and we report here their first application to measurements of oxygen diffusion in silicates, specifically grain boundary diffusion in forsterite.

In the absence of experimental evidence concerning slab versus grain boundary diffusion, we should attempt some estimate based on experience elsewhere. Atkinson and Taylor [1981] have studied a diffusion problem which has some aspects that are similar to our own, nickel diffusion in nickel oxide. They were interested in the relative contributions of bulk and grain boundary diffusion to the oxidation of nickel. They point out that there are normally two conditions for an experiment to yield information about grain boundary diffusion: (1) the mean penetration through the bulk, $(D_V t)^{1/2}$, should be small compared with the separation between the boundaries, i.e., there is no interaction between grains; and (2) the width of the boundary should be small compared with the mean penetration through the bulk, i.e., the amount of diffusant in the boundaries is a small fraction of the total. The first condition is satisfied in our work, but the second is not. On this basis, it appears that we should expect slab diffusion analysis to be appropriate to our data.

In order to properly evaluate the data we need to know: (1) the chemical makeup of the material in the boundary, (2) the state of the material, and (3) the dislocation density around the boundary. In the bulk region near the boundary, we need to know diffusion rates in dislocations in the close packed oxide structures of forsterite. Measurements in MgO have revealed rapid diffusion in dislocations [Holt and Condit, 1966], but quantitative values have not been obtained. Yund et al. [1981] have reported data for dislocation-assisted diffusion of oxygen in albite, but the overall effect is to increase the bulk diffusion coefficient by less than an order of magnitude.

The technique used in this work allows direct observation of diffusion along imperfection paths in a specimen by tracer activation followed by autoradiography. In this work we have illustrated the importance of this in several respects. At the highest temperature, 1374°C, there was localized melting at the grain boundaries, and enhanced diffusion was associated with the presence of localized melt regions. At the lowest temperature, 1121°C, penetration of oxygen along cracks in the specimen was observed. The diffusion coefficient which we measured at 1325°C appears to be characteristic of the solid material in the grain boundary region. It seems plausible that the mixture of phases in the region is likely to contain a high concentration of dislocations and multiphase interfaces which should provide more paths for transport than would be expected in pure single phases of any of the components. Indeed, the diffusion coefficient seems a bit higher than one might expect from other measurements. In the reaction between silica and magnesium oxide to give Mg_2SiO_4 [Borchardt and Schmalzried, 1971] the rate of reaction appears to be governed by grain boundary diffusion. The value which they derive is a little lower than ours. Their materials were purer and the absence of large amounts of second phase material at their boundaries may account for this difference.

Discussion of the Oxygen Tracer Technique

The value of our technique is that it allows the use of oxygen as a radiotracer to measure its concentration and position by autoradiography. An autoradiograph collects data from over the full area of the specimen with a resolution of about 5 μm and afterward it can be examined at leisure. The protons only activate the tracer within a few microns of the surface which further enhances the value of the technique. The limitation in the case of the oxygen-18 tracer is that the natural background of oxygen concentration in all oxides is 0.2%. This means that diffusion down single dislocations will probably be undetectable, but diffusion along clusters of dislocations has been found [Holt and Condit, 1966]. Bulk diffusion can only be measured when the mean penetration of oxygen is at least 2 μm. The penetrations in bulk diffusion studies in olivine are typically much less than this and therefore other measurement techniques are more useful [Reddy et al., 1980]. However, oxygen diffusion in beryllium oxide has been measured in this way with coefficients as small as 6×10^{-17} m^2/s with a mean penetration distance of 10 μm [Bradhurst and de Bruin, 1969]. Its greatest value may turn out to be in observing grain boundary and second phase diffusion mechanisms. It should also find application in studies of solid-solid reactions [Holt, 1967] where oxygen diffusion may be the rate limiting process. Migration of oxygen-18 from a labeled speck of mineral material within a matrix under reaction and creep conditions would reveal much information about transport paths which is not otherwise available.

Acknowledgements. Many people have contributed to development of the techniques discussed in this paper. The LLNL Van de Graaff Facility headed by Richard Fortner with Marvin Williamson and Tarry Schmidt have provided service and advice during the irradiations. The photometric measurements and interpretations have been greatly assisted by Daniel Dietrich, Henry Finn, and Carl Frerking. Summer students Ralph Wheeler and Allison Anderson Connor labored conscientiously through much of the data collection. This research has been funded by the Office of Basic Energy Sciences, Department of Energy. This work was performed under the auspices of the U.S. Department of Energy by the Lawrence Livermore National Laboratory under contract W-7405-ENG-48.

References

Atkinson, A., and R. I. Taylor, The diffusion of ^{63}Ni along grain boundaries in nickel oxide, Philos. Mag. Part A, 43, 979-998, 1981.

Borchardt, G., and H. Schmalzried, Silicate formation in the solid state, Z. Phys. Chem. N.F., 74, 265-283, 1971.

Bradhurst, D. H., and H. J. de Bruin, The self-diffusion of oxygen in beryllium oxide by proton activation of oxygen-18, J. Aust. Ceram. Soc., 5(1), 21-27, 1969.

Calvert, J. M., D. J. Derry, and D. G. Lee, Oxygen diffusion studies using nuclear reactions, J. Phys. D., 7, 940-953, 1974.

Derby, W. S., R. L. Herrick, M. E. Hummell, E. C. Lee, and R. D. Neifert, SOCKITOME, an Interactive Data Processing Code, UCID-17733, Lawrence Livermore National Laboratory, Livermore, Calif., Jan., 1980.

Fisher, J. C., Calculation of diffusion penetration curves for surface and grain boundary diffusion, J. Appl. Phys., 22, 74-77, 1951.

Holt, J. B., Role of oxygen in solid-solid reactions, in Sintering and Related Phenomena, edited by G. Kuczynski, N. A. Hooton, and C. F. Gibbon, pp. 169-190, Gordon and Breach, New York, 1967.

Holt, J. B., and R. H. Condit, Oxygen diffusion in surface defects on MgO as revealed by proton activation, in Materials Science Research, vol 3, edited by W. W. Kriegel and H. Palmour, III, pp. 13-29, Plenum, New York, 1966.

Mark, H., and C. Goodman, Angular distribution of neutrons from ^{18}O (p,n) ^{18}F, Phys. Rev., 101, 768-771, 1956.

McKenzie, D. R., A. W. Searcy, J. B. Holt, and R. H. Condit, Oxygen grain-boundary diffusion in MgO, J. Am. Ceram. Soc., 54, 188-190, 1971.

Passioura, J. B., Quantitative autoradiography in the presence of crossfire, in Microautoradiography and Electron Probe Analysis, edited by U. Luttge, pp. 50-59, Springer-Verlag, New York, 1972.

Paul, W., and H. Steinwedel, Interaction of electrons with matter, in Beta-and-Gamma-Ray Spectroscopy, edited by K. Siegbahn, pp. 1-23, North-Holland, Amsterdam, 1955.

Reddy, K. P. R., S. M. Oh, L. D. Major, Jr., and A. R. Cooper, Oxygen diffusion in forsterite, J. Geophys. Res., 85(B1), 322-326, 1980.

Rogers, A. W., Techniques of Autoradiography, 2nd ed., Elsevier, New York, 1973.

Shewmon, P. G., Diffusion in Solids, McGraw-Hill, New York, 1963.

Yund, R. A., B. M. Smith, and J. Tullis, Dislocation-assisted diffusion of oxygen in albite, Phys. Chem. Miner., 7, 185-189, 1981.

AN APPROACH TO ANALYZING DIFFUSION IN OLIVINE

Ralph H. Condit

Lawrence Livermore National Laboratory
Livermore, California 94550

Abstract. Olivine is pictured as a hexagonal close packing of oxygen ions which provide a matrix for cation diffusion. The network of paths which a cation may take when jumping between normal and interstitial sites is described, and an approach to the analysis of diffusion in this network is suggested. In intermediate diffusion configurations several point defects may be generated which have preferred distances and orientations relative to one another. A succession of these configurations then lead to a final one with the migrating atom in a new site. It is proposed that oxygen diffusion may be by cation-anion divacancies. Two published sets of data on silicon diffusion coefficients differ by four orders of magnitude and it is suggested that this may result from differences in dissolved silica. Measurement of correlation coefficients would make an important contribution to determining actual mechanisms of cation and silicon diffusion.

Introduction

A considerable amount of data has been gathered on diffusion in olivine $(Mg,Fe)_2SiO_4$ and forsterite, Mg_2SiO_4, the pure Mg end member of the phase. One purpose of this paper is to assemble the essential features of this data which may aid in understanding diffusion mechanisms. A summary of some of the main features is shown in Figure 1. For a detailed tabulation of data, the reader is referred to Freer [1981] and Morioka [1980, 1981]. A second purpose is to suggest a way to relate observed diffusion rates to a detailed picture of how atoms jump between normally occupied sites, although no mathematical treatment is attempted. This would have the practical value of providing a lead for going from measured material transport rates to predictions about transport under conditions of temperature and pressure which are not accessible in the laboratory but would be of geological interest.

The situation in olivine is more complicated than in the much-studied alkali halides. There are many more possible jumps between normal (stable) and interstitial (metastable) sites, and the paths connecting them make a network which branches at many points. We will try in this paper to indicate what data is necessary to identify networks and to suggest likely paths. We do so in the belief that even such a vague statement of the right problem is better than a precise formulation of the wrong problem, i.e., to assume that a simple vacancy or interstitial mechanism describes the essentials of actual processes.

The Hexagonal Close-Packed Matrix

Olivine may be regarded as a hexagonal close packing of oxygen anions with the cations nestled in the interstices between them. Therefore we will first discuss some general properties of the hexagonal close-packed (HCP) matrix and then outline a view of the olivine structure using this approach. In Figure 2 the interstices in the HCP lattice are shown with the atoms removed. This attempts to be the best simple representation of the structure that one would see if a HCP stack of metal balls were cast in plastic and the balls were then dissolved away. For each large atom making up this matrix there is one octahedral site, o-site, in which an interstitial atom (if present) would be surrounded by six of the large atoms, and there are two tetrahedral sites, t-sites, in which an interstitial would be surrounded by four large atoms. In our diagram we have shown the o-sites as having a cubic shape. The six cube faces are planes tangent between the interstitial atom and the large atoms around it. The faces of the tetrahedra are also tangent planes, four planes with the four surrounding large atoms.

If the large atoms are in ideal "hard sphere" packing, the sizes of these interstices are such that the radius ratio for an atom in the o-site to the large atoms can be no larger than 0.41.

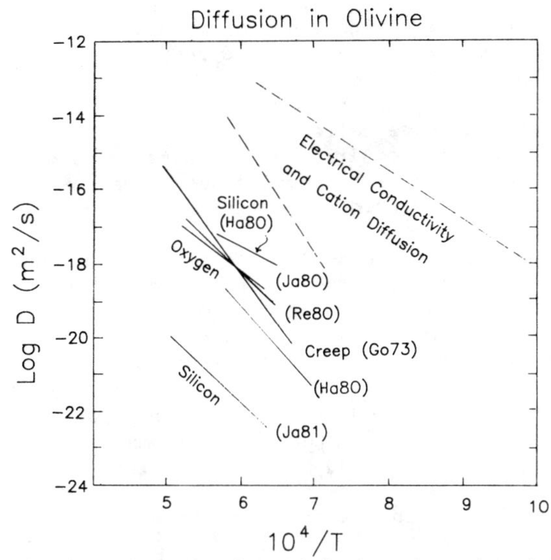

Figure 1. Diffusion in olivine. Diffusion data for olivine are summarized in an Arrhenius plot. The measurements of electrical conductivity which can be translated into cation diffusion and cation diffusion are too numerous to plot individually. Data are taken from Freer [1981], Morioka [1980, 1981], and Duba et al. [1974]. Oxygen diffusion has been measured by Reddy et al. [1980] and Jaoul et al. [1980] and Hallwig et al. [1980] and has been deduced from dislocation climb by Goetze and Kohlstedt [1973]. Silicon diffusion has been measured by Jaoul et al. [1981] and Hallwig et al. [1980].

In a t-site it can be no larger than 0.22. To pass through an orifice between three large atoms it can be no larger than 0.15.

In forsterite we observe from Figure 1 that the divalent cations diffuse more rapidly than oxygen by several orders of magnitude. The data on diffusion of silicon is somewhat perplexing and will be discussed separately below but does not alter our basic picture of oxygens providing a HCP matrix. The slow anion diffusion has a counterpart in other HCP structures, FeS [Condit et al., 1974] and BeO, the wurtzite structure [Condit, 1974], where cation diffusion is by a vacancy mechanism. This supports a picture of a relatively inert anion matrix through which the smaller cations may move. We may suppose that cations do not make such jumps as require them to pass through a tangent point between two anions. Thus, a jump between o-sites in the HCP basal plane might be expected to require an o-t-o sequence of jumps. These jumps are through the orifice between three anions and call for less elastic strain of the anion matrix lattice than would a direct o-o jump in the basal plane. This is further supported in FeS and BeO by the fact that the activation energy for diffusion is the same for the a- and c-directions; the rate limiting jump in these networks is probably the same. This would not be true if direct o-o basal plane jumps were part of the a-diffusion mechanism.

We propose, then, that the preponderance of jumps are between interstices and that these interstices are arranged in a network of a sort shown in Figure 2. We can then discuss a- and c-diffusion in terms of jumps along the lines which connect these interstices. The o-sites are arranged in chains in the HCP c-direction. The t-sites occur in pairs between these chains. The

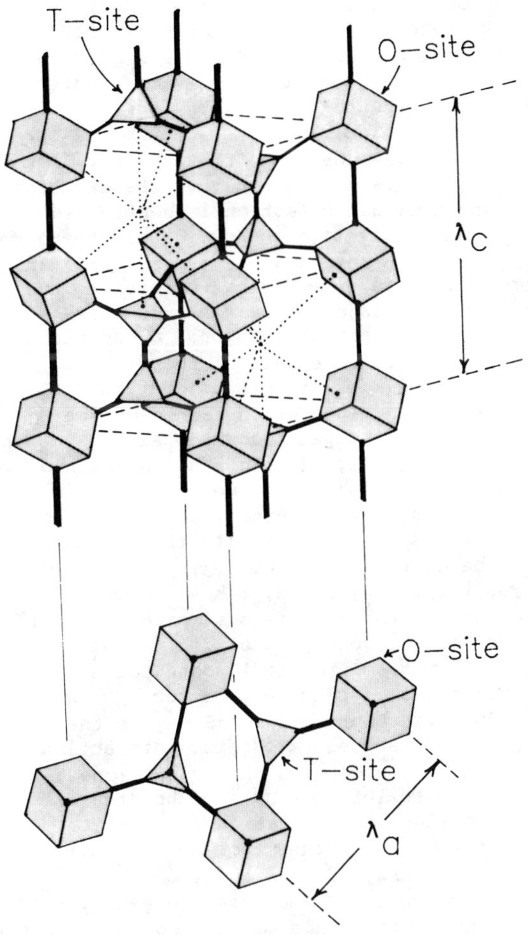

Figure 2. The HCP structure. The interstices in a hexagonal close packed (HCP) unit cell are shown in their approximate shape. There are two types of interstices, tetrahedral (t) shown here as tetrahedra and octahedral (o) shown here as cubes. The flat faces in both sites are the planes of contact between the matrix sphere and an atom if present in the interstitial site. The positions of the large atoms must be imagined in this figures, but their centers would be where the dotted lines converge. In a simple HCP unit cell there are two matrix atoms, two o-sites, and four t-sites.

o-sites neighboring a given o-site include the ring of six around it in the basal plane plus the two in line with it along the HCP c-axis plus the twelve which lie directly above and below the ring of six making a total of twenty. The last twelve o-sites are not nearest neighbors in distance from the origin site, but they are connected to it by an o-t-t-o path; this same type of path can take an atom from one o-site to either of the two o-sites in line with it along the c-axis.

In FeS having the NiAs structure cations are on o-sites, but o-o direct jumps are not favored. From the Dc/Da ratio and its dependence on cation vacancy concentration, it appears that this is not related to the correlation effect which is discussed below. The activation energies for diffusion are the same in the a- and c-directions, and this suggests that the energy barriers for an o-t-o- (a-direction) and o-t-t-o jump (c-direction) are the same, that the t-t jump barrier energy is small, and that the rate for diffusion in either direction is controlled by the o-t jump activation energy. While there may not appear to be any reason for predicting such a preference, it is interesting to refer to the Ni_2In structure [Kjekshus and Pearson, 1964]. The indium atoms are in hexagonal close packing in this material. One half of the Ni atoms are in the o-sites (similar to NiAs) but the remainder are in sites which represent the midpoint in the t-t pair shown in Figure 2. This suggests that passage through this point and even occupancy of this site does not require high energy compared with alternatives such as the o-o jump, at least in a system with the cation sublattice of the NiAs type.

In FeS the ratio, Dc/Da, makes it appear that cations prefer to take diagonal paths which lead from one normally occupied o-site to another which is displaced from it in both the a- and c-directions, a displacement vector c/2, a. A reason for this behavior in FeS may be the nature of the elastic and electrostatic interactions between the two vacancies on o-sites (one being the site the Fe left behind and the other its destination) and the Fe cation in a t-site between them in the transition configuration. The minimum energy of this three body group would place "like charges," the vacant o-sites, as far apart as possible. This would tend to align the three defects into a more or less straight line with the iron occupying t-site in the middle. Thus, a "straight" o-t-t-o jump is preferred over a "loop back" o-t-t-o jump which would give the same net displacement as an o-o jump.

In the wurtzite structures, BeO and ZnO, the cations normally occupy t-sites, but activiation energies for diffusion in the a- and c-directions are again the same. This suggests that an extended series of jumps in the c-direction along an -o-o-o- chain is still unfavored. There also is evidence from the Dc/Da ratio that an atom cannot traverse a possible path between normally occupied t-sites unless an atom on an adjoining site moves aside. This would reduce the repulsion between cations, one (in transit) on an o-site and one in an immediately neighboring t-site. We have called this a "key-lock" mechanism; it requires the stepping aside of one atom to open a path before a second can pass through. Afterwards, the first atom then falls back into place.

From these two examples, it appears in some anion matrices that the cation jump mechanism depends on more than a random walk of the cation alone. The intermediate, discrete configurations must be articulated. These clusters might be called "diffusion molecules" to suggest that there are preferred orientations and spacings between vacancies, interstitial atoms, and momentarily displaced neighboring atoms.

The Olivine Structure

In olivine the atomic arrangement may be depicted in several ways, see Figure 3. The hexagonal close packing of anions is in the b-c plane of this structure. This packing is not perfect HCP because the cations are a bit larger than the interstices would allow by a hard sphere model. There are three types of oxygen atom sites distinguished by their coordination with the divalent cations and silicon ions. All of the oxygens may be regarded as parts of tetrahedra surrounding silicons: in one of these, designated 03, the oxygen ion is directly above or below silicon atom in the a-direction while in the other sites, 01 and 02, the oxygens are at one of the three remaining positions and are more closely associated with the divalent cations. Divalent cations normally occupy two types of sites designated M1 and M2. The M1 sites are aligned in rows extending in the C-direction. The M2 sites may be viewed as spurs extending to the sides of the M1 chain. In olivine only 1/2 of the o-sites are normally occupied, 1/4 as M1 and 1/4 as M2 sites. There are two types of normally unoccupied o-sites. One of these lies directly between two silicon atoms while the other lies between a silicon atom and an unoccupied t-site. There are six types of t-sites, one being normally occupied by silicon.

We propose that in olivine a cation having jumped from an o- to a t-site might just as plausibly move in the olivine b- or c-directions as in the a-direction. Diffusion may take place in the b-c direction by o-t-o jumps, and in the a-direction by o-t-t-o jumps, but not o-o jumps. Figure 2 reveals that the octahedral interstices in a HCP anion matrix line up as a chain along the HCP c-direction; in olivine this is the a-direction. In olivine this might lead us to expect rapid diffusion in the a-direction. This is contrary to observation. Moreover, the activation energies for divalent cation diffusion in the a- and b-directions are the same and so we propose exclusion of o-o jumps. Thus, we cannot

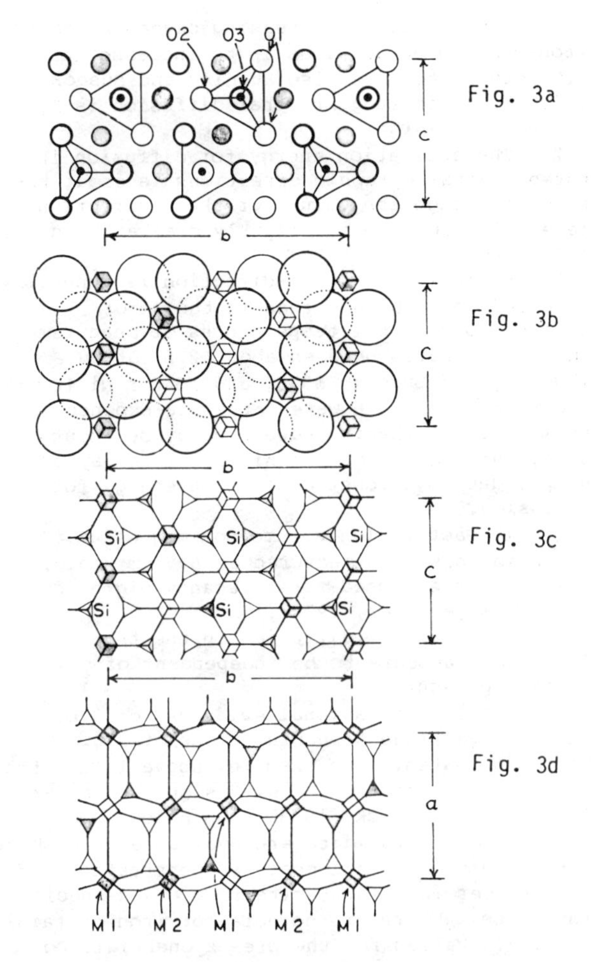

Figure 3. The olivine structure (a) This plan follows Wells [1975]. Small black circles represent Si, shaded circles Mg, and open circles O. The oxygen tetrahedra around Si atoms are indicated by the connecting lines. The three types of oxygen sites, O1, O2, and O3 are indicated; (b) the same structure as in Figure 3a is shown with the difference that the real relative size of the oxygen ions is more closely represented to reveal the HCP character. Those oxygens which are shown are x/4 above and x/4 below the b-c plane. The octahedral interstices between the O atoms are shown as end-on views of cubes. They lie in the x=0, 1 plane. Only those containing Mg ions in the x=1 plane are shaded; (c) the same view of the structure as in Figure 3b with the difference that the oxygen atoms have been removed to show the complete interstice network. The tetrahedral interstices are added. Those sites containing Si atoms visible at the b-c plane are shaded; (d) this is the same manner of presentation as Figure 3c but rotated to give a side view of the lattice. All o- and t-sites which are filled as viewed in the c-direction are shaded. The M1 and M2 cation sites are indicated. M1 sites are found on y=0, 1/2, 1 planes and M2 sites are found on y=1/4, 3/4 planes.

conclude that because inspection reveals a chain of sites in the a- or c-direction, o-o-o or M1, that this provides any explanation for rapid diffusion in that direction.

In the olivine structure, Mg_2SiO_4, the magnesium ions can be replaced by numerous others [Wyckoff, 1968]. Morioka [1980, 1981] has measured the diffusion of Mg, Ni, Mn, Co, and Ca. When iron or similar variable valence ions substitute for part of the magnesium, the stoiciometry of the compound depends on the oxygen partial pressure. We can describe this using the Kröger notation as the reaction,

$$2Fe_M + 1/2\ O_2 \rightleftarrows 2Fe_M^{\cdot} + V_M'' + O_O \qquad (1)$$

i.e., two ferrous ions on normal cation sites, M, become ferric ions, a new M-site is created which is vacant, and a new oxygen site is created in which the former oxygen gas atom now resides. By this mechanism the vacancy concentration on cation sites may be controlled. Actually, the olivine structure does not seem able to dissolve MgO or FeO, and reaction (1) must be accompanied by exsolution of some cation oxide.

The energies required to form vacancies on various sites in olivine have been calculated by Lasaga [1980]. He concludes that cation vacancies should appear more readily in M1 than in M2 sites. The preference of other divalent cations for M1 versus M2 sites has also been investigated, and a body of literature is developing on the subject, see for example Will and Nover [1979], Bish [1981] and Taftø and Spence [1982]. It would be valuable to have data on order-disorder processes as a function of temperature, since these may have a marked effect on diffusion rates.

Silicon atoms rest at the centers of oxygen tetrahedra and all silicon sites are equivalent, see Figure 3a drawn after Wells [1975]. Each oxygen ion can be assigned an association with only one silicon atom; none are shared between two silicons. While the discussion of divalent cation diffusion has treated the oxygen sublattice as HCP, an inquiry into silicon diffusion must recognize possible integrity of the SiO_4 unit. Its reality is suggested by the fact that the Si-O bond distance does not increase with temperature to the same extent as does the divalent cation-oxygen distance [Lager and Meagher, 1978]. This suggests the possibility that an SiO_4 unit or a SiO_3 unit (if one oxygen site is vacant) could rotate as a unit. This could contribute to oxygen diffusion. If one oxygen is absent, the SiO_3 might have a pyramidal structure, partly preserving the O-Si-O bond angles characteristic of SiO_4.

The olivine phase appears capable of dissolving a little silica [Pluschkell and Engell, 1968]. The mechanism by which this may occur has been discussed by Schmalzried [1978] who suggests that vacancies on Si sites or Mg substituted on Si sites are unlikely, and by Stocker [1978] who suggests that the reactions accompanying dissolution may be

$$2SiO_2 \rightleftarrows 2\,V_{Mg}'' + Si_I^{\cdot\cdot\cdot\cdot} + Si_{Si} + 4\,O_O \qquad (2)$$

or

$$SiO_2 \rightleftarrows 2V_{Mg}'' + Si_{Si} + 2\,V_O^{\cdot\cdot} + 2\,O_O \qquad (3)$$

Reaction (2) creates Mg vacancies and Si interstitials while (3) produces vacancies on both Mg and O sublattices. Electrical conductivity increases with SiO_2 content. Since the cations appear to be the current carriers, the increase in cation vacancies by either (2) or (3) corresponds with the data [Pluschkell and Engell, 1968]. However, silicon diffusion rates will depend more strongly on deviations from stoichiometry by (2) than by (3). If there are excess Si atoms in the lattice, Si on interstitial sites, they might pick sites which preserve the Si-O-Si bond angles characteristic of silicon compounds. These configurations must be determined if we are to understand silicon diffusion, and we are aware of no data on this at present. This point is discussed below in the section on silicon diffusion.

Material Transport Data

A principal feature of diffusion in forsterite (compiled by Freer [1981] and Morioka [1981]) is that the cations, Mg, Fe, and so on are much more mobile than oxygen. Morioka has correlated rates with: (1) cation size, the larger the ionic radius of the cation the less mobile; and (2) the concentration of point defects associated with the cations, higher vacancy concentration, allows higher mobility.

Data on electrical conductivity in olivine can be translated into values for diffusion coefficients using evidence that the current carriers are cations [Pluschkell and Engell, 1968]. The values obtained using the Nernst-Einstein equation (see Shewmon [1963] for example) fall in the same general region on the Arrhenius plot, Figure 1, as diffusion coefficients derived in other ways. Conductivity in iron-containing olivine increases also with oxygen partial pressure [Duba et al., 1974]. This would be expected for cation diffusion enhanced by increasing vacancy concentration according to equation (1). The conductivity in the c-direction is greater than in the a- or b-directions by a factor of about 2.5.

Direct tracer self-diffusion measurements are available for magnesium and calcium [Morioka, 1981]. All other diffusion data have been derived from interdiffusion couples where chemical gradients may perturb the self-diffusion values because of a superimposed net vacancy flow. However, the following features appear common for divalent cation diffusion:

1. The effect of increasing oxygen pressure in iron-containing olivines is to increase the diffusion rate (see also Buening and Buseck [1973]). This suggests that diffusion is by a vacancy mechanism.
2. The activation energy for diffusion in the three crystallographic directions is about the same, although the value for the c-direction is generally found to be slightly smaller than the other two.
3. Diffusion in the c-direction is generally observed to be higher than in the a- or b-directions at the temperatures of measurement. This ratio falls between about 2.5 for Mg [Buening and Buseck, 1973] and 15 for Ni [Clark and Long, 1971]. However, the pre-exponential term in the Arrhenius equation may be about the same; the reason for the higher value may be simply due to a lower activation energy for diffusion.
4. The activation volume for magnesium diffusion has been measured as 5.5 cm^3/mole, a value to be expected for a vacancy migration mechanism [Misener, 1974].
5. The ratio of diffusion in the three directions appears to be independent of oxygen partial pressure.
6. There is occasional evidence for a break in the straight line Arrhenius plots of cation diffusion rates. For example, above 1125°C the activation energy for Mg diffusion is greater [Buening and Buseck, 1973] than below.

Oxygen diffusion data are available from three investigators who all measured penetrations of O-18 tracer into forsterite from a gas-specimen interface. There are no data for iron-containing olivine. Values for the pre-exponential, Do, and energy values, E, in the Arrhenius equation are:

Reddy et al. [1980]

E = 415 kJ/mole   Do = 1.06 x 10-6 m2/s

Hallwig et al. [1980]

E = 475 kJ/mole   Do = 1.6 x 10-4 m2/s

Jaoul et al. [1980]

E = 320 kJ/mole   Do = 5.6 x 10-7 m2/s

The apparent features are:

1. The activation energy for diffusion is the same for the three crystallographic directions within experimental error.
2. The pre-exponential coefficients for diffusion in the a- and c-directions are about the same but are less than for the b-direction by a factor of perhaps 1.5.
3. Oxygen diffusion apppears to be oxygen pressure independent in the forsterite studied [Jaoul et al., 1980].
4. The oxygen diffusion coefficient deduced from dislocation climb [Goetze and Kohlstedt,

1973] agrees well with that measured by isotopic exchange.

Data on silicon diffusion are available from two groups of investigators. Their experimental techniques were similar; diffusion was measured from a thin deposit of Mg_2SiO_4 enriched in Si-30. The penetration of this tracer into the substrate forsterite was measured using an ion probe and specimens were sectioned by ion milling. The results are in striking disagreement. Values reported at the same temperature differ by over four orders of magnitude. The Arrhenius values for the pre-exponential, Do, and energy, E, are:

Jaoul et al. [1981]

E = 377 kJ/mol   Do = 1.5 x 10-10 m2/s

Hallwig et al. [1980]

E = 176 kJ/mol   Do = 8.5 x 10-13 m2/s

Jaoul et al. draw the following additional conclusions from their work: (1) silicon diffusion seems to be isotropic; (2) silicon diffusion is not a function of oxygen partial pressure in forsterite.

Discrepancies

It may appear difficult to reconcile some of the available data with our present understanding of atomic diffusion in ceramic and mineral materials. In olivine the following points seem to be puzzling:

1. The differences between measured silicon diffusion coefficients in forsterite present one of the more unusual disagreements in the study of solid state kinetics. In both groups, Jaoul and Hallwig, the experimental technique seems to be under good control and so one is inclined to wonder if there may not be differences in the materials. Tsukahara [1976] predicted diffusion rates for Si, Mg, and O in olivine based on the ion sizes and the sizes of the orifice through which the ion needs to jump in order to transit from one site to another. His predictions are in relatively good agreement with the observed magnitudes for Mg and O. There is poor agreement for Si as measured by Hallwig (the measured Do is too low, although the values in the measured temperature range agree), and by Jaoul (the values are low).
2. Dislocation climb provides us with an oxygen diffusion coefficient which agrees well with self-diffusion numbers. If Si is faster than O, then O would be rate limiting. However, if silicon is slower, we would expect this slowest mover to limit the rate of dislocation climb.
3. The measured activation energy for diffusion of oxygen is less than we might expect based on the calculation of energy for formation and migration of an oxygen vacancy [Lasaga, 1980].
4. The ratios of cation diffusion in the a-, b-, and c-directions need to be explained.
5. The occasionally observed breaks in the straight-line Arrhenius dependence of cation diffusion rates on temperature need to be explained.

The Correlation Factor

An important concept for understanding diffusion is the correlation factor, generally designated, f. An introduction is given by Shewmon [1963] and is more extensively discussed by Manning [1968]. This factor is a measure of the probability that the direction of a specific tracer atom jump following a given jump will not be related to the direction of that original jump. If, to pick an extreme example, the atom sites are arranged in a chain, then a vacancy will have only two filled sites next to it. An atom can change places with a vacancy but its next jump will reverse this displacement as the random walk of the vacancy takes it in the other direction. The correlation factor is zero; the atom may jump often but these jumps will not take it anywhere in the long run. In such a geometry the only way in which a specified tracer atom can interchange positions with a neighbor is for one of them to step out of the chain, let the second pass, and then fall back into the chain. In such a case diffusion along the chain will not be governed by the jumps within the chain but by jumps out from the chain which allow position interchange. More commonly, diffusion is not one-dimensional and f is between 0.5 and 1.0.

The correlation coefficient may be measured in several ways. One of these is to compare the diffusion coefficient derived from ion migration in an electric field with the value obtained by self diffusion. Since an electric field favors jumps in the direction of the field, the transport measured by it is not subject to the same correlation effect as is self-diffusion [Chemla, 1956]. Such measurements should be quite feasible for divalent cations. For silicon, the correlation coefficient should be measurable if it is very small, i.e., if the ionic transport number is large compared with self-diffusion transport. For oxygen, the small diffusion coefficient and a near-unity correlation factor would make this type of experiment more difficult.

Under conditions of material transport driven by chemical gradients such as interdiffusion of Mg and Fe or in solid state reactions, there may be a net flux of vacancies which will wash out the correlation effects. Therefore, it is possible that in some solid state reactions diffusion might appear to be more rapid than one would determine from a self-diffusion measurement.

Discussion of Cation Diffusion

It has been calculated by Lasaga [1980] that the energy to form a vacancy on an M1 site is less than on an M2 site by 200 to 300 kJ/mol. This has been offered as a suggestion for rapid diffusion in the c-direction in olivine, since the M1 sites form chains along this axis. However, as pointed out above in the discussion of correlation effects, this need not relate directly to fast diffusion in the given direction.

The only computational attempt at explaining divalent cation diffusion in olivine in terms of individual jump frequencies was carried through by Ohashi and Finger [1974]. They counted three different types of jumps between octahedral cation sites which they designated: ν_1 parallel to the a-axis (what we have said could be a "loop back" o-t-t-o jump or an o-o jump), ν_2 an interlayer jump with the vector components a/2, b/4, and c/4, in other words, a jump from one o-site to another which is neither coplanar in the b-c plane with the original or directly in line with it along the olivine a-axis (a "diagonal" jump which we would describe as belonging to a "straight" o-t-t-o sequence), and ν_3 a jump in the b-c plane (which we would describe as an o-t-o jump). We claim that their analysis neglects the component jumps: the fact that (1) the transition from one o-site to another may include distinguishable o-t and t-o steps, each with its own activation energy, (2) there will be a change of direction in going from o-t to t-o, and (3) there is an option at the t-site of going toward either one of three o-sites surrounding it (one of which would be the site from which it originated). This neglect also leads to a small miscount of the number of o-sites which may be available from any given o-site. This number sums to 20 sites rather than ten around M1 and twelve around M2 as they state.

Nevertheless, Ohashi and Finger obtain interesting results which may not be entirely undermined by their approximations. They carry through a random walk calculation in which D_a/D_c and D_b/D_c are calculated as functions of the ratios between ν_1, ν_2, and ν_3. They conclude that ν_2 is the dominant jump. If this is indeed the case, its dominance needs to be rationalized in terms of the "diffusion molecular" configurations of the vacancies, transiting cation in some t-site, and other neighboring atoms as has been described for iron diffusion in FeS [Condit et al., 1974]. It is interesting to note that in that structure, too, "diagonal" jumps between o-sites appeared to dominate.

An experimental program which would differentiate between the various possibilites would include measurement of the correlation factor for diffusion in the three directions in single crystals. Such anisotropy of ionic conductivity as has been measured is not greatly different from diffusion coefficient anisotropy, and this suggests that correlation coefficients should range in value between about 0.5 and 1.0. It would also be interesting to see how the presence of different cation species affects the diffusion of others. From studies of ordering between M1 and M2 sites [Taftø and Spence, 1982] it appears that Ni prefers M1 sites and Mn prefers M2 sites. They might be appropriate dopants to select. The occasionally observed breaks in the straight line Arrhenius plot for Fe-Mg inter-diffusion may be manifestations of an order-disordering between the cation sub-lattices.

Discussion of Oxygen Diffusion

The fact that the activation energy for diffusion in the three directions is the same for oxygen indicates that the same rate limiting step is common to them all, i.e., that the same energy barrier is common to the formation of the metastable configuration which may then result in a jump in any of the three directions. It appears plausible that a vacancy on a large anion site should be accompanied by an associated cation vacancy. Thus, the mobile species may well be cation-anion divacancies. The greater mobility of cations suggests that the migration of divacancies should be limited by the mobility of anions, not cations. The activation volume for creep has been reported as 13.4 cm^3/mole, a little larger than the 11.6 cm^3/mole one would expect for an oxygen vacancy alone [Ross et al., 1979]. While this discrepancy may be within experimental error, it is tempting to speculate that the measured activation volume also includes the volume of a cation vacancy, 5.5 cm^3/mole [Misener, 1974] and that the mobile species is a cation-anion divacancy. The activation energy for diffusion of oxygen is less than expected from calculation [Lasaga, 1980] and may result from a low energy for migration of the divacancy. In a simple hexagonal close packed lattice when jumps to neighboring sites are all equally probable, there is little anisotropy. Another possibility is that the SiO_4^{-4} or SiO_3^{-2} units may rotate at high temperature and thereby randomize the distribution of tracer atoms between O_1, O_2, and O_3 sites. The "diffusional molecule" configuration energies need to be calculated to see which predicts the observed Db/Da ratio of about 1.5.

Since there are three types of oxygen sites, one should allow for the possibility that diffusion of oxygen might occur through only one sublattice and that interchange between this and the other two might be slow. This would assume no rotation of a SiO_4^{-4} unit. This becomes apparent when we consider a mean penetration of, say, 0.3 μm, a distance typical of some oxygen diffusion studies. The number of jumps to achieve this by random walk is about 10^6. We must know if equilibration between two subsets of oxygen sites will take place in less than this number of jumps. Calculations of Lasaga [1980] suggest that the energies to form an oxygen vacancy in the three sites may differ by about 82 kJ/mol. This means that at 1200°C the rates of jump from these sites might differ by a factor of 800. In a million jumps, therefore, there will be enough jumps to assure good local equilibrium exchange. However, if the energies were to differ by twice this amount, 165 kJ/mol., the difference in frequencies of jump from the two sites would be 600,000, a significant fraction of a million, and there would not be time for local equilibration between the two types of sites. The difference for vacancy formation in M1 and M2 is estimated to range between 200 and 300 kJ/mol., and if oxygen diffusion is by divacancies, it is conceivable that equilibration would not occur. This could be determined by absolute measurements of oxygen exchange at the gas-solid interface. A slow interchange between the sublattices should also result in distortion of the penetration profile. Such distortion, if observed, would normally be attributed to dislocation diffusion, but the possibility of sublattice interchange should be examined in future experiments. A barrier to easy interchange of oxygen between gas and solid may also occur and would produce an opposite distortion of a penetration profile [Reddy et al., 1980].

Discussion of Silicon Diffusion

In trying to explain the difference between the silicon diffusion data of Jaoul et. al. and Hallwig et. al. [1980], we will propose that variation in stoichiometry makes large differences in silicon diffusion rates; that the Jaoul material was free of SiO_2 while the Hallwig material contained some SiO_2 in solution. Olivine can dissolve a little silica [Pluschkell and Engell, 1968] which suggests that such dissolution could be accomplained by introduction of interstitial silicon, see equation 2. Diffusion mechanisms in the presence or absence of interstitial silicon could be expected to differ.

The measured diffusion coefficient of silicon in stoichiometric olivine may be small. Silicon atom jumps may be frequent but with a correlation factor of 10^{-4}. It is interesting that the activation energy measured by Jaoul for silicon diffusion is very similar to that for oxygen, because this suggests that the rate limiting step for silicon diffusion may be some process involving oxygen ion jumps.

Correlation effects would be less important if interstitial silicon were present, because an interstitial defect in the lattice would not remember its point of origin in the just-preceeding jump in the same way that a silicon-oxygen divacancy jump would. If the mechanism involves an interstitialcy jump in which an interstitial atom displaces a second from its normal site to itself become interstitial, this argument might have less strength.

Nomenclature

The infancy of this type of analysis can be surmized from the lack of a generally accepted nomenclature. The term, interstitial, for example, has a clear meaning for describing a carbon atom in an iron lattice. The free energies of the carbon atom in any of the normally occupied intersitial sites are all the same. In olivine, however there are many octahedral interstices between oxygen ions. Some are normally occupied, the M1 and M2 sites, but some are normally unoccupied. What may be a normal site for one atom, Si on a t-site, would be an interstitial site for another, Mg on a t-site. Azaroff [1961] at one time suggested the term, voidal diffusion, for the type of jump sequences we have discussed, but that has not gained currency. A way for coping with this may be to elaborate the symbolism of Kröger to indicate reactions. Thus,

$$Mg_{M1} + V_t \rightleftarrows V_{M1}'' + Mg_t^{\cdot\cdot} \quad (4)$$

would indicate the "reaction" in which a Mg ion moves from one M1 site into a t-site. The situation can get more involved as we postulate divacancies, $(V_{O3}^{\cdot\cdot})(V_{M1}'')$, or clusters of three defects as suggested for silicon diffusion in stoichiometric forsterite, $(V_{M1}'')(V_{O3}^{\cdot\cdot})(Si_{ti}^{\cdot\cdot\cdot\cdot})$, where ti indicates some t-site not normally occupied by silicon. These are the "diffusion molecules" suggested earlier.

Summary and Conclusions

We have reviewed material transport in olivine with attention to some apparent discrepancies between expectations and observations. We anticipate that most of these can be resolved by undertaking a detailed analysis of the network of paths which atoms follow in going from one normal site to the next. The intermediate clusters of atoms on normal sites, vacancies, and interstitial atoms will likely pass through a series of discrete configurations which might be called "diffusion molecules." Such an analysis must take into account jump correlation and we propose that this may modify the effective jump frequency by orders of magnitude in the case of silicon. This would explain the apparent discrepancy between the slow self-diffusion of silicon as measured by Jaoul et al. [1980] compared with oxygen and the fact that oxygen diffusion seems to be the rate limiting step for dislocation climb. One way for measuring the correlation factor would be to compare self-diffusion and ionic transport numbers. This should be done for divalent cations and silicon transport in the three principal crystallographic directions in olivine.

Order-disorder processes between the M1 and M2 cation sub-lattices at diffusion temperatures should also be studied for their effects on diffusion. To carry out a serious theoretical program will require a computational effort to select (1) plausible diffusion molecule configurations and (2) calculate energies of transition between them. As mentioned in the introduction, the intermediate configurations may transit to other configurations and the network of transit options may be laid out like an electrical network with the articulated "molecules" being the nodes and the transitions being the branches. The nodes in this network need not represent atoms on one or another site but configurations of atoms. The branches would represent possible transitions from one to another molecule configuration.

Acknowledgment. Work performed under the auspices of the U.S. Department of Energy by the Lawrence Livermore National Laboratory under contract number W-7405-ENG-48.

References

Azaroff, L. V., Role of crystal structure in diffusion, I, Diffusion paths in closest packed crystals, J. Appl. Phys., 32, 1658-1662, 1961.

Bish, D. L., Cation ordering in synthetic and natural Ni-Mg olivine, Am. Mineral., 66, 770-776, 1981.

Buening, D. K., and P. R. Buseck, Interdiffusion of the cations, Mg and Fe, in olivines, J. Geophys. Res., 78, 6852-6862, 1973.

Chemla, M., Diffusion of radioactive ions in crystals, Ann. Phys. Ser., 13, 1, 959-999, 1956.

Clark, A. M., and J. V. P. Long, The anisotropic diffusion of nickel in olivine, in Diffusion Processes, Proceedings of the Thomas Graham Memorial Symposium, Univ. Strathclyde Vol. 2, edited by J. N. Sherwood, A. V. Chadwick, W. M. Muir, and F. L. Swinton, pp. 511-521, Gordon and Breach, New York, 1971.

Condit, R. H., Diffusion path networks in the wurtzite lattice, in Defects and Transport in Oxides, edited by M. S. Seltzer and R. I. Jaffee, pp. 303-313, Plenum, New York, 1974.

Condit, R. H., R. R. Hobbins, and C. E. Birchenall, Self-diffusion or iron and sulfur in ferrous sulfide, Oxid. Met., 8, 409-455, 1974.

Duba, A., H. C. Heard, and R. N. Schock, Electrical conductivity of olivine at high pressure and under controlled oxygen fugacities, J. Geophys. Res., 79, 1667-1673, 1974.

Freer, R., Diffusion in silicate minerals and glasses: A data digest and guide to the literature, Contrib. Mineral. Petrol., 76, 440-454, 1981.

Goetze, C., and D. L. Kohlstedt, Laboratory study of dislocation climb and diffusion in olivine, J. Geophys. Res., 78, 5961-5971, 1973.

Hallwig, D., R. Schachtner, and H. G. Sockel, Diffusion of magnesium, silicon, and oxygen in Mg_2SiO_4, and formation of the compound in the solid state, in Science of Ceramics, vol. 10, edited by H. Hausner, pp. 385-393, Ber. Deutsche Keram. Ges., Bad Honnef/Rhein, 1980.

Jaoul, O., C. Froidevaux, W. B. Durham, and M. Michaut, Oxygen self-diffusion in forsterite: Implications for the high-temeperature creep mechanism, Earth Planet. Sci. Lett., 47, 391-397, 1980.

Jaoul, O., M. Poumellec, C. Froidevaux, and A. Havette, Silicon diffusion in forsterite: A new constraint for understanding mantle deformation, in Analasticity in the Earth, Geodynamics Ser., vol 4, edited by F. D. Stacey, M. S. Paterson, and A. Nicholas, pp 95-100, AGU, Washington, D. C., 1981,

Kjekshus, A., and W. B. Pearson, Phases with the nickel arsenide and closely-related structures, in Progress in Solid State Chemistry, vol. 1, edited by H. Reiss, pp. 83-174, Pergamon, New York, 1964.

Lager, G. A., and E. P. Meagher, High temperature structural study of six olivines, Am. Mineral., 63, 365-377, 1978.

Lasaga, A. C., Defect calculations in silicates, Am. Mineral., 65, 1237-1248, 1980.

Manning, J. R., Diffusion Kinetics for Atoms in Crystals, Van Nostrand, Princeton, N. J., 1968.

Misener, D. J., Cationic diffusion in olivine to 1400°C and 35 kbar, in Geochemical Transport and Kinetics, Spec. Publ. 634, edited by A. W. Hoffman, B. J. Giletti, H. S. Yoder, Jr., and R. A. Yund, pp. 1117-1129, Carnegie Institution of Washington, Washington, D. C., 1974.

Morioka, M., Cation diffusion in olivine, I, Cobalt and magnesium, Geochim. Cosmochim. Acta, 44, 759-762, 1980.

Morioka, M., Cation diffusion in olivine, II, Ni-Mg, Mn-Mg, Mg, and Ca, Geochim. Cosmochim. Acta, 45, 1573-1580, 1981.

Ohashi, Y., and L. W. Finger, Diffusion anisotropy in olivine model calculations, in Annual Report of the Director, Geophysical Laboratory, Publ. 1655, H. S. Yoder, Jr., Director, pp. 403-405, Carnegie Institution of Washington, Washington, D. C., 1973-74.

Plushkell, W. Von, and H. J. Engell, Ionic and electronic conductivity in magnesium orthosilicate, Ber. Deutsche Keram. Ges., 45, 388-394, 1968.

Reddy, K. P. R., S. M. Oh, L. D. Major, Jr., and A. R. Cooper, Oxygen diffusion in forsterite, J. Geophys. Res., 85(B1), 322-326, 1980.

Ross, J. V., H. G. Ave'Lallemant, and N. L. Carter, Activation volume for creep in the upper mantle, Science, 203, 261-263, 1979.

Schmalzried, H., Reactivity and point defects of double oxides with emphasis on simple silicates, Phys. Chem. Miner., 2, 279-294, 1978.

Shewmon, P. G., Diffusion in Solids, McGraw-Hill, New York, 1963.

Stocker, R. L., Influence of oxygen pressure on

defect concentrations in olivine with a fixed cationic ratio, Phys. Earth Planet. Inter., 17, 118-129, 1978.

Taftø, J., and J. C. H. Spence, Crystal site location of iron and tracer elements in a magnesium-iron olivine with a new crystallographic technique, Science, 218, 49-51, 1982

Tsukahara, H., Diffusion and diffusion creep in olivine and ultrabasic rocks, J. Phys. Earth, 24, 89-103, 1976.

Wells, A. F., Structural Inorganic Chemistry, 4th ed., p. 812, Clarendon, Oxford, 1975.

Will, G., and G. Nover, Influence of oxygen partial pressure on the Fe/Mg distribution in olivines, Phys. Chem. Miner., 4, 199-208, 1979.

Wyckoff, R. W. G., Crystal Structures, 2nd ed., vol. 4, pp. 160-163, Interscience, New York, 1968.

TRACE ELEMENT DIFFUSION IN OLIVINE: MECHANISM AND A POSSIBLE IMPLICATION TO NATURAL SILICATE SYSTEMS

Masana Morioka

Radioisotope Centre, The University of Tokyo, Yayoi, Bunkyo-ku
Tokyo 113, Japan

Kazuhiro Suzuki

Department of Earth Sciences, Nagoya University, Furocho, Chigusa-ku
Nagoya 464, Japan

Hiroshi Nagasawa

Department of Chemistry, Gakushuin University, Mejiro, Toshima-ku
Tokyo 171, Japan

Abstract. Variation of diffusion coefficient (D) of divalent cations with ionic radius (IR) and its implications for the natural systems were discussed on the basis of the observed D's in the single crystal Mg- and Mn-olivines. The similar trend of variation observed for D-IR diagrams for the Mg- and Mn-olivines and periclase (MgO) suggests that the shape of the D-IR patterns is determined by the crystal structure and IR. Some of the distributions of the rare earth elements (La and Ce) observed in natural systems are consistent with the crystal-structure-controlled trend of D-IR variation.

Introduction

The mechanism of distribution of minor and trace elements in silicate systems has been a major interest in geochemistry. Since the later 1960's, the mechanism of equilibrium partitioning of trace elements among silicate phases has become much clearer than ever [e.g., Jensen, 1973; Matsui et al., 1977]. This development sheds light on the origin and history of the igneous rocks by means of minor and trace element determinations, producing many fruitful results of which the lunar rock studies were examples. Recent developments in sensitive analytical tools for determination of elemental distributions, however, revealed that much of the distributions of minor and trace elements on a microscopic scale have been determined by nonequilibrium processes; diffusion of cations through crystal lattice of rock-forming minerals apparently being one of the most important.

Although numbers of diffusion coefficients in silicate minerals have been determined [e.g., Buening and Buseck, 1973; Clark and Long, 1971; Foland, 1974; Misener, 1974], the mechanism of diffusion of cations, particularly of minor and trace elements, is not well known. If these diffusion mechanisms were established, further development of application of minor and trace element geochemistry to the petrological problems could be expected.

We have studied the diffusion mechanism of cations in olivine crystals [Morioka, 1980, 1981], by systematically changing diffusing cation species using end-member components. Morioka [1981] suggested that diffusion of Ni, Mg, Co, Fe, Mn and Ca in olivine can be explained as the defect-mediated interdiffusion (namely, diffusion in which ions occupying crystal sites diffuse by a crystal defect). In this paper we further discuss the D-IR relationship and possible implications of the relationship for the distribution of trace elements in the natural systems.

Determination of Diffusion Coefficients

Interdiffusion of Ni-Mg, Co-Mg and Mn-Mg were determined by coupled annealing of single crystal end-members at 1100-1300 °C and measurement of diffusion profiles with a GEOL, JXA-5 electron microprobe analyzer [Morioka, 1980, 1981]. D's of Mg and Ca in Mg-olivine, and those of Ni, Co, Mn, Ca, Sr and Ba in Mn-olivine were determined by a tracer diffusion technique using stable isotope tracers and using Ca and radioactive tracers, respectively [Morioka, 1980, 1981, 1983].

The values of D in olivine discussed in this

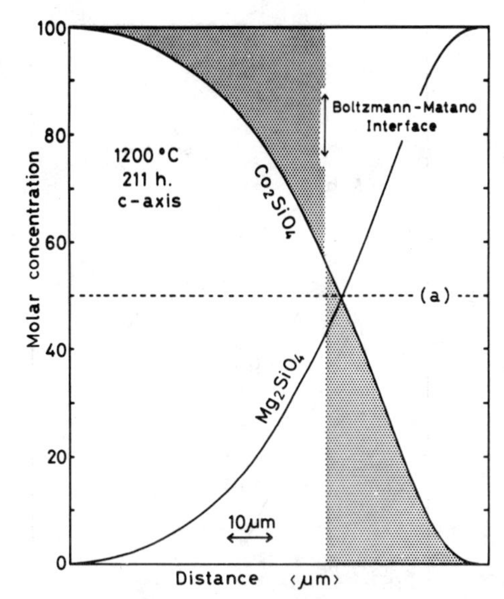

Fig. 1. Concentration profile of Co_2SiO_4-Mg_2SiO_4 system. Symmetry with respect to the horizontal line at 50% (a) indicates interdiffusion between Co and Mg (or 1 to 1 substitution between Co and Mg in diffusion) holds for all concentration range involved. Asymmetry with respect to the estimated original interface (Boltzmann-Matano interface) indicates typical concentration dependent diffusion.

paper are those for direction perpendicular to (001) plane, or along the *c*-axis.

Cation Diffusion in Olivine Under Experimental Conditions

Typical concentration profiles of interdiffusion (Figure 1) were obtained by heating two pure end-member olivine components, e.g. Mg_2SiO_4 and Co_2SiO_4 crystals, in contact with each other for 1100 - 1300 °C. These patterns which are asymmetric with respect to the estimated initial interface (indicated by Boltzmann-Matano interface) indicate that diffusivities are dependent on the concentration of the diffusing cations. From these concentration profiles, D's were calculated using the Boltzmann-Matano equation [e.g., Jost, 1960].

The calculated D's show concentration dependence expressed as

$$D = D_0 \exp(-\alpha C_{Mg}) \qquad (1)$$

where α is a constant, and C_{Mg} is the concentration of Mg_2SiO_4 [Morioka, 1981]. Figure 2 shows a typical D-C_{Mg} plot for Co_2SiO_4-Mg_2SiO_4 system. For the purpose of comparison of concentration dependent D's of different pairs, Morioka [1980, 1981] calculated D's in pure Mg_2SiO_4 by extrapolation, namely, D's at infinite dilution of the cations in Mg_2SiO_4. By using the extrapolated values, we can compare D's of cations in the same medium.

Figure 3 shows variation of D with the ionic radius of diffusing cations. The pattern for Mg-olivine at T = 1200 °C was constructed from the values observed at 1200 °C by Morioka [1980, 1981] and Buening and Buseck [1973] (Fe), and the value of Mg estimated by extrapolation from those observed at different temperatures in log D - 1/T plot. The values of D are approximately equal for Ni and Mg, increase from Mg to Fe and then decrease towards Ca and larger cations. This tendency is explained by Morioka [1981] as a combination of two effects, i.e., (1) the space effect which limits the larger cation to move slower than a smaller cation due to spacial hindrance, and (2) defect density around cations increase with $R_M - R_{Mg}$ (R_M and R_{Mg} denotes ionic radius of divalent cation M and Mg, respectively) which provides the cation the chance of a jump from one site to another. The decreasing trend from Fe to Ca and larger cations indicates the first effect is dominant in this ionic radius range, while the increasing trend from Mg to Fe suggests that the second effect is dominant in this range.

D's for Mn_2SiO_4 are about two orders of magnitude higher than those for Mg_2SiO_4. However, D's for Mn_2SiO_4 show a trend of variation with ionic radius which is similar to that observed for Mg_2SiO_4, increase from Mg to Co, decrease from Mn to Ca and larger cations.

D's for MgO, periclase [Kingery et al., 1976], also show a similar trend of variation (Figure 3). Periclase has a cubic structure similar to that of NaCl, halite, with Mg and O ions occupying the site of Na and Cl ions, respectively [e.g., Deer

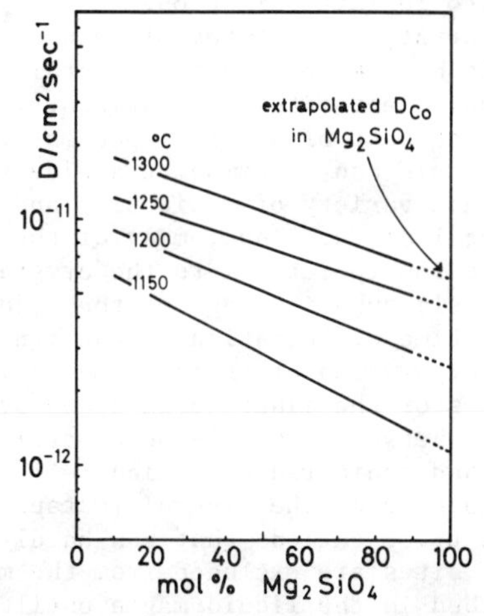

Fig. 2. Interdiffusion coefficients calculated by using Boltzmann-Matano equation and graphical method [Morioka, 1980] for Co_2SiO_4-Mg_2SiO_4 system along the *c*-axis plotted against molar concentration of Mg_2SiO_4. The extrapolated value is used for comparison of D at infinite dilution. The values are from Morioka [1980].

et al., 1965]. On the basis of melting relations of periclase with the end-member oxides, NiO, CoO and FeO [Phase Diagrams for Ceramists, 1964], Mg is most favorably accepted among the divalent cations by the cation site in the crystal structure, as is the case for Mg_2SiO_4. Thus, Fe, Mn, Zn, and Co with the ionic radii which are a little larger than the size of the cation site in periclase have higher diffusivities compared with Mg, the major cation.

In contrast to the characteristic D-IR patterns for the minerals, silicate melts with less rigid structures show much smaller variation in the D-IR relationships (Figure 3). The D-IR patterns observed for the olivines and periclase indicate D-IR relationships are determined by the crystal structure and the size of the diffusing ions.

Diffusion of Trace Elements in Minerals

The mechanism of diffusion of trace elements in rock-forming minerals in igneous rocks is not well known. However, if trace elements occupy crystal sites in the minerals substituting for the major cations, the mechanism of diffusion of the trace elements would be identical to that of interdiffusion under infinite dilution observed for the olivine systems, and diffusivities of trace elements would show regularities which are similar to those observed for the olivine systems. Although no direct comparison of observed tracer diffusivity and interdiffusivity extrapolated to infinite dilution has been reported for rock-forming minerals, one example supporting the above idea is known for a CoO-NiO system [Stiglich et al., 1973] in which tracer diffusivities are equal to the interdiffusivities extrapolated to infinite dilution.

Because crystal structures of most rock-forming minerals have more than two crystallochemically nonequivalent sites for cations, igneous rocks which consist of several species of rock-forming minerals can accommodate a wide range of cations with a variety of ionic size and charge in the crystal sites. Thus, many of the trace elements are incorporated into the crystal sites of the minerals substituting for the major cations at the time of formation of the igneous rocks. Incorporation of trace elements in the crystal sites of the minerals is shown by the regularities between mineral-magma partition coefficients and ionic radius, which reflect the crystal structure of the mineral [Matsui et al., 1977]. The trace cations that can hardly enter the crystal sites are excluded from the minerals, being enriched in the liquid magma until a mineral phase that accepts the trace cation starts to crystallize. Based on these observations, most trace elements are likely to occupy crystal sites of minerals substituting for the major cations rather than to be incorporated as clusters or aggregates at the grain boundaries or crystal defects, etc.

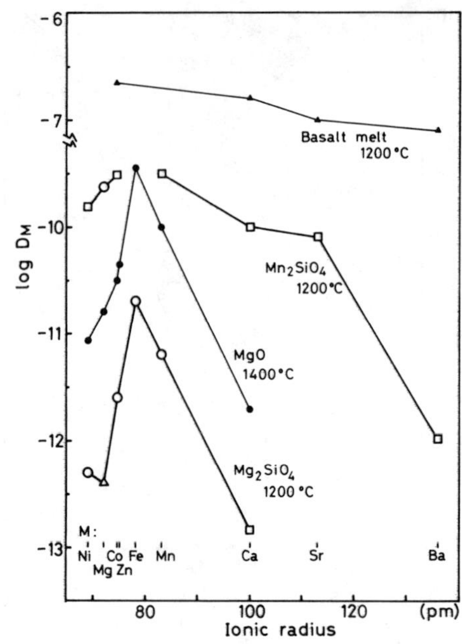

Fig. 3. Diffusion coefficient-ionic radius (D-IR) diagram for Mg- and Mn-olivines, periclase (MgO) and basalt. Open circle, interdiffusion coefficient extrapolated to infinite dilution; square, determined by tracer diffusion technique; open triangle, estimated by extrapolation on log D-(1/T) plot; closed circle, tracer diffusion coefficients for periclase taken from Kingery et al. [1976]; solid triangle, tracer diffusion coefficients for basalt melt taken from Hofmann and Magaritz [1977].

The physicochemical state of incorporation of minor cations in silicate minerals was examined by Morris [1975] using electron paramagnetic resonance (EPR). He showed that Eu^{3+} and Gd^{3+} were incorporated at Ca^{2+} structural sites in $CaMgSi_2O_6$ and other Ca-bearing minerals, while they were incorporated as aggregates or clusters of these ions in Mg_2SiO_4 and Clino-$MgSiO_3$.

From these observations it is concluded that most of the minor and trace elements are incorporated at crystal sites in the minerals in igneous rocks. Thus, understanding interdiffusion mechanism is important for discussing the behavior of minor and trace elements in igneous minerals.

Trace Element Diffusion in Natural Systems

Distribution of trace elements determined by modern instruments and techniques have shown that trace elements often show heterogeneous distributions or zoning in individual mineral grains [e.g., Philpotts and Schnetzler, 1970; Schnetzler and Philpotts, 1970; Suzuki, 1981]. These heterogeneities are often explained by kinetic effects such as change of partition coefficients due to the change in physicochemical conditions during

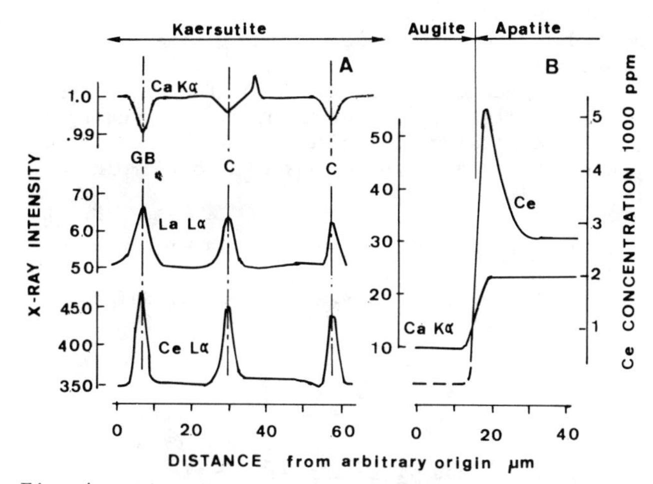

Fig. 4. Distribution of rare earth elements (La and Ce) in (a) a kaersutite and (b) an augite-apatite system. La, Ca and Ce concentrations in Figure 4a are shown as intensities of La L_{α}, Ca K_{α} and Ce L_{α} X-rays in 100 sec, respectively. Ce concentrations in augite were too low to resolve diffusion profile, and thus are shown as broken line. GB and C indicate a grain boundary and a cleavage, respectively.

crystal growth [Lagache and Carron, 1982], or slow diffusion rate of cations in the magma from which the minerals crystallized [Albarede and Bottinga, 1972; Magaritz and Hofmann, 1978]. Although many of the heterogeneous distributions of trace elements in silicate minerals in igneous rocks can be explained as the results of kinetic effects at the time of formation of the minerals, some of the distributions can be explained by diffusion in solids during slow cooling stage followed the formation of the crystal phase or later thermal events.

Such redistribution of the elements in mineral phases by diffusion depends both on D's and the solubilities of elements in the crystals. The fact that Fe and Mn which have fairly high solubilities in olivine have much higher D's compared with those of more soluble cations (e.g., Ni) and of much less soluble cations (e.g., Ca, Sr) could produce distributions of elements in olivine which are quite different from those expected from equilibrium distribution. Exclusion of the elements from olivine crystal, for example, starts with the removal of Fe and Mn at higher rates followed by slow but much more complete removal of Ca, Sr and Ba which have much lower solubilities. Incorporation of the elements into olivine from surrounding phases takes place with the initial enrichment of Fe and Mn followed by slow but steady increase of Ni and Co which have higher solubilities but lower diffusion rates in olivine. Similarly, the rare earth element abundances in natural rocks which are a useful geochemical tool for igneous petrology could be modified by the diffusion-partition controlled processes.

Recently Suzuki [1981] measured the distributions of La and Ce in a hornblendite from Ngong area, Kenya and found that these rare earth elements are enriched in cleavages or cracks in augite and kaersutitic hornblende, and in apatite inclusion in augite near the boundary with augite. The observed distributions of La and Ce can be explained by diffusion coupled with an increasing tendency of augite and kaersutite to exclude La and Ce, which do not fit well with the crystal sites of the minerals, with decreasing temperature.

Figure 4a shows variation of La and Ce concentrations in a kaersutite crystal [Suzuki, 1981]. The observed enrichment of La and Ce at the cleavages can be explained as a result of exclusion of these rare earth elements from the crystal to the cleavages by diffusion at the time of

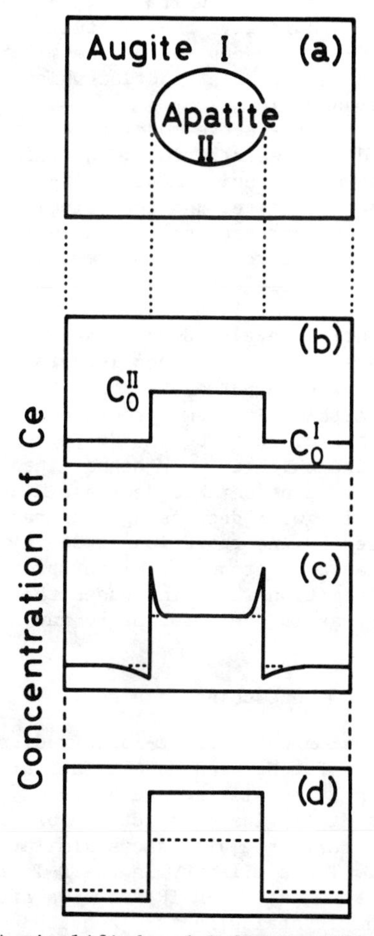

Fig. 5. A simplified model for Ce diffusion for augite-apatite system. (a) A cross-section of the augite-apatite system. Figure 5b, 5c and 5d indicate concentration profile of Ce along the line passes through the center of the apatite inclusion. (b) Initial concentration profile. (c) Concentration profile during heating at T, and at time t. (d) Final concentration profile at equilibration (t = ∞).

some secondary thermal event, where microcrystals or aggregates could easily grow up.

Figure 4b shows more typically a possible diffusion profile in an augite-apatite system, which can be explained by a simplified two-stage model which is shown in Figure 5: The augite and apatite crystals have originally crystallized at a temperature T_0, under chemical equilibrium. Because of the chemical equilibrium, concentrations of Ce in augite (C_0^I) and in apatite (C_0^{II}) are uniform within the individual crystals, and (C_0^I/C_0^{II}) = K_0 at T_0. The crystals were cooled rapidly, thus holding the initial concentration distributions (b). The crystals were then heated to T for relatively longer period of time by a secondary thermal event. If T is significantly lower than T_0, augite crystal tends to exclude Ce which is accepted by apatite crystal, or $K < K_0$. As a consequence, Ce becomes enriched in apatite near the boundary with augite (c). Heating for infinite time would give the eventual distribution at T (d). Slow cooling of the minerals could result in a Ce distribution which is similar to that shown in (c).

In order for the above model to be true for the observed Ce concentrations, the diffusion coefficient of Ce in augite must be significantly higher than in apatite, and the solubility of Ce in augite must decrease substantially with decrease in temperature. The observed mineral/magma partition coefficients of Ce for augite (e.g., 0.166 [Onuma et al., 1968]) are much smaller than those in apatite (e.g., 18-52.2 [Nagasawa, 1970]). This indicates that Ce is much less favorably accepted by augite compared with apatite. Although it is not certain, it seems probable that the energy of substitution of Ce for Ca in the crystal site in augite which is much higher than that in apatite causes considerable decrease in solubility of Ce in augite with decreasing temperature. The strain caused by incorporation of Ce in Ca site in augite may also produce a higher probability of defect formation which increases the diffusivity of Ce as is the case for Fe and Mn in olivine.

Concluding Remarks

1. Diffusion coefficient-ionic radius (D-IR) relationships for Mg- and Mn-olivines are similar, although D's in Mn-olivine are uniformly higher than D's in Mg-olivine by about two orders of magnitude. Periclase also shows similar D-IR relationship. The similarity in the D-IR relationships suggests that the D-IR patterns are determined by the crystal structure and ionic radius.

2. Similarity of the D-IR patterns for Mg- and Mn-olivines which include D's for tracer diffusion and for interdiffusion extrapolated to infinite dilution, together with other observations, indicates that the mechanism of tracer diffusion observed in olivine system is identical to the interdiffusion of the same ion at infinite dilution. This suggests that interdiffusion is an important process which determined the trace element distributions in rock-forming minerals in igneous rocks.

3. Enrichment of La and Ce at the cleavages of kaersutite and that of Ce near the boundary of apatite inclusion in augite may be the results of exclusion of La and/or Ce from the mafic minerals by diffusion.

Acknowledgments. We thank H. Takei for permitting us to use his furnace facilities for single crystal growth and M. Takakuwa, M. Shigeno and R. Ito for assistance in tracer diffusion measurements.

References

Albarede, F., and Y. Bottinga, Kinetic disequilibrium in trace element partitioning between phenocrysts and host lava, Geochim. Cosmochim. Acta, 36, 141-156, 1972.

Buening, D. K., and P. R. Buseck, Fe-Mg lattice diffusion in olivine, J. Geophys. Res., 78, 6852-6862, 1973.

Clark, A. M. and J. V. P. Long, The anisotropic diffusion of nickel in olivine, in Thomas Graham Memorial Symposium on Diffusion Processes, pp. 511-521, Gordon and Breach, New York, 1971.

Deer, W. A., B. A. Howie, and J. Zussman, Rock Forming Minerals, vol. 5, p. 1, Longmans, London, 1965.

Foland, K. A., Alkali diffusion in orthoclase, in Geochemical Transport and Kinetics, Publ. 634, edited by A. Hofmann, B. Gilletti, H. Yoder,Jr, and R. Yund, pp. 77-98, Carnegie Institution of Washington, Washington, D. C., 1974.

Hofmann, A. W., and M. Magaritz, Diffusion of Ca, Sr, Ba, and Co in a basalt melt: Implications for the geochemistry of the mantle, J. Geophys. Res., 82, 5432-5440, 1977.

Jensen, B. B., Patterns of trace element partitioning, Geochim. Cosmochim. Acta, 37, 2227-2242, 1973.

Jost, W., Diffusion in Solids, Liquids, Gases, pp. 31-32, Academic, New York, 1960.

Kingery, W. D., H. K. Bowen, and D. R. Uhlmann, Introduction to Ceramics, pp. 248-249, John Wiley, New York, 1976.

Lagache, M., and J. P. Carron, Zonation des elements en traces au cours de la croissance des cristaux dans les bains silicates: L'exemple de Rb, Cs, Sr et Ba das le systeme Qz-Ab-Or-H_2O, Geochim. Cosmochim. Acta, 46, 2151-2158, 1982.

Magaritz, M., and A. W. Hofmann, Diffusion of Sr, Ba and Na in obsidian, Geochim. Cosmochim. Acta, 42, 595-605, 1978.

Matsui, Y., N. Onuma, H. Nagasawa, H. Higuchi, and S. Banno, Crystallo-chemical control in trace element partitioning, Bull. Soc. Fr. Mineral. Cristallogr., 100, 317-324, 1977.

Misener, D. J., Cationic diffusion in olivine to 1400 °C and 35 kb, in Geochemical Transport and Kinetics, Publ. 634, edited by A. Hofmann, B. Gilletti, H. Yoder,Jr, and R. Yund, pp. 117-129,

Carnegie Institution of Washington, Washington, D. C., 1974.

Morioka, M., Cation diffusion in olivine, I, Cobalt and magnesium, Geochim. Cosmochim. Acta, 44, 759-762, 1980.

Morioka, M., Cation diffusion in olivine, II, Ni-Mg, Mn-Mg, Mg and Ca, Geochim. Cosmochim. Acta, 45, 1573-1580, 1981.

Morioka, M., Cation diffusion in olivine, III, Mn_2SiO_4 system, Geochim. Cosmochim. Acta, in press, 1983.

Morris, R. V., Electron paramagnetic resonance study of the site preferences of Gd^{3+} and Eu^{3+} in polycrystalline silicate and aluminate minerals, Geochim. Cosmochim. Acta, 39, 621-634, 1975.

Nagasawa, H., Rare-earth concentrations in zircons and apatites and their host dacites and granites, Earth Planet. Sci. Lett., 9, 359-364, 1970.

Onuma, N., H. Higuchi, H. Wakita, and H. Nagasawa, Trace element partition between two pyroxenes and the host lava, Earth Planet. Sci. Lett., 5, 47-51, 1968.

Phase Diagrams for Ceramists, edited by M. K. Reser, pp. 52, 54, and 110, American Ceramic Society, Columbus, Ohio, 1964.

Philpotts, J. A., and C. C. Schnetzler, Phenocryst-matrix partition coefficient for K, Rb, Sr and Ba, with applications to anorthosite and basalt genesis, Geochim. Cosmochim. Acta, 34, 307-322, 1970.

Schnetzler, C. C., and J. A. Philpotts, Partition coefficients of rare earth elements and barium between igneous matrix material and rock forming mineral phenocrysts, II, Geochim. Cosmochim. Acta, 34, 331-340, 1970.

Stiglich, J. J., Jr., J. B. Cohen, and D. H. Whitmore, Interdiffusion in CoO-NiO solid-solution, J. Am. Ceram. Soc., 56, 119-125, 1973.

Suzuki, K., Grain boundary concentration of rare earth elements in a hornblende cumulate, Geochem. J., 15, 295-303, 1981.

EXTENDED DEFECTS AND VACANCY NON-STOICHIOMETRY IN ROCK-FORMING MINERALS

David R. Veblen

Department of Earth and Planetary Sciences, The Johns Hopkins University
Baltimore, Maryland 21218

Abstract. Extended defects in minerals, both linear and planar, can have geochemically and geophysically important consequences. Whereas point defects can be investigated fruitfully with theoretical methods, transmission electron microscopy provides an important experimental tool for studying extended defects. In at least some cases, there may be a dynamic chemical relationship between point defect concentrations and the length or area of dislocations or planar defects in a crystal. Many solid-state reactions take place by mechanisms involving the nucleation and propagation of dislocations or planar defects. Extended defects also may alter the diffusion properties of minerals, especially important in the case of replacement reactions. Substantial non-stoichiometry resulting from the substitution of large proportions of vacancies on cation sites is a common feature in a wide variety of important rock-forming minerals and may provide an opportunity for studying the detailed structural properties of vacancies.

Introduction

It is clear from other papers in this collection that point defects can exercise an important controlling influence over a variety of geologically important processes. In the interest of breadth, in this contribution I will point out that extended defects (linear and planar) also may play a critical role in certain geological phenomena. Indeed, in some cases it is likely that there exists a chemical interplay between point and extended defects, and the extended defects cannot be ignored if one is to understand fully the point defect behavior.

This paper is not intended to be a thorough review of extended defects or their functions in mineralogical processes. It portrays instead a rather personal view of a few aspects of defect chemistry in rock-forming minerals. No attempt has been made to cover many important aspects of extended defects, such as the role that dislocations play in deformation processes. Likewise, the effects of planar defects on major and minor element distributions have been considered elsewhere and are not discussed here [Veblen and Buseck, 1981; Buseck and Veblen, 1978].

This paper should not be construed as a suggestion that point defects are not important in the earth sciences. It has been established clearly that point defects are important for chemical diffusion processes, electrical conductivity of minerals, mechanical properties, and processes such as exsolution (precipitation) that depend on intracrystalline diffusion. Indeed, the final section of the paper points out that large concentrations of vacancies are not rare and unimportant, but rather that non-stoichiometry based on vacancy substitution is a common phenomenon in a wide range of rock-forming minerals.

Methods of Study

Many aspects of the structure and chemistry of point defects in solids can be treated theoretically (reviews of some of these elegant treatments are included elsewhere in this volume; see, e.g., Kröger; Anderson). This is largely because point defects generally are structurally and chemically rather simple and because in many cases they represent equilibrium phenomena in crystals. Thus, their atomic structures, concentrations, and thermodynamic effects are theoretically accessible. There are also experimental methods for studying point defects, including diffusion experiments, electrical conductivity measurements, and spectroscopic studies. All of these methods are, however, in a way indirect, and diffusion and electrical conductivity studies in particular are prone to experimental difficulties. Some types of point defects may be observable via diffraction contrast in conventional transmission electron microscopy (TEM) studies [e.g., Crump and Mitchell, 1963], but the electron bombardment in such experiments can create and move point defects, and the interpretation of such results can be ambiguous. Likewise, subtle variations of intensity in images obtained with high-resolution TEM (HRTEM) methods may be related to the exact distributions of various atomic species and/or vacancies on

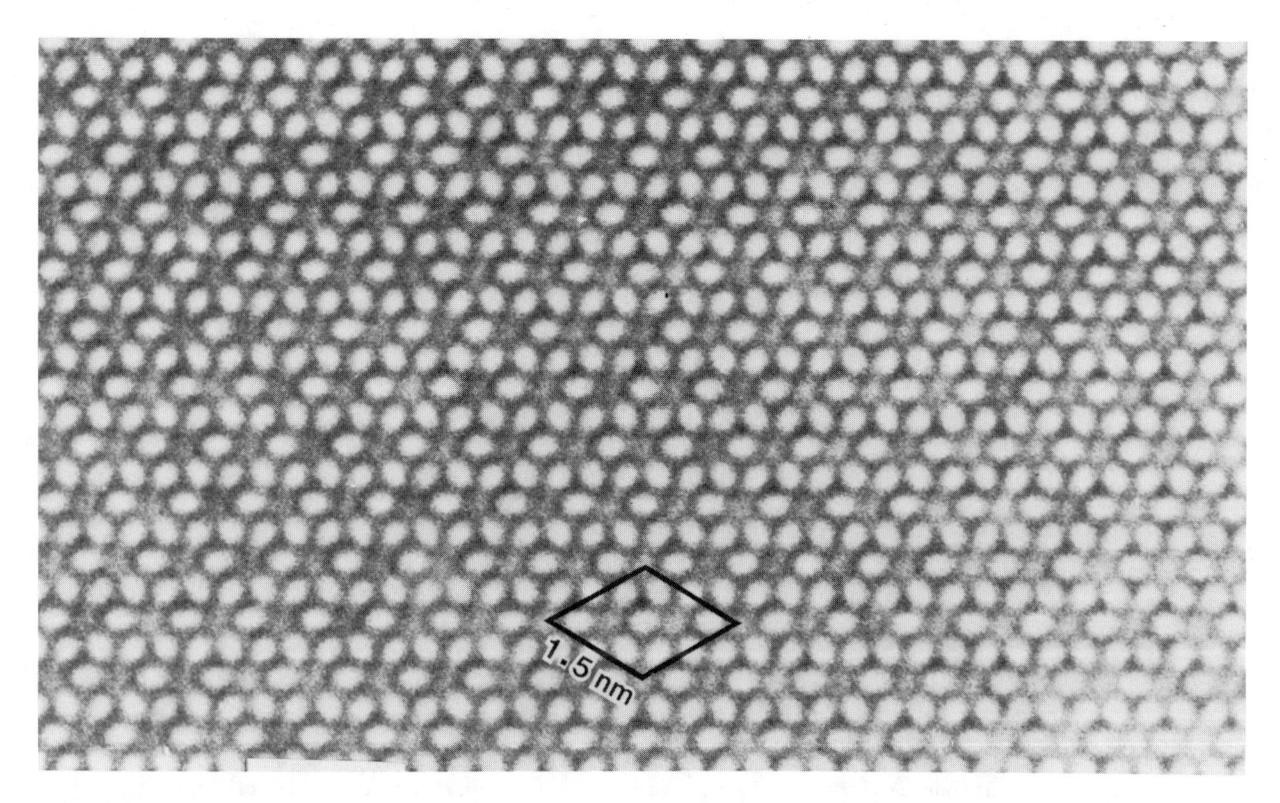

Fig. 1. High-resolution transmission electron micrograph of the tourmaline structure. Non-periodic variations in the intensities of the centers of the hexagonal "flowers" may result from inhomogeneous distributions of cations and/or vacancies [see Iijima et al., 1973]. These intensity variations can best be seen by looking at the figure from a distance at a low angle. The unit cell is outlined. Photograph courtesy of Sumio Iijima, Meijo University, Japan.

specific crystallographic sites; such variations can be seen in the HRTEM image of tourmaline shown in Figure 1 [Iijima et al., 1973]. Again, however, quantitative or detailed structural interpretation of such results are not yet possible.

Compared to point defects, extended defects are extremely difficult to treat theoretically [Lasaga, 1981]. They tend to be structurally and chemically much more complex, and they do not represent equilibrium phenomena. The detailed structures of dislocation cores, for example, are still not understood. On the other hand, experimental methods for the direct observation of extended defects with TEM are well-established. Reliable conventional TEM methods for the observation of dislocations, stacking faults, and antiphase boundaries have long been established [Hirsch et al., 1965], and, as a result, we know a great deal about their behavior in a wide variety of crystalline materials [Hirth and Lothe, 1982]. Similarly, the past ten years have seen a blossoming of the HRTEM technique, with which crystal structures and some types of extended defects can be more-or-less directly imaged under appropriate experimental conditions [Spence, 1980]. This method is particularly valuable for determining the structures and distributions of various types of intergrowth defects and for clarifying the structures of fine-grained materials. Most of the figures that follow were obtained with the HRTEM method.

Chemical Relationships Between Vacancies and Extended Defects

It is well known that a chemical relationship can exist between vacancies and dislocations in crystals. Perhaps the most dramatic demonstration of this balance is afforded by the quenching of certain alloys and halides from high temperature, followed by annealing: vacancies condense out of the bulk structure onto dislocations that were initially relatively straight, causing them to climb into helices, as shown in Figure 2 (see, for example, Eikum and Thomas [1964]). In this case, the vacancies become unstable at the lower temperature. Because they are kinetically precluded from diffusing out of the crystal, they instead contribute to an increased dislocation length. This example graphically shows that there can be a relationship between the point defect concentration and the length of dislocations present in a crystal. In the most simple-

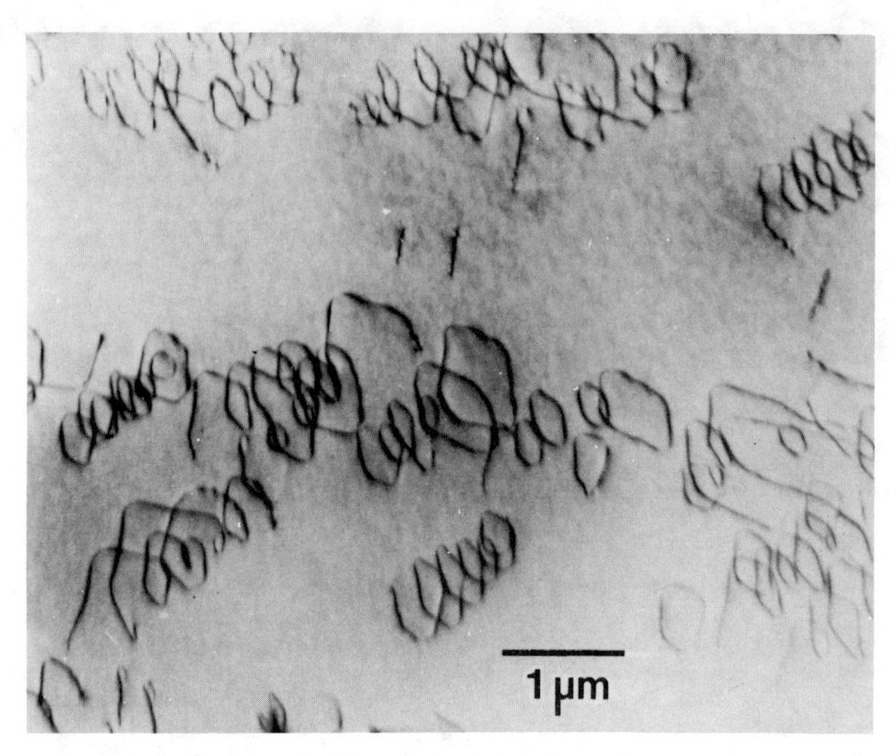

Fig. 2. Dislocations in a Mg-Al alloy. The helical morphology is due to climb that results from precipitation of vacancies onto the dislocations during cooling from high temperature. Photograph courtesy of Gareth Thomas, University of California, Berkeley.

minded view, both vacancies and dislocations represent "open space" within a crystal, and this space can be transferred back and forth between them. Similar helical dislocations have been observed in olivine [Ricoult, 1978] and cuprite [Veblen and Post, 1983]. Although these may have resulted from crystal growth or some other process, they at least raise the possibility of vacancy condensation on dislocations in minerals.

Another similar demonstration of vacancy condensation involves the formation of dislocation loops when some crystalline materials are cooled from high temperature. Indeed, the counting and measurement of such loops in TEM images is an established technique for experimentally determining point defect concentrations in quenched and irradiated materials [Hirsch et al., 1965, p. 427].

It is also possible, at least in theory, for there to exist a chemical relationship between vacancies and planar defects. Consider, for example, the following reaction for an idealized high-vacancy aluminous pyroxene composition, which is found as a component of some high-pressure natural pyroxenes:

$$4M_{1.75}\square_{0.25}(Si,Al)_2O_6 \rightarrow M_7(Si,Al)_8O_{24}$$

(pyroxene) (oxyamphibole)

where M refers to octahedrally coordinated cations and $\square$ to a vacancy. This reaction suggests that upon quenching from high pressure and temperature, lamellae of oxyamphibole might form in the pyroxene by vacancy condensation. Indeed, in one example from a kimberlite, narrow amphibole lamellae have been observed (Figure 3; Veblen and Buseck [1981]). Although an alteration origin for these cannot be ruled out, it is certainly possible that these lamellae formed as a metastable manifestation of the instability of vacancies in the pyroxene that occurred during emplacement of the kimberlite.

The possibility and consequences of a dynamic chemical relationship between point defects and extended defects in minerals have not been given a great deal of consideration by geologists in the past, although the interplay between vacancies and dislocations is clearly recognized in some studies involving dislocation climb [e.g., Goetze and Kohlstedt, 1973]. Such relationships clearly exist in many crystalline systems and could prove to be of general importance in annealing and deformation processes in the earth.

The Role of Extended Defects in Solid-State Reactions

The mechanisms of solid-state reactions commonly involve the formation and propagation of extended defects. In addition, the presence of extended defects or extreme structural disorder

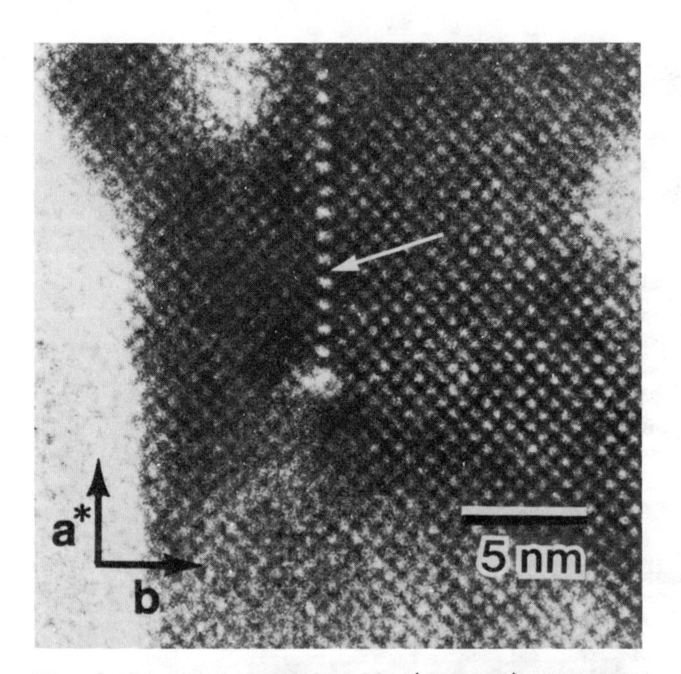

Fig. 3. An amphibole lamella (arrowed) that is one double silicate chain wide terminating in a high-vacancy, high-Al pyroxene from a kimberlite. The formation of such lamellae could result from condensation of vacancies during the quench from high pressure. From Veblen and Buseck [1981].

is typical of minerals that have undergone only partial reaction to another phase. Observation of these defects has become an important tool for understanding how phase transformations and replacement reactions occur in minerals.

The formation and movement of dislocations constitutes an important mechanism for solid-state transformation of one crystal structure into another. For example, reactions from one pyroxene polymorph to another, driven either by deformation or by hydrostatic thermodynamic instability, can occur by this process [Coe and Kirby, 1975; McLaren and Etheridge, 1976; Iijima and Buseck, 1975]. In the case of transformations between enstatite polymorphs, partial dislocations form and move through the structure in the (100) planes, leaving (100) stacking faults behind them. Incompletely reacted crystals contain these planar defects, the distribution of which may contain information on the processes that led to their formation. Allowed to run to completion, this reaction process will result in a new ordered structure.

Planar defects also can be of critical importance for reactions in which the chemical composition of the crystal is changed. Perhaps the most extensively studied examples of these phenomena are those exhibited during hydration reactions among chain and sheet silicates (see Veblen [1981], for a recent review; also see Nakajima and Ribbe [1981], Akai [1982], and

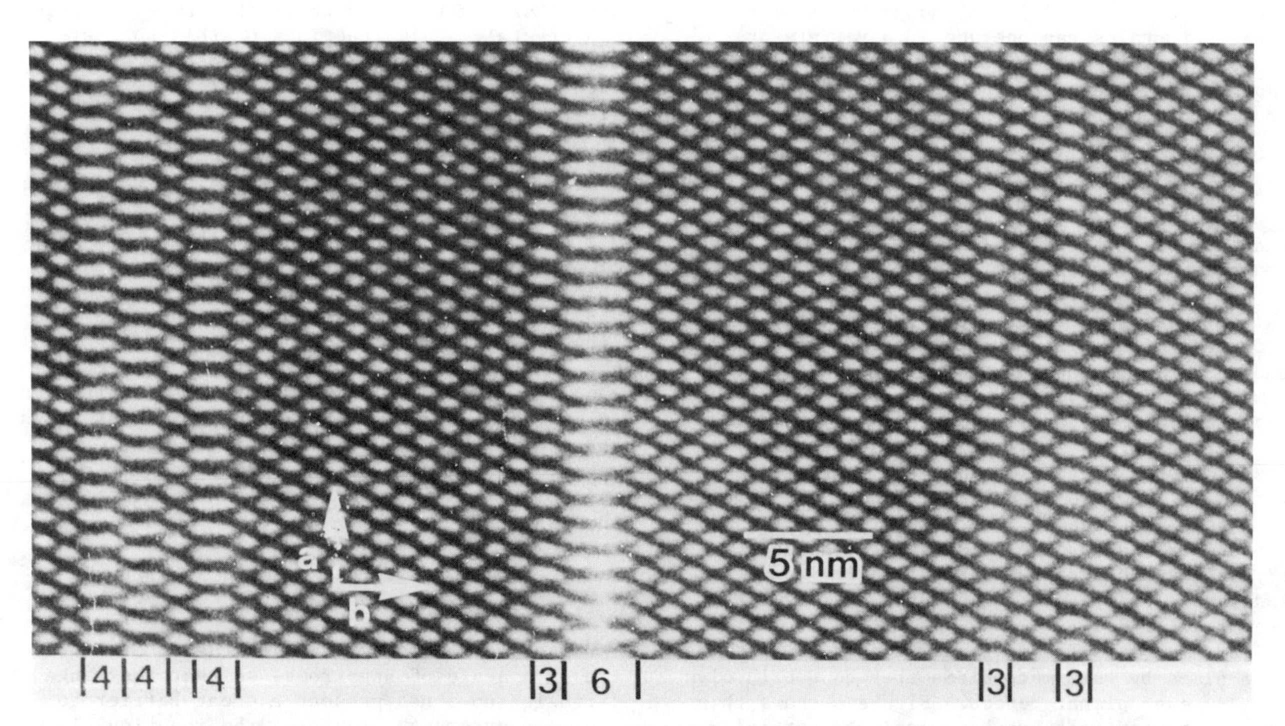

Fig. 4. Planar defects in anthophyllite. The defects consist of intercalated structure with triple, quadruple, and sextuple silicate chains, in contrast with the double-chain structure of the amphibole anthophyllite. From Veblen [1981].

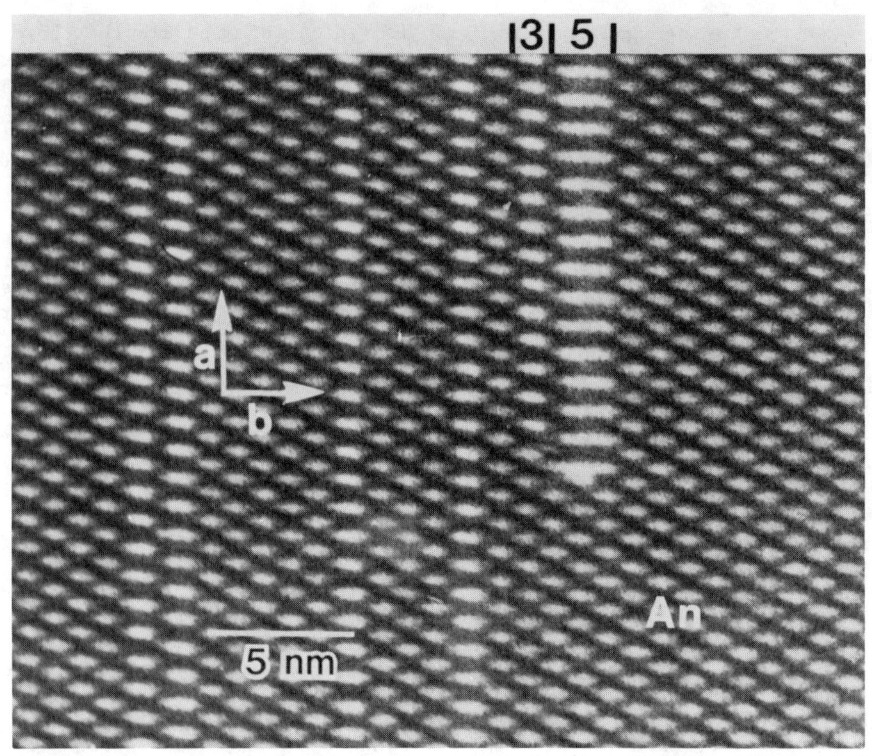

Fig. 5. The cooperative termination of a quintuple-chain defect and a triple-chain defect in disordered anthophyllite.

Cressey et al. [1982]). Although the same structural principles can operate in a variety of other reacting systems, we will examine a few aspects of these hydration reactions, since they provide a good example of the role of planar defects.

One common mechanism for hydration reactions in chain silicates involves the nucleation and growth of lamellae of wide-chain structure in a reactant with narrower silicate chains. Figure 4 shows a number of planar defects, produced by such a reaction, in the amphibole anthophyllite, which contains double silicate chains; the defects in this figure contain triple, quadruple, and sextuple chains, as indicated. In the simplest case, such defects form individually or cooperatively by nucleation, followed by growth at their terminations; the cooperative termination of a quintuple-chain and a triple-chain defect is shown in Figure 5. Planar defects of this sort that change the chemical composition of the crystal have been called Wadsley defects, crystallographic shear planes [Wadsley and Andersson, 1970], and polysomatic defects [after Thompson, 1978].

In some cases, these replacement reactions take place by bulk mechanisms that do not involve this type of planar defect. Figure 6 shows this for a pyroxene that has been replaced partially by amphibole. Rather than being devoured by narrow lamellae, the pyroxene is replaced along a broad, planar reaction front that sweeps through the crystal. Bulk reactions of this sort are also observed in some cases for other reactions, such as amphibole hydration. Likewise, pyroxene alternatively can be replaced by either amphibole or triple-chain silicate via mechanisms involving nucleation and growth of planar defects.

Several important generalizations about reactions in minerals have come out of electron microscopic observations. Most obviously, it is clear that extended defects play an important role in many reactions. What is less obvious is the fact that the energies and propagation kinetics of specific defect types may subtly control the reaction behavior of a system. This may result in multiple reaction paths or in different mechanisms for a given reaction, which depend on the specific geological conditions. In some cases, the extended defect structures may provide an explanation for the formation of metastable phases in rocks.

The Role of Extended Defects in Diffusion

In polymorphic reactions that occur by extended defect mechanisms, diffusion distances are on a scale of angstroms, so that the range of diffusion does not present a great barrier to reaction progress. Instead, the reaction kinetics are controlled predominantly by the nucleation and propagation rates of the defects.

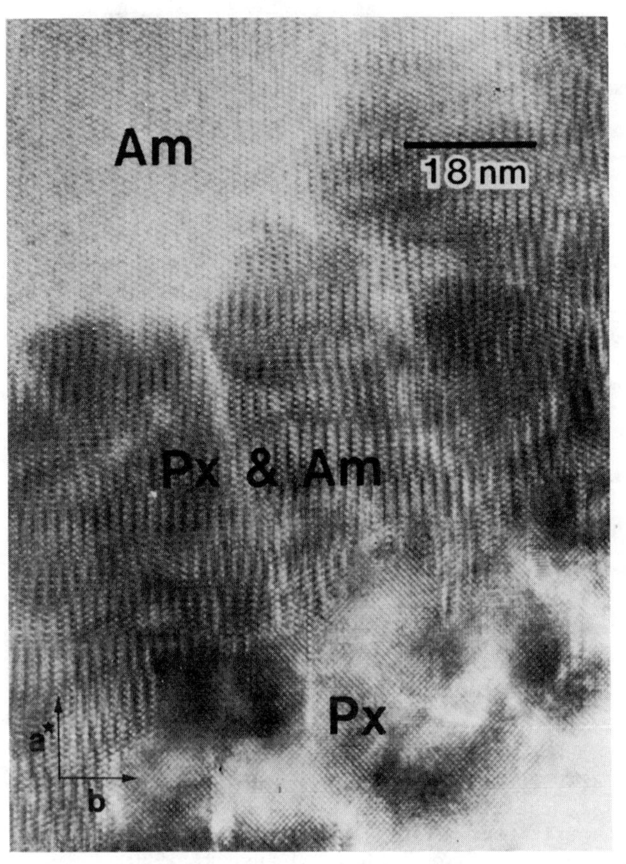

Fig. 6. A reaction front in a crystal of pyroxene that has been replaced partially by amphibole. The boundary between the structures is tilted with respect to the viewing direction, producing a moire effect. From Veblen and Buseck [1981].

This picture may be quite different, however, for replacement reactions in which the chemical composition of the crystal changes. In these cases, examples of which were given in the last section, chemical transport between the crystal and its environment is necessary, and it is expected that diffusion processes at least in part will control the kinetics of reaction. As is well known, the diffusion rates in typical rock-forming minerals are extremely sluggish, compared to metallic systems, for example. The question arises as to how some of these replacement reactions can proceed at all on reasonable time scales, given the diffusion constraints. The answer is probably that in reactions of these types, the transport mechanism is not normal "lattice" diffusion.

In replacement reactions involving the growth of planar defects, there typically is a structural tunnel at the termination of the propagating defect, as shown in Figure 7 for the case of amphibole replacement by sextuple-chain silicate [Veblen and Buseck, 1980]. These tunnels can be very large, even compared to those in many zeolites and other ordered structures in which rapid pipe diffusion occurs (also called interstitial or ultrafast diffusion; see Peterson [1968]). For example, the tunnels of Figure 7 have free apertures of at least 0.48nm, while zeolites with free apertures of only 0.26nm exhibit rapid diffusion of cations as large as Rb [Breck, 1974]. Substantial diffusion can occur in zeolites on time scales of minutes or seconds, even at room temperature; and in tunnel structures, diffusion coefficients for species that are transported by the interstitial mechanism may be highly anisotropic and as much as 10^{12} times higher than for species tranported by "lattice" diffusion [Dyson et al., 1967]. It thus seems very likely that chemical transport associated with these replacement reactions is dominated by pipe diffusion along channels that are localized exactly at the site of the reaction (given the sizes of some of these tunnels, this perhaps should be referred to as "culvert diffusion"!). In this view, the bulk diffusion properties of the crystal, which in general are controlled by vacancy-hopping and other "lattice" diffusion mechanisms, may be totally irrelevant to transport during the reaction. It is instead the structures of the extended defect terminations that control diffusion and are of importance for understanding the reaction kinetics. In addition, in reactions that occur by the migration of interfaces (e.g., Figure 6), enhanced diffusion along the interface is undoubtedly of great importance.

Pipe diffusion also can operate along the cores of dislocations. One study of geological interest is that of Yund et al. [1981], on pipe diffusion of oxygen along dislocations in albite. It was found that a high dislocation density of 5×10^{9} cm^{-2} enhanced the overall diffusion coefficient by less than an order of magnitude, compared with that for undeformed albite. Thus, it was concluded that in most geologically relevant situations, pipe diffusion along dislocations will not alter drastically the diffusion properties of feldspars, and perhaps other minerals as well. Yund et al. did indicate, however, that pipe diffusion may be important for mineralogical changes that are controlled by nucleation.

Diffusion along extended defects, such as dislocations, may be most important mineralogically because they provide a chemical short circuit between the interior of a crystal and its environment. Thus, dislocations could cause nucleation of heterogeneous reactions within crystals, rather than just at grain boundaries. As an example, it may well be that the alteration clouding that occurs pervasively throughout feldspar crystals in many different types of rocks occurs by the nucleation of alteration products along dislocation cores.

A particularly good example of the drastic effects that dislocations can have on reactions is the study of Boland and Duba [this volume] on

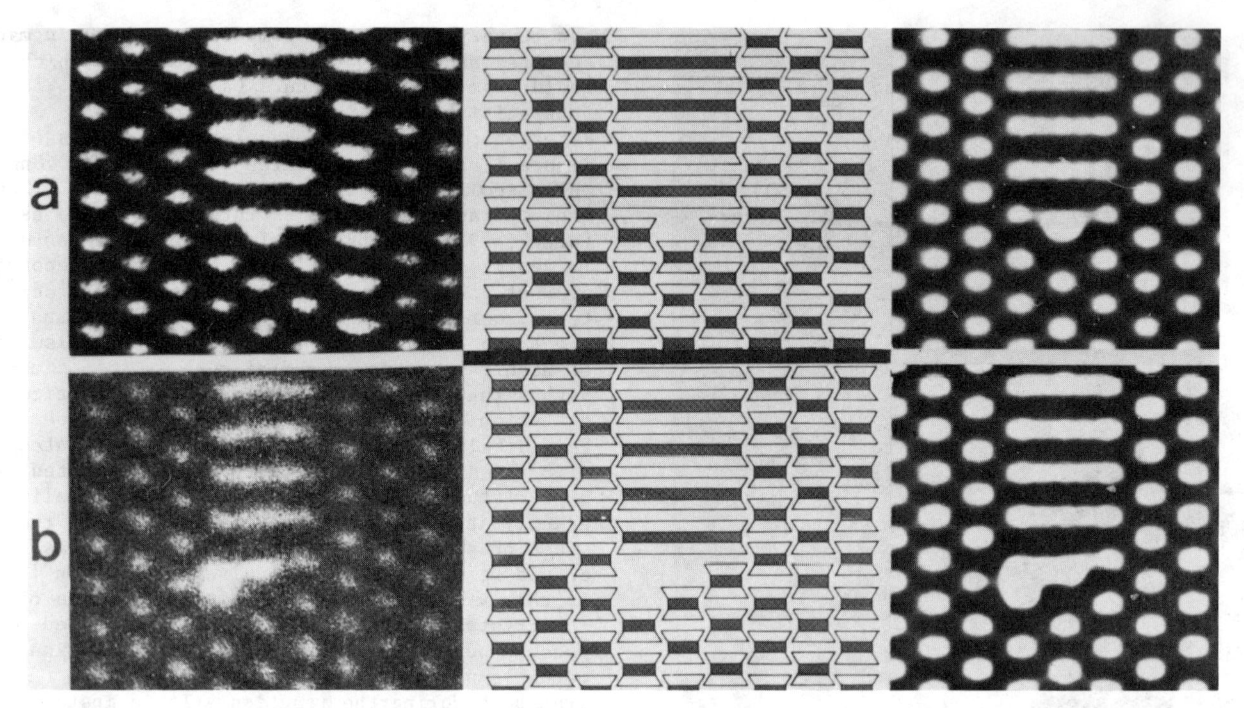

Fig. 7. Two different types of large structural tunnels at the terminations of sextuple-chain lamellae in the amphibole anthophyllite. On the left are experimental HRTEM images; in the center are models for the termination structures (given in the I-beam representation, see Veblen et al. [1977], for example); and on the right are simulated images that were computed using the models in the center. Scale is indicated by the amphibole I-beams, which are 0.9nm wide. The triple-chain lamella adjacent to the sextuple structure in Figure 7a is ignored in the model. From Veblen and Buseck [1980].

reduction of olivine. It was found that near mechanically treated surfaces, where dislocation density was high, the number of metallic precipitates formed by reduction was four orders of magnitude greater than in relatively dislocation-free olivine. In addition, there were differences in the types of reactions that occurred in the high- and low-dislocation areas.

Vacancy Non-Stoichiometry in Rock-Forming Minerals

When asked to name a non-stoichiometric mineral based on large concentrations of vacancies, most mineralogists will mention something like pyrrhotite ($Fe_{1-x}S$) or wüstite ($Fe_{1-x}O$). This is quite natural, since the formulae of these minerals typically are written with a variable number of Fe, and it is understood that the "x" gives the proportion of vacancies on the Fe sites. What is often overlooked by those not intimately involved with the crystal chemistry is that nonstoichiometry based on substantial proportions of vacancies is an extremely common phenomenon in rock-forming minerals; the nonstoichiometry is masked in many cases simply by our tendency to write end-member formulae not exhibiting this property. Those interested in the detailed structural properties of vacancies in rock-forming minerals could possibly learn a great deal by careful examination of the site geometries in mineral series exhibiting varying degrees of this type of non-stoichiometry.

In the following paragraphs I will note briefly some of the important minerals and mineral groups that can exhibit significant non-stoichiometry based on vacant cation sites. As will become apparent, whether we consider a mineral to be stoichiometric or not often depends on our definition of the ideal mineral structure. For example, the pyrrhotites are clearly non-stoichiometric with respect to the troilite composition (FeS), but there are certain discreet pyrrhotite compositions at which ordered structures occur; if one wishes, these discrete compositions can be written stoichiometrically, with integral numbers of Fe and S. However, the important point here is that all of the pyrrhotites contain vacancies in sites that can otherwise be occupied by Fe, and for our present purpose we will consider them to be examples of vacancy non-stoichiometry.

The sheet silicates exhibit a variety of non-stoichiometric phenomena. The interlayer sites, between the tetrahedral-octahedral-tetrahedral layers, are generally shown in written formulae as being completely vacant (talc, pyrophyllite) or completely filled with monovalent (K, Na) or divalent (Ca, Ba) cations (the micas and brittle micas). Yet, there are some glaring exceptions to these formulae among the rock formers of the group. For example, the interlayer sites of the sodium trioctahedral mica wonesite [Spear et al., 1981] are about half vacant (immiscibility at low temperatures results in exsolution of talc and Na-biotite [Veblen, 1983]). Likewise, talc compositions with as much as 35% Na in the interlayer sites have been reported [Schreyer et al., 1980]. Perhaps more important is the fact that trioctahedral micas very commonly contain more limited numbers of vacancies in their interlayer sites than wonesite [Foster, 1960; Hawthorne and Černý, 1982]. The full range of interlayer contents, from full to empty, is exhibited by the clay mineral sheet silicates. In fact, it is this partial occupancy that results in the expandable hydration properties of smectites. Thus, common mud can contain large amounts of a non-stoichiometric sheet silicate based on high vacancy concentrations.

Non-stoichiometry also can be important in the octahedral sheets of rock-forming sheet silicates. These minerals are generally split into the trioctahedral types, with all of the octahedrally-coordinated sites occupied, and the dioctahedral types, which are supposed to contain exactly one-third vacancies in the octahedral sheets. Although unusual, marked deviations from these extreme types can occur, and small deviations are common [Hawthorne and Černý, 1982; Zussman, 1979].

As in the sheet silicates, the amphiboles possess a large site that can either be vacant or contain alkali cations, usually Na. Although amphibole end-member formulae typically are written with this site either completely empty or completely full, any aficionado of amphiboles knows that these extremes are seldom fulfilled: amphiboles are almost without exception examples of vacancy non-stoichiometry. Indeed, it is of historical interest that the importance of vacancies for structural crystal chemistry probably was first recognized in amphiboles by Warren [1930].

Although pyroxenes generally are treated as stoichiometric minerals, it has been demonstrated in recent years that high-pressure aluminous pyroxenes can contain substantial proportions of vacancies in their octahedral sites (see discussion in section on "chemical relationship between vacancies and extended defects," above). This vacancy non-stoichiometry has been observed in both synthetic pyroxenes [O'Hara and Yoder, 1967; Wood and Henderson, 1978; Gasparik and Lindsley, 1980] and in pyroxenes from kimberlites [Smyth, 1980]. Petrologists should be aware of the possibility of pyroxene non-stoichiometry, especially at high pressures, where a high-vacancy pyroxene component may replace an alternative aluminous phase assemblage.

Like the chain and sheet silicates, some of the framework silicates can contain significant numbers of vacancies, and silica polymorphs can contain appreciable numbers of alkali cations in sites that are vacant in the pure SiO_2 system. For example, Longhi and Hays [1979] showed that in the $CaAl_2Si_2O_8$-SiO_2 system, the feldspar anorthite can contain up to 9% vacancies in the Ca sites (an excess of SiO_2), and cristobalite can contain about 5% of the anorthite component (non-stoichiometric addition of Ca into voids and Al into the framework of the structure). Natural tridymites and cristobalites can contain appreciable amounts of a $NaAlSiO_4$ component, and stuffed derivatives of the silica polymorphs [Buerger, 1954], such as nepheline and kalsilite, are likely candidates for non-stoichiometric substitution of vacancies in the alkali sites.

Finally, at least some island silicates (orthosilicates) can exhibit important degrees of vacancy non-stoichiometry. Mullite is a classic case, with oxygen vacancies on one site and limited numbers of Al ions substituting on another, mostly vacant site [Burnham, 1963; Ribbe, 1980]. Olivine from at least several terrestrial environments can contain substantial numbers of vacancies in the octahedrally-coordinated sites, at least up to the composition of laihunite, approximately $\square Fe_3Si_2O_8$ [Kitamura et al., 1983].

From the above discussion, it should be clear that the existence of substantial numbers of vacancies (or substitution of appreciable numbers of cations into "normally" vacant sites) is a widespread phenomenon and produces non-stoichiometry in a variety of important rock-forming silicates. This behavior appears to be the most common for relatively large, low-valence cations, but it also occurs for octahedral cations and anions. Although only pyrrhotite and wüstite were mentioned, this type of non-stoichiometry is also important in sulfides, oxides, and many other chemical groups. While vacancies are commonly treated merely as "defects," it should be realized that they can occur in numbers large enough to drastically alter the chemical compositions of many minerals. Likewise, structure refinement of minerals with very high vacancy concentrations might provide some insight into the detailed structures of vacancies in minerals where they occur in trace numbers.

Conclusions

In this paper, a number of points have been made concerning extended defects in the earth sciences and the role of vacancies in some rock-forming minerals. These points can be summarized as follows:

1. Point defects can be treated successfully in a theoretical framework, but their direct experimental observation is very difficult. Extended defects, on the other hand, are theo-

retically intractable, but a great deal can be learned about them through direct observation in the transmission electron microscope.

2. There can be a chemical interplay between point and extended defects in a crystal: in some cases, vacancies can be annihilated to produce dislocations, and in others, condensation of vacancies can produce planar defects.

3. Both dislocations and planar defects can play important roles in solid-state reactions.

4. The chemical diffusion necessary for some solid-state reactions may take place primarily by interstitial mechanisms along linear or planar defects. Pipe diffusion along dislocations may be important for reactions involving nucleation of new phases inside crystals.

5. Vacancies do not occur only as low-concentration defects. Indeed, in many rock-forming minerals they can have concentrations high enough to grossly affect the mineral chemistry. In some cases, complete solid solution even can exist between end members having all vacancies vs. complete occupancy by cations on a crystallographic site.

Acknowledgements. I thank Sumio Iijima for providing Figure 1 and Gareth Thomas for providing Figure 2. I also thank Peter R. Buseck and Timothy L. Grove for their reviews. This work was supported by the National Science Foundation, grant EAR83-06861.

References

Akai, J., Polymerization process of biopyribole in metasomatism at the Akatani ore deposit, Japan, Contrib. Mineral. Petrol., 80, 117-131, 1982.

Anderson, A. B., Point defects in crystals: A quantum chemical methodology and its applications, this volume.

Boland, J. N., A. Duba, The defect-mechanisms for the solid-state reduction of olivine, this volume.

Breck, D. W., Zeolite molecular sieves, John Wiley, New York, 1974.

Buerger, M. J., The stuffed derivatives of the silica structures, Am. Mineral., 39, 600-614, 1954.

Burnham, C. W., Crystal structure of mullite, Year Book Carnegie Inst. Washington, 62, 223-227, 1963.

Buseck, P. R. and D. R. Veblen, Trace elements, crystal defects, and high-resolution electron microscopy, Geochim. Cosmochim. Acta, 42, 669-678, 1978.

Coe, R. S. and S. H. Kirby, The orthoenstatite to clinoenstatite transformation by shearing and reversion by annealing: Mechanism and potential applications, Contrib. Mineral. Petrol., 52, 29-55, 1975.

Cressey, B. A., E. J. W. Whittaker, and J. L. Hutchison, Morphology and alteration of asbestiform grunerite and anthophyllite, Mineral. Mag., 46, 77-87, 1982.

Crump, J. C., III and J. W. Mitchell, Hexagonal networks of linear imperfections in single crystals of cadmium, Philos. Mag., 8, 59-69, 1963.

Dyson, B. F., T. Anthony, and D. Turnbull, Interstitial diffusion of copper in tin, J. Appl. Phys., 38, 3408, 1967.

Eikum, A. and G. Thomas, Precipitation and dislocation nucleation in quench-aged Al-Mg alloys, Acta Mettall., 12, 537-545, 1964.

Foster, M. D., Interpretation of the composition of trioctahedral micas, U.S. Geol. Surv. Prof. Pap., 354-B, 11-49, 1960.

Gasparik, T. and D. H. Lindsley, Phase equilibria at high pressure of pyroxenes containing monovalent and trivalent ions, in Pyroxenes, Mineral. Soc. of Am. Rev. in Mineral., vol. 7, edited by C. T. Prewit, pp. 309-339, Mineralogical Society of America, Washington, D.C., 1980.

Goetze, C. and D. L. Kohlstedt, Laboratory study of dislocation climb and diffusion in olivine, J. Geophys. Res., 78, 5961-5971, 1973.

Hawthorne, F. C. and P. Černý, The mica group, Short Course Handb. Mineral. Assoc. Can., 8, 63-98, 1982.

Hirsch, P., A. Howie, R. B. Nicholson, D. W., Pashley, and M. J. Whelan, Electron Microscopy of Thin Crystals, Butterworths, London, 1965.

Hirth, J. P. and J. Lothe, Theory of Dislocations, 2nd ed., John Wiley, New York, 1982.

Iijima, S. and P. R. Buseck, High resolution electron microscopy of enstatite I: Twinning, polymorphism, and polytypism, Am. Mineral., 60, 758-770, 1975.

Iijima, S., J. M. Cowley and G. Donnay, High resolution electron microscopy of tourmaline crystals, Tschermaks Mineral. Petrogr. Mitt., 20, 216-224, 1973.

Kitamura, M., B. Shen, S. Banno, and N. Morimoto, Fine textures of laihunite--A nonstoichiometric distorted olivine-type mineral, Am. Mineral., 68, in press, 1983.

Kröger, F. A., Point defects in solids: Physics, chemistry, and thermodynamics, this volume.

Lasaga, A. C., The atomistic basis of kinetics: Defects in minerals, in Kinetics of Geochemical Processes, Mineral. Soc. of Am. Rev. in Mineral., vol. 8, edited by A. C. Lasaga and R. J. Kirkpatrick, pp. 261-319, Mineralogical Society of America, Washington, D.C., 1981.

Longhi, J. and J. F. Hays, Phase equilibria and solid solution along the join $CaAl_2Si_2O_8$-SiO_2, Am. J. Sci., 279, 876-890, 1979.

McLaren, A. C. and M. A. Etheridge, A transmission electron microscope study of naturally deformed orthopyroxene, Contrib. Mineral. Petrol., 57, 163-177, 1976.

Nakajima, Y. and P. H. Ribbe, Texture and structural interpretation of the alteration of pyroxene to other biopyriboles Contrib.

Mineral. Petrol., 78, 230-239, 1981.
O'Hara, M. J. and H. S. Yoder, Formation and fractionation of basic magma at high pressures, Scott. J. Geol., 3, 67-117, 1967.
Peterson, N. L., Diffusion in metals, in Solid State Physics, vol. 22, edited by F. Seitz, D. Turnbull, and H. Ehrenreich, pp. 409-512, Academic, New York, 1968.
Ribbe, P. H., Aluminum silicate polymorphs (and mullite), in Orthosilicates, Mineral. Soc. of Am. Rev. in Mineral., vol. 5, edited by P. H. Ribbe, pp. 189-214, Mineralogical Society of America, Washington, D.C., 1980.
Ricoult, D. L., Ricuit experimental de l'olivine, thése du doctorat, 3me cycle, Nantes Univ., Nantes, France, 1978.
Schreyer, W., K. Abraham, and H. Kulke, Natural sodium phlogopite coexisting with potassium phlogopite and sodian aluminian talc in a metamorphic evaporite sequence from Derrag, Tell Atlas, Algeria, Contrib. Mineral. Petrol., 74, 223-233, 1980
Smyth, J. R., Cation vacancies and the crystal chemistry of breakdown reactions in kimberlitic omphacites, Am. Mineral., 65, 1185-1191, 1980.
Spear, F. S., R. M. Hazen, and D. Rumble, III, Wonesite: A new rock forming silicate from the Post Pond Volcanics, Vermont, Am. Mineral., 66, 100-105, 1981.
Spence, J. C. H., Experimental High-Resolution Electron Microscopy, 370 pp., Oxford University Press, New York, 1980.
Thompson, J. B., Jr., Biopyriboles and polysomatic series, Am. Mineral., 63, 239-249, 1978.
Veblen, D. R., Non-classical pyriboles and polysomatic reactions in biopyriboles, in Amphiboles and Other Hydrous Pyriboles--Mineralogy, Mineral. Soc. of Am. Rev. in Mineral., vol. 9A, edited by D. R. Veblen, pp. 189-236, Mineralogical Society of America, Washington, D.C., 1981.
Veblen, D. R., Exsolution and crystal chemistry of the sodium mica wonesite, Am. Mineral., 68, 554-565, 1983.
Veblen, D. R. and P. R. Buseck, Microstructures and reaction mechanisms in biopyriboles, Am. Mineral., 65, 599-623, 1980.
Veblen, D. R. and P. R. Buseck, Hydrous pyriboles and sheet silicates in pyroxenes and uralites: Intergrowth microstructures and reaction mechanisms, Am. Mineral., 66, 1107-1134, 1981.
Veblen, D. R., P. R. Buseck, and C. W. Burnham, Asbestiform chain silicates: New minerals and stuctural groups, Science, 198, 359-365, 1977.
Veblen, D. R. and J. E. Post, A TEM study of fibrous cuprite (chalcotrichite): Microstructures and growth mechanisms, Am. Mineral., 68, 790-803, 1983.
Wadsley, A. D. and S. Andersson, Crystallographic shear, and the niobium oxides and oxide fluorides in the composition region MX_x, 2.4<x<2.7, in Perspectives in Structural Chemistry III, edited by J. D. Dunitz and J. A. Ibers, pp. 1-58, John Wiley, New York, 1970.
Warren, B. E., The crystal structure and chemical composition of the monoclinic amphiboles, Z. Kristallogr., 72, 493-517, 1930.
Wood, B. J. and C. M. B. Henderson, Composition and unit cell parameters of synthetic non-stoichiometric tschermakitic clinopyroxenes, Am. Mineral., 63, 66-72, 1978.
Yund, R. A., B. M. Smith, and J. Tullis, Dislocation-assisted diffusion of oxygen in albite, Phys. Chem. Miner., 7, 185-189, 1981.
Zussman, J., The crystal chemistry of the micas, Bull. Minéral., 102, 5-13, 1979.

DIFFUSIONAL CREEP PHENOMENA IN POLYCRYSTALLINE OXIDES

Ronald S. Gordon

University of Utah, Salt Lake City, Utah 84112

Abstract. Examples are given for the kinetic limitation of diffusional creep rates in polycrystalline oxides by (1) cation lattice diffusion, (2) cation grain boundary diffusion, (3) anion grain boundary diffusion and (4) interfacial effects at grain boundaries. Well defined examples of diffusional creep (Nabarro-Herring, Coble and combinations thereof) have been identified in polycrystalline MgO, Al_2O_3 and TiO_2. The effects of aliovalent substitutional dopants (e.g. Fe in MgO; Fe, Mn, Ti, Fe-Ti, and Mn-Ti in Al_2O_3; and Ta in TiO_2) on processes which are important in diffusional creep are presented. With the exception of Ta additions to TiO_2, all dopants increase creep rates by enhancing cation lattice diffusion, cation grain boundary diffusion or both. In the case of TiO_2, tantalum doping suppresses lattice diffusion. When lattice diffusion processes are considerably enhanced (e.g. TiO_2 and Ti-doped Al_2O_3) and cation grain boundary diffusion is limited, diffusional creep which is rate-limited by interfacial defect reactions at grain boundaries is dominant. Divalent impurities in Al_2O_3 such as Fe^{2+} and Mn^{2+} considerably enhance aluminum grain boundary diffusion and facilitate defect creation/ annihilation reactions at grain boundaries.

Introduction

The earth's mantle consists to a large degree of crystalline mineral matter with a relatively coarse grain size. It is at a sufficiently high temperature such that creep can occur in response to any non-hydrostatic stress component which may be present. In contrast to the creep of polycrystalline ceramics at high temperatures under atmospheric conditions, creep of minerals in the earth's mantle may be strongly influenced by the effect of hydrostatic pressure on diffusivity. Depending on the sign and magnitude of the activation volume, the ion diffusivity both in the lattice and in the grain boundary will be different than those under atmospheric conditions.

Since the grain size is expected to be fairly large in geologic materials, diffusional creep contributions may be small and dislocation or non-viscous modes of deformation might be important. However, as will be discussed in this paper, creep which is controlled by grain boundary diffusion could be important even under conditions of coarse grain size. Also considering the geologic time scale, diffusional creep even in coarse grained materials may be significant over very long periods of time.

Another mechanism, other than conventional lattice and grain boundary diffusion, which might be important in rock deformation is a "pressure solution" or "solution-precipitation" mechanism of diffusional creep. In this case, transport through an intergrannular liquid phase (i.e. aqueous, glass or molten rock) will be much more rapid than that which is possible by "solid- state" grain boundary diffusion.

In this paper, the basic principles of diffusional creep in polycrystalline ionic solids will be reviewed, emphazing solid state diffusion mechanisms. For a discussion of the "pressure solution" type of mechanism, the reader is referred elsewhere [Burton, 1977]. Creep data in three model oxide systems (MgO, Al_2O_3 and TiO_2) will be reviewed to illustrate the effects of soluble aliovalent dopants, oxygen fugacity (partial pressure), and grain size. Two of the materials (MgO and Al_2O_3) are wide band gap oxides in which the concentrations of intrinsic electronic and lattice defects are small. The third material, TiO_2, is an intermediate band gap oxide in which intrinsic or extrinsic electronic defect concentrations can be dominant. The roles of lattice diffusion, grain boundary diffusion and interfacial defect reactions at grain boundaries are discussed. Detailed implications of these results for deformation in geologic systems are left to the experts.

Theory

In general the diffusional creep of a polycrystalline ceramic solid, $A_\alpha B_\beta$, follows a relation of the following form [Gordon, 1973, 1975]:

$$\dot{\varepsilon} = 44\Omega_V \sigma D^C / kT(GS)^3 \qquad (1)$$

in which $\dot{\varepsilon}$ is the steady state strain rate, σ is the stress, (GS) is the grain size, Ω_V is the molecular volume of $A_\alpha B_\beta$ and D^C is a complex mass transport parameter.

TABLE 1. Theoretical Grain Size Dependencies in Diffusional Creep

Grain Size Exponent, m	Controlling Mechanism
1	defect creation/annihilation at grain boundaries [Burton, 1977]
2	cation and/or anion lattice diffusion
3	cation and/or anion grain boundary diffusion
3	"pressure-solution" transport in an intergranular liquid, glassy or molten phase

Depending on the mechanism of mass transport, the apparent grain size exponent (i.e. m in $\dot{\varepsilon} \alpha (GS)^{-m}$) can take on values between 1 and 3. In Table 1, a summary is given for the various mechanisms of diffusional creep. It is not uncommon in experimental data for an intermediate grain size exponent (e.g. $2 < m < 3$) to be encountered indicating a mixed deformation mechanism. In the general situation, three different processes can control diffusional creep in a polycrystalline compound, $A_\alpha B_\beta$, in which the transport of the cation is coupled with that of the anion (Ikuma, 1981): (1) lattice diffusion, (2) grain boundary diffusion and (3) interfacial defect reactions at grain boundaries. Assuming for simplicity that the rates of the interfacial defect reactions vary linearly with stress and that they act in series only with lattice diffusion processes (some mechanisms for interfacial kinetics involve a σ^2 dependence [Ikuma and Gordon, 1981]; also interfacial processes can in principle act in series with boundary diffusion [Ikuma and Gordon, 1981]), Ikuma and Gordon [1981] derived the following equation for D^C:

$$(D^C)^{-1} = \frac{\alpha\left[\frac{(GS)D_A^\ell}{\pi} + \frac{(GS)^2K_A}{44}\right]}{\frac{(GS)^3D_A^\ell K_A}{44\pi} + \delta_A D_A^b\left[\frac{(GS)D_A^\ell}{\pi} + \frac{(GS)^2K_A}{44}\right]} + \frac{\beta\left[\frac{(GS)D_B^\ell}{\pi} + \frac{(GS)^2K_B}{44}\right]}{\frac{(GS)^3D_B^\ell K_B}{44\pi} + (\delta_B D_B^b)\left[\frac{(GS)D_B^\ell}{\pi} + \frac{(GS)^2K_B}{44}\right]} \tag{2}$$

Here, D_A^ℓ and D_A^b are the $A^{\beta+}$ ion lattice and grain boundary diffusivities, D_B^ℓ and D_B^b are the $B^{\alpha-}$ ion lattice and grain boundary diffusivities, δ_A and δ_B are the effective widths of the regions of enhanced diffusion near the grain boundaries for $A^{\beta+}$ and $B^{\alpha-}$ ions, and K_A and K_B are the interfacial rate constants for anion and cation vacancy (or interstitial) creation or annihilation, respectively, at the grain boundaries.

In general, it is clear that the diffusion coefficients which are extracted from diffusional creep data are complex quantities which involve several basic mass transport parameters. Impurities which are present in solution can significantly alter lattice and grain boundary diffusivities and hence influence the value of D^C. Furthermore D^C is a strong function of the grain size.

For purposes of illustration, let us examine several limiting cases for this generalized diffusional creep theory. First in the limit of rapid kinetics for interfacial defect reactions, (2) can be simplified to terms involving the four basic diffusion parameters, D_A^ℓ, $\delta_A D_A^b$, D_B^ℓ, $\delta_B D_B^b$.

$$(D^C)^{-1} = \frac{\alpha}{\frac{(GS)}{\pi}D_A^\ell + \delta_A D_A^b} + \frac{\beta}{\frac{(GS)}{\pi}D_B^\ell + \delta_B D_B^b} \tag{3}$$

Equation (3) predicts that the overall diffusional creep rate will be controlled by the slowest moving ionic species diffusing over its fastest path.

Kinetics controlled by grain boundary diffusion. At very small grain sizes and low temperatures, $(GS/\pi)D_A^\ell \ll \delta_A D_A^b$ and $(GS/\pi)D_B^\ell \ll \delta_B D_B^b$. In this limit, (1) can be written as

$$\dot{\varepsilon} = \frac{44\Omega_V\sigma}{\alpha kT(GS)^3}\left[\frac{\delta_A D_A^b}{1 + \frac{\beta}{\alpha}\frac{\delta_B D_B^b}{\delta_A D_A^b}}\right] \tag{4}$$

and Coble [1963] creep, which is rate-limited by the slowest moving species in the grain boundary, should be observed. Coble creep has been reported [Terwilliger et al., 1970] for the deformation of Fe-doped MgO (0.05-0.27 cation %) for grain sizes between 7 and 23 μm at temperatures $\leq$ 1300°C. In this case, diffusional creep rates were controlled by cation (magnesium) grain boundary diffusion.

The only other well defined case of Coble creep was observed in Fe(2 cation %) - doped Al_2O_3 (17-100 μm) tested under reducing conditions ($P_{O_2} \approx 10^{-1}$ Pa) where a substantial portion of the iron was in the divalent state [Lessing and Gordon, 1977].

Kinetics controlled by cation lattice diffusion. At larger grain sizes, higher temperatures, and under conditions of enhanced anion grain boundary diffusion, $(GS/\pi)D_A^\ell >> \delta_A D_A^b$, $\delta_B D_B^b >> (GS/\pi)D_B^\ell$ and $(GS/\pi)D_A^\ell << \delta_B D_B^b$. In this limit, (1) can be written

$$\dot{\varepsilon} = 44\Omega_V \sigma D_A^\ell / \alpha \pi kT(GS)^2 \tag{5}$$

and Nabarro-Herring [Nabarro, 1948; Herring, 1950] creep, which is rate-limited by cation lattice diffusion, should be dominant. Abundant examples of Nabarro-Herring creep controlled by cation lattice diffusion have been observed [Tremper et al., 1974; Hodge and Gordon, 1978; Lessing and Gordon, 1977; Ikuma and Gordon, 1983; Philpot et al., 1983] in polycrystalline MgO, Al_2O_3 and TiO_2.

Mixed kinetics. Frequently [Hodge and Gordon, 1978; Lessing and Gordon, 1977; Ikuma and Gordon, 1983] the apparent grain size exponent lies between values characteristic of Nabarro-Herring (m=2) and Coble creep (m=3). If anion grain boundary diffusion is rapid (typically observed in polycrystalline MgO and Al_2O_3), then diffusional creep will be rate-limited by a combination of cation grain boundary and cation lattice diffusion with 2 < m < 3, i.e.

$$\dot{\varepsilon} = 44\Omega_V \sigma D_A^\ell / \pi kT\alpha(GS)^2 + 44\Omega_V \sigma \delta_A D_A^b / kT\alpha(GS)^3 \tag{6}$$

Any one of three conditions are important for the enhancement of contributions due to cation grain boundary diffusion: (1) small grain size, (2) low temperatures, (3) specific dopants (e.g. F^- in MgO, Fe^{2+} in Al_2O_3, Mn^{2+} in Al_2O_3).

Finally when the cation lattice diffusivity is sufficiently enhanced by appropriate dopants (e.g. Fe^{3+} in MgO), the grain size is reasonably coarse, and the temperature is sufficiently high, diffusional creep kinetics can be controlled in part by anion grain boundary diffusion [Tremper et al., 1974]. The conditions appropriate for this situation are as follows: $(GS/\pi)D_A^\ell >> \delta_A D_A^b$ and $(GS/\pi)D_B^\ell << \delta_B D_B^b$. The general creep equation then becomes

$$\dot{\varepsilon} = \frac{44\Omega_V \sigma}{\alpha \pi kT(GS)^2} \left[\frac{D_A^\ell}{1 + \frac{\beta(GS)D_A^\ell}{\alpha \pi \delta_B D_B^b}} \right] \tag{7}$$

Mixed kinetics (2 < m < 3) have been observed [Tremper et al., 1974] in Fe-doped MgO (2.65-5.30%). In this system, cation lattice diffusion is significantly enhanced by the presence of trivalent iron in solid solution through the creation of cation lattice vacancies. If cation lattice diffusion becomes too fast, then it no longer is entirely rate-limiting and oxygen grain boundary diffusion becomes rate-limiting in part. In the case of Al_2O_3 doped with iron or manganese, while enhancing cation lattice diffusion to some degree, doping significantly enhanced cation grain boundary diffusion. Oxygen grain boundary diffusion was always too fast relative to aluminum lattice or grain boundary diffusion to be rate-limiting.

Kinetics influenced by interfacial reactions. Let us briefly examine a few situations in which interfacial reactions may be important. First in the limit of very rapid anion grain boundary diffusion (good assumption for the three model materials analyzed in this paper), the following relation can be written for D^C:

$$D^C = \frac{(GS)^3 D_A^\ell K_A / 44\pi\alpha}{(GS)D_A^\ell/\pi + (GS)^2 K_A/44} + \delta_A D_A^b/\alpha \tag{8}$$

Equation (8) has several interesting limits.

Limit I: Rapid Cation Interface Reaction

$$(GS)D_A^\ell/\pi <<< (GS)^2 K_A/44 \tag{9}$$

In this case

$$D^C = \frac{1}{\alpha}\left[\frac{(GS)D_A^\ell}{\pi} + \delta_A D_A^b\right] \tag{10}$$

and diffusional creep is just a competition between cation lattice and cation grain bounday diffusion.

Limit II: Slow Cation Interface Reaction

$$(GS)D_A^\ell/\pi >> (GS)^2 K_A/44 \tag{11}$$

In this case,

$$D^C = \frac{1}{\alpha}\left[\frac{(GS)^2 K_A}{44} + \delta_A D_A^b\right] \tag{12}$$

and

$$\dot{\varepsilon} = \frac{44\Omega_V \sigma}{kT\alpha(GS)}\left[\frac{K_A}{44} + \frac{\delta_A D_A^b}{(GS)^2}\right] \tag{13}$$

Limit IIA: Rapid Cation Grain Boundary Diffusion

In this case,

$$\dot{\varepsilon} = \frac{44\Omega_V \sigma \delta_A D_A^b}{\alpha kT(GS)^3} \tag{14}$$

and diffusional creep is controlled by cation grain boundary diffusion.

Limit IIB: Slow Cation Grain Boundary Diffusion

In this case,

$$\dot{\varepsilon} = \frac{\Omega_V K_A \sigma}{\alpha kT(GS)} \tag{15}$$

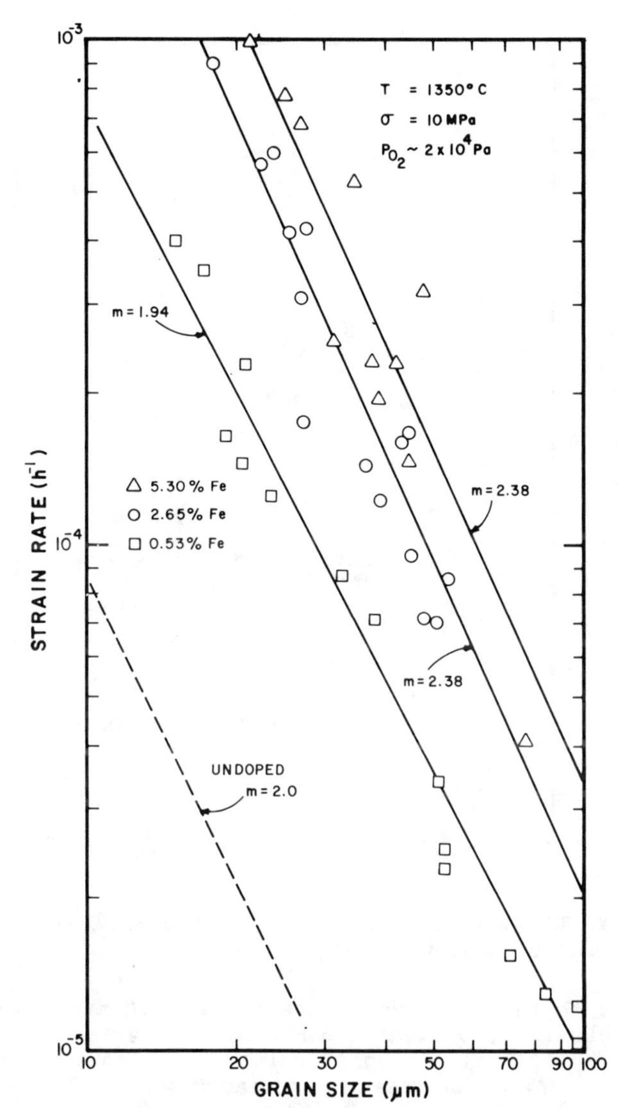

Fig. 1. Effect of grain size on the diffusional creep rate at 1350°C of polycrystalline MgO, pure and doped with iron.

and diffusional creep is controlled by the cation interfacial defect reaction in the limit of small grain sizes.

Limit III: Slow Cation Grain Boundary Diffusion

In this case,

$$\dot{\varepsilon} = \frac{\Omega_V \sigma}{\pi \alpha kT(GS)} \left[\frac{D_A^\ell K_A}{D_A^\ell/\pi + (GS)K_A/44} \right] \quad (16)$$

Limit IIIA: Small Grain Size and/or High D_A^ℓ

In this case,

$$D_A^\ell/\pi \gg (GS)K_A/44 \quad (17)$$

$$\dot{\varepsilon} = \frac{\Omega_V \sigma K_A}{\alpha kT\ (GS)} \quad (18)$$

and diffusional creep is again interface reaction controlled. The important point to be made here is that interface controlled creep kinetics are not restricted to the small grain size limit. They can be controlling in intermediate to large grain size materials providing cation lattice diffusion is very fast and cation grain boundary diffusion is slow. The creep of Ti-doped (and Mg-Ti compositions) Al_2O_3 [Ikuma and Gordon, 1981] apparently satisfied these conditions (i.e. intermediate grain size, very high D_A^ℓ and apparently low $\delta_A D_A^b$). At these compositions, the grain size dependence was much smaller (m ~1). The titanium dopant was so effective in enhancing aluminum lattice diffusion that the interfacial defect reaction at the grain boundaries became rate-limiting. When Fe or Mn was present in combination with Ti, cation grain boundary diffusion was significantly enhanced and the interfacial reactions were no longer a problem. It is possible that the interfacial rate constants for each diffusing species depend on the magnitude of the corresponding grain boundary diffusion coefficient. Interfacial control of diffusional creep has also been observed for undoped, polycrystalline TiO_2 [Philpot et al., 1983].

In summary, diffusional creep in polycrystalline ceramic materials is expected to be important under the following conditions: (1) low stresses below those required for the initiation of plastic deformation by dislocation motion (typically in complex ceramic systems (particularly minerals) slip is expected to be generally more difficult and the Von Mises criterion for homogeneous polycrystalline plasticity may not be obeyed), (2) small grain sizes in most cases, (3) high grain boundary and/or lattice diffusivity which can be influenced by the oxygen fugacity and aliovalent dopants in solid solution.

Diffusional Creep of Model Polycrystalline Oxides

In this section, typical creep data will be reviewed for three oxide systems to illustrate the various diffusional creep regimes which are important in polycrystalline ceramic materials. The effects of grain size, oxygen fugacity, and aliovalent dopants will be discussed. These variables are perhaps the most important in understanding the relative contributions of lattice and grain boundary diffusion processes to overall creep deformation. Little emphasis will be given to the effect of temperature, since apparent activation energies are not particularly sensitive indicators of the diffusion mechanism.

All results, which will be reported, were derived from steady state and viscous (i.e. $\dot{\varepsilon}\alpha\sigma$) creep data taken under conditions of dead-load, four-point bending on specimens with densities in excess of 98% of theoretical. Details concerning the preparation and testing of specimens are given elsewhere [Tremper et al., 1974; Hollenberg and Gordon, 1973; Lessing and Gordon, 1977; Ikuma and Gordon, 1982, 1983; Philpot et al., 1983].

MgO, pure and doped with iron. In Figure 1, the effect of grain size is shown for the creep of both undoped and iron-doped polycrystalline MgO for grain sizes up to about 100 μm (all grain sizes reported herein are linear intercept averages multiplied by the factor 1.5). The creep data for the undoped material were obtained by extrapolation from measurements taken on material with grain sizes below 10 μm [Hodge and Gordon, 1978]. Doping polycrystalline MgO with 0.53 cation % iron increased the diffusional creep rate by a factor of ten, thereby increasing the ranges of grain size and stress over which diffusional creep is dominant. The reciprocal square grain size dependence indicates that the creep of undoped and iron-doped (0.53 cation %) MgO is controlled by a lattice diffusion process. Diffusion coefficients which were extracted from the creep data at 1350°C are plotted in Fig. 2. They are consistent with magnesium lattice diffusion (magnesium tracer diffusion in undoped MgO is estimated to be about 1.2×10^{-12} cm^2/sec based on measurements [Wuensch et al., 1973] taken at higher temperatures). The enhancement in the creep rate due to iron doping is believed to be

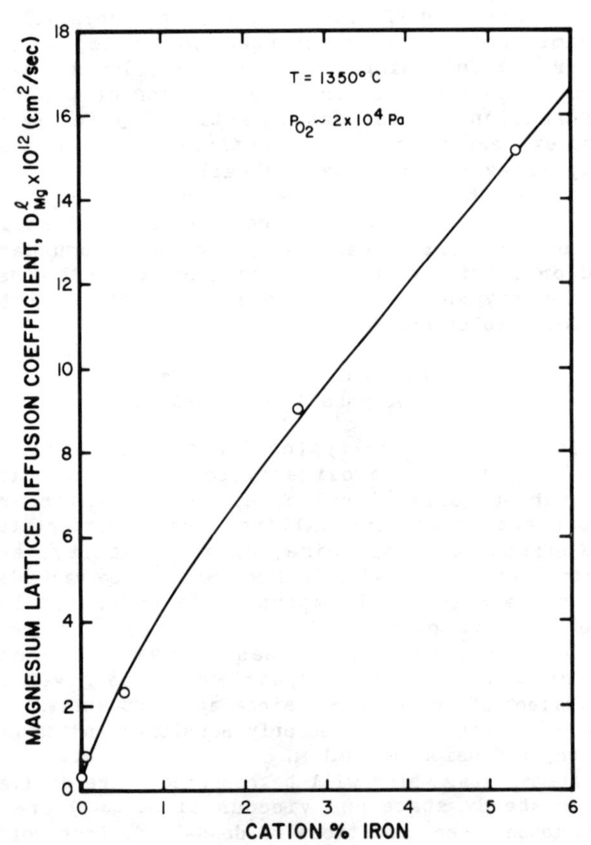

Fig. 2. Effect of iron dopant concentration on magnesium lattice diffusivity at 1350°C in MgO.

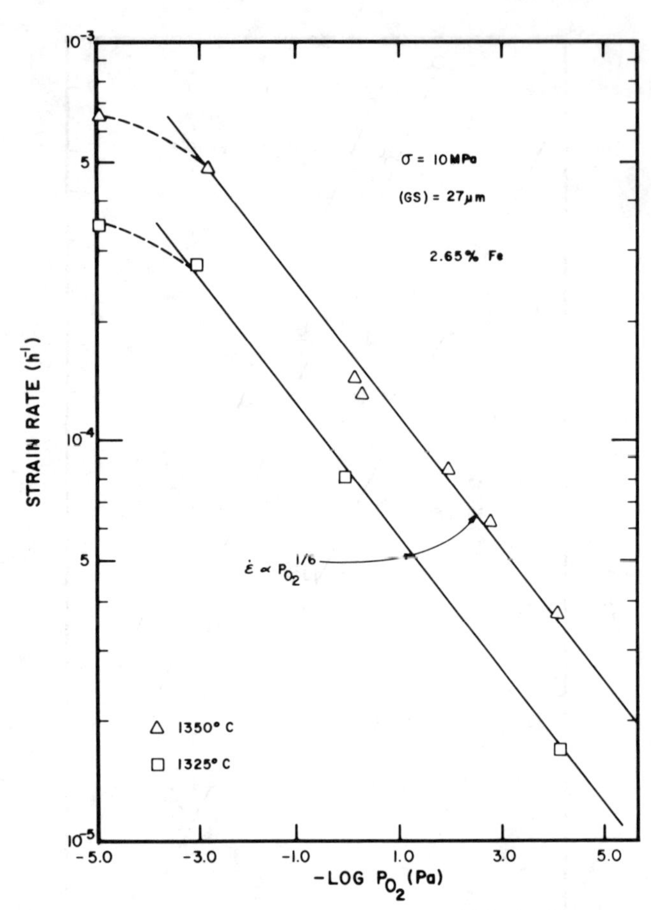

Fig. 3. Effect of oxygen fugacity on diffusional creep rate of iron-doped, polycrystalline MgO.

due to the increased concentration of magnesium vacancies (V''_{Mg}) which compensate trivalent iron in substitutional solid solution. [Tremper et al., 1974]. The relative concentrations of trivalent [$Fe^{\cdot}_{Mg}$] and divalent [Fe^{*}_{Mg}] iron can be controlled by the oxygen fugacity as is illustrated in Fig. 3. A reduction in the oxygen fugacity to 10^{-5} Pa led to over an order of magnitude reduction in the creep rate due to an increased concentration of divalent iron and a corresponding decrease in the cation vacancy concentration [V''_{Mg}]. This effect can be represented by the following quasi-chemical defect reaction:

$$V''_{Mg} + O^{*}_{O} + 2Fe^{\cdot}_{Mg} = 2Fe^{*}_{Mg} + 1/2\ O_{2(g)} \quad (19)$$

The mixed grain size dependence (i.e. m=2.34) at higher iron dopant concentrations (2.65 and 5.3 cation %) and the lower creep activation energies (339 versus 490 kJ/mole) in oxidizing atmospheres has been interpreted [Tremper et al., 1974] as a mixed creep mechanism involving both magnesium lattice and oxygen grain boundary diffusion.

Oxygen grain boundary diffusion ($\delta_O D^{o}_{O}$ ~1.6-3.2 x 10^{-14} cm^3/sec at 1350°C) is projected to be dominant in situations where the concentration

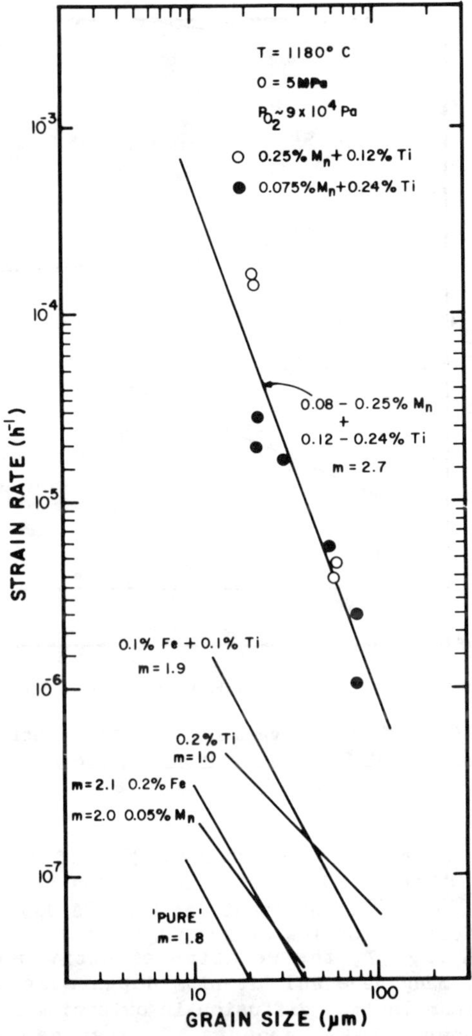

Fig. 4. Effect of grain size on diffusional creep rate at 1180°C of polycrystalline Al_2O_3 doped with transition metal impurities.

of trivalent iron is large and the temperature and grain size are increased [Tremper et al., 1974; Hodge et al., 1977]. On the contrary, magnesium grain boundary diffusion at 1350°C is not projected (Hodge et al., 1977) to be dominant at 1350°C unless the grain size is well below 10μm ($\delta_{Mg}D^{b}_{Mg}$ ~6 x 10^{-16} cm^3/sec).

Al_2O_3, pure and doped with transition metal dopants. Solid solution impurities can have a pronounced effect on the diffusional creep rate of polycrystalline Al_2O_3. In Fig. 4 the effect of grain size at 1180°C is shown for the creep of polycrystalline Al_2O_3 doped simultaneously with small amounts of Mn (0.08-0.25 cation %) and Ti (0.12-0.24 cation %). The data shown in Fig. 4 [Ikuma and Gordon, 1982] for the undoped Al_2O_3 and material doped with Mn, Fe, Ti or Fe-Ti were extrapolated from measurements at higher temperatures (~1400-1450°C). Small amounts of Mn and Ti dopants enhanced the diffusional creep rate over three orders of magnitude leading to significant mass transport rates at a temperature approximately 200° lower than that in undoped material where comparable transport rates are encountered. The rapid transport at low temperatures is believed [Ikuma and Gordon, 1982] to be due to significant enhancement in both aluminum grain boundary diffusion ($\delta_{Al}D^{b}_{Al}$ ≈4.5 x 10^{-16} cm^3/sec at 1180°C) and aluminum lattice diffusion (D^{ℓ}_{Al} ~3.4 X 10^{-13} cm^2/sec at 1180°C). In undoped Al_2O_3, the aluminum lattice and grain boundary diffusion parameters at 1180°C are ~2x$10^{-16}$$cm^2$/sec and <<$10^{-18}$ cm^3/sec, respectively. It is possible that some type of boundary phase may be responsible for the enhanced grain boundary transport and accelerated creep rates in the Mn-Ti doped aluminas.

Grain size effects at higher temperatures (1450°C) are shown in Fig. 5. All lines represent least square fits to the data (Lessing and Gordon, 1977 Ikuma and Gordon, 1983]. Only a portion of the actual data are shown to maintain clarity of presentation. Several interesting observations can be drawn from these data. With the exception of titanium present as a single dopant and the iron dopant (2 cation %) at very low oxygen fugacities,

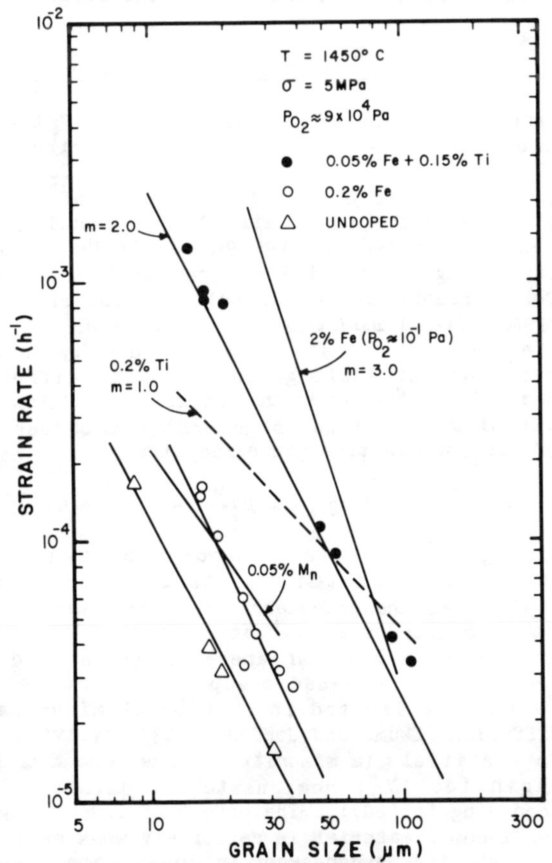

Fig. 5. Effect of grain size on diffusional creep rate at 1450°C of polycrystalline Al_2O_3 doped with transition metal impurities.

all creep data are consistent with Nabarro-Herring creep (m ~2) controlled by aluminum lattice diffusion. Fe^{2+}, Mn^{2+} and Ti^{4+} present in substitutional solid solution enhance cation diffusion by the creation of additional aluminum lattice interstitials ($Al_I^{\bullet\bullet\bullet}$ or vacancies (V_{Al}'''). Small amounts (~0.2 cation %) of iron and Ti, present simultaneously in solid solution, are much more effective in enhancing diffusional creep than either Mn or Fe present as a single dopant (Ikuma Gordon, 1983). When titanium is present as a single dopant, a much weaker grain size dependence (i.e. $\dot{\varepsilon}\alpha(GS)^{-1}$) was encountered [Ikuma and Gordon, 1981] indicating the presence of interfacial controlled creep which becomes dominant in the limit of small grain sizes. Substantial enhancement in deformation rate was encountered in the creep of coarse-grained titanium--doped Al_2O_3. It is believed that the concentration of Ti^{4+} (i.e. $Ti_{Al}^{\bullet}$) is sufficiently high leading to a large concentration of compensating aluminum ion vacancies (V_{Al}'''). Under these conditions and in the absence of significant aluminum grain boundary diffusion, the creation of lattice defects at grain boundaries can limit the overall mass transport rate. In the case of iron-doped material (e.g. 2 cation %), the concentration of divalent iron [Fe_{Al}'] can be increased when the oxygen fugacity is lowered, i.e.

$$O_O^* + 2Fe_{Al}^* = 2Fe_{Al}' + V_O^{\bullet\bullet} + 1/2O_{2(g)} \qquad (20)$$

Under these conditions, the kinetics of transport are consistent with a strong contribution of aluminum grain boundary diffusion and Coble Creep (i.e. m~3) is observed. The role of Fe^{2+} (and even Mn^{2+} for that matter) in enhancing $\delta_{Al}D_{Al}^b$ (>10^{-14} cm^3/sec at 1450°C for 2% Fe at reduced oxygen fugacities) is believed to be responsible for the enhanced creep rates in the mixed dopant systems [Ikuma and Gordon, 1983). In these situations, the creep rate does not become limited by interfacial reactions at small grain sizes and diffusional creep is controlled by the concentration of tetravalent titanium ($Ti_{Al}^{\bullet}$) which governs the concentration of aluminum lattice vacancies, i.e.

$$6\ Ti_{Al}^* + 3/2\ O_{2(g)} = 2V_{Al}''' + 3O_O^* + 6Ti_{Al}^{\bullet} \qquad (21)$$

In Fig. 6, the effects of oxygen fugacity are shown for selected dopants in polycrystalline Al_2O_3. As indicated earlier, a decrease in oxygen fugacity increases the concentration of divalent iron (Fe_{Al}') and manganese (Mn_{Al}') according to (20). The increased creep rates are believed to be rate limited in part by aluminum lattice diffusion [Ikuma and Gordon, 1983] involving cation interstitials (a minority lattice defect if oxygen vacancies ($V_O^{\bullet\bullet}$) compensate the divalent dopant according to (20)). The effect of oxygen fugacity on Ti-doped material is relatively weak even though a substantial enhancement in creep rate (compared to undoped material) was observed. Interfacial control of the mass transport kinetics is consistent with this effect. In the Fe-Ti mixed dopant,

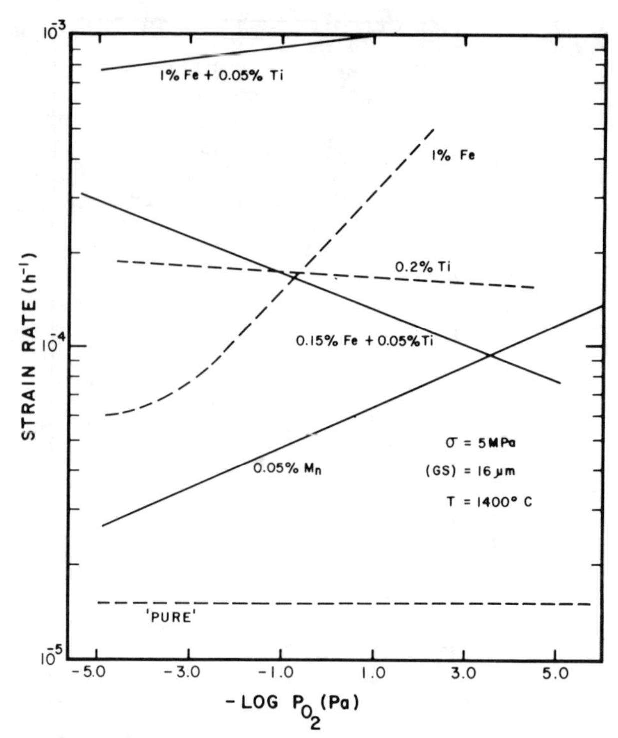

Fig. 6. Effect of oxygen fugacity on diffusional creep rate at 1400°C of polycrystalline Al_2O_3 doped with transition metal impurities.

Ti^{4+} (i.e. $Ti_{Al}^{\bullet}$) dominates the defect chemistry according to (21) until the concentration of Fe^{2+} (i.e. Fe_{Al}') is dominant (at high Fe dopant concentrations and/or low oxygen fugacities).

In Fig. 7, the relative effectiveness of the iron, manganese and titanium dopants in enhancing aluminum lattice diffusion in oxidizing atmospheres is shown. Small amounts of manganese are about 3 to 4 times more effective than comparable amounts of iron. The most effective single dopant is titanium. Values of D_{Al}^{ℓ}, estimated (lower bounds) from creep data in the limit of coarse grain size (~100μm) where the contribution of lattice diffusion is the highest, indicate over an order of magnitude enhancement in the aluminum lattice diffusivity for dopant levels ≤0.2% Ti. In the Fe-Ti mixed dopant, Ti^{4+} is dominant until the total iron concentration exceeds about 2%. Less than 2% of the total iron is divalent; whereas approximately 80% of the titanium is tetravalent [Ikuma and Gordon, 1983].

The Mn-Ti dopant pair (<0.4 cation % total impurities) is the most effective in enhancing aluminum lattice diffusion (over three orders of magnitude and off the scale in Fig. 7). In all situations oxygen grain boundary diffusion in polycrystalline alumina is believed to be very fast and never rate limiting [Ikuma and Gordon, 1983].

TiO_2, pure and doped with tantalum. In TiO_2, an intermediate band gap semiconductor, the defect

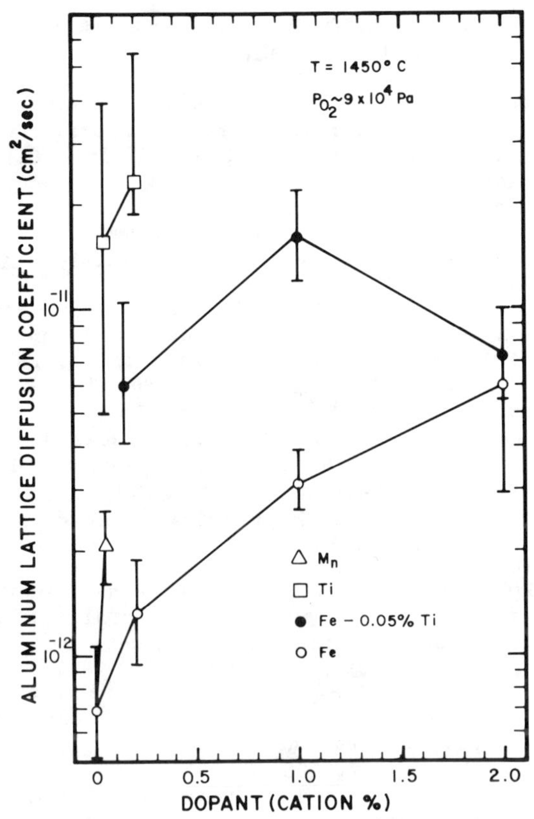

Fig. 7. The effect of transition metal dopants on the aluminum lattice diffusivity in Al_2O_3 at 1450°C.

structure is dominated to a large degree by the concentration of intrinsic or extrinsic conduction band electrons. Depending upon the donor concentration, temperature, and oxygen fugacity, extrinsic and intrinsic conduction band electrons can be compensated by substitutional donors (e.g. $Ta^{\bullet}_{Ti}$) and by intrinsic lattice defects (e.g. $V_O^{\bullet\bullet}$, $Ti_I^{\bullet\bullet\bullet}$, $Ti_I^{\bullet\bullet\bullet\bullet}$), respectively.

In the case of MgO and Al_2O_3, intrinsic electronic and atomic defect concentrations are very small at most temperatures due to very large band gaps and high defect formation energies. Hence, the concentration of both the majority and minority lattice defects are controlled extrinsically by dopants. Due to enhanced oxygen grain boundary transport in MgO and Al_2O_3, aliovalent dopants normally enhance lattice diffusion by increasing the concentration of either majority or minority lattice defects.

In TiO_2, the concentration of minority lattice defects can be either suppressed or enhanced by soluble aliovalent dopants (e.g. $Ta^{\bullet}_{Ti}$) or changes in the oxygen fugacity. For example, doping with tantalum suppresses the concentrations of oxygen vacancies ($V_O^{\bullet\bullet}$) and titanium interstitials ($Ti_I^{\bullet\bullet\bullet}$, $Ti_I^{\bullet\bullet\bullet\bullet}$), while enhancing the concentration of titanium vacancies (V_{Ti}''''). In undoped rutile, the concentrations of lattice defects such as anion vacancies and cation interstitials are relatively high (e.g. ~10^{-4}–10^{-5} atom fraction at 1400°C). At a fixed dopant concentration of tantalum, a decrease in the oxygen fugacity will lead to increases in the concentrations of both oxygen vacancies and titanium interstitials, according to the following quasi-chemical defect reactions:

$$O_O^{*} = V_O^{\bullet\bullet} + 2e' + 1/2\, O_{2(g)} \qquad (22)$$

$$Ti_{Ti}^{*} + 2O_O^{*} = Ti_I^{\bullet\bullet\bullet\bullet} + 4e' + O_{2(g)} \qquad (23)$$

Increases in the concentrations of these defects should lead to enhanced mass transport and faster creep rates.

In Fig. 8 representative data at 1200°C are presented for the diffusional creep of polycrystalline TiO_2, pure and doped with 1 cation % Ta. Tantalum doping suppressed the creep rate in air by one to two orders of magnitude depending on the grain size. This drop in creep rate is due to the suppression of anion vacancies or cation interstitials. At very low oxygen fugacities, the concentrations of these defects are enhanced and creep rates in doped material are increased and are equivalent to those in undoped material.

Extensive experiments [Philpot et al., 1983]

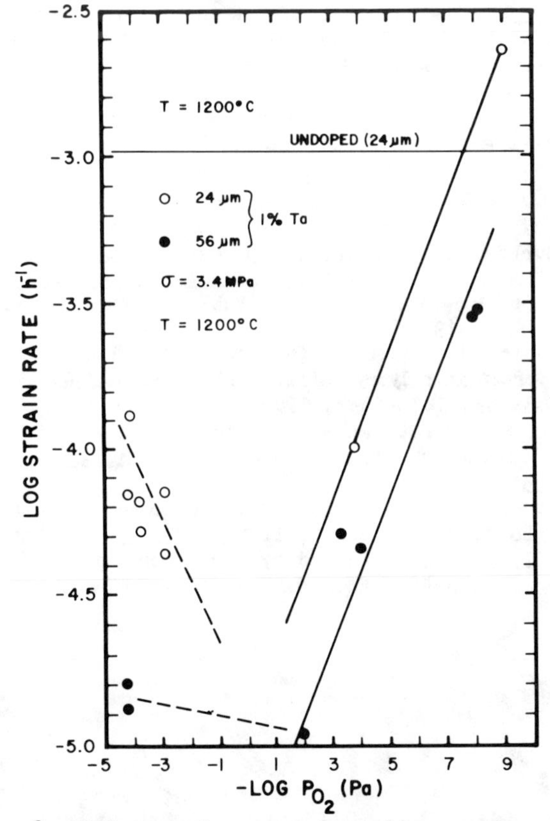

Fig. 8. Effects of oxygen fugacity, grain size and tantalum dopant on the diffusional creep of polycrystalline rutile at 1200°C.

over a range of grain sizes (5-50μm) in both oxidizing and reducing (~10^{-9} Pa) atmospheres indicate that the creep of tantalum-doped rutile obeys the Nabarro-Herring relation (i.e. m=2). Hence, mass transport is rate limited by a lattice diffusion mechanism involving oxygen vacancies, titanium interstitials or titanium vacancies depending on the dopant concentration and oxygen fugacity.

The diffusional creep rate of undoped TiO_2 is a weak inverse function of the grain size and independent of the oxygen fugacity. It is believed that because of the high intrinsic lattice defect concentrations, lattice diffusion is too fast to be accommodated by defect creation and/or annihilation processes at grain boundaries. Hence the mass transport kinetics are controlled by interfacial effects.

Summary

Aliovalent dopants in solid solution, oxygen fugacity and grain size are the most sensitive variables to investigate when one is trying to analyze and quantify the various mass transport processes which are important in the diffusional creep of polycrystalline oxide ceramics.

Acknowledgements. This work was supported by the Department of Energy under Contract EY-76-S-02-1591. The author acknowledges the contributions of his former students, H. K. Bowen, R. A. Giddings, J. D. Hodge, G. W. Hollenberg, Y. Ikuma, P. A. Lessing, K. Philpot, G. R. Terwilliger, and R. T. Tremper.

References

Burton, B., Diffusional Creep of Polycrystalline Materials, Diff. and Defect Monogr. Ser., no. 5, pp. 106-109, Trans Tech Publications, Bay Village, Ohio, 1977.

Coble, R. L., Model for boundary diffusion controlled creep in polycrystalline materials, J. Appl. Phys., 34(6), 1679-1682, 1963.

Gordon, R. S., Mass transport in the diffusional creep of ionic solids, J. Am. Ceram. Soc., 56(3) 147-152, 1973.

Gordon, R. S., Ambipolar diffusion and its application to diffusion creep, in Mass Transport Phenomena in Ceramics, edited by A. R. Cooper and A. H. Heuer, pp. 445-464, Plenum, New York, 1975.

Herring, C., Diffusional viscosity of a polycrystalline solid, J. Appl. Phys., 21(5), 437-445, 1950.

Hodge, J. D., and R. S. Gordon, Grain growth and creep in polycrystalline magnesium oxide fabricated with and without a LiF additive, Ceramurgia Inter., 4(1), 17-20, 1978.

Hodge, J. D., P. A. Lessing, R. S. Gordon, Creep mapping in a polycrystalline ceramic: Application to magnesium oxide and magnesiowustite, J. Mater. Sci., 12(8), 1598-1604, 1977.

Hollenberg, G. W., and R. S. Gordon, Effect of oxygen partial pressure on the creep of polycrystalline Al_2O_3 doped with Cr, Fe, or Ti, J. Am. Ceram. Soc., 56(3), 140-147, 1973.

Ikuma, Y., and R. S. Gordon, Role of interfacial defect creation-annihilation processes at grain boundaries on the diffusional creep of polycrystalline alumina, in Surfaces and Interfaces in Ceramic and Ceramic Metal Systems, edited by J. Pask, and A. Evans, pp. 283-294, Plenum, New York, 1981.

Ikuma, Y., and R. S. Gordon, Enhancement of the diffusional creep of polycrystalline Al_2O_3 by simultaneous doping with manganese and titanium, J. Mater. Sci., 17, 2961-2967, 1982.

Ikuma, Y., and R. S. Gordon, Effect of doping simultaneously with iron and titanium on creep of polycrystalline Al_2O_3, J. Am. Ceram. Soc., 66(2), 139-147, 1983.

Lessing, P. A., and R. S. Gordon, Creep of polycrystalline alumina, pure and doped with transition metal impurities, J. Mater Sci., 12, 2291-2302, 1977.

Nabarro, F. R. N., Report of Conference on Strength of Solids, pp. 75-90, Physical Society, London, 1948.

Philpot, K., Y. Ikuma, G. R. Miller, and R. S. Gordon, High temperature steady-state diffusional creep of polycrystalline rutile, pure and doped with tantalum, J. Mater. Sci., 18, 1698-1708, 1983.

Terwilliger, G. R., H. K. Bowen, and R. S. Gordon, Creep of polycrystalline $MgO-Fe_2O_3$ solid solutions at high temperatures, J. Am. Ceram. Soc., 53(5), 241-251, 1970.

Tremper, R. T., R. A. Giddings, J. D. Hodge, and R. S. Gordon, Creep of polycrystalline MgO--$FeO-Fe_2O_3$ solid solutions, J. Am. Ceram. Soc., 57(10), 421-428, 1974.

Wuensch, B. J., W. C. Steel, and T. Vasilos, Cation self diffusion in single crystal MgO, J. Chem. Phys. 58(12), 5258-5266, 1973.

WATER-RELATED DIFFUSION AND DEFORMATION EFFECTS IN QUARTZ AT PRESSURES OF 1500 AND 300 MPA

S. J. Mackwell and M. S. Paterson

Research School of Earth Sciences, Australian National University
Canberra 2601, Australia

Abstract. Natural and synthetic quartz specimens have been hydrothermally treated at 1500 MPa, 900°C and the diffusion profile of the water-related species determined by serial sectioning and infrared absorption measurements. The diffusion coefficient D is found to be around 10^{-12} to 10^{-11} m^2s^{-1} and the solubility of the order of 1000 $H/10^6Si$ at 1500 MPa, 900°C. The solubility therefore appears not to have increased greatly from 300 MPa to 1500 MPa whereas D has increased about six orders of magnitude, thereby explaining why quartz has been successfully weakened by diffusing water into it at 1500 MPa but not at 300 MPa. It has also been found that heat treatment of hydroxyl-rich synthetic quartz at 1500 MPa, 900°C, with and without added water, leads to a large increase in strength, suggesting that some sort of equilibration or re-arrangement of the hydroxyl has occurred and throwing into question the significance of the low flow stresses measured on synthetic quartz at low pressures.

Introduction

There is now a considerable literature concerning the mechanical weakening of single crystals of quartz as a result of some sort of bulk interaction with water, to which the term "hydrolytic weakening" is commonly applied; a resumé of the experimental studies has been given by Paterson and Kekulawala [1979]. It may be presumed that the effects involve water-related defects within the crystals. However, the precise nature of the defects is not clear, although there has been some speculation about the mechanism of the weakening [Griggs, 1967; Hirsch, 1981; Hobbs, 1981]. The aim of the present work is to obtain more information about the kinetics of incorporation of "water" in quartz and on its mechanical effects, especially as a function of the confining pressure, in order to constrain more narrowly the possible nature of the water-related defect and its properties.

The original observations on hydrolytic weakening in quartz were made in experiments in which both the heat treatment that introduced the "water" into the quartz and the mechanical test that demonstrated the resultant weakening were carried out in the same run at 1500 MPa confining pressure and at temperatures around 900°C, using a solid-medium apparatus [Griggs and Blacic, 1964]. Attempts to produce the same effects in experiments at confining pressures in the range 300 to 500 MPa in a gas-medium apparatus have been unsuccessful, evidently because of much lower diffusivity of the water-related defects at the lower pressures [Kekulawala et al., 1978]. Very little further work has been done with the original type of hydrolytic weakening experiment (see, however, Hobbs and Tullis [1979]). Almost all other studies subsequent to the original experiments and purporting to relate to hydrolytic weakening have been carried out on crystals already containing a water-related species, characterized by a broad 3 μm infrared absorption, which has been incorporated during the growth of the crystals ("wet" synthetic crystals and amethyst). Such crystals show a pronounced weakening above a certain temperature [Griggs, 1967].

There has been a widespread presumption in the literature that the two weakening phenomena are identical in kind. In order to explore the validity of this view and to introduce more experimental flexibility, the approach has recently been taken of studying the properties of "hydrolytically weakened" quartz at 300 MPa confining pressure after exposing the quartz to water at 1500 MPa [Mainprice and Paterson, 1983]. The present paper is a progress report concerning such experiments in which, in particular, the diffusion coefficient at 1500 MPa, 900°C of the water-related species, identified by its infrared absorption, has been determined approximately and the strength at 300 MPa confining pressure of specimens heat treated at 1500 MPa has been explored. (For convenience, in this paper we shall use the term "water" to refer to any water-related species that may be involved in the phenomena under study but is not specifically identified, and the term "hydroxyl" to refer to

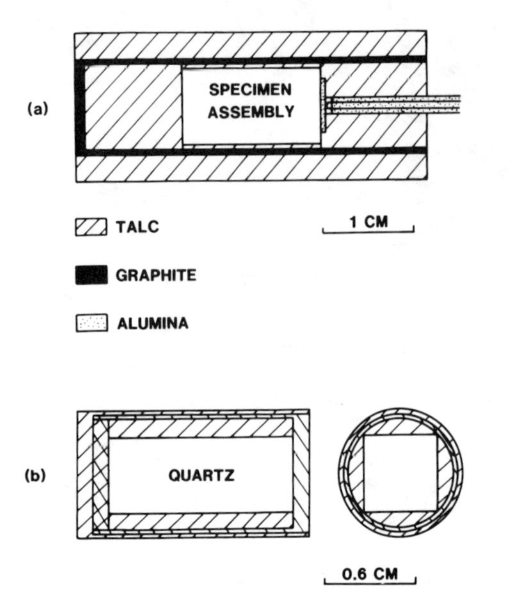

Fig. 1. (a) High pressure assembly used in the 1500 MPa heat treatment experiments in the solid-medium apparatus. (b) Detail of the copper capsule enclosing the specimen.

any species identified by infrared absorption in the 3 μm region, measured relative to that in a standard "dry" crystal).

Specimen Material And Experimental Details

Specimens have been taken from two quartz crystals, one natural and one synthetic. The natural crystal, N2, was an exceptionally large, optically clear, colourless rock crystal of unknown provenance, with well developed crystallographic faces and original dimensions of the order of 150 mm; infrared analysis gave an initial hydroxyl concentration of about 20 H/10^6Si associated mainly with a sharp absorption band at about 3370 cm^{-1}. The synthetic crystal, W4, was also optically clear and colourless, the same crystal as used and described by Hobbs et al. [1972]; all the present specimens were taken from the same Z-growth horizon and were fairly homogeneous in hydroxyl concentration, around 100-200 H/10^6Si associated mainly with a broad absorption band.

Specimens in the form of 5 x 5 x 12mm prisms were cut from both crystals using a diamond saw. The N2 specimens had the long axis normal to an $\underline{m}$(10$\bar{1}$0) plane and the lateral faces normal to $\underline{c}$ and $\underline{a}$ crystallographic axes. The W4 specimens had the long axis at 15° to an $\underline{m}$ plane and 90° to the $\underline{c}$ axis (the sense of rotation from $\perp$ $\underline{m}$ is unknown since it cannot be distinguished by the Laue procedure used for orienting the crystals); one pair of lateral faces was normal to the $\underline{c}$ axis and the other pair was inclined at 15° to an $\underline{m}$ plane and contained the $\underline{c}$ axis. In all cases the lateral faces were finely polished using 0.05 μm alumina powder and, unless otherwise mentioned, they were subsequently etched for two hours in a 90%HF, 10% HNO_3 plus MgO etchant at room temperature. No growth twins were revealed by the etching.

Heat treatment of the specimens was carried out at 1500 MPa pressure and 900°C in a solid-medium apparatus of Boyd and England [1960] type, using an assembly with talc confining medium and graphite furnace, illustrated in Fig. 1. The long overlap in the design of the copper capsule enclosing the specimen aimed at ensuring initial sealing of the capsule under the applied pressure. Except in the case of the W4 specimen heat treated dry, 5 μℓ of de-ionized water was added to the capsules before closure, this amount being accommodated in the clearances between the quartz and copper parts. In the heat treatment runs the pressure-temperature path up to 1500 MPa, 900°C and back again was chosen so as to keep the quartz in the alpha field while maximizing the temperature at any pressure in order to minimize departure from nonhydrostatic condition. After the heat treatment the capsules appeared to be well sealed by the welding of the outer cylindrical wall to the inner one during the run. In the later runs, in which the practice was adopted of puncturing the capsule carefully after the run by filing a knick in it, a drop of residual water was seen to exude in all cases except one, confirming that in general the water was retained throughout the heat treatment. The specimen was recovered by dissolving away the copper with dilute nitric acid, after which it was either used for determination of the diffusion profile or subjected to a mechanical test at 300 MPa, 900°C. Infrared spectra were obtained between each step in the above procedures.

In spite of the attempts to minimize nonhydrostatic stresses, all heat treated specimens were cracked on a submillimeter to millimeter scale. The cracks were mainly oriented normal to the cylindrical axis of the high pressure assembly, suggesting extension cracking during cooling or pressure release.

The infrared spectra were measured with a Pye Unicam SP3-200 ratio recording spectrophotometer, operated in absorbance mode, with a beam aperture of about 2 x 5mm and an attenuator in the reference beam adjusted to give a zero reading without the specimen. Unless otherwise specified, the spectra were measured with unpolarized radiation propagated parallel to the $\underline{c}$ axis and were expressed as absorption coefficient K versus wavenumber $\tilde{\nu}$. In view of the presence of the cracks, the heat treated specimens were carefully dried before measurement by holding under a diffusion pump vacuum of about 10^{-3} Pa for at least two hours and then impregnated under rotary pump vacuum with hexachlorobutadiene (C_4Cl_6), a liquid of similar refractive index to quartz and of negligible absorption in the infrared region of interest here, in order to minimize light scattering or reflection from the cracks;

without the impregnation the background scattering tended to be at least twice that shown in Fig. 2a at 4000-3800 cm^{-1} and to slope downwards appreciably towards lower wavenumbers, suggesting the presence of a component of Rayleigh scattering not present after impregnation.

In order to compare the spectra from virgin and heat treated specimens in respect of hydroxyl absorption (Fig. 2a), the additional background for the latter specimens was subtracted by matching the two spectra at 4000-3800 cm^{-1} and below 2600 cm^{-1} and assuming the additional background to vary linearly in between. The slope of the additional background thereby inferred for heat treated specimens was very small, probably independent of wavelength within experimental error and suggestive of an absorption by opaque material in the cracks or of losses by internal reflection at the cracks, which are predominantly subparallel to the beam. The additional background would then arise from an effective screening of the beam by the projected fractional area f of the reflecting cracks and opaque material, according to the relation $f = 1-10^{-\bar{K}\tau}$ where $\bar{K}$ is the apparent absorption coefficient due to the screening for a specimen thickness τ. The additional background for the heat treated specimen in Fig. 2a corresponds to about 40 percent screening of the beam, roughly consistent with the visual estimate of the projected area of the cracking.

For the absolute determination of the absorption spectrum attributed to hydroxyl in any specimen, the measured spectrum for our standard "dry" synthetic quartz crystal A6-13 (Fig. 2a) was subtracted from the spectrum of the specimen after matching the two spectra at 4000-3800 cm^{-2} and below 2600 cm^{-1} in order to eliminate differences in background absorption, as described in the previous paragraph.

The relationship between the absorption coefficient $K(\tilde{\nu})$ at wavenumber $\tilde{\nu}$ and the hydroxyl concentration c was assumed to be that given by Paterson [1982], namely,

$$c = \frac{1}{150} \int \frac{K(\tilde{\nu})}{\gamma(3780 - \tilde{\nu})} d\tilde{\nu} \text{ in mol H/}\ell \qquad (1)$$

Here K and $\tilde{\nu}$ are given in cm^{-1} and γ is the anisotropy factor, taken to be 1/3 for the broad band absorption.

The diffusion profile parallel to the c axis was determined in a sequence of steps in each of which a layer of 50-200 μm thickness was ground from a face parallel to the basal plane, the face repolished, the specimen re-impregnated, and the absorption spectrum re-measured as described above, the 2 x 5 mm beam being directed to pass through the central portion of the face of the specimen. The sequential removal of layers was generally continued from one side of the specimen until a total thickness of about 1.5 mm had been removed and then a similar sequence of layers was removed from the other side of the crystal, permitting, in effect, two determinations of the diffusion profile.

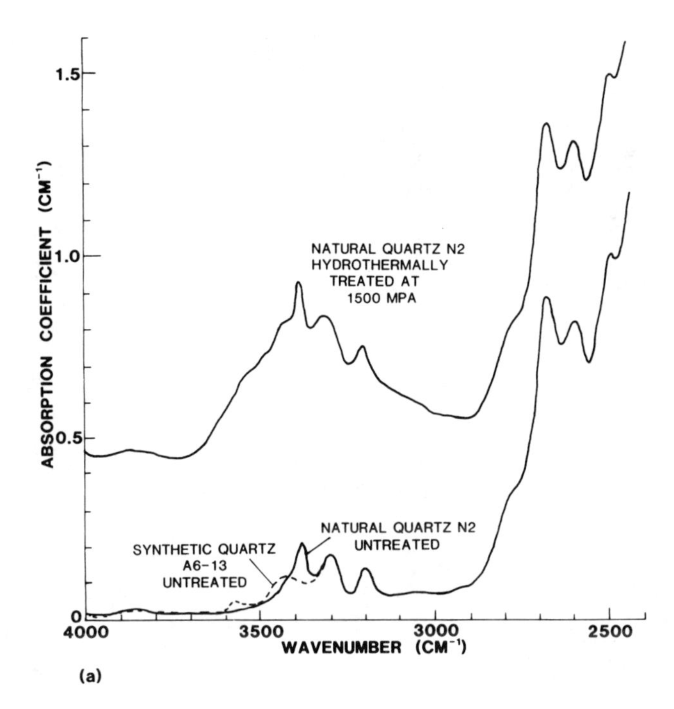

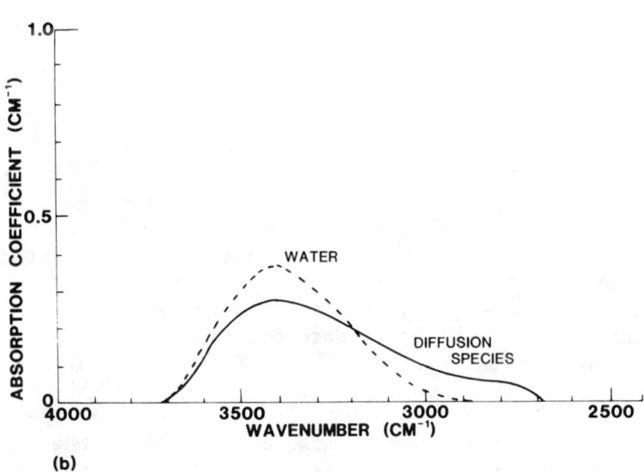

Fig. 2. (a) Room temperature infrared absorption spectrum of a specimen of natural quartz N2 subsequent to heat treatment at 1500 MPa ("wet" surface layer removed). Spectra of untreated N2 and the "dry" synthetic quartz A6-13 are included for comparison. (b) Difference spectrum of the diffusion species derived by subtracting the room temperature infrared absorption spectrum of untreated N2 from that for the specimen hydrothermally treated at 1500 MPa ("wet" surface layer removed). A spectrum of water is included for comparison.

The deformation experiments at 300 MPa confining pressure were carried out in a gas-medium deformation apparatus [Paterson, 1970] at 900°C, $10^{-5}s^{-1}$ strain rate. The rectilinear specimens

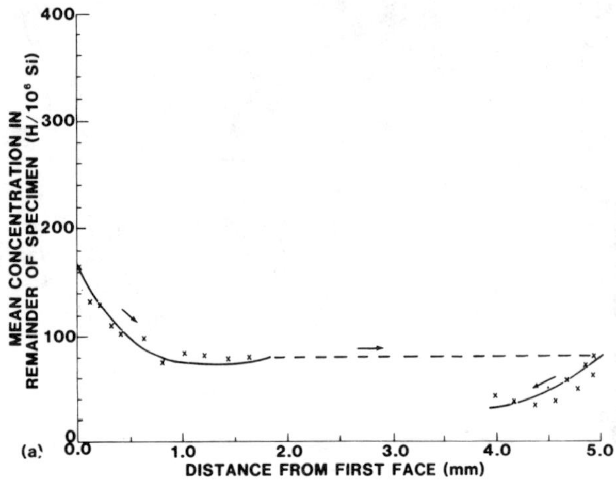

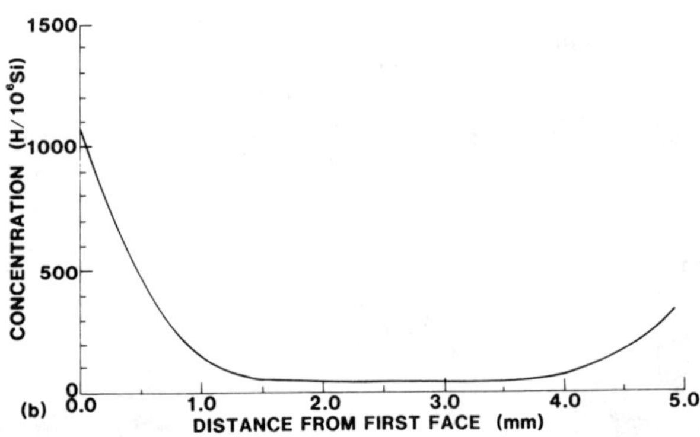

Fig. 3. (a) Mean concentration of introduced hydroxyl remaining in the specimen as a function of distance from the c face first sectioned, determined from the serial sectioning of specimen N2-22 (crosses), and the least squares fit to that data using (5) to determine the diffusion parameters (solid line). (b) Concentration profile of introduced hydroxyl in specimen N2-22 determined from the diffusion parameters calculated using the least squares fit of (5) to the mean concentration data in Figure 3a. The relatively low concentration at the second face may represent water starvation at that face, tending to bias c_s to lower and D to higher values.

were inserted in copper sleeves of 10 mm outside diameter, with copper inserts to fill the spaces between specimen and sleeve, and the assembly was sealed mechanically in a copper jacket of 0.25 mm wall thickness [Paterson et al., 1982, Figure 1a]. The load supported by the copper was allowed for in calculating the stress but was very small compared with that supported by the quartz.

Diffusion of Hydroxyl in Quartz at 1500 MPa, 900°C

The procedure described in the previous section has been used to determine diffusion profiles of the unidentified mobile species that produces locally the entity characterized by the infrared absorption that we assume to be OH-stretching absorption; for brevity we shall refer to the profiles as representing hydroxyl diffusion. The diffusion profiles have been measured only in the direction parallel to the c axis and, in view of the known strong anisotropy of oxygen tracer and sodium diffusion in quartz [Choudhury et al., 1965; Frischat, 1970; B.J. Giletti, personal communication, 1982; P.F. Dennis, personal communication, 1982], it has so far been assumed that the hydroxyl diffusion is similarly anisotropic so that the measured profiles represent essentially one-dimensional diffusion.

Considerable difficulty has been experienced in achieving consistency in the determination of the diffusion profiles. The most important source of variation appears to be associated with the situation at the surface of the specimens. Thus, the total amount of hydroxyl entering a specimen in a given amount of time was found to be doubled when the surface was etched instead of being left in its polished state; on the other hand, when specimens were heat treated in talc instead of in pure water (for example, by using talc filler pieces instead of copper, Figure 1b), a considerably reduced amount of hydroxyl was found to enter the specimens, although this may be partly due to the setting up of a diffusion profile in the talc itself due to water exhaustion. Even amongst etched specimens there is some variability, the source of which has not yet been identified although it is possible that it still has to do with the effectiveness of the etching or that it is due to water starvation due to contact with the copper.

Very high hydroxyl concentrations are commonly found in a 20-30 μm layer at the surface. This is thought to be associated with a layer of silica re-deposited on the surface during cooling and pressure release and it has therefore been disregarded in analysing the diffusion profile.

Fig. 3 shows typical results for an etched specimen of natural quartz, presented in Fig. 3a as the mean concentration $\bar{c}(x_A,x_B)$ of introduced hydroxyl in the remaining portion of the specimen at each stage of thickness reduction, where X_a is the thickness that has been removed from the first face (A) and x_B that removed from the second face (B). This quantity $\bar{c}$ is derived directly from the measured absorption spectrum after subtracting the spectrum of the virgin material, displaced so as to match the background levels at 4000-3800 cm^{-1} and below 2600 cm^{-1} as described in the previous section; the resulting difference spectrum, $\Delta K(\tilde{\nu})$, characteristic of the species being profiled, is of the form shown in Fig. 2b (essentially the same spectrum is obtained at 78K, showing that no significant amount of liquid water, such as in fluid inclusions, is contributing to the absorption). The derivation of $\bar{c}$ from $\Delta K(\tilde{\nu})$ is done with the aid of the relation (1) which, changing from mol H/ℓ to H/10^6Si and putting $\gamma = 1/3$ gives

$$\bar{c}(x_A, x_B) = 452 \int \frac{\Delta K(\tilde{\nu})}{3780 - \tilde{\nu}} d\tilde{\nu} \text{ in H}/10^6\text{Si} \quad (2)$$

The difference in successive $\bar{c}$ values can be used to obtain the hydroxyl concentration in the layer just removed, which gives directly the diffusion profile when plotted against the distance of the layer from the surface; the diffusion profile can then be fitted by a model profile in order to obtain the diffusion coefficient and other parameters. Alternatively, a more indirect fitting procedure can be used whereby the model profile is first integrated and then fitted to the plot of $\bar{c}$ versus thickness removed. The latter procedure has the disadvantage that it involves assuming initially the form of the model profile on both sides of the specimen, whereas in the first procedure the profiles on each side can be analysed independently, but the experimental scatter is less emphasized in the $\bar{c}$ plot compared with the higher degree of scatter that appears in the diffusion profile itself through the differencing procedure. Of course, the same inherent errors are present in both and the two procedures are equivalent.

Since in these experiments the diffusion profiles on opposite sides of the specimen die out before overlapping at the centre, we have chosen the Fick's law solution for a semi-infinite slab with fixed concentration at the surface as the model for fitting to the observations. The solution [Carslaw and Jaeger, 1959, p. 60; Crank, 1975, p. 32], adapted to the finite slab, is

$$c(x) = c_0 + (c_S - c_0)\,\mathrm{erfc}\,\frac{x}{2\sqrt{Dt}} + (c_S - c_0)\,\mathrm{erfc}\,\frac{X - x}{2\sqrt{Dt}} \quad (3)$$

where $c(x)$ is the concentration at a distance x from the reference face, c_0 is a superimposed uniform concentration, c_S is the maximum or fixed concentration at the two surfaces, X is the initial thickness of the slab, D is the diffusion coefficient, t the elapsed time and $\mathrm{erfc}(x/2\sqrt{Dt})$, $\mathrm{erfc}[(X - x)/2\sqrt{Dt}]$ are the complementary error functions referring to the profiles on the two sides, only one of which is ever significantly different from zero at any given x. Then we have

$$\bar{c}(x_A, x_B) = \frac{1}{X - x_A - x_B} \int_{x_A}^{X - x_B} c(x)\,dx \quad (4)$$

as the expression for the mean concentration in the remaining thickness, corresponding to the observations (2). Substituting (3) into (4), carrying through the integration and neglecting small terms under the condition that x_A and x_B are less than $X/2$ gives the model value

$$\bar{c}(x_A, x_B) = c_0 + \frac{2\sqrt{Dt}(c_S - c_0)}{\sqrt{\pi}(X - x_A - x_B)} \left\{ \exp\frac{-x_A^2}{4Dt} + \exp\frac{-x_B^2}{4Dt} - \frac{x_A\sqrt{\pi}}{2\sqrt{Dt}}\,\mathrm{erfc}\,\frac{x_A}{2\sqrt{Dt}} - \frac{x_B\sqrt{\pi}}{2\sqrt{Dt}}\,\mathrm{erfc}\,\frac{x_B}{2\sqrt{Dt}} \right\} \quad (5)$$

By obtaining an optimum fit of (5) to the observations (2), or of (3) to the experimental diffusion profile derived from the observations, we can obtain a set of values of D, c_S and c_0. With these values, the best-fit diffusion profile can be recalculated using (3) as shown in Figure 3b. (Actually in this figure the right hand branch appears to be anomalous relative to the other measurements, probably due to water starvation, and the analysis may not strictly apply to this branch). The goodness of fit has been measured by the standard deviation σ of the experimental values from the model values (for example, (2) from (5)), and the uncertainty in each of the derived parameters D, c_S and c_0 has been obtained by determining the amount by which it would have to depart from its optimum value, while holding the remaining parameters at their optimum values, in order to double the standard deviation σ.

The values of D, c_S and c_0 thus obtained from the relatively few runs in which the amount of cracking or other experimental interference was sufficiently small to permit the uncertainties to be less than the values of the parameters themselves are given in Table 1. Although the procedure of using the difference spectra $\Delta K(\tilde{\nu})$ eliminates the effect of any hydroxyl initially present and might lead one to expect that c_0 would be zero, the data could not be fitted optimally with $c_0 = 0$. The finite values of c_0 are thought to be associated probably with water penetration along cracks since the values tend to be lower in specimens with lower crack densities. The presence of this quasi-uniform distribution is thought not to affect seriously the conclusions about D and c_S.

The value of D for natural quartz at 1500 MPa, 900°C is thus found to be around 10^{-12} to 10^{-11} m^2s^{-1}. This result broadly agrees with the value of D of about $5.10^{-12} m^2 s^{-1}$ deduced by Blacic [1981] for the same conditions based on recognizing a "halo" or mantle of deformation effects in thin section in deformed specimens, equating the thickness of this "halo" to the effective depth of penetration L of the water, and using $L = \sqrt{10Dt}$; note that Blacic's value refers to a direction of 45° to the c axis (see Blacic [1975] for the orientations).

The considerable range of values of surface concentration c_S may reflect the presence of a variable surface barrier, discussed earlier, or possible starvation of water in some cases, as

TABLE 1. Diffusion Parameters

Specimen Number	Diffusion Coefficient D, $10^{-12}m^2s^{-1}$	Surface Concentration c_s, $H/10^6Si$	Superimposed Uniform Concentration c_0, $H/10^6Si$	Remarks
N2-4	2 ± 1	1600 ± 600	100 ± 25	water retention not checked
N2-22	13 ± 4	1080 ± 350 360 ± 120	30 ± 10	failed water retention test; water penetration different on the two sides, fitted with same D
W4-3	2 ± 2	460 ± 400	60 ± 15	synthetic quartz with initial OH content 120 $H/10^6Si$, not included in the values of c_s and c_0; water retained

well as the difficulty of defining precisely the position of the surface when a thin siliceous layer is deposited during cooling. Taking these potential factors into account, the maximum values of c_S would then represent the best value for the solubility of hydroxyl in quartz at 1500 MPa, 900°C, namely, of the order of 1000 $H/10^6Si$.

The results for wet synthetic quartz W4 are similar to those for the natural quartz, suggesting that, within the present accuracy, neither the initial hydroxyl content nor variations in any other impurity content have grossly affected the diffusion coefficient or solubility. Further,

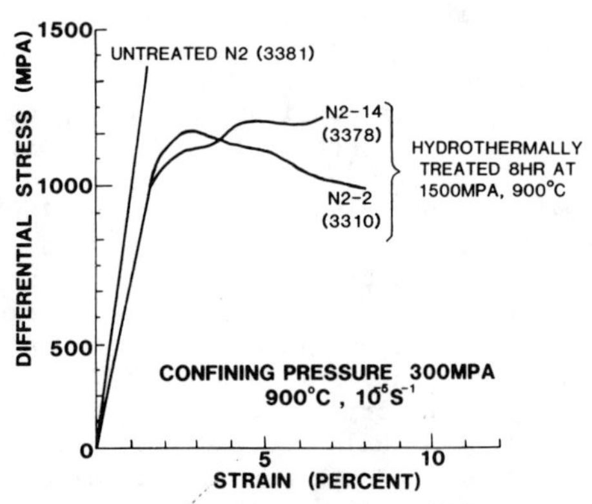

Fig. 4. Stress-strain curves at 300 MPa confining pressure, 900°C and $10^{-5}s^{-1}$ strain rate for specimens of natural quartz N2 hydrothermally treated at 1500 MPa for 8 hours. The stress-strain curve for an untreated N2 specimen tested under the same conditions is included for comparison.

the circumstance that a Fick's law solution with constant D can be fitted to the data for both quartz samples gives no support to the suggestion made by Paterson and Kekulawala [1979] that the diffusion coefficient for water in quartz may be concentration dependent.

Stress-Strain Behavior at 300 MPa, 900°C

Natural Quartz Heat Treated in Water at 1500 MPa

Two deformation experiments were carried out at 300 MPa confining pressure and 900°C on specimens of natural quartz N2 that had been heat treated with water at 1500 MPa, 900°C for 8 hours. The stress-strain curves are shown in Fig. 4, together with the stress-strain curve for a virgin specimen. The latter showed no evidence of plastic yielding up to a differential stress of 1500 MPa, at which the test was terminated in view of previous experience that fracture occurs under these conditions at stresses not much higher. In contrast, both heat treated specimens yielded at about 1000 MPa and the flow stress rose to around 1200 MPa after a few percent strain (the subsequent fall in stress for N2-2 may be an artifact arising from cracking of the alumina endpieces).

The distribution of hydroxyl in the heat treated specimens would be expected to be similar to that depicted in Figure 3b; that is, apart from the superimposed quasi-uniform concentration c_0, discussed previously, there should be a diffusion halo extending to a depth of about 1mm from each c face at concentrations above about 20 $H/10^6Si$, the minimum level needed for hydrolytic weakening to be evident at 900°C in synthetic quartz (see Paterson and Kekulawala [1979] for formulae for the hydrolytic weakening temperature from the work of Griggs and Blacic and of Hobbs et al.). If the specimen can then be regarded as consisting effectively of two phases, an unweakened core of 3mm thickness with an initial flow stress greater than 1500 MPa and a pair of weakened layers of 2 mm total thickness, the latter would need to have an initial flow stress of less than 250 MPa in order to account for the observed initial flow stress of about 1000 MPa. However, microscopical observations on a thin section cut in the plane containing the specimen axis and c crystallographic axis show no evidence for such a heterogeneity; the deformation features, mainly deformation bands with traces parallel to the a axis (coinciding with the trace of the m slip planes) and a widespread distribution of fine microcracking (Fig. 5), did not suggest any distinction between core and water-diffused mantle and, in particular, the zoning or "halo" referred to by Blacic [1981] could not be recognized. Since, for strain compatibility, the total strains undergone in the core and mantle zones must have been equal, either the identity of microstructure reflects the equality of strain in spite of different stress levels or it reflects some additional homogenization process such as a local re-distribution of the quasi-homogeneous superimposed hydroxyl concentration (thought initially to be associated with the cracking) or a re-distribution by core diffusion along dislocations as they penetrated the central domain during the deformation.

Fig. 5. Optical photomicrograph of a thin section of specimen N2-14 deformed at 300 MPa confining pressure, 900°C and $10^{-5}s^{-1}$ strain rate. The compression direction, normal to the m prism plane, is vertical and the c axis, in the direction of which the diffusion is thought principally to occur, is horizontal. The scale bar represents 1.0 mm. Crossed polarizers.

Wet Synthetic Quartz Heat Treated at 1500 MPa

Two deformation tests were carried out on specimens of the synthetic crystal W4 after heat treatment with additional water for 8 hours at 1500 MPa, 900°C, giving the uppermost stress-strain curves in Fig. 6. The infrared difference spectrum for the heat treated quartz relative to the dry synthetic quartz A6-13 (Figure 2a) is shown in Figure 7a, as well as the difference spectrum for the virgin specimen. From the figure it is seen that the total hydroxyl absorption has increased during the heat treatment, the increase being strongest in the vicinity of 3400 cm^{-1}. It is therefore remarkable that the heat treated specimens had much higher flow stresses than the virgin synthetic quartz (lowest curve in Figure 6) in spite of the total hydroxyl content being increased; the strength of the heat treated synthetic specimens was, in fact, comparable to that of the heat treated natural specimens.

A further deformation test was carried out on another specimen of synthetic quartz W4 that was similarly heat treated at 1500 MPa, 900°C but without the addition of water to the copper capsule and for a period of four hours. The infrared difference spectrum relative to A6-13 is shown in Figure 7b, from which it is seen that there has been some modification of the spectrum; also, in spite of no water being added to the capsule, there again appears to have been some increase in hydroxyl content, although not as much as for the specimens heat treated with added water. Again there is an increase in strength

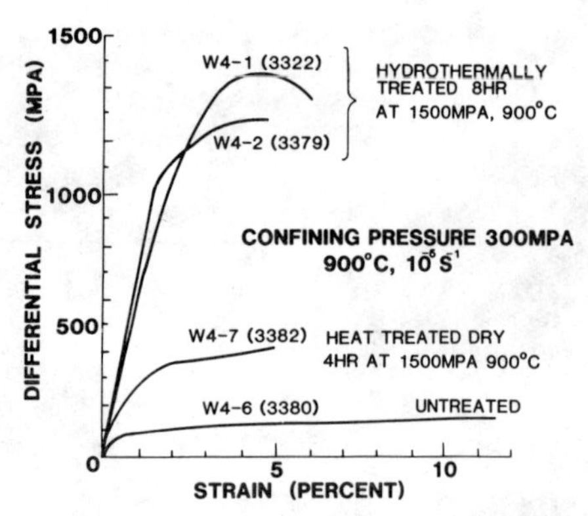

Fig. 6. Stress-strain curves at 300 MPa confining pressure, 900°C and $10^{-5}s^{-1}$ strain rate for specimens of synthetic quartz W4. Two specimens, 3322 and 3379, were tested subsequent to hydrothermal treatment at 1500 MPa for 8 hours, while a third, 3382, was tested after "dry" heat treatment at 1500 MPa for 4 hours. The stress-strain curve for an untreated W4 specimen deformed under the same conditions, 3380, is included for comparison.

(intermediate curve in Figure 6) which is also of intermediate degree (strict comparison is not possible because of the different durations of heat treatment). The apparent small increase in hydroxyl in this specimen may possibly be related to a high activity of hydrogen in the pressure medium due to interaction between the water released in dehydration and the graphite furnace element, since the copper capsule is probably permeable to hydrogen on the timescale of the experiments.

The microstructures of the heat treated synthetic quartz specimens were broadly similar to those of the heat treated natural specimens, although some local differences occurred such as the appearance of zones of criss-crossed short lamellae with traces about ± 45° to the specimen axis.

Discussion

Diffusion Coefficient and Solubility

It has been strikingly evident for some time that on the laboratory timescale the degree of water penetration into quartz during hydrothermal treatment is vastly greater when the pressure is around 1500 MPa than at 300-500 MPa [Blacic, 1975; Kekulawala et al., 1978]. The present results indicate that the reason for this difference lies mainly in the greatly different diffusion coefficients for the effective water-related species rather than in the solubilities. Thus, on the one hand, the diffusion coefficient at 1500 MPa, 900°C has now been established to be of the order of 10^{-12} to 10^{-11} m^2s^{-1} compared with a probable value at 300 MPa, 900°C of less than 10^{-18} m^2s^{-1}, perhaps of the order of 10^{-19} m^2s^{-1}, as estimated from the kinetics of bubble growth and the rate of re-weakening after low-pressure precipitation (Kekulawala et al. [1981]; note, however, that the corresponding value of D for tracer oxygen under hydrothermal conditions at 100 MPa, 900°C, which might be expected to be similar, is around 10^{-17} m^2s^{-1} according to B.J. Giletti (personal communication, 1982) and P.F. Dennis (personal communication, 1982)). On the other hand, the present study suggests a solubility at 1500 MPa, 900°C of not much more than 1000 H/10^6Si compared with a solubility at 300 MPa, 900°C thought to be around 300 H/10^6Si, based on non-precipitation in synthetic crystals during heat treatment [Kekulawala et al., 1981]. Thus, the change in diffusion coefficient in this

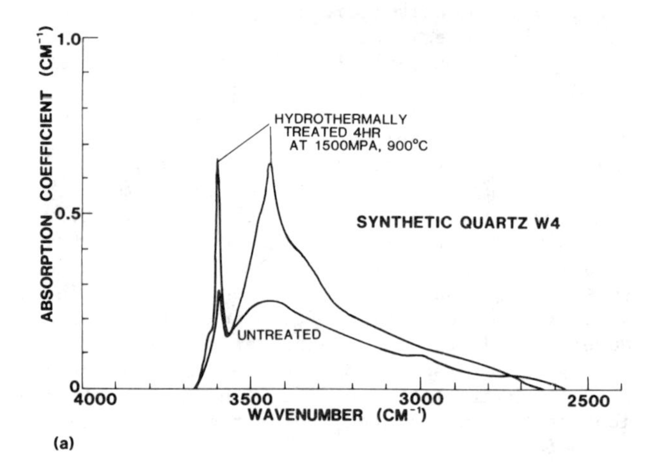

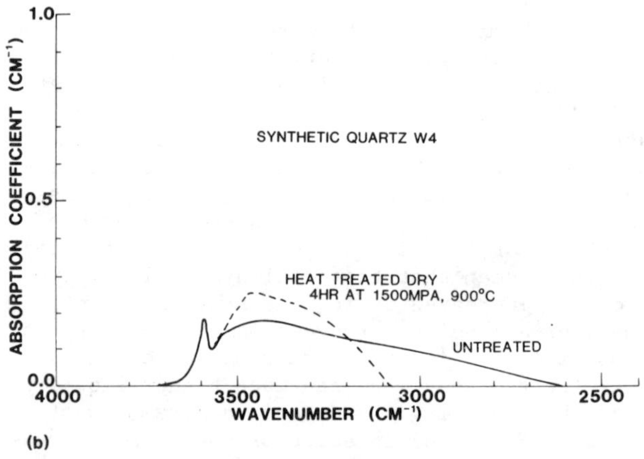

Fig. 7. Room temperature infrared absorption spectra after subtraction of the reference "dry" A6-13 spectrum and correction for background scattering: (a) for a specimen of synthetic quartz W4 before and after heat treatment at 1500 MPa for 4 hours ("wet" surface layer removed); (b) for a specimen of synthetic quartz W4 before and after "dry" heat treatment at 1500 MPa for 4 hours (surface layer removed).

pressure range appears to be around six orders of magnitude compared with less than an order of magnitude change in solubility. This change in diffusion coefficient implies that in the time needed for an effective penetration of the water of 1 mm at 1500 MPa, 900°C, the effective penetration at 300 MPa would be less than 1 μm, thereby explaining the failure to achieve weakening in 300 MPa experiments.

The sign and magnitude of the infrared pressure effect do not conform to the usual effect in solids, which is a relatively small decrease in diffusion coefficient with increase in pressure. Since it is presumably meaningless to interpret the large increase with pressure in terms of a large negative activation volume, the effect suggests that there is an increase in the pre-exponential Arrhenius term D_0 or a decrease in activation energy Q or changes in both (the change in activation energy in going from alpha to beta quartz, as determined by B.J. Giletti (personal communication, 1982) is inadequate to explain the difference between 1500 MPa and 300 MPa behavior purely in terms of the phase change).

The relatively small change in hydroxyl solubility also suggests that the greatly accelerated diffusion that occurs at 1500 MPa is not attributable to a corresponding increase in the concentration of the diffusing species, even though a large increase in concentration of species such as H_3O^+ might have been expected as a result of equilibration with the free water phase which is much more dissociated at 1500 MPa (the ionization constant at 1500 MPa, 900°C being 10^4 times that at 300 MPa, 900°C according to Holzapfel [1969]). Rather, the accelerated diffusion is presumably related to a greatly increased mobility of the diffusing species, probably along the relatively open channels parallel to the c axis in the quartz structure, which would be responsible for the anisotropy of the diffusion. This increase in mobility may be somewhat analogous to that involved in the transition to superionic conductivity for lithium in beta-eucryptite with increase in temperature [von Alpen et al., 1977]; beta-eucryptite has a structure derived from that of quartz. Although increase in pressure usually has the opposite effect to increase in temperature, the present situation may be exceptional in respect of the influence of pressure, as indicated by the existence of negative coefficients of expansion in beta-eucryptite and beta-quartz [Schulz, 1974; Mayer, 1960]. The analogy is further supported by the value of $D \sim 10^{-12}$ to 10^{-11} m^2s^{-1} at 1500 MPa, 900°C being within the band of values, 10^{-9} to 10^{-12} m^2s^{-1}, quoted as characteristic of superionic conductors [Mundy, 1979].

Strength

The data on the stress-strain behavior at 300 MPa, 900°C of the natural and synthetic quartz after 1500 MPa heat treatment are still very limited. However, if the results for the present orientations can be taken as typical, some general trends are suggested.

The most striking effect is that the 1500 MPa heat treatment, both without and with added water, has considerably strengthened the synthetic quartz, in spite of increases in the mean hydroxyl concentration in the specimens. The resulting flow stresses are in fact rather similar to those for natural vein quartz [Paterson and Kekulawala, 1979] and for quartzites [Heard and Carter, 1968; Mainprice and Paterson, 1983]. It is therefore tempting to speculate that this level, around 500 to 1000 MPa differential stress at 300 MPa confining pressure and 900°C, is typical of quartz that has been in some way "equilibrated" or "normalized" in respect of its water content. However, owing to the ambiguity of interpreting the behavior of the heterogeneous specimens of the heat treated natural quartz, it is not yet clear whether the flow stress of the natural quartz with diffused-in water also falls into this category of strength.

The circumstance that the untreated wet synthetic quartz has much lower flow stresses under the same testing conditions suggests that it has not been "equilibrated" in the same sense as the 1500 MPa heat treated specimens, in spite of the apparent reversibility at 300 MPa of the strengthening due to precipitation at atmospheric pressure reported by Kekulawala et al. [1981], an effect which now appears in some conflict with the present results and which needs further work to resolve. The difference between the 300 and 1500 MPa cases may reflect a different structural arrangement of the water incorporated during the relatively fast growth of the wet specimens, as well as sluggishness in the reorganization of this arrangement at 300 MPa. The need to go to 1500 MPa to bring about the reorganization suggests that this process is dependent on diffusion occurring on the millimeter scale, such as would be required to transport some component to the surface of the specimen, although it is not clear why such a component would not precipitate locally if it were present unstably.

In conclusion, there is now some doubt about the applicability of stress-strain measurements made in the laboratory on wet synthetic crystals to the behavior of quartz under conditions more favorable to equilibration. It is also evident that there can be considerable variations in the strength of hydroxyl-bearing quartz without there being, in most cases, any obviously correlated changes in the nature of the infrared absorption spectrum in the 3 μm region, although in all cases so far studied a broad infrared absorption band has always been present whenever it has been possible to deform specimens at differential stresses below 1500 MPa at 900°C. Finally, in view of the marked influence of pressure on the diffusion coefficient, it may be expected that a similar pressure effect will be seen in diffusion-limited creep behavior.

Acknowledgments. We thank Paul Dennis and Bruno Giletti for making data on oxygen diffusion available prior to publication; John Fitzgerald, David Kohlstedt, Jean-Paul Poirier, and many other colleagues for helpful discussions, and Ted Ringwood for access to the Boyd-England apparatus. Bill Hibberson, Graeme Horwood, and Paul Willis gave valuable experimental help.

References

Blacic, J. D., Plastic-deformation mechanisms in quartz: The effect of water, Tectonophysics, 27, 271-294, 1975.

Blacic, J. D., Water diffusion in quartz at high pressure: Tectonic implications, Geophys. Res. Lett., 8, 721-723, 1981.

Boyd, F. R., and J. L. England, Apparatus for phase-equilibrium measurements at pressures up to 50 kilobars and temperatures up to 1750°C, J. Geophys. Res., 65, 741-748, 1960.

Carslaw, H. S., and J. C. Jaeger, Conduction of Heat in Solids, Clarendon, Oxford, 1959.

Choudhury, A., D. W. Palmer, G. Amsel, H. Curien and P. Baruch, Study of oxygen diffusion in quartz by using the nuclear reaction $O^{18}(p,\alpha)N^{15}$, Solid State Commun., 3, 119-122, 1965.

Crank, J., The Mathematics of Diffusion, Clarendon, Oxford, 1975.

Frischat, G. H., Sodium diffusion in natural quartz crystals, J. Am. Ceram. Soc., 53, 357, 1970.

Griggs, D. T., Hydrolytic weakening of quartz and other silicates, Geophys. J.R. Astron. Soc., 14, 19-31, 1967.

Griggs, D. T., and J. D. Blacic, The strength of quartz in the ductile regime (abstract), EoS Trans. AGU, 45, 102, 1964.

Heard, H. C., and N. L. Carter, Experimentally induced "natural" intragranular flow in quartz and quartzite, Am. J. Sci., 266, 1-42, 1968.

Hirsch, P. B., Plastic deformation and electronic mechanisms in semiconductors and insulators, J. Phys. Colloq. C3, 42, C3-149 to C3-159, 1981.

Hobbs, B. E., The influence of metamorphic environment upon the deformation of minerals, Tectonophysics, 78, 335-383, 1981.

Hobbs, B. E., and T. E. Tullis, The influence of pressure on hydrolytic weakening in quartz (abstract), EoS Trans. AGU, 60, 370, 1979.

Hobbs, B. E., A. C. McLaren, and M. S. Paterson, Plasticity of single crystals of synthetic quartz, in Flow and Fracture of Rocks, Geophys. Monogr. Ser., vol. 16, edited by H. C. Heard, I. Y. Borg., N. L. Carter, and C. B. Raleigh, pp. 29-53, AGU, Washington, D. C., 1972.

Holzapfel, W. B., Effect of pressure and temperature on the conductivity and ionic dissociation of water up to 100 kbar and 1,000°C, J. Chem. Phys., 50, 4424-4428, 1969.

Kekulawala, K. R. S. S., M. S. Paterson, and J. N. Boland, Hydrolytic weakening in quartz, Tectonophysics, 46, T1-T6, 1978.

Kekulawala, K. R. S. S., M. S. Paterson, and J. N. Boland, An experimental study of the role of water in quartz deformation, in Mechanical Behavior of Crustal Rocks, Geophys. Monogr. Ser., vol. 24, edited by N. L. Carter, M. Friedman, J. M. Logan, and D. W. Stearns, pp. 49-60, AGU, Washington, D. C., 1981.

Mainprice, D. H., and M. S. Paterson, Experimental studies of the role of water in the plasticity of quartzites, J. Geophys. Res., in press, 1983.

Mayer, G., Recherches expérimentales sur une transformation du quartz, Rapp. CEA R Fr. Commis. Energ. At., 1330, 101 pp., 1960.

Mundy, J. N., Diffusion and ionic conductivity in solid electrolytes, in Fast Ion Transport in Solids, Electrodes and Electrolytes, edited by P. Vashista, J. N. Mundy, and G. K. Shenoy, pp. 159-164, Elsevier, New York, 1979.

Paterson, M. S., A high-pressure, high-temperature apparatus for rock deformation, Int. J. Rock Mech. Min. Sci., 7, 517-526, 1970.

Paterson, M. S., The determination of hydroxyl by infrared absorption in quartz, silicate glasses and similar materials, Bull. Minéral., 105, 20-29, 1982.

Paterson, M. S., and K. R. S. S. Kekulawala, The role of water in quartz deformation, Bull. Minéral., 102, 92-98, 1979.

Paterson, M. S., P. N. Chopra, and G.R. Horwood, The jacketing of specimens in high-temperature high-pressure rock-deformation experiments, High Temp. High Pressures, 14, 315-318, 1982.

Schulz, H., Thermal expansion of beta eucryptite, J. Am. Ceram. Soc., 57, 313-318, 1974.

von Alpen, U., E. Schönherr, H. Schulz, and G. H. Talat, β-eucryptite--A one-dimensional Li-ionic conductor, Electrochim. Acta, 22, 805-807, 1977.

THE HYDROLYTIC WEAKENING EFFECT IN QUARTZ

B. E. Hobbs

Department of Earth Sciences, Monash University
Clayton, Victoria 3168, Australia

Abstract. Experiments on single crystals of quartz have shown that an order of magnitude increase in the fugacity of H_2O is associated with about an order of magnitude decrease in the flow strength at a given temperature and pressure. The classical interpretation of this hydrolytic weakening effect is that H_2O groups are incorporated into the quartz structure as Si-OH.HO-Si groups. Then, in order to move a dislocation, OH.HO bonds need to be broken rather than Si-O bonds. The rate controlling process is envisaged as the diffusion of the (OH)-defect to or with the dislocation core. This paper discusses the manner in which charged hydrogen- or hydroxyl-defects alter the concentrations of other charged defects such as kinks and jogs on dislocations or vacancies and interstitials and so have an influence on the deformation rate. As an example, an increase in the concentration of negatively charged (OH)-defects leads to an increase in the concentration of positively charged kinks on dislocations thus increasing the strain rate. Other deformation mechanisms involving diffusion of oxygen and silicon with or without climb of dislocations or motion of kinks are also investigated and are shown to be capable of explaining the observed effect. This defect chemistry interpretation is consistent with the classical interpretation but also proposes other mechanisms where the direct diffusion of (OH)-defects plays no role in the process. As an example, an increase in the concentration of negatively charged (OH)-defects increases both the concentration of positively charged jogs and positively charged silicon interstitials in such a way as to explain the magnitude of the hydrolytic weakening effect. As such, the rate controlling process is the climb of dislocations controlled by silicon diffusion, not the diffusion of (OH)-defects. Although several different mechanisms are capable of explaining the hydrolytic weakening effect, many have different dependencies upon the activity of oxygen so that properly designed experiments are capable of establishing which mechanism actually operates.

1. Introduction

In 1964 Griggs and Blacic discovered that the introduction of small amounts of (OH) into the structure of crystalline quartz results in an order of magnitude decrease in the plastic flow stress for a given imposed strain rate. The effect has subsequently been studied by a number of workers including Griggs and Blacic [1965], Griggs [1967], Hobbs et al. [1972], Baëta and Ashbee [1970a, b], Balderman [1974], Blacic [1975], Morrison-Smith et al. [1976], Kirby [1975, 1983], Kirby and McCormick [1979], Kirby et al. [1977], Kekulawala et al. [1978, 1981], and Linker and Kirby [1981]. The effect is known as hydrolytic weakening.

It was also shown that the introduction of trace amounts of (OH) has profound effect on the recrystallization kinetics of quartz [Carter et al., 1964; Hobbs, 1968; Green et al., 1970; Tullis et al., 1973] so that apparently the incorporation of (OH) not only facilitates dislocation motion and/or generation but facilitates direct diffusive processes as well. This is supported by an increase in oxygen diffusivity in "wet" as opposed to "dry" environments [Dennis, 1983; Yund, 1983].

The interpretation of the effect has traditionally been that the (OH) is incorporated into the quartz structure as a Si-OH.HO-Si defect so that an H_2O group is placed in the structure at an already occupied oxygen site [see Griggs, 1967]. Then, in order for deformation to proceed it is only necessary to break an OH.HO bond rather than an Si-O bond. Griggs [1967] proposed that this process facilitated the motion of kinks on dislocation lines whereas McLaren and Retchford [1969] proposed that the process facilitated the climb of jogs. These interpretations suppose that (OH) is able to diffuse to or with dislocations and so "hydrolyse" Si-O-Si groups immediately adjacent to the moving dislocation. The diffusion of (OH) through the quartz structure is therefore envisaged as the rate controlling step in the deformation process.

The purpose of this paper is to present an alternative explanation of the hydrolytic weakening effect in quartz in terms of the influence that (OH)-defects have on the concentrations of those defects that control dislocation motion. Thus, in the interpretation presented here, diffusion of (OH) through the quartz structure

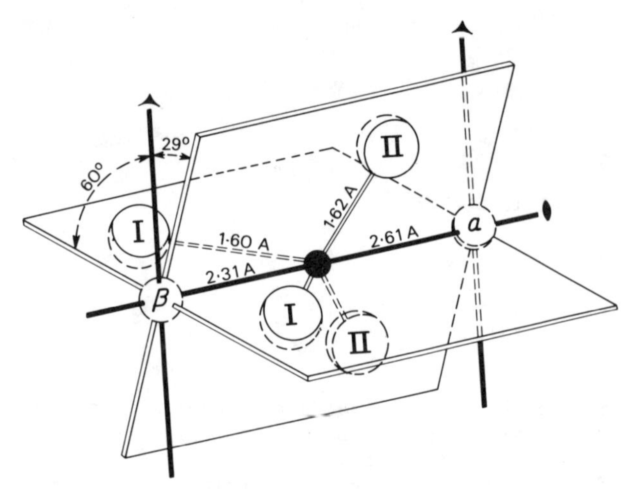

Fig. 1. Crystal structure of α-SiO_2. The silicon atom is black and the oxygen atoms are labelled I and II. α and β are interstitial positions lying at the intersection of the three fold c-axes and two fold a-axes; they are positions commonly occupied by H, Li and Na. (After Griscom [1978].)

need not play any role in facilitating deformation although such a mechanism is still compatible with what is presented here. The enhancement of charged vacancy and interstitial concentrations by the presence of (OH)-defects is postulated as one potentially important control on the hydrolytic weakening effect. A second important control involves the enhancement of charged jog and kink concentrations by the presence of (OH)-defects. The essence of the argument is presented in Hobbs [1980] whereas the argument involving kinks alone is presented by Hirsch [1981] and Hobbs [1983a].

In what follows, the electronic band structure of α-quartz is considered (section 2) and then the structures of the common point and line defects in quartz are reviewed (sections 3 and 4). Then follows a discussion of the defect chemistry of pure quartz (section 5) and of the SiO_2- H_2O system (section 6). This is followed by a consideration of the hydrolytic weakening process (section 7).

A review of the principles used in this paper is given by Hobbs [1981]. The defect symbolism used is due to Kröger [1974]. Except where stated it is assumed that the defect concentrations are small and that there is no interaction between point defects to form clusters and other associations.

In this paper, the arguments are presented for pure quartz and for quartz that contains only (OH)-defects as constituting the main dopant. Both natural and synthetic quartz commonly contain Al, Na and Li as impurities in significant levels and the arguments for this situation are presented in Hobbs [1983a].

2. The Electronic Structure of α-SiO_2

Alpha-quartz is built from slightly distorted SiO_4 tetrahedra connected together so that each oxygen is shared by two tetrahedra, the Si-O-Si bond angle being ca. 144°. The crystal structure is trigonal belonging to the two enantimorphous space groups $P3_121$ or $P3_221$. There are three silicons and six oxygens per unit cell; each silicon lies on one of the three two-fold a-axes; however, the oxygens do not lie on symmetry axes but in each tetrahedron consist of two pairs. The oxygens in one pair, O(I) in Figure 1, lie 1.598 Å from the central silicon whilst those in the other pair, O(II), lie 1.616 Å from the silicon. The three fold symmetry axes pass through channels which contain, at places such as α and β in Figure 1, sites that can accommodate interstitial cations such as H, Li and Na [see Kats, 1962, p. 258].

Pantelides and Harrison [1976] have shown that the electronic structure of SiO_2 can be considered in terms of a bonding unit consisting of an oyxgen and one sp^3 hybrid orbital from each of two adjacent silicon atoms (see also Harrison [1980, pp. 261-2701]). The oxygen 2p orbitals and the silicon sp^3 hybrid orbitals are shown in Figure 2. The following interactions take place to form the five bond orbitals shown in Figure 3.

1. The oxygen p_z-orbital, directed closest to the Si-O direction, interacts with the anti-symmetric pair of silicon sp^3 hybrids to form strong bonding and strong anti-bonding bond orbitals labelled B_z and A_z in Figure 3.
2. The oxygen p_x-orbital (Fig. 2), interacts with the symmetric pair of silicon sp^3 hybrids to form weak bonding and weak anti-bonding bond orbitals labelled B_x and A_x in Figure 3.
3. There is no interaction between the p_y-orbitals and the sp^3 orbitals and the p_y-orbital becomes a non-bonding bond orbital, B_y. Its energy remains close to the O-2p level.

The six electrons available for bonding fill the levels up to the non-bonding B_y level whilst the anti-bonding states are unoccupied. The B_y level and below therefore constitute the valence band of SiO_2 whereas the anti-bonding states constitute the conduction band. The band gap is fixed more or less by the energy difference between the O-2p level and the Si-sp^3 level although Laughlin et al. [1979] indicate that interaction between the Si-sp^3 orbitals and the O-2s orbital, which essentially forms a narrow band well below the valence band (Fig. 3), imparts some character to the lower parts of the conduction band and thus contributes to the large size of the band gap.

The calculations of Laughlin et al. [1979] give details of the energy bands and these are in broad agreement with many earlier calculations of the band structure (see references 1 to 12 in Laughlin et al. [1979]). The band gap is taken as 9.2 eV to fit X-ray emission and X-ray photoemission data. The gap is indirect with two fold degeneracies in

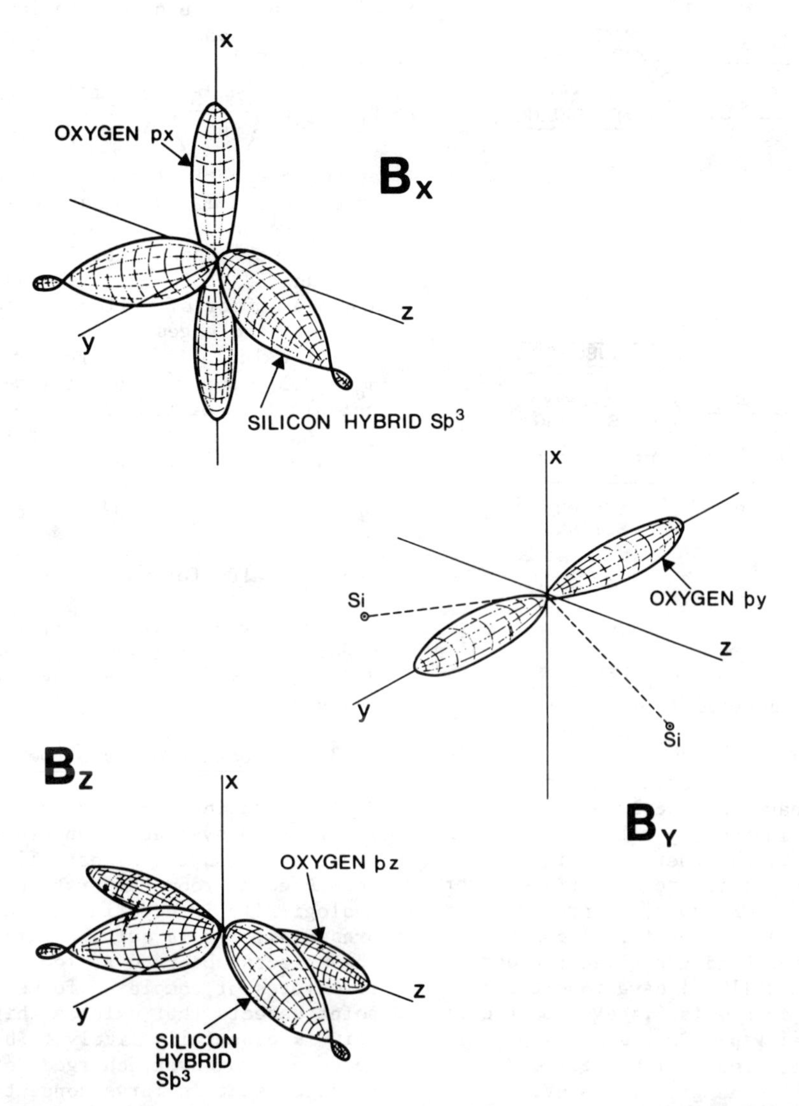

Fig. 2. Orbital interactions in α-SiO_2. The interactions pictured give rise to the weakly bonding and antibonding (B_x, A_x), non-bonding (B_Y) and strong bonding and antibonding (B_z, A_z) bond orbitals labelled in Figure 3.

the conduction band edge at K and A in the Brillouin Zone and at Γ, H and L in the valence band edge.

None of the band gap calculations mentioned above predict or explain why the Si-O-Si bond angle should be 144°; in fact they predict [see Pantelides and Harrison, 1976, p. 2685; Harrison, 1980, pp. 275-277] that the bond angle should be 180°. A number of explanations are offered for this discrepancy [see Harrison, 1980] without a satisfactory solution. O'Keeffe and Hyde [1978, 1981] propose that non-bonding interactions between silicons are important but neglected in the band calculations and that it is these interactions that are essentially responsible for establishing the 144° bond angle. This same suggestion is made by Pantelides and Harrison [1976] but so far the effect has not been incorporated into the band calculations. The details of the band structure as calculated however seem to be consistent to first order with a range of spectroscopic data (see Laughlin et al. [1979] and references 1 to 12 therein).

The band gap of 9.2 eV mentioned above is the optical band gap and is approximately equivalent (except for the energy of an exciton; see Mott [1981]) to the absorption edge for quartz. The energy that is important for defect concentration calculations is the thermal band gap or the thermal energy required to excite an electron from

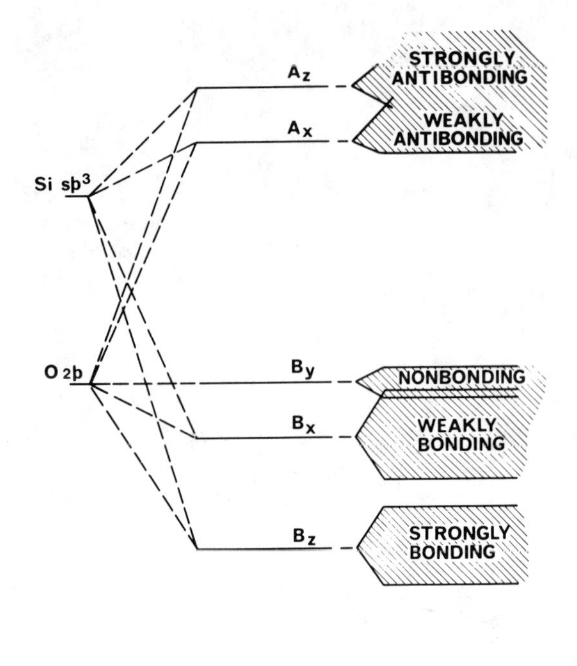

Fig. 3. Energy levels in α-SiO_2.

the top of the valence band to the base of the conduction band. Because of the Frank-Condin Principle, the thermal band gap will be less than the optical band gap by the energy required to move atoms to equilibrium lattice positions after the thermal transition. However, at present there is no theoretical and little empirical basis for arriving at the thermal band gap given the optical band gap. Morin et al. [1977] have reported that the thermal band gap in MgO is 6.4 eV compared to 7.7 eV for the optical gap. On the basis of this observation a 20% reduction is taken for quartz also to give a thermal band gap of 7.4 eV. Clearly the matter is not very satisfactory and experimental determinations of the thermal gap are required. The band gap will vary with temperature and pressure but the normal situation is for these two effects to be in opposing senses. Moreover the changes with temperature and pressure are small compared to the size of the gap and in view of the uncertainties associated with the value of the thermal gap, these effects are neglected here.

Thermal excitation of an electron from the valence to the conduction band may be expressed by the reaction

$$\text{neutral crystal} \rightleftarrows e' + h^{\bullet} \qquad (1)$$

for which the law of mass action gives

$$K_i = np \qquad (2)$$

where K_i is the equilibrium constant for reaction (1), n is the concentration (cm^{-3}) of electrons, e', and p is the concentration (cm^{-3}) of holes in the valence band, $h^{\bullet}$. K_i is given by [see Kröger, 1974, p. 77]

$$K_i = 4g_c g_v \left(\frac{2\pi m k T}{h^2}\right)^3 \left(\frac{m_e^* m_h^*}{m^2}\right)^{3/2} \exp\left(-E_T/kT\right) cm^{-6} \qquad (3)$$

assuming spherical bands. Here m is the mass of an electron, m_e^* and m_h^* are the effective masses of electrons and holes, k is Boltzmann's constant, h is Planck's constant, T is the absolute temperature and E_T is the thermal band gap. g_c and g_v are the number of valleys in the conduction and valence band edges.

The calculations of Fowler et al. [1978] give $m_e^* = 0.5$ m and $m_h^* = 5$ m for α-quartz and the band calculations of Laughlin et al. [1979, Fig. 6] give $g_v = 2$ and $g_c = 2$. E_T is taken to be 7.4 eV. Equation (3) now becomes

$$K_i = 3.67 \times 10^3 \, T^3 \exp(-7.4/kT) cm^{-6} \qquad (4)$$

and its value for various values of T is given in Table 1.

Thus, even at 1200°C, K_i remains small, and predicts from equation (2) a concentration of only 2.2×10^8 cm^{-3} of electrons and holes at this temperature for n = p.

3. Structure of Point Defects in α-Quartz

Although there is a considerable amount of information available concerning the point defects present in quartz, almost all studies have been conducted at room temperature or below and not at geologically interesting temperatures and pressures. It is quite probable therefore that the defects observed at low temperatures will be associates or complexes formed from much simpler point defects that exist at high temperatures. This is especially likely to be so for (OH)-type defects. Moreover, charged defects are not likely to be present in large concentrations at equilibrium at low temperatures whereas significant concentrations of such defects may exist at high temperatures. The main point defects that have been recognized at low temperatures and that are likely to be geologically important are considered below.

TABLE 1. Values of the Equilibrium Constant K_i for α-Quartz

T^o, K	K_i, cm^{-6}
873	3.8×10^{-2}
1073	6.8×10^6
1273	3.3×10^{12}
1473	5.0×10^{16}

Oxygen Vacancies

The oxygen vacancy in α-quartz has been widely studied. In the radiation literature this defect is called the E'-center. Yip and Fowler [1975] have presented a convincing model for this defect although there has been some discussion of the precise model for the defect by Haberlandt and Ritschl [1980]. The defect consists of a single oxygen vacancy with a single electron that spends most of its time in one of the two sp^3 orbitals of adjacent silicon atoms. At room temperature therefore this defect is $V_o^{\bullet}$ in the symbolism of Kröger [1974]. The silicon closer to the vacancy, Si(I), relaxes slightly towards the vacancy whilst the other silicon, Si(II), relaxes away from the vacancy to become almost coplanar with the three oxygens that surround it. This is shown diagrammatically in Figure 4. Two E'-centers are identified, namely E'_1 and E'_2. E'_1 is an oxygen vacancy with the structure shown in Figure 4 but with an electron occupying the sp^3 orbital of Si(I). Early work by Weeks [1963] had suggested that E'_2 consisted of an electron trapped on an sp^3 silicon orbital with an adjacent oxygen and another silicon site vacant. Other workers [see Lell et al., 1966] had suggested that this could even be a vacant SiO_2 group with an electron trapped on a nearby silicon. However Feigl and Anderson [1970] have shown that the E'_2 center has essentially the same structure as E'_1 and it is now identified [Griscom, 1978] as a single oxygen vacancy in which an electron is trapped in an sp^3 non bonding orbital of an Si(II) atom. Both of these defects are associated with an optical absorption band in the vicinity of 6 eV and it had been thought that this was the energy level of the defect below the conduction band [see Bennett and Roth, 1971; Hobbs, 1981]. However Griscom [1979], Griscom and Fowler [1980], and Schirmer [1980] have proposed that the absorption band just below 6 eV is due to charge transfer between Si(I) and Si(II) in the oxygen vacancy (Fig. 5a).

Herman et al. [1980] have calculated the energy

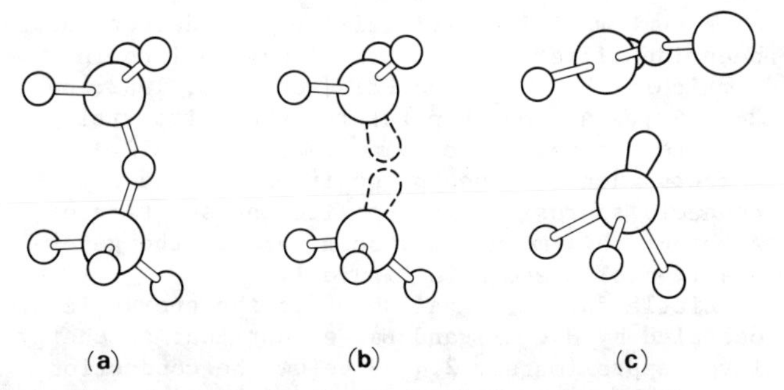

Fig. 4. Structure of the V_o defect in α-SiO_2. (a) Undistorted structure; small circles, oxygens; large circles, silicons. (b) Oxygen atom removed, structure undistorted. (c) Oxygen atom removed and structure relaxed so that one SiO_3 group becomes coplanar. (After Feigl et al. [1974].)

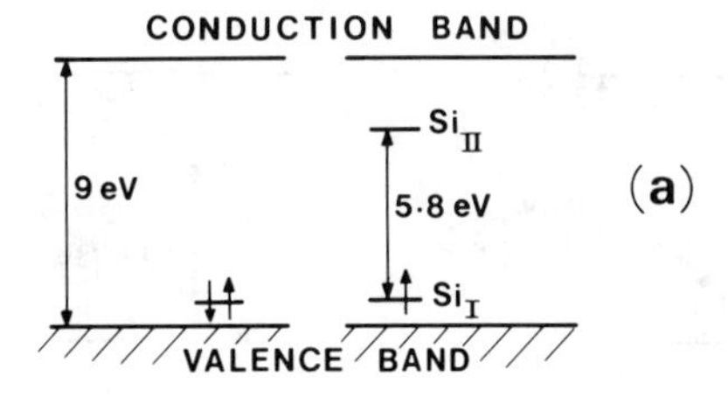

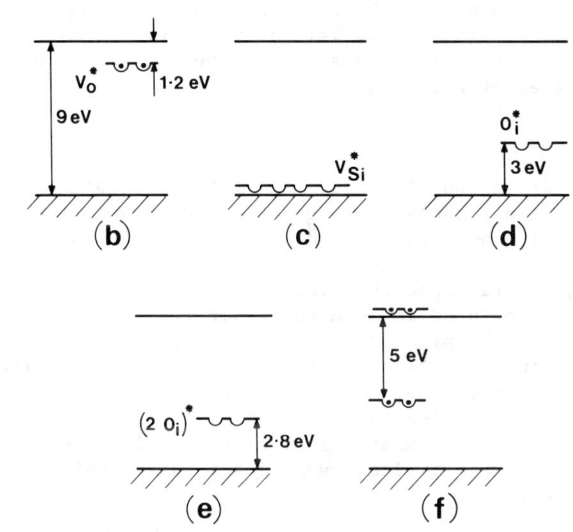

Fig. 5. (a) Energy levels associated with the neutral oxygen vacancy and with the relaxed oxygen vacancy. (After Griscom [1979].) (b) Energy level for the V_o defect based on the calculations of Herman et al. [1980]. (c) Energy level for the V_{Si} defect based on the calculations of Herman et al. [1980]. (d) Energy level for the O_i defect based on the calculations of Herman et al. [1980]. (e) Energy level for the peroxy defect suggested by Griscom [1979]. (f) Energy levels for the silicon interstitial based on the calculations of Herman et al. [1980].

levels associated with a neutral oxygen vacancy and propose a doubly occupied level 1.2 eV below the conduction band minimum as shown in Figure 5b.

Ciraci and Erkoc [1981] have also calculated that the oxygen vacancy introduces an energy level about one eV below the conduction band edge.

Silicon Vacancies

As pointed out by Herman et al. [1980], the removal of a neutral silicon from SiO_2 means the removal of four electrons from the topmost two energy levels in the valence band. Since these uppermost levels belong to oxygen 2p non-bonding orbitals, the removal of a neutral silicon has very little effect on these levels. The two vacant levels become degenerate with the top of the balence band as shown in Figure 5c.

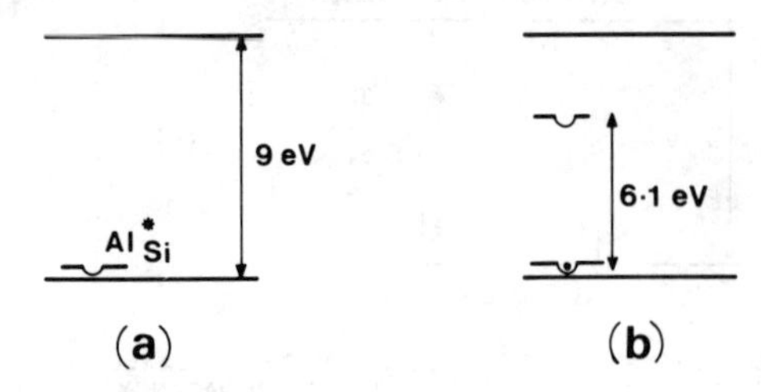

Fig. 6. (a) Energy level associated with the Al_{Si} defect based on the calculations of Herman et al. [1980]. (b) Energy levels associated with the (Al_{Si}, H_i) defect based on the calculations of Herman et al. [1980].

Ciraci and Erkoc [1981] calculate that the silicon vacancy introduces levels between the non-bonding and weakly bonding bands in the valence band.

In principle then the neutral silicon vacancy, V^*_{Si}, could be ionized to become V''''_{Si} at 0°K. However in practice these four holes probably remain close to the vacancy to maintain electrical neutrality, the vacancy and four holes forming a bound state [see Adler, 1975, pp. 246-247]. As yet it is not clear if this is the situation or whether the silicon vacancy acts as a shallow acceptor.

Oxygen Interstitials

Herman et al. have calculated that the introduction of a neutral oxygen interstitial creates an occupied interstitial O-2s band 11.3 eV below the valence band maximum, two occupied interstitial O-2p bands 2.1 and 2.3 above the valence band maximum and an empty O-2p band 3.0 eV above the valence band maximum as shown in Figure 5d.

A related defect, the peroxy defect, is discussed by Griscom [1979]. This is formed by the incorporation of oxygen to convert a Si-O-Si group to a Si-O-O-Si group. Griscom [1979] envisages that the common point defects in α-quartz are Frenkel defect pairs formed from vacant oxygen sites and peroxy defects. He proposes a defect level 2.8 eV above the valence band edge arising from the peroxy defect as shown in Figure 5e and tentatively correlated this with the "blue" luminescence found in many forms of SiO_2.

Silicon Interstitials

Introduction of a neutral silicon interstitial only slightly perturbs the already filled valence band levels but there is now a filled interstitial band 4 eV above the valence band maximum. The remaining two electrons occupy a band close to or in the conduction band edge (see Fig. 5f). Other unfilled interstitial bands arising from the Si-3p levels occur high in the conduction band [Herman et al., 1980].

The Aluminum Defect Center

Aluminum tends to be the common impurity in natural quartz where it may be present in concentrations of about 3000 Al per 10^6Si. Recent reviews concerning the defect are given by Weil [1975], Griscom [1978, 1979], and Koumvakalis [1980]. Aluminum substitutes for silicon in the quartz structure with a hole situated mainly in the non bonding 2p orbital of an adjacent oxygen atom [see Nuttall and Weil, 1976]. In natural quartz electrical neutrality is maintained at least at low temperatures by formation of associated complexes of the form $(Al'_{Si}, H^{\bullet}_i)^*$, $(Al'_{Si}, Li^{\bullet}_i)^*$ and $(Al'_{Si}, Na^{\bullet}_i)^*$. After irradiation by X-rays, γ-rays or electrons, Al-bearing quartz becomes "smoky." The precise nature of the optical absorption spectrum and its relation to the Al-center is still a matter for further study [Griscom, 1978, 1979], but Koumvakalis [1980] has reported a direct correlation between the Al^*_{Si} center and an absorption peak at ∿ 2.5 eV after irradiation.

Herman et al. [1980] point out that substitution of a neutral Al-atom for a neutral Si-atom removes one electron from the top of the valence band composed of non bonding O-2p orbitals. Their calculations cannot distinguish between a delocalized hole at the top of the valence band and a shallow acceptor level (Fig. 6a).

The addition of an interstitial hydrogen along with the Al-center adds one electron to fill the valence band and creates an empty hydrogen level 6.1 eV above the valence band maximum [Herman et al. 1980] as shown in Figure 6b.

Other impurities such as Fe^{3+} and Ga^{3+} also substitute for Si in quartz and the Fe-center has been extensively studied [Hassan and Cohen, 1974; Cohen and Hassan, 1974; Scala and Hutton, 1976]. Scala and Hutton conclude that Fe^{3+} can be incorporated into the quartz structure in two ways; Fe^{3+} tends to substitute for Si but also enters the structure interstitially.

Interstitial Impurity Defects

A number of interstitial impurity defects have been identified in α-quartz. They include in particular H_i, Li_i, and Na_i [see Kats, 1962, pp. 248-260]. At low temperatures these impurities tend to act as charge compensators for Al'_{Si} defects when they occupy positions in the c-axis channels approximately opposite the substituted Al-atom; this means they occur around the general position of α and β in Figure 1.

Little data are available for the energy levels occupied by H_i, Li_i and Na_i except that an energy level approximately 2.4 eV below the conduction band is known to be associated with the incorporation of Na into SiO_2. These three impurities are readily moved by an electric field and hence are easily ionized. Presumably they act as relatively shallow donors and form energy levels close to the conduction band.

Hydrogen Associated Defects

Many hydrogen defects give rise to absorption in the near infrared, particularly in the general region of 3 μm^{-1} wavenumber [see Kats, 1962]. The situation is complicated by the large number of absorption peaks, broad regions of structureless absorption and by the observation that individual details of the absorption spectrum may not be duplicated from one quartz specimen to the next. For a detailed discussion of the complexities in the infrared spectrum of quartz see Wood [1960].

However, several clarifying observations have been made:

1. Kats [1962] showed that some hydrogen is incorporated interstitially, generally in the form of charge compensators for Al'_{Si} or with other interstitials such as $Li^{\bullet}_{i}$ in more complicated configurations. As with other interstitials, $H^{\bullet}_{i}$ seems to be incorporated at sites such as α and β in Figure 1, opposite the Al_{Si} defect and in the c-axis channel. Several variants are possible in the general vicinity of α and β and all such defects give rise to sharp absorption bands at low temperatures. However, not all sharp bands have been identified with specific defects [see Chakraborty and Lehmann, 1976]; much of the complexity presumably arises from the hydrogen being able to occupy a number of sites near α and β and from the hydrogen being associated with a number of other impurities such as Al, Fe, Li, and Na in particular.

2. In crystals that have been heat treated at room pressure, part of the broad absorption band around 3.2 μm^{-1} wavenumber is converted to an ice absorption spectrum at liquid nitrogen temperatures [see Paterson and Kekulawala, 1979]. This indicates that at room temperature at least part of the spectrum results from a water phase present either as very small bubbles or in structurally disoriented regions in the crystal [Aines and Rossman, 1983]. Some authors such as Bambauer et al. [1969] have suggested that the milkiness that forms in (OH)-rich quartz upon heating to moderate temperatures consists of H_2O precipitating in the form of bubbles and have proposed the reaction

$$\text{Si-OH:HO-Si} \rightleftharpoons H_2O + \text{Si-O-Si} \tag{5}$$

Such a reaction is well documented in silica glass [see Greaves, 1978, and references therein].

3. The hydrolytic weakening effect appears to correlate with that part of the infrared spectrum that remains broad and structureless even down to liquid helium temperatures [Kekulawala et al., 1978, 1981; Paterson and Kekulawala, 1979]. It is not clear what gives rise to this absorption; Walrafen and Luongo [1975] have identified (OH) groups oriented randomly in {0001} and perhaps these contribute to this part of the absorption.

Although there are still some uncertainties associated with the quantitative determination of (OH) in quartz from infrared data [Paterson, 1982], the amounts so far reported in single crystals that show the weakening effect range from about 300 (OH) per 10^6Si to about 4000 (OH) per 10^6Si, the weakening effect being strongly dependent on the (OH) concentration [see Griggs, 1967; Hobbs et al., 1972]. An average concentration of 1000 (OH) per 10^6Si is equivalent to one (OH) group in 0.3 x 10^3 unit cells or an average of one (OH) group every ca. 21 Si atoms in any direction. It is this low concentration that so drastically influences mechanical and recrystallization behavior. Alternatively, with a dislocation density of say $10^9 cm^{-2}$, ca. 10^{16}(OH) groups cm^{-3} are required to place one (OH) group in each unit cell of dislocation line. Since a crystal with an average (OH) content of 1000 (OH) per 10^6Si contains ca. 10^{19}(OH) groups cm^{-3}, it is difficult to see why the effect should be so strongly concentration dependent on the basis of the Griggs [1967] model.

The precise nature of the (OH)-defects in quartz is still not clear. Two models have been suggested; that due to Griggs [1967] is interpreted here as consisting of an H_2O molecule incorporated at an existing oxygen site either interstitially or as two hydrogens associated with a peroxy defect. The other model, documented by Nuttall and Weil [1980], consists of three or four hydrogen atoms substituted for a silicon. This latter defect exists at low temperatures with a trapped hole suggesting that it has an energy level at or very close to the top of the valence band. Such an energy level is also to be expected from arguments identical to those used for the V^{*}_{Si} or Al'_{Si} defects in quartz [see Herman et al., 1980]: removal of a neutral silicon atom removes four electrons from the topmost levels of the valence band (the degenerate oxygen non bonding 2p levels), addition of three hydrogen atoms adds three electrons leaving one vacant level close to or degenerate with the top of the valence band. Addition of four electrons creates a neutral defect but $(4H)_{Si}$ could still act as an acceptor because of the different "electron affinity" the defect would have compared to silicon.

The $(3H)'_{Si}$ or $(4H)'_{Si}$ defects are perhaps to be favored since (1) as pointed out by McLaren et al. [1983], they are embryonic water bubbles, (2) small clusters of these defects could give rise to the broad structureless infrared absorption, and (3) they could act mechanically in the same manner as the classical Griggs defect so that now OH.HO bonds need to be broken at a dislocation core instead of O-Si-O bonds in order to facilitate dislocation motion.

No defect energy data are available for the Griggs defect other than the calculations by Bennett and Roth [1971] which suggest that this defect acts as a relatively shallow acceptor in quartz.

4. Line Defects in Quartz

The introduction of dislocations into an otherwise perfect crystal structure creates a periodic

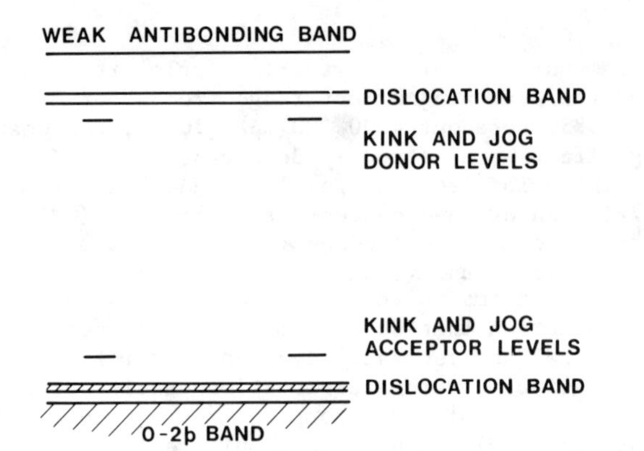

Fig. 7. Dislocation bands and kink and jog levels in α-SiO_2. The precise positions of these levels are unknown; the donor and acceptor levels may in fact lie much closer to the O-2p and weak antibonding bands respectively.

array of lattice distortions and (for edge dislocations) of dangling bonds. Since these defects are periodic electron energy bands may be generated within the band gap [see Read, 1954; Heine, 1966; Labusch and Schroter, 1980; Hirsch, 1979, 1981]. Lattice distortions are likely to be associated with dislocation bands close to the valence and conduction bands since such distortions lead to changes in bond length. Dangling bonds are likely to be associated with bands midway between strong bonding and antibonding states [see Labusch and Schroter, 1980]. In materials such as Ge and Si energy bands due to such effects are generated within the band gap; the lowermost one, close to the valence band, is filled at 0°K and acts as a donor band whereas the highest band, just below the conduction band, is empty at 0°K and acts as an acceptor. With the Fermi level at mid gap, a screw dislocation tends to be neutral. An edge dislocation in Ge and Si introduces a partly filled band within the band gap [see Labusch and Schroter, 1980]. Kinks on dislocation lines being localized disturbances in the otherwise periodic structure of the dislocation line are expected to introduce localized levels (both acceptors and donors) within the band gap [Hirsch, 1979, 1981]. The same is expected of jogs in dislocation lines.

No calculations or experimental observations are yet available for quartz but results similar in principle to those for Ge and Si are likely to be true. An edge dislocation in SiO_2 presumably has dangling Si- and O-bonds and there is lattice dilation associated with the end of the extra half plane [see McLaren et al., 1971]; a screw dislocation has only lattice distortion. The states to be expected therefore are those associated with perturbations in the vicinity of oxygen 2p orbitals and of silicon sp^3 hybrid orbitals. Since the highest energy oxygen 2p orbital is non bonding little change in energy is expected whereas a band split off from the conduction band is likely (Fig. 7). States midway between strongly bonding and antibonding levels are likely to be close to the top of the valence band; Calabrese and Fowler [1978] for instance place the center of the strongly bonding band in a α-quartz at about -16 eV (zero being taken at the lowermost point in the conduction band) and the mid point of the strongly antibonding band at about +6 eV. Thus states midway between strongly bonding and antibonding would be around -5 eV which is one or two eV above the valence band.

Although no calculations are available for line defects in quartz, calculations and experimental observations of surface states are available [see Ciraci and Ellialtioglu, 1982]. The results are in broad agreement with what has been said above: surfaces composed only of oxygens give rise to states in the valence band whereas surfaces composed only of silicons give rise to states in the band gap (in this case a half filled band approximately 2 eV below the conduction band edge).

The proposed energy levels associated with dislocations in quartz are shown in Figure 7. Jogs and kinks are proposed to be associated with localized energy levels within the band gap acting as both donors and acceptors.

If neutral kinks, K*, become ionized according to

$$\begin{aligned} K^* &\rightleftharpoons K^{\bullet} + e' \\ K^* &\rightleftharpoons K' + h^{\bullet} \end{aligned} \tag{6}$$

and neutral jogs, J*, become ionized according to

$$\begin{aligned} J^* &\rightleftharpoons J^{\bullet} + e' \\ J^* &\rightleftharpoons J' + h^{\bullet} \end{aligned} \tag{7}$$

then the concentrations of charged jogs and kinks are given by

$$\begin{aligned} [K\cdot] &= K_{K\cdot}\ [K^*]\ n^{-1} \\ [K'] &= K_{K'}\ [K^*]\ p^{-1} \\ [J\cdot] &= K_{J\cdot}\ [J^*]\ n^{-1} \\ [J'] &= K_{J'}\ [J^*]\ p^{-1} \end{aligned} \tag{8}$$

where $K_{K\cdot}$, $K_{K'}$, $K_{J\cdot}$, $K_{J'}$ are the equilibrium constants for reactions (6a), (6b), (7a), (7b) respectively.

5. The Defect Chemistry of Pure Quartz

In pure quartz it is assumed that the only point defects present at high temperatures are oxygen vacancies and interstitials and silicon vacancies and interstitials. Pure quartz would appear to belong to that class of materials that remains close to stoichiometric over relatively large ranges of the imposed activities of oxygen

TABLE 2. Reactions Expressing the Incorporation of Point Defects Into Pure Quartz

Reaction	Equilibrium Constant	Law of Mass Action
Incorporation of oxygen: $O_2 = 2O_o + V_{Si}$	K_o	$[V_{Si}] = K_o a_{O_2}$
Incorporation of silicon: $Si = Si_{Si} + 2V_o$	K_{Si}	$[V_o] = K_{Si}^{1/2} a_{Si}^{1/2} = K_{SiO_2}^{1/2} K_{Si}^{1/2} a_{SiO_2}^{1/2} a_{O_2}^{-1/2}$
Breakdown of SiO_2: $SiO_2 = Si + O_2$	K_{SiO_2}	$a_{Si} a_{O_2} = K_{SiO_2} a_{SiO_2}$
Frenkel oxygen disorder: $O_o = V_o + O_i$	K_{Fo}	$[O_i] = K_{Fo} K_{SiO_2}^{-1/2} K_{Si}^{-1/2} a_{SiO_2}^{-1/2} a_{O_2}^{1/2}$
Frenkel silicon disorder: $Si_{Si} = V_{Si} + Si_i$	K_{FSi}	$[Si_i] = K_{FSi} K_o^{-1} a_{O_2}^{-1}$
Excitation of electron from valence band to conduction band: neutral lattice = $e' + h^{\cdot}$	Ki	np = Ki
Ionization of silicon vacancy: $V_{Si} = V_{Si}^{''''} + 4h^{\cdot}$	$K_{Si'}$	$[V_{Si}^{''''}] = K_o K_{Si'} a_{O_2} p^{-4}$
Ionization of oxygen vacancy: $V_o = V_o^{\cdot\cdot} + 2e'$	$K_{o^{\cdot}}$	$[V_o^{\cdot\cdot}] = K_{Si}^{1/2} K_{o^{\cdot}} K_{SiO_2}^{1/2} a_{SiO_2}^{1/2} a_{O_2}^{-1/2} n^{-2}$
Ionization of silicon interstitial: $Si_i = Si_i^{\cdot\cdot\cdot\cdot} + 4e'$	$K_{Si^{\cdot}}$	$[Si_i^{\cdot\cdot\cdot\cdot}] = K_{Si^{\cdot}} K_{FSi} K_o^{-1} a_{O_2}^{-1} n^{-4}$
Ionization of oxygen interstitial: $O_i = O_i^{''} + 2h^{\cdot}$	K_o'	$[O_i^{''}] = K_{Si}^{-1/2} K_{o'} K_{Fo} K_{SiO_2}^{-1/2} a_{SiO_2}^{-1/2} a_{O_2}^{1/2} p^{-2}$

and silicon. This means that the concentrations of uncharged point defects must remain low over this range of activities and that if charged point defects are present then they must be present in proportions to maintain both electrical neutrality and stoichiometry over this range.

The reactions expressing the incorporation of point defects into pure quartz are set out in Table 2 together with the equilibrium constants for each reaction and the results of applying the law of mass action to each reaction. This table contains five equations involving the six quantities $[V_{Si}^{''''}]$, $[V_o^{\cdot\cdot}]$, $[Si_i^{\cdot\cdot\cdot\cdot}]$, $[O_i^{''}]$, n and p. The equation for electrical neutrality

$$4[V_{Si}^{''''}] + 2[O_i^{''}] + n = 4[Si_i^{\cdot\cdot\cdot\cdot}] + 2[V_o^{\cdot\cdot}] + p \quad (9)$$

supplies one more equation which then enables expressions to be written down for each of the six quantities in terms of the activity of oxygen, a_{O2}.

However, because quartz seems to be stoichiometric over large ranges in a_{O2} we will be interested in one of the following electrical neutrality conditions:

$$[V_o^{\cdot\cdot}] \approx [O_i^{''}] \quad (10a)$$

$$[V_{Si}^{''''}] \approx [Si_i^{\cdot\cdot\cdot\cdot}] \quad (10b)$$

$$[V_o^{\cdot\cdot}] \approx 2[V_{Si}^{''''}] \quad (10c)$$

$$[O_i^{''}] \approx 2[Si_i^{\cdot\cdot\cdot\cdot}] \quad (10d)$$

Conditions (10a) and (10b) correspond to Frenkel disorder on the oxygen and silicon sublattices respectively whereas conditions (10c) and (10d) correspond to Schottky disorder. Expressions for the concentrations of the six charged point defects in pure quartz may now be derived for each of the neutrality conditions given in (10); these expressions are summarized in Table 3. Notice that within these neutrality ranges expressed in (10) only n and p vary with a_{O2}; the other charged point defect concentrations are independent of a_{O2}. The variation of defect concentration with a_{O2} is summarized for each of the conditions in (10) in Figure 8.

At present it is not possible to say which of the conditions in (10) is relevant to quartz. As mentioned above, oxygen vacancies are well known in quartz [see Griscom, 1979] so that either $[V_o^{\cdot\cdot}]$ $[V_o^{\cdot\cdot}] = [O_i^{''}]$ or $[V_o^{\cdot\cdot}] = 2[V_{Si}^{''''}]$ is probably the condition that controls stoichiometry in

TABLE 3. Dependence of Charged Defect Concentrations Upon a_{O2} for Various Stoichiometric Neutrality Ranges

Defect	Neutrality Range			
	$[V_o^{\bullet\bullet}] = [O_i'']$	$[V_o^{\bullet\bullet}] = 2[V_{Si}'''']$	$[Si_i^{\bullet\bullet\bullet\bullet}] = [V_{Si}'''']$	$2[Si^{\bullet\bullet\bullet\bullet}] = [O_i'']$
$[V_o^{\bullet\bullet}]$	0	0	0	0
$[V_{Si}'''']$	0	0	0	0
$[O_i'']$	0	0	0	0
$[Si_i^{\bullet\bullet\bullet\bullet}]$	0	0	0	0
n	- 1/4	- 1/4	- 1/4	- 1/4
p	+ 1/4	+ 1/4	+ 1/4	+ 1/4

quartz; Griscom [1979] favors $[V_o^{\bullet\bullet}] = [O_i'']$, that is, Frenkel disorder on the oxygen sublattice. Experiments are needed to settle the matter.

Notice that each of the four neutrality conditions (10) imply different rate controlling diffusion mechanisms in quartz. For $[V_o^{\bullet\bullet}] = [O_i'']$ with $[Si_i^{\bullet\bullet\bullet\bullet}] > [V_{Si}'''']$ as shown in Figure 8a, silicon defects are minority species and, assuming all defects have the same mobility, silicon can diffuse faster by an interstitial rather than by a vacancy mechanism. Thus diffusion by a silicon interstitial mechanism is the rate controlling mechanism. Similar arguments lead to an oxygen vacancy mechanism as rate controlling for the condition $[V_{Si}''''] = [Si_i^{\bullet\bullet\bullet\bullet}]$, $[V_o^{\bullet\bullet}] > [O_i'']$ as in Figure 8b, to a vacancy silicon mechanism for $[V_o^{\bullet\bullet}] = 2[V_{Si}'''']$, $[Si_i^{\bullet\bullet\bullet\bullet}] > [O_i'']$ as in Figure 8c and to an interstitial silicon mechanism for $[O_i''] = 2[Si_i^{\bullet\bullet\bullet\bullet}]$, $[V_o^{\bullet\bullet}] > [V_{Si}'''']$ as in Figure 8d. Since it appears that oxygen diffusion in quartz is faster than silicon, condition (10b), namely, $[V_{Si}''''] = [Si_i^{\bullet\bullet\bullet\bullet}]$ seems to be ruled out as a likely neutrality condition.

6. The H_2O-SiO_2 System in the Solid State

Consider the situation where pure SiO_2 is in equilibrium with H_2O vapor. We suppose that the fugacity of oxygen, f_{O_2}, is controlled by the environment (that is, by a suitable buffer) and the pressure, P, and temperature, T, of the system are prescribed. Under these conditions, the fugacities of H_2 and of H_2O are fixed and are given by [see Edgar, 1973]

$$f_{H_2} = (P\,\gamma_{H_2O}\,\gamma_{H_2})/(K_{H_2O}\,f_O^{1/2}\,\gamma_{H_2} + \gamma_{H_2O}) \quad (11)$$

$$f_{H_2O} = (P\,K_{H_2O}\,f_{O_2}^{1/2}\,\gamma_{H_2}\,\gamma_{H_2O})/(K_{H_2O}\,f_{O_2}^{1/2}\,\gamma_{H_2} + \gamma_{H_2O}) \quad (12)$$

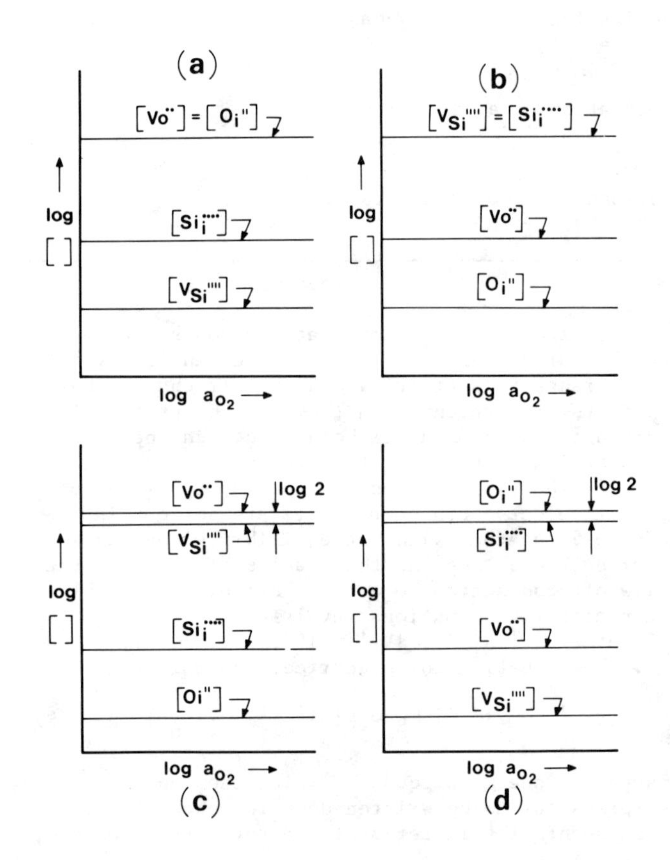

Fig. 8. Four possibilities for plots of Log (defect concentration) against log a_{O_2} for stoichiometric, pure quartz. Figures 8a and 8b represent Frenkel disorder on the oxygen and silicon sublattices, whereas Figures 8c and 8d represent Schottky disorder.

where γ_{H_2}, γ_{H_2O} are the fugacity coefficients for H_2 and H_2O and K_{H_2O} is the equilibrium constant for the reaction.

$$H_2 + \tfrac{1}{2}O_2 \rightleftharpoons H_2O \qquad (13)$$

Equations (11) and (12) assume that f_{O_2} is small compared to both f_{H_2} and f_{H_2O}, a situation commonly true in experimental and metamorphic conditions.

The Griggs Defect $(HOH)_i$

We assume that hydrogen is incorporated into the quartz structure according to the following two reactions:

$$H_2 \rightleftharpoons 2H_i \qquad (14)$$

$$H_2O \rightleftharpoons (HOH)_i \qquad (15)$$

The defect described by (15) is the Si-OH.HO-Si defect proposed by Griggs [1967]. Note that with a concentration of 1000 (OH) per 10^6 Si only one in 10^3 unit cells contains such a defect so that any distortion associated with the interstitial incorporation of the defect is spread over a relatively large volume.

Application of the law of mass action to (14) and (15) gives

$$[H_i] = K_H^{\frac{1}{2}} a_{H_2}^{\frac{1}{2}} \qquad (16)$$

$$[(HOH)_i] = K_{HOH}\, a_{H_2O} \qquad (17)$$

where K_H, K_{HOH} are the equilibrium constants for (14) and (15).

Ionization of these defects is expressed by

$$H_i \rightleftharpoons H_i^{\bullet} + e' \qquad (18)$$

$$(HOH)_i \rightleftharpoons (HOH)_i' + h^{\bullet} \qquad (19)$$

from which are obtained

$$[H_i^{\bullet}] = K_H^{\frac{1}{2}} K_{H^{\bullet}}\, a_{H_2}^{\frac{1}{2}}\, n^{-1} \qquad (20)$$

$$[(HOH)_i'] = K_{HOH} K_{HOH'}\, a_{H_2O}\, p^{-1} \qquad (21)$$

where $K_{H^{\bullet}}$, $H_{HOH'}$ are the equilibrium constants for reactions (18) and (19).

In order to demonstrate the effect of adding H_2O to pure SiO_2 one of the neutrality fields (10) that controls stoichiometry in pure quartz is selected and the argument developed in detail. Similar arguments follow for the other three conditions in (10). We choose $[V_O^{\bullet\bullet}] \approx 2[V_{Si}'''']$.

Expressions for the intrinsic point defect concentrations $[V_O^{\bullet\bullet}]$, $[V_{Si}'''']$, $[Si_i^{\bullet\bullet\bullet\bullet}]$ $[O_i'']$ may be read from Table 2. These four equations together with (2), (20) and (21) and the expression for electrical neutrality:

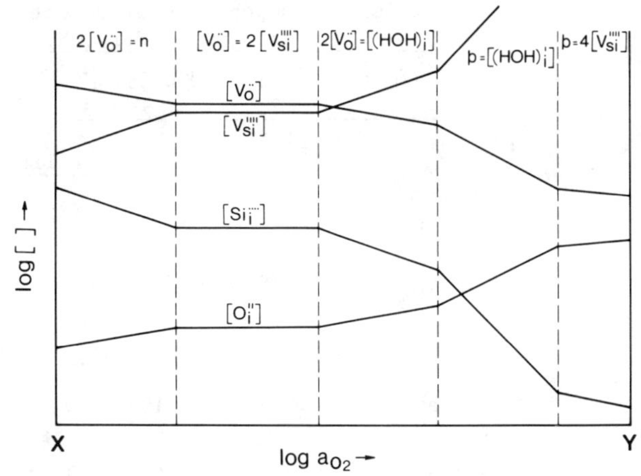

Fig. 9. Variation of defect concentration with activity of oxygen for the situation where $[V_O^{\bullet\bullet}] = 2[V_{Si}'''']$ is the stoichiometric condition in pure quartz. At relatively high a_{O_2} the Griggs defect $(HOH)_i'$ is assumed to participate in the neutrality condition. This is the section X-Y across the phase diagram shown in Figure 10.

$$2[V_O^{\bullet\bullet}] + [H_i^{\bullet}] + p = 4[V_{Si}''''] + [(HOH)_i'] + n \qquad (22)$$

supply eight equations which involve the eight quantities $[V_O^{\bullet\bullet}]$, $[V_{Si}'''']$, $[Si_i^{\bullet\bullet\bullet\bullet}]$, $[O_i'']$, n, p, $[H_i^{\bullet}]$ and $[(HOH)_i']$.

We can now proceed to construct phase diagrams showing the variation of the concentrations of point defects with a_{O_2} at various a_{H_2O} (Fig. 9). This diagram is constructed at a value of a_{H_2O} where $(HOH)_i'$ participates in maintaining charge neutrality and is divided into a number of fields where only two charged defects are responsible for charge neutrality, the other defects in (22) being present in minority concentrations [see Brouwer, 1954; Kröger, 1974, pp. 152-157]. Table 4 presents expressions for all charged point defects in the neutrality fields of Figure 9.

Figure 9 however comprises one section across the $a_{H_2O} - a_{O_2}$ phase diagram shown in Figure 10. This diagram is constructed by the method outlined by Van Gool [1966, pp. 101-103]; the boundary between $[H_i^{\bullet}] = 4[V_{Si}'''']$ and $[H_i^{\bullet}] = [(HOH)_i']$ being found for instance by calculating $4[V_{Si}'''']$ as a function of a_{O_2} and a_{H_2O} using $[H_i^{\bullet}] = [(HOH)_i']$, then calculating $[(HOH)_i']$ as a function of a_{O_2} and a_{H_2O} using $[H_i^{\bullet}] = 4[V_{Si}'''']$. At the boundary these two expressions are equal. A number of points arise from Figure 10.

1. Figure 9 comprises a section along a line such as X - Y in Figure 10.

2. Expression (12) shows that it is not possible to vary f_{H_2O} and f_{O_2} independently of each other at constant P and T and so long as H_2, O_2 and H_2O are the only vapor phases present. In order to experimentally construct sections such as A-B or C-D on Figure 10 at constant f_{O_2} it would be necessary to vary P at constant T or to introduce another inert vapor phase which has negligible

TABLE 4. Dependence of Defect Concentrations Upon Activities of Water and Oxygen for (OH) Defect Formed by Griggs Defect

Defect	$[H_i^{\bullet}]$	$[(HOH)_i^{\bullet}]$	$[V_o^{\bullet\bullet}]$	$[O_i'']$	$[V_{Si}'''']$	$[Si_i^{\bullet\bullet\bullet\bullet}]$	n	p
Neutrality Range $n = [H_i^{\bullet}]$								
a_{H_2O}	+1/4	+5/4	-1/2	+1/2	+1	-1	+1/4	-1/4
a_{O_2}	-1/8	-1/8	-1/4	+1/4	+1/2	-1/2	-1/8	+1/8
Neutrality Range $[H_i^{\bullet}] = 4[V_{Si}'''']$								
a_{H_2O}	+2/5	+11/10	-1/5	+1/5	+2/5	-2/5	+1/10	-1/10
a_{O_2}	0	-1/4	0	0	0	0	-1/4	+1/4
Neutrality Range $[H_i^{\bullet}] = [(HOH)_i']$								
a_{H_2O}	+3/4	+3/4	+1/2	-1/2	-1	+1	-1/4	+1/4
a_{O_2}	-1/8	-1/8	-1/4	+1/4	+1/2	-1/2	-1/8	+1/8
Neutrality Range $2[V_o^{\bullet\bullet}] = [(HOH)_i']$								
a_{H_2O}	+5/6	+2/3	+2/3	-2/3	-4/3	+4/3	-1/3	+1/3
a_{O_2}	-1/12	-1/6	-1/6	+1/6	+1/3	-1/3	-1/6	+1/6
Neutrality Range $[(HOH)_i'] = p$								
a_{H_2O}	+1	+1/2	+1	-1	-2	+2	-1/2	+1/2
a_{O_2}	-1/4	0	-1/2	+1/2	+1	-1	0	0

solubility in SiO_2. Similar comments are true for any other section line across Figure 10 other than those that are compatible with expression (12).

3. Figure 10 is valid so long as other charged defects such as O_i'' and $Si_i^{\bullet\bullet\bullet\bullet}$ remain minority defects.

4. Since the field $[V_o^{\bullet\bullet}] = 2[V_{Si}'''']$ in Figure 10 probably covers a range of a_{O_2} that may be physically difficult to attain (see section 4 and Hobbs [1981]), the area of Figure 10 that is likely to be of geological interest is only the central part. To change f_{H_2O} or f_{H_2} significantly at constant T and f_{O_2} when H_2O, H_2 and O_2 are the only vapor phases present is also difficult; a change in pressure from 300 MPa to 1500 MPa at $f_{O_2} = 10^{-11}$ MPa and 1200°K results in about an order of magnitude increase in f_{H_2O} (data from Shaw and Wones [1964] for γH_2, from Burnham et al. [1969] for γ_{H_2O}, and from the JANAF Thermochemical Tables [1971] for K_{H_2O}). Under these conditions, f_{H_2O} changes from about 270 MPa at P = 300 MPa to about 3.5×10^3 MPa at P = 1500 MPa.

A section across Figure 10 at relatively high a_{O_2} is shown in Figure 11 (section line A-B). The effect of passing from the neutrality range $[V_o^{\bullet\bullet}] = 2[V_{Si}'''']$ characteristic of pure quartz into the range $2[V_o^{\bullet\bullet}] = [(HOH)_i']$ where the $(HOH)_i'$ defect is responsible for maintaining charge neutrality is to decrease the concentration of acceptor defects, O_i'' and V_{Si}'''', and to increase the concentration of donor defects, $V_o^{\bullet\bullet}$ and $Si_i^{\bullet\bullet\bullet\bullet}$.

The fastest moving defect species of the minority component changes as f_{H_2O} increases. V_{Si}'''' is the rate controlling defect in pure quartz and this decreases in concentration as f_{H_2O} increases implying that the diffusion coefficient decreases. However, this is soon overtaken by a rapid increase in $[Si_i^{\bullet\bullet\bullet\bullet}]$ which becomes rate controlling. Thus if deformation in quartz is rate controlled by diffusion then the rate controlling mechanism in pure quartz would be a silicon vacancy mechanism which would continue as f_{H_2O} was first increased resulting in fact in a strengthening effect. For further increases in

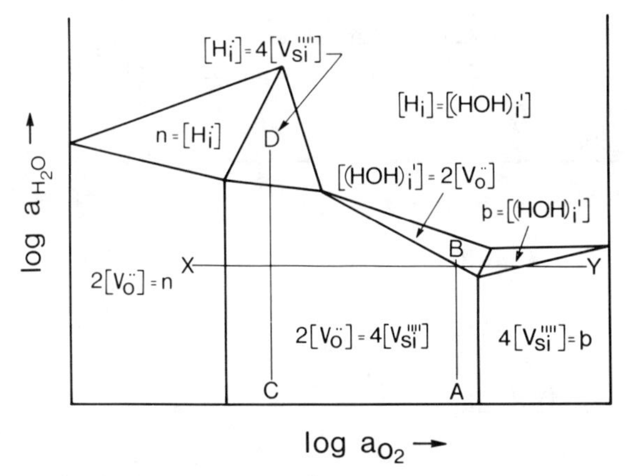

Fig. 10. Phase diagram for the SiO_2-H_2O system assuming hydrogen is incorporated in SiO_2 interstitially and as the Griggs defect $(HOH)_i'$. The Section X-Y at constant a_{H_2O} is presented in Figure 9, whereas the section A-B and C-D at relatively high and low a_{O_2} are presented in Figures 11 and 12.

f_{H_2O} a silicon interstitial mechanism would become rate controlling leading to a weakening effect. The weakening effect over the range where $Si_i^{\cdot\cdot\cdot}$ is rate controlling is quite dramatic with more than an order of magnitude increase in the diffusion coefficient for an order of magnitude increase in f_{H_2O}.

Figure 12 is a section (C-D) across Figure 10 at relatively low f_{O_2}. The defect responsible for maintaining charge neutrality as f_{H_2O} increases is now $H_i^{\cdot}$ and the effect of introducing this defect is to decrease the concentrations of all donor defects such as $V_O^{\cdot\cdot}$ and $Si_i^{\cdot\cdot\cdot}$ whilst increasing the concentrations of all acceptor defects such as V_{Si}'''' and O_i''. The rate controlling defect remains V_{Si}'''' for a small increase in f_{H_2O} but

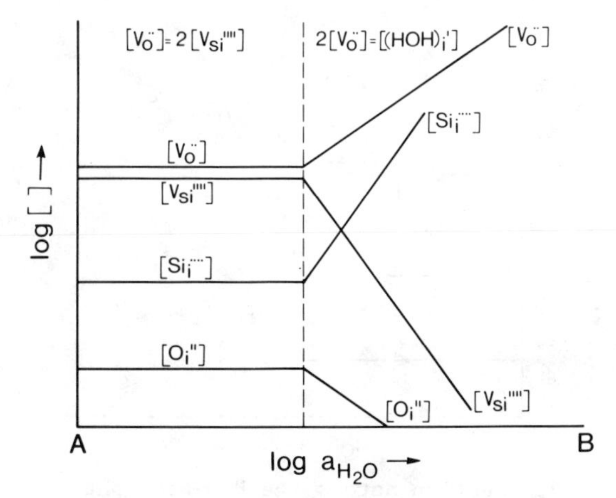

Fig. 11. Section A-B across the phase diagram given in Figure 10. Activity of oxygen is relatively high.

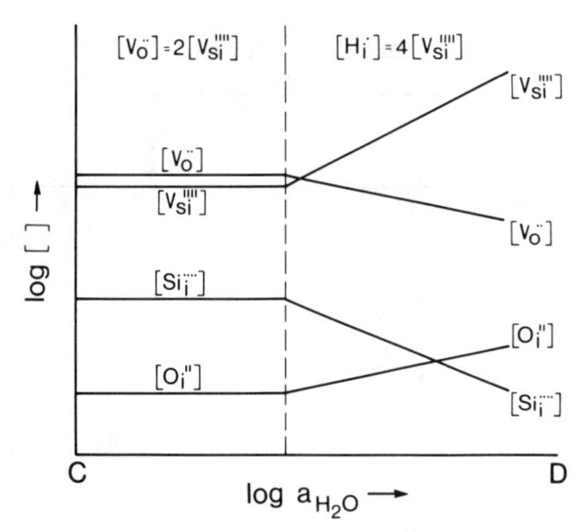

Fig. 12. Section C-D across the phase diagram given in Figure 10. Activity of oxygen is relatively low.

then becomes $V_O^{\cdot\cdot}$ and then finally O_i'' at high f_{H_2O}. Thus the rate controlling mechanism passes from a silicon vacancy mechanism at low f_{H_2O} (where there is a slight weakening effect) to an oxygen vacancy mechanism resulting in a gradual strengthening effect to finally an oxygen interstitial mechanism at high f_{H_2O} when the material begins to weaken. The effect of $H_i^{\cdot}$ overall is to strengthen quartz; over physically accessible ranges of f_{H_2O} it is doubtful if significant weakening due to the oxygen interstitial mechanism would become important.

The $(3H)_{Si}$ Defect

Incorporation of the $(3H)_{Si}$-defect is expressed by

$$3H_2 \rightleftharpoons 2(3H)_{Si} + 4V_o \tag{23}$$

for which the Law of Mass Action gives, using Table 2,

$$[(3H)_{Si}] = K_{3H}^{1/2} K_{Si}^{-1} K_{SiO_2}^{-1} a_{SiO_2}^{-1} a_{H_2}^{3/2} a_{O_2} \tag{24}$$

Ionization of this defect is expressed by

$$(3H)_{Si} \rightleftharpoons (3H)_{Si}' + h^{\cdot} \tag{25}$$

from which

$$[(3H)_{Si}'] = K_{3H'} K_{3H}^{1/2} K_{Si}^{-1} K_{SiO_2}^{-1} a_{SiO_2}^{-1} a_{H_2}^{3/2} a_{O_2} p^{-1} \tag{26}$$

K_{3H} and $K_{3H'}$ are the equilibrium constants for reactions (23) and (25).

In addition, hydrogen may be incorporated interstitially according to reaction (14) to give concentrations of neutral and charged hydrogen interstitials expressed by (16) and (20).

The dependence of various charged defects upon

TABLE 5. Dependence of Defect Concentrations Upon Activities of Water and Oxygen for (OH) Defect Formed by $(3H)_{Si}$ Substitution

Defect	$[H_i^{\cdot}]$	$[(3H_{Si})']$	$[V_o^{\cdot\cdot}]$	$[O_i'']$	$[V_{Si}'''']$	$[Si_i^{\cdot\cdot\cdot\cdot}]$	n	p
			Neutrality Range $[H_i^{\cdot}] = n$					
a_{H_2O}	+1/4	+7/4	-1/2	+1/2	+1	-1	+1/4	-1/4
a_{O_2}	-1/8	+1/8	-1/4	+1/4	+1/2	-1/2	-1/8	+1/8
			Neutrality Range $[H_i^{\cdot}] = 4[V_{Si}'''']$					
a_{H_2O}	+2/5	+8/5	-1/5	+1/5	+2/5	-2/5	+1/10	-1/10
a_{O_2}	0	0	0	0	0	0	-1/4	+1/4
			Neutrality Range $[H_i^{\cdot}] = [(3H)_{Si}']$					
a_{H_2O}	+1	+1	+1	-1	-2	+2	-1/2	+1/2
a_{O_2}	0	0	0	0	0	0	-1/4	+1/4
			Neutrality Range $[(3H)_{Si}'] = 2[V_o^{\cdot\cdot}]$					
a_{H_2O}	+1	+1	+1	-1	-2	+2	-1/2	+1/2
a_{O_2}	0	0	0	0	0	0	-1/4	+1/4
			Neutrality Range $[(3H)_{Si}'] = p$					
a_{H_2O}	+5/4	+3/4	+3/2	-3/2	-3	+3	-3/4	+3/4
a_{O_2}	-1/8	+1/8	-1/4	+1/4	+1/2	-1/2	-1/8	+1/8

the activities of water and of oxygen may now be derived as with the Griggs defects and these dependencies are given in Table 5. A similar table is easily constructed for the $(4H)_{Si}$ defect but is not included here.

The phase diagram for incorporation of the $(3H)_{Si}$ defect into quartz where $[V_o^{\cdot\cdot}] = 2[V_{Si}'''']$ is the neutrality condition for stoichiometric, pure quartz is similar to that shown in Figure 10 except that the $[(HOH)_i^{\cdot}] = 2[V_o^{\cdot\cdot}]$, $[H_i^{\cdot}] = [(HOH)_i^{\cdot}]$ and $p = [(HOH)_i^{\cdot}]$ fields are replaced by $[(3H)_{Si}'] = 2[V_o^{\cdot\cdot}]$, $[H_i^{\cdot}] = [(3H)_{Si}']$ and $p = [(3H)_{Si}']$ respectively. A section across such a phase diagram for high activities of oxygen is shown in Figure 13. A section at low a_{O_2} is identical to Figure 12. Notice that the variation of defect concentration with a_{H_2O} is stronger for the $(3H)_{Si}$ defect (Figure 13) than for the Griggs defect (Figure 11). Conclusions identical to those drawn from Figure 11 and 12 where the Griggs

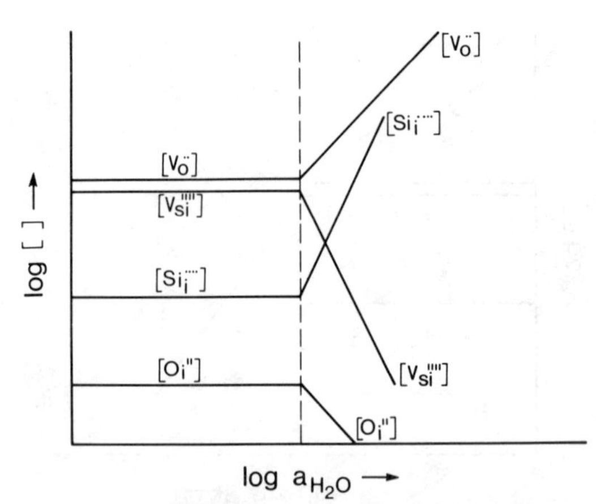

Fig. 13. Section across the H_2O-SiO_2 phase diagram where hydrogen is assumed to be incorporated interstitially and as the $(3H)_{Si}$ defect. Activity of oxygen is relatively high.

defect is important may be made for the $(3H)_{Si}$ defect as well.

7. Discussion

The classical interpretation of the hydrolytic weakening effect involves the migration of (OH) defects to dislocations where they facilitate the migration of kinks on dislocation lines [Griggs, 1967]. The rate controlling species is envisaged as the diffusion of the (OH) species through the quartz structure. In the following discussion this process and a number of others are investigated from the point of view of defect chemistry. All of the processes examined here are capable of explaining a decrease in strength with an increase in f_{H_2O}. It remains for carefully designed experiments to determine the precise dependence of creep rate upon f_{H_2O} so that the relevant mechanism may be delineated.

Deformation Rate Controlled by Kinks Alone

If the deformation rate is controlled by kinks alone and does not depend on an (OH) atmosphere diffusing to or with the kink then the effect must be associated solely with an increase in kink concentration with increase in f_{H_2O} [Hirth and Lothe, 1968; Hirsch, 1979, 1981; Hobbs, 1983b].

From expression (8) and Table 5, the kink concentrations for various neutrality conditions of interest are given in Table 6.

TABLE 6. Kink Concentration Dependence on a_{O_2}, a_{H_2O} for Two Neutrality Conditions

Neutrality Condition	Kink	a_{H_2O}	a_{O_2}
$2[V_o^{\bullet\bullet}] = [(3H)'_{Si}]$	$K^\bullet$	+1/2	+1/4
	K'	-1/2	-1/4
$[H_i^\bullet] = 4[V''''_{Si}]$	$K^\bullet$	+1/10	-2/5
	K'	-1/10	+2/5

Deformation Controlled by Kink Motion With Diffusion of (OH) to or With the Kink

This is the classical Griggs [1967] mechanism for hydrolytic weakening. The dependence of strain rate on a_{H_2O} and a_{O_2} is now given by multiplying the kink concentration by the diffusivity of the (OH) defect. We consider two assumptions here: (1) that the diffusivity of the (OH) defect is proportional to the concentration of $(3H)'_{Si}$, and (2) that diffusion of the $(3H)'_{Si}$ defect takes place by a silicon vacancy mechanism and hence the diffusivity of the (OH) defect is proportional to $[V''''_{Si}]$. These two possibilities are taken into account in Table 7.

Deformation Rate Controlled by Diffusion

The view presented here is that the presence of (OH)-defects in the quartz structure increases the concentration of charged oxygen vacancies and silicon interstitials so that the concentration, and hence the diffusivity, of the rate controlling atomic species is increased. Diffusion of (OH) through the structure plays no role. This leads directly to an increase in strain rate at constant stress since the strain rate is proportional to the diffusion coefficient of the rate controlling species for diffusion-controlled creep. The physical reason for this effect is that the (OH)-defect responsible for hydrolytic weakening in quartz appears to act as a relatively shallow acceptor and hence is readily ionized compared to intrinsic point defects such as $V_o^{\bullet\bullet}$, O''_i, and $Si_i^{\bullet\bullet\bullet}$. The introduction of (OH)-defects into quartz therefore increases the concentration of donor defects such as $V_o^{\bullet\bullet}$ and $Si_i^{\bullet\bullet\bullet}$ whilst decreasing the concentration of acceptor defects such as V''''_{Si} and O''_i.

As has been pointed out, we do not know what point defects are responsible for maintaining stoichiometry and charge balance over the range of a_{O_2} where quartz is stoichiometric although the demonstrated high concentrations of oxygen vacancies make $[V_o^{\bullet\bullet}] = [O''_i]$ or $[V_o^{\bullet\bullet}] = 2[V''''_{Si}]$ prime candidates; for the latter neutrality condition increase in (OH)-defect concentration is first associated with a decrease in silicon diffusion by a vacancy mechanism before a rapidly increasing interstitial mechanism becomes dominant (see Figure 13).

If deformation of quartz occurs solely by a diffusion mechanism then the steady state creep rate, $\dot{\varepsilon}$, which is proportional to the diffusion coefficient for the rate controlling diffusing species will depend on a_{H_2O} in different ways depending on the stoichiometric condition and the magnitude of a_{H_2O}. Various dependences on a_{O_2} are also possible depending on the experimental set-up. These possibilities are summarized in Tables 4 and 5. Clearly, careful experiments should be able to delineate the stoichiometric condition, the mechanisms of diffusion and the identity of the (OH)-defect responsible for hydrolytic weakening.

Deformation Rate Controlled by the Climb of Dislocations

If climb of jogs is rate controlling then the strain rate is proportional to the product of the jog concentration and the concentration of rate controlling diffusive species [see Hobbs, 1983b]. Thus, the dependence of strain rate upon the thermodynamic activities of chemical components is found by multiplying the individual jog and point defect dependencies. For the neutrality range, $[(3H)'_{Si}] = 2[V_o^{\bullet\bullet}]$, the strain rate dependence for various jog-deformation mechanisms is tabulated in

TABLE 7. Strain Rate Dependence on a_{O2}, a_{H2O} for Two Neutrality Conditions for Deformation by the Classical Griggs [1967] mechanism

Neutrality Condition	Deformation Mechanism	a_{H_2O}	a_{O_2}
Diffusivity of (OH)-Defect Proportional to $[(3H)'_{Si}]$			
$2[V_o^{\bullet\bullet}] = [(3H)'_{Si}]$	$K^{\bullet}$, $(3H)'_{Si}$	+3/2	+1/4
	K', $(3H)'_{Si}$	+1/2	-1/4
$[H_i^{\bullet}] = 4[V_{Si}'''']$	$K^{\bullet}$, $(3H)'_{Si}$	+17/10	-2/5
	K', $(3H)'_{Si}$	+3/2	+2/5
Diffusivity of (OH)-Defect Proportional to $[V_{Si}'''']$			
$2[V_o^{\bullet\bullet}] = [(3H)'_{Si}]$	$K^{\bullet}$, $(3H)'_{Si}$	-3/2	+1/4
	K', $(3H)'_{Si}$	-5/2	-1/4
$[H_i^{\bullet}] = 4[V_{Si}'''']$	$K^{\bullet}$, $(3H)'_{Si}$	+1/2	-2/5
	K', $(3H)'_{Si}$	+3/10	+2/5

Table 8. The combinations that give strong positive dependence on a_{H2O} are: (1) positive jogs with a charged oxygen vacancy diffusion mechanism leading to a strain rate proportional to $a_{O_2}^{1/4}$, $a_{H_2O}^{3/2}$; (2) positive jogs with a charged silicon interstitial diffusion mechanism leading to a strain rate proportional to $a_{O_2}^{1/4}$, $a_{H_2O}^{5/2}$; (3) negative jogs with a charged silicon interstitial diffusion mechanism leading to a strain rate proportional to $a_{O_2}^{-1/4}$, $a_{H_2O}^{3/2}$.

Inspection of Figure 13 shows that for this neutrality range, a silicon insterstitial diffusion mechanism is expected to be rate controlling and hence (2) and (3) are favored with (2) giving the strongest variation of strain rate with a_{H2O}. Clearly, suitable experiments could lead to a decision since different oxygen dependencies are operative.

Identical conclusions to these follow if the Griggs defect rather than the $(3H)_{Si}$ is assumed to be responsible for hydrolytic weakening although the precise dependence of strain rate upon a_{O_2} and a_{H_2O} is different. The point is that the diffusion of the (OH)-defect need not be rate controlling; the influence of (OH) can be to increase the concentration of other defects responsible for deformation and hence to enhance the strain rate for a given stress.

The above discussion assumes that the changes in strain rate that accompany changes in f_{O_2} and f_{H_2O} arise solely from changes in the concentrations of rate controlling defects. The possibility that the changes in strain rate arise from changes in the mobilities of the rate controlling defects or from combinations of the concentrations and mobilities is neglected. The reason for this is that it is difficult to treat the mobilities explicitly from a defect chemistry point of view although the influence of defect chemistry upon the mobilities of defects is well known. Some discussion of this aspect of the problem in quartz is given in Hobbs [1981, 1983a]. Linker and Kirby [1981] have shown that the creep properties of (OH)-bearing quartz are strongly anisotropic, as is diffusion; such behavior can only arise from anisotropy in mobility but the precise manner in which such behavior is controlled by defect chemistry is yet to be understood. It remains to be established by experiment whether concentration or mobility terms or a combination of the two are all important in the hydrolytic weakening process.

8. Summary and Conclusions

This paper has reviewed what is known of the electronic and defect structure of pure quartz and has considered the influence that hydrogen related defects have upon its mechanical properties. The general thesis is that these hydrogen related defects can act as donors (in the form of interstitial hydrogens, H_i) or as acceptors (in the form of the classical Griggs defect, $(HOH)'_i$, or as three or four hydrogens substituted for a silicon, $(3H)'_{Si}$ or $(4H)'_{Si}$). As such, increasing the concentration of $H^{\bullet}_i$ increases the concentrations of acceptors such as V''''_{Si}, O''_i and dislocation kinks and jogs, K' and J'; the concentrations of donors such as $V^{\bullet\bullet}_o$, $Si^{\bullet\bullet\bullet\bullet}_i$, $K^{\bullet}$ and $J^{\bullet}$ are decreased. The reverse statements are true for increasing concentrations of the acceptor hydrogen-related defects. Thus, since the strain rate is related in some way to the concentrations of some of these defects, depending on the precise mechanism of deformation, changing the concentration of hydrogen related defects will change the strain rate for a given stress, temperature and pressure.

This is proposed as the origin of the hydrolytic weakening effect.

A number of possible mechanisms of deformation have been proposed in this paper:

Deformation Controlled by Migration of Kinks Alone. See Table 6. For the situations considered the dependence of strain rate upon a_{H_2O} is not strong enough to explain the observed effect.

Deformation Controlled by Kink Motion with Diffusion of (OH) to or with the Kink - The Classical Griggs Mechanism. See Table 7. For the situations considered only the $[K^{\bullet}, (3H)'_{Si}]$ mechanism gives a dependence of strain rate upon a_{H_2O} high enough to explain the effect (with an exponent of 1.5 or 1.7). This however would have to be for a situation where the diffusion of (OH) is concentration dependent.

Deformation Controlled Solely by Diffusion. See Tables 4 and 5. The magnitude of the effect on diffusion alone, which is presumably what is observed in the recrystallization situation, is readily explained but many possibilities exist depending on the stoichiometry condition in pure quartz. Notice that for many situations the $(3H)'_{Si}$ defect results in much stronger dependencies of creep rate upon a_{H_2O} than does the Griggs defect. The oxygen dependencies are also quite different.

Deformation Controlled by the Climb of Dislocations. See Table 8. For the situation considered the magnitude of the effect is consistent with $[J^{\bullet}, V^{\bullet\bullet\bullet}_o]$, $[J^{\bullet}, Si^{\bullet\bullet\bullet}_i]$ and $[J', Si'''_i]$ mechanisms giving exponents for a_{H_2O} of 1.5 or 2.5.

Clearly the various possibilities can only be checked by experiment but in doing so it is essential to note that not only a dependence on a_{H_2O} is involved but a dependence on a_{O_2} as well. This is an additional experimental difficulty but has an advantage also in that it enables distinctions between some of the mechanisms to be made.

Finally, this paper deals only with the situation where the quartz is pure except for the addition of hydrogen related defects. The situation where other dopants such as Al, Na, Li are present is considered in Hobbs [1983a].

The overall conclusion to be reached is that as yet there is not enough data to distinguish between a Griggs type mechanism where (OH)-defects diffuse with or to the dislocation and other mechanisms where diffusion of oxygen or silicon to jogs is rate controlling. Both are capable of giving the observed strong dependencies of strain rate upon a_{H_2O}. It is possible of course that both mechanisms do operate, the Griggs type mechanism at fast strain rates or low temperatures and the climb/diffusion mechanisms at slow strain rates or high temperatures.

TABLE 8. Dependence of Strain Rate Upon a_{O_2} and a_{H_2O} for Various Rate Controlling Processes

Mechanism	a_{O_2}	a_{H_2O}
$K^{\bullet}$, $J^{\bullet}$	+1/4	+1/2
$K^{\bullet}$, J'	-1/4	-1/2
$J^{\bullet}$, $V^{\bullet\bullet}_o$	+1/4	+3/2
J', $V^{\bullet\bullet}_o$	-1/4	+1/2
$J^{\bullet}$, V''''_{Si}	+1/4	-3/2
J', V''''_{Si}	-1/4	-5/2
$J^{\bullet}$, O''_i	+1/4	-1/2
J', O''_i	-1/4	-3/2
$J^{\bullet}$, $Si^{\bullet\bullet\bullet\bullet}_i$	+1/4	+5/2
J', $Si^{\bullet\bullet\bullet\bullet}_i$	-1/4	+3/2

Neutrality condition is $[(3H)'_{Si}] = 2[V^{\bullet\bullet}_o]$.

Acknowledgements. Much of this paper was written whilst the author was a visiting professor at the University of California, Davis. Thanks go to the Geology Department at U.C. Davis for the opportunity to sit and write and special thanks go to Alison Ord, Harry Green, Rob Twiss, Peter Vaughn, and Barclay Kamb for helpful discussions. The manuscript was substantially improved by comments from Steve Kirby.

References

Adler, D., The imperfect solid-transport properties, in Treatise on Solid State Chemistry, vol. 2, Defects in Solids, pp. 237-332, Plenum, New York, 1975.

Aines, R. D., and G. R. Rossman, Water in minerals? A peak in the infrared, J. Geophys. Res., in press, 1983.

Baëta, R. D., and K. H. G. Ashbee, Mechanical deformation of quartz, I, Constant strain rate

compression experiments, Philos. Mag., 22, 604-624, 1970a.
Baëta, R. D., and K. H. G. Ashbee, Mechanical deformation of quartz, II, Stress relation and thermal activation parameters, Philos. Mag., 22, 625-635, 1970b.
Balderman, M. A., The effect of strain rate and temperature on the yield point of hydrolytically weakened synthetic quartz, J. Geophys. Res., 79, 1647, 1974.
Bambauer, H. U., G. O. Brunner, and F. Laves, Light scattering of heat-treated quartz in relation to hydrogen containing defects, Am. Mineral., 54, 718-724, 1969.
Bennett, A. J., and L. M. Roth, Electronic structure of defect centres in SiO_2, J. Phys. Chem. Solids, 32, 1251-1261, 1971.
Blacic, J. D., Plastic deformation mechanisms in quartz: The effect of water, Tectonophysics, 27, 271-294, 1975.
Brouwer, G., A general asymptotic solution of reaction equations common in solid-state chemistry, Philips Res. Rep., 9, 366-376, 1954.
Burnham, C. W., J. R. Holloway, and N. F. Davis, Thermodynamic properties of water to 1000°C and 10,000 bars, Spec. Pap. Geol. Soc. Am., 132, 96 pp., 1969.
Calabrese, E., and W. B. Fowler, Electronic energy-band structure of α-quartz, Phys. Rev. B, 18, 2888-2896, 1978.
Carter, N. L., J. M. Christie, and D. T. Griggs, Experimental deformation and recrystallization of quartz, J. Geol., 72, 687-733, 1964.
Chakraborty, D., and G. Lehmann, On the structures and orientations of hydrogen defects in natural and synthetic quartz crystals, Phys. Status Solidi A, 34, 467-474, 1976.
Ciraci, S., and S. Ellialtioglu, Surface electronic structure of silicon dioxide, Phys. Rev., B, 25, 4019-4030, 1982.
Ciraci, S., S. Erkoc, Electronic energy structure of vacancy and divacancy in SiO_2, Solid State Commun., 40, 801-803, 1981.
Cohen, A. J., and F. Hassan, Ferrous and ferric ions in synthetic α-quartz and natural amethyst, Am. Mineral., 59, 719-728, 1974.
Dennis, P. F., Oxygen self diffusion in quartz under hydrothermal conditions, Is hydrogen an important defect species at low water pressures (< 100 MPa)?, J. Geophys. Res., in press, 1983.
Edgar, A. D., Experimental Petrology: Basic Principles and Techniques, Oxford at the Clarendon Press, London, 1973.
Feigl, F. J., and J. H. Anderson, Defects in crystalline quartz: Electron paramagnetic resonance of E' vacancy centres associated with germanium impurities, J. Phys. Chem. Solids, 31, 575-596, 1970.
Feigl, F. J., W. B. Fowler, and K. L. Yip, Oxygen vacancy model for the E_1' center in SiO_2, Solid State Commun., 14, 225-229, 1974.
Fowler, W. B., P. M. Schneider, and E. Calabrese, Band structures and electronic properties of SiO_2, in The Physics of SiO_2 and Its Interfaces, edited by S. T. Pantelids, pp. 70-74, Pergamon, New York, 1978.
Greaves, G. N., A model for point defects in silica, in The Physics of SiO_2 and Its Interfaces, edited by S. T. Pantelides, pp. 268-272, Pergamon, New York, 1978.
Green, H. W., D. T. Griggs, and J. M. Christie, Syntectonic and annealing recrystallization of fine-grained quartz aggregates, in Experimental and Natural Rock Deformation, edited by P. Paulitsch, Springer-Verlag, Berlin-Heidelberg, pp. 272-335 (1970).
Griggs, D. T., Hydrolytic weakening of quartz and other silicates, Geophys. J. R. Astron. Soc., 14, 19-32, 1967.
Griggs, D. T., and J. D. Blacic, The strength of quartz in the ductile regime (abstract), Eos Trans. AGU, 45, 102-103, 1964.
Griggs, D. T., and J. D. Blacic, Quartz: Anomalous weakness of synthetic crystals, Science, 147, 292-295, 1965.
Griscom, D. L., Defects and impurities in α-quartz and fused silica, in The Physics of SiO_2 and Its Interfaces, edited by S. T. Pantelides, pp. 232-252, Pergamon, New York, 1978.
Griscom, D. L., Point defects and radiation damage processes in α-quartz, Proc. Annu. Freq. Control Symp. 33rd, 98-109, 1979.
Griscom, D. L., and W. B. Fowler, Electron-transfer model for E'-center optical absorption in SiO_2, in Physics of MOS Insulators, edited by G. Lucovsky, S. T. Pantelides, and F. L. Galaenar, pp. 97-101, Pergamon, New York, 1980.
Haberlandt, H., and F. Ritschl, CNDO/2 calculations of α-quartz electronic properties of the perfect cluster and EPR parameters of the O^- vacancy, Phys. Status Solidi B, 100, 503-508, 1980.
Harrison, W. A., Electronic structure and the properties of solids, 582 pp., W. H. Freeman, San Francisco, Calif., 1980.
Hassan, F., and A. J. Cohen, Biaxial color centers in amethyst quartz, Am. Mineral., 59, 709-718, 1974.
Heine, V., Dangling bonds and dislocations in semiconductors, Phys. Rev., 146, 568-570, 1966.
Herman, F., D. J. Henderson, and R. V. Kasowski, Electronic structure of vacancies and interstitials in SiO_2, in Physics of MOS Insulators, edited by G. Lucovsky, S. T. Pantelides, and F. L. Galeenar. pp. 107-111, Pergamon, New York, 1980.
Hirsch, P. B., A mechanism for the effect of doping on dislocation mobility, J. Phys. Paris, 40, C6-117 to C6-121, 1979.
Hirsch, P. B., Plastic deformation and electronic mechanisms in semiconductors and insulators, J. Phys. Paris, 42, C3-149 to C3-159, 1981.
Hirth, J. P., and J. Lothe, Theory of Dislocations, 780 pp., McGraw-Hill, New York, 1968.
Hobbs, B. E., Recrystallization of single crystals of quartz, Tectonophysics, 6, 353-401, 1968.
Hobbs, B. E., A new interpretation of the hydrolytic weakening effect (abstract), Eos Trans. AGU, 61, 1129, 1980.

Hobbs, B. E., The influence of metamorphic environment upon the deformation of minerals, Tectonophysics, 78, 335-383, 1981.
Hobbs, B. E., Point defect chemistry of minerals under a hydrothermal environment, J. Geophys. Res., in press, 1983a.
Hobbs, B.E., Constraints on the mechanism of deformation of olivine imposed by defect chemistry, Tectonophysics, 92, 35-69, 1983b.
Hobbs, B. E., A. C. McLaren, and M. S. Paterson, Plasticity of single crystals of synthetic quartz, in Flow and Fracture of Rocks, Geophys. Monogr. Ser., vol. 16, edited by H. C. Heard, I. Y. Borg, N. I. Carter, and C. B. Raleigh, pp. 29-53, AGU, Washington, D. C., 1972.
JANAF Thermochemical Tables, Ref. Data Ser., vol. 37, 1141 pp., National Bureau of Standards, Washington, D. C., 1971.
Kats, A. Hydrogen in α-quartz, Philips Res. Rep., 17, 133-279, 1962.
Kekulawala, K. R. S. S., M. S. Paterson, and J. N. Boland, Hydrolytic weakening in quartz, Tectonophysics, 46, T1-T6, 1978.
Kekulawala, K. R. S. S., M. S. Paterson, and J. N. Boland, An experimental study of the role of water in quartz deformation, in Mechanical Behavior of Crustal Rocks: The Handin Volume, Geophys. Monogr. Ser., vol. 24, edited by N. L. Carter, M. Friedman, J. Logan, and D. Stearns, AGU, Washington, D. C., 1981.
Kirby, S. H., Creep of synthetic alpha quartz, Ph.D. thesis, 193 pp., Univ. of Calif., Los Angeles, 1975.
Kirby, S. H., Hydrogen-bonded hydroxyl in synthetic quartz: Analysis, mode of incorporation, and role in hydrolytic weakening, J. Geophys. Res., in press, 1983.
Kirby, S. H., and J. W. McCormick, Creep of hydrolytically weakened synthetic quartz crystals oriented to promote $\{2\bar{1}\bar{1}0\}$ <0001> slip: A brief summary of work to date, Bull. Mineral., 102, 124-137, 1979.
Kirby, S. H., J. W. McCormick, and M. Linker, The effect of water concentration on creep rates of hydrolytically-weakened synthetic quartz single crystals (abstract), Eos Trans. AGU, 58, 1239, 1977.
Koumvakalis, N., Defects in crystalline SiO_2: Optical absorption of the aluminum-associated hole center, J. Appl. Phys., 51, 5528-5532, 1980.
Kröger, F. A., The Chemistry of Imperfect Crystals, vol. 2, 988 pp., Elsevier, New York, 1974.
Labusch, R., and W. Schroter, Electrical properties of dislocations in semiconductors, in Dislocations in Solids, edited by F. R. N. Nabarro, pp. 127-191, North-Holland, Amsterdam, 1980.
Laughlin, R. B., J. D. Joannopoulos, and D. J. Chadi, Bulk electronic structure of SiO_2, Phys. Rev., B, 20, 5228-5237, 1979.
Lell, E., N. J. Kreidl, and J. R. Hensler, Radiation effects in quartz, silica and glasses, Prog. Ceram. Sci., 4, 1-93, 1966.
Linker, F. L., and S. H. Kirby, Anisotropy in the rheology of hydrolytically weakened synthetic quartz crystals, in Mechanical Behavior of Crustal Rocks: The Handin Volume, Geophys. Monogr. Ser., vol. 24, edited by N. L. Carter, M. Friedman, J. Logan and D. Stearns, AGU, Washington, D. C. 1981.
McLaren, A. C., J. A. Retchford, Transmission electron microscope study of the dislocation in plastically deformed synthetic quartz, Phys. Status Solidi A, 33, 657-668, 1969.
McLaren, A. C., C. F. Osborne, and L. A. Saunders, X-ray topographic study of dislocations in synthetic quartz, Phys. Status Solidi A, 4, 235-247, 1971.
McLaren, A. C., R. F. Cook, S. T. Hyde, and R. C. Tobin, The mechanisms of the formation and growth of water bubbles and associated dislocation loops in synthetic quartz, Phys. Chem. Miner., 9, 79-94, 1983.
Morin, F. J., J. R. Oliver, and R. M. Housley, Electrical properties of forsterite Mg_2SiO_4, Phys. Rev. B, 16, 4434-4445, 1977.
Morrison-Smith, D. J., M. S. Paterson, and B. E. Hobbs, An electron microscope study of plastic deformation in single crystals of synthetic quartz, Tectonophysics, 33, 43-79, 1976.
Mott, N. F., Electronic properties of vitreous silicon dioxide, in The Physics of SiO_2 and Its Interfaces, edited by S. T. Pantelides, pp. 1-13, Pergamon, New York, 1978.
Nuttall, R. H. D., and J. A. Weil, Double-hole aluminum center in α-quartz, Solid State Commun., 19, 141-142, 1976.
Nuttall, R. H. D., and J. A. Weil, Two hydrogenic trapped-hole species in α-quartz, Solid State Commun., 33, 99-102, 1980.
O'Keeffe, M., and B. G. Hyde, On Si-O-Si configurations in silicates, Acta Crystallogr. Sect. B, 34, 27-32, 1978.
O'Keeffe, M., and B. G. Hyde, The role of non bonded forces in crystals, in Structure and Bonding in Crystals, vol. 1, pp. 227-254, Academic, New York, 1981.
Pantelides, S. T., and W. A. Harrison, Electronic structure, spectra and properties of 4:2 coordinated materials, I, Crystalline and amorphous SiO_2 and Geo_2, Phys. Rev. B, 13, 2667-2791, 1976.
Paterson, M. S., The determination of hydroxyl by infrared absorption in quartz, silicate glasses and similar materials, Bull. Mineral., 105, 20-29, 1982.
Paterson, M. S., and K. R. S. S. Kekulawala, The role of water in quartz deformation, Bull. Mineral., 102, 92-98, 1979.
Read, W. T., Jr., Theory of dislocations in germanium, Philos. Mag., 45, 775-796, 1954.
Scala, C. M., and D. R. Hutton, Site assignment of Fe^{3+} in α-quartz, Phys. Status Solidi B, 73, K115-K117, 1976.
Schirmer, O. F., Small polaron aspects of defects in oxide materials, J. Phys. Paris, 41, C6-479 to C6-484, 1980.

Shaw, H. R., and D. R. Wones, Fugacity coefficients for hydrogen gas between 0° and 1000°C for pressures to 3000 atm, Am. J. Sci., 262, 981-929, 1964.

Tullis, J., J. M. Christie, and D. T. Griggs, Microstructures and preferred orientations of experimentally deformed quartzites, Geol. Soc. Am. Bull., 84, 297-314, 1973.

Van Gool, W., Principles of Defect Chemistry of Crystalline Solids, 148 pp., Academic, New York, 1966.

Walrafen, G. E., and J. P. Luongo, Raman and infrared investigations of OH^- and H_2O in hydrothermal α-quartz, Spex Speaker, 20(3), 1-6, 1975.

Weeks, R. A., Paramagnetic spectra of E_2' centers in crystalline quartz, Phys. Rev., 130, 570-576, 1963.

Weil, J. A., The aluminum centers in α-quartz, Radiat. Eff., 26, 261-265, 1975.

Wood, D. L., Infrared absorption of defects in quartz, J. Phys. Chem. Solids, 13, 326-336, 1960.

Yip, K. L., and W. B. Fowler, Electronic structure of E_1' centers in SiO_2, Phys. Rev. B, 11, 2327-2338, 1975.

Yund, R., Lattice diffusion in silicates under hydrothermal conditions, J. Geophys. Res., in press, 1983.

EXPERIMENTAL EVIDENCE FOR THE EFFECT OF CHEMICAL ENVIRONMENT UPON THE CREEP RATE OF OLIVINE

D. L. Ricoult and D. L. Kohlstedt

Cornell University, Department of Materials Science and Engineering
Ithaca, New York 14853

Abstract. Natural iron-bearing olivine crystals have been deformed at one atmosphere under controlled partial pressures of oxygen and activities of the constituent oxide phases. The creep rate varies as $p_{O_2}^{1/6}$ for olivine crystals not buffered by an oxide phase and as $p_{O_2}^{0}$ for crystals buffered by an oxide phase. At a given stress, temperature and oxygen partial pressure, the samples buffered with $(Mg,Fe)SiO_3$ deform ten times faster than those buffered with $(Mg,Fe)O$; that is, the strain rate increases as $a_{en}^{1.2}$. A point defect model consistent with the experimental results is presented here. It is suggested that the creep rate of iron-bearing olivine is controlled by the nucleation and displacement of positively charged jogs along the dislocation lines. In turn, the motion of the jogs is limited by the rate of self-diffusion of oxygen by a vacancy mechanism. Similarly, forsterite single crystals, annealed in either MgO powder or $MgSiO_3$ powder, have been deformed at constant stress. The recorded strain rate is independent of both the oxygen partial pressure and the activity of the constituent oxides. These results for forsterite are also consistent with a model in which the controlling creep mechanism is the motion of positively charged jogs along dislocations, coupled with oxygen diffusion via a vacancy mechanism.

1. Introduction

The creep behavior of natural polycrystals of olivine [Raleigh, 1968; Carter and AveLallemant, 1970; Blacic, 1972; Post, 1977; Zeuch and Green, 1979; Chopra and Paterson, 1981] as well as natural single crystals of olivine [Kohlstedt and Goetze, 1974; Kohlstedt et al., 1976; Durham and Goetze, 1977; Kohlstedt and Hornack, 1981] and of synthetic forsterite [Durham et al., 1979; Durham and Goetze, 1977; Darot and Gueguen, 1981] has been extensively investigated to determine high-temperature flow laws which might be used to model convective flow in the earth's upper mantle. Yet, major discrepancies still exist between the results obtained by the various research groups. Because of the substantial differences in the apparatuses used (one-atmosphere, gas-medium, solid-medium rigs), one major difference from one laboratory to the next is the chemical environment to which the specimens were exposed. The present paper systematically explores the effects of chemical environments and associated point defect concentrations on the deformation behavior of olivine.

The Gibbs phase rule states that four independent state variables must be fixed to thermodynamically define a ternary system such as forsterite, Mg_2SiO_4. These state variables can be pressure, temperature, partial pressure of oxygen (p_{O_2}) and activity of one of the constituent oxides (a_{ox}). In a quaternary system such as iron-bearing olivine, $(Mg,Fe)_2SiO_4$, five parameters must be specified; experimentally, the fifth parameter is the Fe/Mg ratio. Most investigations have examined the effects of temperature and/or pressure on the creep rate of olivine. In addition, Hornack [1978], Jaoul et al. [1980], and Kohlstedt and Hornack [1981] examined the effects of p_{O_2}; by changing the partial pressure of oxygen as they deformed olivine single crystals, these authors established that the creep rate varies as $p_{O_2}^{1/6}$. Their experiments were carried out with the cationic, (Mg+Fe)/Si, ratio fixed. In no instance was the activity of one of the constituent oxides fixed.

For forsterite, if we assume that the Mg/Si ratio can depart slightly from the stoichiometric value of 2, a solid solubility field can be envisioned. The compositional extent of the solid solution field may be very small [Schmalzried, 1981]. The electrical conductivity measurements of Pluschkell and Engell [1968] indicate a solubility of one mole percent of silica in forsterite at 1400°C and a lesser solubility for magnesium oxide. The oxide activity can be fixed by buffering forsterite with either MgO or $MgSiO_3$. When the crystal is not buffered by an oxide, the Mg/Si ratio is fixed by the composition of the sample itself and is assumed to be constant. For this assumption to hold, there must be no preferential vaporization of either Mg or Si [Smyth, 1976]. In this case, the activity of MgO changes with temperature, pressure and partial pressure of oxygen. (Abelard and Baumard [1982] have devised a graphical representation for describing the variation of the activity of MgO as a function of p_{O_2} when the cationic ratio is constant.)

In the course of this study, we have deformed forsterite crystals buffered either with MgO or $MgSiO_3$ and olivine crystals buffered either with (Mg,Fe)O or $(Mg,Fe)SiO_3$ to determine the dependence of the creep rate on the activity of one of the constituent oxide phases, as well as on the partial pressure of oxygen. If diffusion plays a role in the creep process, the strain rate, $\dot{\varepsilon}$, is proportional to a diffusion coefficient which in turn is proportional to the concentration of the slowest diffusing defect species. The defect concentrations are functions of the thermodynamic parameters defined earlier (in particular the partial pressure or oxygen and the activity of one constituent oxide). Consequently, the strain rate can be expressed as [Kohlstedt and Ricoult, 1984]

$$\dot{\varepsilon} = A\, \sigma^{n}\, p_{O_2}^{m}\, a_{ox}^{p} \exp -(Q/RT) \qquad (1)$$

where A is a constant, σ is the applied stress, a_{ox} is the activity of one oxide phase, Q is the activation energy of the controlling mechanism, R is the gas constant and T is the absolute temperature. It is the main emphasis of the present investigation to measure m and p, that is, to examine the role of the partial pressure of oxygen and of the activity of the constituent oxides on the creep rate. The long range goal of such experiments is to determine the controlling mechanism for creep, by defining the point defect species, the diffusion of which rate limits the creep process.

The point defect chemistry of olivine has been the subject of several studies [Smyth and Stocker, 1975; Stocker, 1978a, b; Stocker and Smyth, 1978; Hobbs, 1983; Nakamura and Schmalzried, 1983, 1984]. Stocker [1978b] treated the case of a fixed cationic ratio (i.e., unbuffered samples). Likewise, Stocker and Smyth [1978] calculated the effect of enstatite activity as well as oxygen partial pressure on the point defect chemistry of olivine for various charge neutrality conditions. We will largely use the calculations of these authors to interpret our experimental results. The Kroger-Vink notation [Kroger, 1964] for point defects will be used throughout this work.

2. Experimental Procedures

2.1. Samples

Both forsterite and iron-bearing olivine samples were used. The forsterite single crystals were grown from the melt by Union Carbide. A mass spectrometric analysis on one of our forsterite single crystals yielded a Mg/Si ratio of 2.02. Similarly, Takei and Kobayashi [1974] reported that forsterite crystals grown by the Czochralski method always contained excess MgO, even when the melt they were grown from was SiO_2-rich. Other experimental evidence leads to the conclusion that synthetic forsterite generally contains excess MgO: Relandeau [1981] reports the presence of MgO precipitates in the synthetic polycrystalline forsterite he used for sintering and creep experiments; Kohlstedt and Ricoult [1984] observed MgO precipitates in forsterite single crystals grown by the floating-zone method. However, transmission electron microscopy revealed no precipitates in our starting material at magnifications up to 150,000X. Less than 80 weight-ppm iron was detected by optical mass spectroscopy and electron microprobe analysis. These two techniques revealed no aluminum, although this element has been reported in the literature as being a common impurity [Hobbs, 1983]. The detection limit for aluminum was 100 ppm by optical spectroscopy and 60 ppm by microprobe analysis. A neutron activation energy analysis revealed a small amount of iridium which was introduced from the iridium crucibles used to grow the crystals.

The single crystals of iron-bearing olivine used for this study came from San Carlos, Arizona. The composition of the samples was $(Fe_{0.10}Mg_{0.90})_2SiO_4$. The following impurities were detected by wet chemical analysis: 0.1-0.3% Ni, 0.03-0.1% Mn and less than 0.01% Al, Co, Cr and V. For completeness, it should be mentioned that small amounts of carbon were analyzed by secondary ion mass spectrometry (SIMS) in both our forsterite and our olivine samples. Using x-ray photoelectron spectroscopy, Oberheuser et al. [1983] also measured 60-180 wt.ppm carbon in olivine single crystals from the same locality.

All samples were oriented with a real-time back-reflection Laue camera [Bilderback, 1979]. The estimated accuracy of the final orientation of the samples, after X-ray orientation and cutting, is ±2°. The crystals were cut into parallelepipeds of approximate dimensions 2.5 mm x 2.5 mm x 5 mm. The long dimension was cut parallel to $[110]_c$, $[101]_c$, or $[011]_c$. As an example, $[101]_c$ refers to a direction 45° to [100] and [001], by analogy to the indices of a similar direction in a crystal with cubic symmetry. The two end-faces were carefully polished with 1 μm alumina powder. The samples were then ultrasonically cleaned in water for two minutes.

2.2. Buffering

Some of the as-prepared forsterite samples were annealed in $MgSiO_3$ powder at 1793°K for 120 hours. An appropriate mixing ratio of CO and CO_2 gases maintained a p_{O_2} of 2×10^{-5} Pa at this temperature. Embedded in the buffering powder inside a molybdenum boat, the samples were annealed in the horizontal alumina tube of an annealing furnace. The enstatite powder was prepared following the gelling method described by Hamilton and Henderson [1968]. A Debye-Scherrer X-ray pattern of the powder confirmed the presence of enstatite together with some forsterite. Because chemical analyses revealed excess MgO in the crystals, only two samples were annealed in MgO powder at 1773°K for 120 hours.

Similarly, olivine crystals were annealed in either MgO or natural enstatite, $(Mg_{0.9} Fe_{0.1}) SiO_3$, powder. The annealing temperature was 1873°K in the former case and, to avoid eutectic melting, 1523°K in the latter case. In both cases, the CO/CO_2 mixing ratio was 0.1 and the annealing time was 100 hours. In this initial set of experiments, the MgO was free of iron. More recently, we have annealed San Carlos crystals in (Mg,Fe)O and $(Mg,Fe)SiO_3$ powders prepared by the gelling method referred to above. For an olivine of composition $(Mg_{0.90}Fe_{0.10})_2 SiO_4$, to maintain the Mg/Fe ratio in the olivine, the appropriate buffers are $(Mg_{0.90}Fe_{0.10})SiO_3$ [Medaris, 1969] and $(Mg_{0.70}Fe_{0.30})O$ [Nafziger and Muan, 1967], assuming that the partitioning of Fe and Mg between coexisting olivine and magnesiowustite is not a strong function of temperature. One series of olivine crystals was annealed in magnesiowustite at 1673°K for 100 hours with a p_{O_2} of 8.1×10^{-7} Pa. Another series of crystals was annealed in enstatite powder at 1618°K for 100 hours in a p_{O_2} of 8.1×10^{-7} Pa. It should be noted that buffering $(Mg_{0.90}Fe_{0.10})_2SiO_4$ with $(Mg_{0.70}Fe_{0.30})O$ powder is not strictly exact, because the partitioning of Fe and Mg between the two coexisting phases is a function of oxygen partial pressure. However, it was not practically possible to vary the buffer composition with each change in oxygen partial pressure; thus oxide powders of a single composition were used throughout this study.

2.3. Deformation

So that small amounts of buffer powders would be present during the deformation, no attempt was made to clean the side-faces of the samples after annealing. The heating was ensured by a free-standing tungsten foil furnace supported by two graphite end-pieces each supported by a molybdenum leg. The samples were deformed in a dead-load apparatus, between carefully polished thoriated-tungsten pistons. The load was applied by stacking lead weights onto a pan fastened to the top piston. Under the assumption that the volume of the sample remained constant during the deformation run, lead shot was added after every percent of recorded strain to keep the nominal stress constant. The frictional forces induced by an O-ring seal set around the top piston were negligible compared to the magnitude of the weights used. The sample shortening was measured by two direct current displacement transducers (DCDT) whose summed output was fed into a chart recorder.

$CO-CO_2$ gas mixtures were flowed through the specimen chamber to control the partial pressure of oxygen; the gas mixing ratio was monitored with flowmeters calibrated with a zirconia oxygen sensor. The total gas flow was maintained constant for any p_{O_2} condition. The presence of tungsten and molybdenum in the system required that the CO/CO_2 gas ratio be kept at values larger than 0.25. Each p_{O_2} condition was held constant for at least one hour to allow the sample to equilibrate with its environment. At least two identically buffered samples were deformed for each orientation of the stress.

3. Results

3.1. Olivine

The experimental conditions of the runs conducted on olivine single-crystals are listed in Tables 1 and 2. For the $[101]_c$ orientation, the recorded strain rates were all normalized to a stress of 35 MPa. The accuracy of the strain rate data is estimated to be better than 5%. The results obtained on olivine crystals buffered with MgO and natural $(Mg,Fe)SiO_3$ powders are reported in Figure 1. The samples buffered against $(Mg,Fe)SiO_3$ deform one order of magnitude faster than those buffered against MgO. The average activation energy is 523±27 KJ/mole for those buffered against (Mg,Fe)O, as reported in Table 3.

Two $[110]_c$ crystals annealed and buffered in $(Mg_{0.90}Fe_{0.10})SiO_3$ powder and two [110] crystals annealed and buffered in $(Mg_{0.70}Fe_{0.30})O$ powder were deformed. The raw experimental data are reported in Table 2 and the strain rate-p_{O_2} results are plotted in Figure 2. Again, at a given stress and temperature, the strain rate is higher for samples buffered in enstatite powder than for those buffered in magnesiowustite. The Gibbs free energy of formation of iron-bearing olivine from mixed oxides is -25.7 KJ/mole [Hobbs, 1983]. A one order of magnitude difference in strain rate between the two sets of buffered samples yields a value for the exponent of the activity of enstatite in Eq. 1 of p = 1.2.

To obtain information about the time required for the samples to equilibrate with the powdered buffers (i.e., the defect relaxation time), one olivine crystal was given a special treatment. In a first step, it was annealed in $(Mg,Fe)SiO_3$ powder. Subsequently, the crystal was deformed while surrounded with (Mg,Fe)O powder. A steady-state strain rate was reached not more than forty minutes after the furnace reached the run temperature, 1675°K. The creep rate remained constant for three hours (1.7% strain in the steady-state regime). This value of strain rate is reported as a star symbol in Figure 2; it is in excellent agreement with the creep rate measured for samples both annealed and buffered in (Mg,Fe)O powder prior to the deformation.

The partial pressure of oxygen was varied over three orders of magnitude during the deformation of some buffered samples. No variation of the

TABLE 1. Summary of Runs Performed on Iron-Bearing Olivine: $[101]_c$ Orientation

Experiment Number	Buffer	Temperature, °C	Stress, MPa	P_{O_2}, atm	Strain Rate, s^{-1}	Normalized Strain Rate*
1	MgO	1600	30	3×10^{-9}	9.8×10^{-6}	1.7×10^{-5}
		1500	30	6×10^{-10}	1.9×10^{-6}	3.4×10^{-6}
2	MgO	1600	37	3×10^{-9}	8.9×10^{-6}	7.2×10^{-6}
			44	3×10^{-9}	2.6×10^{-5}	1.1×10^{-5}
			49	3×10^{-9}	4.2×10^{-5}	1.2×10^{-5}
		1500	49	6×10^{-10}	2.6×10^{-6}	7.5×10^{-7}
			54	6×10^{-10}	6.4×10^{-6}	1.3×10^{-6}
3	$(Mg,Fe)SiO_3$	1600	37	3×10^{-9}	1.4×10^{-4}	1.1×10^{-4}
			43	3×10^{-9}	4.4×10^{-4}	2.1×10^{-4}
		1500	43	6×10^{-10}	6.6×10^{-5}	3.1×10^{-5}
			51	6×10^{-10}	1.8×10^{-4}	4.5×10^{-5}
		1400	51	8×10^{-11}	1.1×10^{-5}	2.7×10^{-6}
			56	8×10^{-11}	1.8×10^{-5}	3.2×10^{-6}
4	$(Mg,Fe)SiO_3$	1600	35	1×10^{-10}	2.5×10^{-4}	2.5×10^{-4}
		1549	35	6×10^{-11}	1.2×10^{-4}	1.2×10^{-4}
		1496	35	2×10^{-11}	3.9×10^{-5}	3.9×10^{-5}
		1451	35	3×10^{-12}	4.6×10^{-6}	4.6×10^{-6}
		1399	35	3×10^{-12}	4.6×10^{-6}	4.6×10^{-6}
		1500	35	2×10^{-11}	4.1×10^{-5}	4.1×10^{-5}

*The strain rates were all normalized to a stress of 35 MPa. We used a stress exponent $n = 3.7$ [Durham and Goetze, 1977].

TABLE 2. Summary of Runs Performed on Iron-Bearing Olivine: $[110]_c$ Orientation

Experiment Number	Buffer	Temperature, °C	Stress, MPa	P_{O_2}, atm	Strain Rate, s^{-1}
1	$(Mg,Fe)O$	1404	90	3.0×10^{-12}	8.8×10^{-7}
		1404	90	2.0×10^{-10}	9.9×10^{-7}
		1404	90	3.0×10^{-12}	9.9×10^{-7}
		1404	90	1.0×10^{-9}	1.0×10^{-6}
2	$(Mg,Fe)O$	1400	90	3.0×10^{-12}	8.5×10^{-7}
		1400	90	1.0×10^{-9}	7.6×10^{-7}
		1400	90	3.0×10^{-12}	8.1×10^{-7}
3	$(Mg,Fe)SiO_3$	1400	90	3.0×10^{-12}	1.2×10^{-5}
		1400	90	1.0×10^{-10}	1.4×10^{-5}
		1400	90	1.0×10^{-9}	8.5×10^{-6}
4	$(Mg,Fe)SiO_3$	1400	90	3.0×10^{-12}	1.1×10^{-5}

The stress was applied along $[110]_c$.

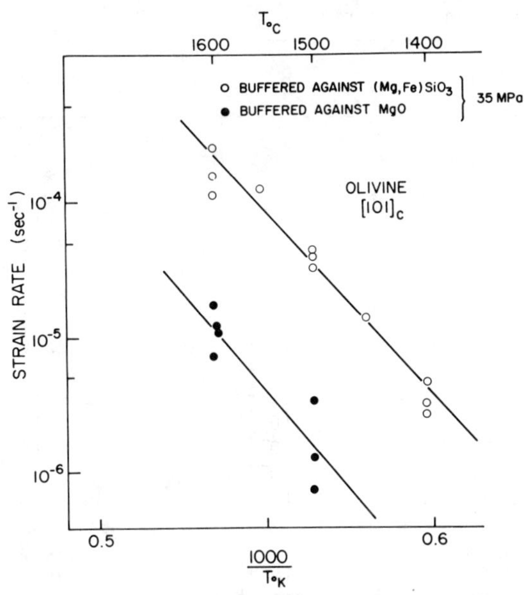

Fig. 1. Strain rate versus inverse temperature for creep of iron bearing olivine single crystals buffered against either MgO or $MgSiO_3$ powders. The stress was applied along $[101]_c$.

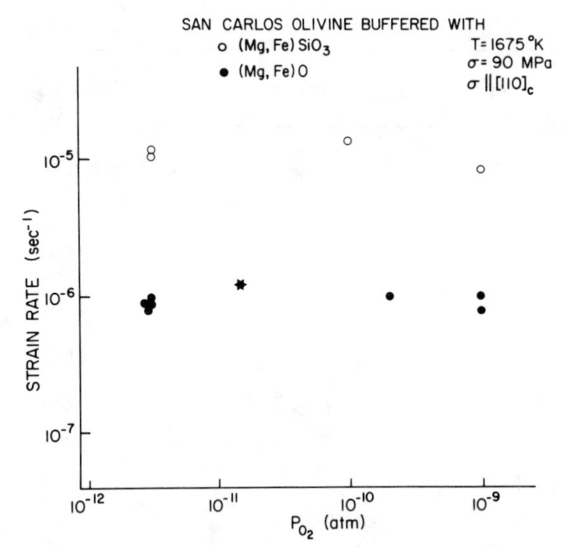

Fig. 2. Strain rate versus partial pressure of oxygen for iron-bearing olivine single crystals buffered against either (Mg,Fe)O or $(Mg,Fe)SiO_3$ powders. The stress was applied along [110] . The star symbol corresponds to a sample first annealed in $(Mg,Fe)SiO_3$ powder and then deformed in (Mg,Fe)O powder.

strain rate with oxygen partial pressure was recorded, in contrast to the $p_{O2}^{1/6}$ dependence observed for unbuffered samples [Hornack, 1978; Jaoul et al., 1980; Kohlstedt and Hornack, 1981]. For an exponent of 1/6, the strain rate changes by a factor of 3.2 when the p_{O2} is changed by three orders of magnitude. Such a change is easily detectable in our creep apparatus. Indeed, we recorded a $p_{O2}^{1/6}$ dependence of the strain rate for unbuffered samples in the present study. After the p_{O2} was changed, a new steady-state strain rate was always reached in less than 30 minutes.

TABLE 3. Determination of the Activation Energy for Creep of Forsterite and Iron-Bearing Olivine

Material	Orientation	Buffer	Stress, MPa	Number of Data Points	Q, KJ/mole
Forsterite	$[101]_c$	none	40	10	627
	$[101]_c$	$MgSiO_3$	40	18	572
	$[101]_c$	all	40	28	585
	$[110]_c$	none	30	11	694
	$[110]_c$	$MgSiO_3$	30	8	782
	$[110]_c$	all	30	19	
	$[011]_c$	none	70	4	723
	$[011]_c$	$MgSiO_3$	70	3	568
	$[011]_c$	all	70	7	644
	$[011]_c$	self+MgO	90	8	815
	$[011]_c$	$MgSiO_3$	90	10	769
	$[011]_c$	all	90	18	786
Olivine	$[101]_c$	MgO	35	7	560
	$[101]_c$	$(Mg,Fe)SiO_3$	35	12	523

TABLE 4. Summary of Runs Performed on Forsterite

Experiment Number	Buffer	Temperature °C	Stress, MPa	p_{O_2}, atm	Strain Rate, s^{-1}
		$[110]_c$ Orientation			
1	none	1500	30	$<10^{-14}$	3.5×10^{-5}
		1603	30	$<10^{-14}$	3.9×10^{-4}
		1500	30	$<10^{-14}$	1.2×10^{-5}
		1400	30	$<10^{-14}$	1.1×10^{-6}
		1500	30	$<10^{-14}$	1.5×10^{-5}
		1600	30	$<10^{-14}$	3.2×10^{-4}
2	none	1401	30	$<10^{-14}$	2.4×10^{-6}
		1449	30	$<10^{-14}$	1.0×10^{-5}
		1500	30	$<10^{-14}$	3.7×10^{-5}
		1549	30	$<10^{-14}$	7.2×10^{-5}
		1603	30	$<10^{-14}$	4.2×10^{-4}
3	$MgSiO_3$	1600	30	$<10^{-14}$	4.9×10^{-4}
		1500	30	$<10^{-14}$	1.9×10^{-4}
		1600	30	$<10^{-14}$	5.6×10^{-4}
		1653	30	$<10^{-14}$	2.2×10^{-3}
4	$MgSiO_3$	1452	30	$<10^{-14}$	3.3×10^{-6}
		1550	30	$<10^{-14}$	3.8×10^{-5}
		1500	30	$<10^{-14}$	1.2×10^{-5}
		1600	30	$<10^{-14}$	1.7×10^{-4}
		$[101]_c$ Orientation			
1	none	1450	40	1.1×10^{-11}	1.6×10^{-6}
		1500	40	3.5×10^{-11}	7.5×10^{-6}
		1604	40	2.5×10^{-10}	1.7×10^{-4}
		1497	40	3.5×10^{-11}	1.2×10^{-5}
		1451	40	1.1×10^{-11}	4.5×10^{-6}
		1600	40	2.5×10^{-10}	1.6×10^{-4}
2	none	1597	40	2.5×10^{-10}	1.4×10^{-4}
		1498	40	3.5×10^{-11}	1.8×10^{-5}
		1596	40	2.5×10^{-10}	9.0×10^{-5}
		1695	40	1.6×10^{-9}	4.0×10^{-4}
3	$MgSiO_3$	1500	40	3.5×10^{-11}	2.2×10^{-6}
		1600	40	2.5×10^{-10}	3.2×10^{-5}
		1404	40	4.0×10^{-12}	3.3×10^{-7}
		1500	40	3.5×10^{-11}	5.2×10^{-6}
		1600	40	2.5×10^{-10}	3.1×10^{-4}
		1700	40	1.6×10^{-9}	3.1×10^{-4}
		1600	40	2.5×10^{-10}	4.6×10^{-5}
		1500	40	3.5×10^{-11}	6.3×10^{-6}
4	$MgSiO_3$	1600	40	$<10^{-14}$	7.7×10^{-5}
		1500	40	$<10^{-14}$	1.8×10^{-5}
		1600	40	$<10^{-14}$	1.7×10^{-4}
		1500	40	$<10^{-14}$	2.8×10^{-5}
		1402	40	$<10^{-14}$	2.0×10^{-6}
5	$MgSiO_3$	1555	40	1.1×10^{-10}	1.4×10^{-5}
		1601	40	2.5×10^{-10}	5.8×10^{-5}
		1703	40	1.6×10^{-9}	3.5×10^{-4}

TABLE 4. (continued)

Experiment Number	Buffer	Temperature °C	Stress, MPa	P_{O_2}, atm	Strain Rate, s^{-1}
		1601	40	$2.5x10^{-10}$	$5.8x10^{-5}$
		1503	40	$3.5x10^{-11}$	$2.6x10^{-6}$
		$[011]_c$ Orientation			
1	none	1650	70	$<10^{-14}$	$4.3x10^{-5}$
		1550	70	$<10^{-14}$	$2.8x10^{-6}$
		1450	70	$<10^{-14}$	$2.4x10^{-7}$
		1600	70	$<10^{-14}$	$1.4x10^{-5}$
2	none	1599	90	$<10^{-14}$	$5.5x10^{-5}$
		1548	90	$<10^{-14}$	$1.0x10^{-5}$
		1500	90	$<10^{-14}$	$3.0x10^{-6}$
		1600	90	$<10^{-14}$	$5.4x10^{-5}$
		1651	90	$<10^{-14}$	$2.7x10^{-4}$
3	$MgSiO_3$	1547	70	$<10^{-14}$	$1.8x10^{-6}$
		1600	70	$<10^{-14}$	$4.1x10^{-6}$
		1648	70	$<10^{-14}$	$1.3x10^{-5}$
4	$MgSiO_3$	1599	90	$<10^{-14}$	$6.4x10^{-5}$
		1501	90	$<10^{-14}$	$3.5x10^{-6}$
		1549	90	$<10^{-14}$	$1.6x10^{-5}$
		1600	90	$<10^{-14}$	$9.8x10^{-5}$
		1650	90	$<10^{-14}$	$4.2x10^{-4}$
5	MgO	1500	90	$2.2x10^{-10}$	$2.5x10^{-6}$
6	MgO	1500	90	10^{-12}	$4.0x10^{-6}$
		1500	90	10^{-9}	$3.8x10^{-6}$
7	$MgSiO_3$	1598	90	$<10^{-14}$	$2.2x10^{-5}$
		1648	90	$<10^{-14}$	$1.0x10^{-4}$
		1599	90	$<10^{-14}$	$3.3x10^{-5}$
		1549	90	$<10^{-14}$	$1.1x10^{-5}$
		1499	90	$<10^{-14}$	$3.5x10^{-6}$
8	$MgSiO_3$	1550	100	10^{-9}	$3.0x10^{-5}$
		1550	100	10^{-12}	$2.8x10^{-5}$
		1550	100	10^{-9}	$2.7x10^{-5}$

3.2. Forsterite

The experimental parameters of the runs on forsterite are reported in Table 4, along with the strain rate values computed from the displacement versus time recordings. The results for forsterite are also presented as strain rate versus temperature plots in Figures 3a, 3b, and 3c for three orientations of the applied stress with respect to the crystal: $[110]_c$, $[101]_c$ and $[011]_c$, respectively. The scatter of the data points is large for the middle orientation. Darot [1980] observed a similar behavior which he attributed to interactions between dislocations of the two slip systems which are simultaneously activated when the stress is applied along $[101]_c$. However, Durham et al. [this volume] have shown that only minor microstructural interactions occur between slip systems in olivine. Possibly the scatter from one run to the next arises due to different amounts of strain produced on one system versus the other.

Linear regressions were performed on the various sets of data points for each of the three orientations. The best linear fits obtained are also shown in Figures 3a, 3b, and 3c. In one case only, the strain rate differs between the two buffered cases (σ = 70MPa, $[011]_c$ orientation). However, this result relies on only a few data points and should be interpreted with cau-

TABLE 5. Summary of p_{O_2} and a_{ox} Dependences of the Creep Rate of Variously Buffered Forsterite and Iron-Bearing Olivine

Sample	Buffer	p_{O_2} Dependence	$\dot{\varepsilon}_{mw}/\dot{\varepsilon}_{en}$	a_{en} Dependence
Forsterite	unbuffered	0		
	MgO	0	1	0
	$MgSiO_3$	0		
Iron-olivine	unbuffered	1/6		
	(Mg,Fe)O	0	≅0.1	1.2
	$(Mg,Fe)SiO_3$	0		

tion. In all other instances, the strain rate is independent of the buffering condition. Two samples annealed and buffered in MgO powder were deformed; their creep rates are indicated by arrows in Figure 3b. These data points fall in the same range as the other points. The activation energy corresponding to each linear fit is presented in Table 3.

A few experiments were run to investigate the effect of changing the partial pressure of oxygen. The p_{O_2} was varied by up to five orders of magnitude. No change in strain rate was recorded for any sample, whether or not it was buffered. The dependence on partial pressure of oxygen and enstatite activity of the strain rate of both forsterite and iron-bearing olivine are summarized in Table 5.

4. Point Defect Model

4.1. Olivine

Our experimental creep results for natural olivine, as summarized in Table 5, cannot be explained by a diffusion-controlled dislocation mechanism involving either a single silicon or a single oxygen defect. As proposed by Gueguen [1979] and later by Hobbs [1983] to account for the discrepancy between the activation energy for creep, Q_C, and that for self-diffusion, Q_D, in forsterite, we have considered the possibility that the formation and motion of jogs or kinks play a major role in determining the velocity of the dislocations.

For dislocation creep, the strain rate can be expressed as a function of the dislocation velocity, v, through the Orowan equation,

$$\dot{\varepsilon} = \rho b v \tag{2}$$

where ρ is the free dislocation density. If the creep rate is controlled by climb of dislocations, then,

$$v = m_J c_J \tag{3}$$

where m_J and c_J are the jog mobility and jog concentration, respectively. Both m_J and c_J increase exponentially with increasing temperature; thus,

$$v = C \exp -[(Q_D + Q_F) / RT] \tag{4}$$

C is a constant and Q_D and Q_F are the activation energies for the diffusion of the rate-controlling species and for the formation of a jog, respectively. From Eqs. 2 and 4,

$$\dot{\varepsilon} \propto \exp -[(Q_D + Q_F) / RT] \tag{5}$$

Consequently, for this model $Q_C > Q_D$. Hobbs [1983] extended the point defect calculations of Stocker and Smyth to include jogs by considering the reactions $J^x = J^{\cdot} + e'$ and $J^x = J' + h^{\cdot}$. At a given temperature, the concentration of neutral jogs is assumed to be constant, so that

$$[J^{\cdot}] \propto [e']^{-1} \tag{6}$$

$$[J'] \propto [h^{\cdot}]^{-1} \tag{7}$$

From Eqs. 2 and 3, the dependence of the strain rate on p_{O_2} or a_{ox} is the product of the individual dependences of the concentration of charged jogs and the concentration of the rate-controlling defect species.

Similarly, the ionization of kinks can be written as $K^x = K^{\cdot} + e'$ and $K^x = K' + h^{\cdot}$. At a given temperature,

$$[K^{\cdot}] \propto [e']^{-1} \tag{8}$$

$$[K'] \propto [h^{\cdot}]^{-1} \tag{9}$$

Since the concentrations of electrons and holes are functions of the oxygen partial pressure and the oxide activity, so are the concentrations of charged kinks. Consequently, if glide of dislocations via the motion of charged kinks is rate-limiting for creep, then the creep rate is expected to vary with p_{O_2} and a_{ox} in the same manner.

From our experiments, the creep rate of buffered olivine is such that

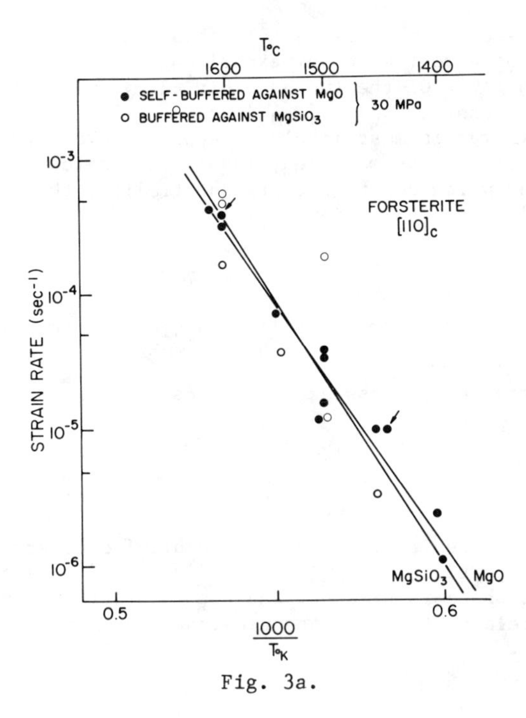

Fig. 3a.

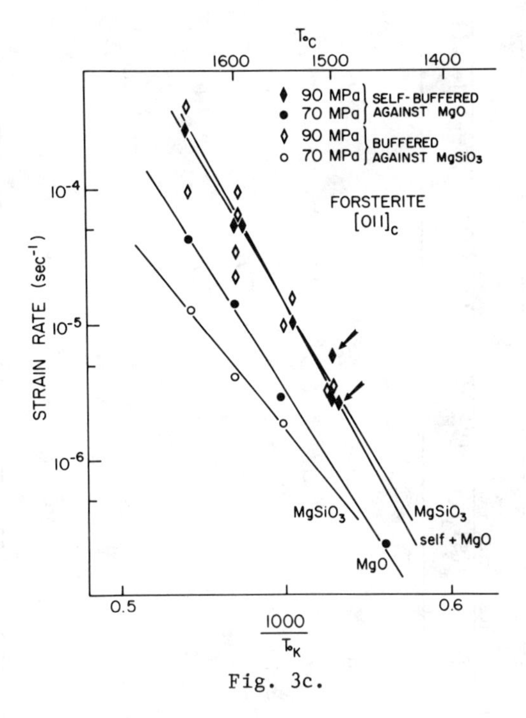

Fig. 3c.

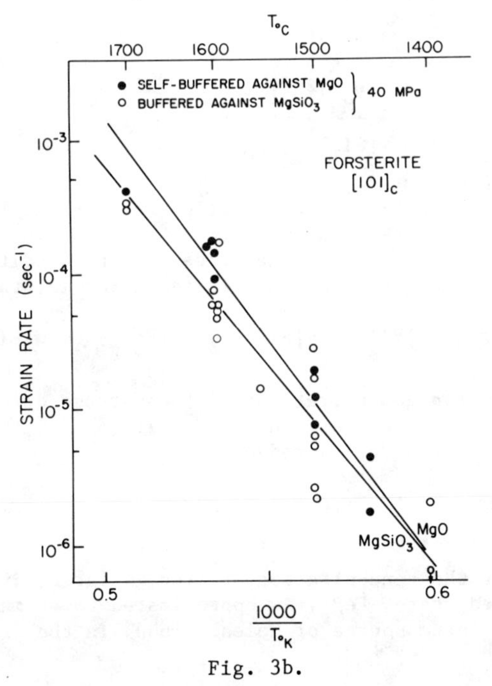

Fig. 3b.

Fig. 3. Strain rate versus inverse temperature for creep of forsterite single crystals. The stress was applied along (a) $[110]_c$, (b) $[101]_c$ and (c) $[011]_c$. In all cases, the buffering temperature was 1793°K.

$$\dot{\varepsilon} \propto p_{O_2}^{0}\, a_{en}^{1.2} \tag{10}$$

In Table 6, we have listed the dependences of the concentrations of kinks and of jog-point defect pairs on p_{O_2} for buffered olivine crystals using various charge neutrality conditions. The strain rate is independent of p_{O_2} in only three instances: (1) $K^{\bullet}$ and K' for the neutrality condition $[Fe^{\bullet}_{Mg}] = [e']$; (2) $(J^{\bullet}, V^{\bullet\bullet}_{O})$ for either $2[V''_{Mg}] = [Fe^{\bullet}_{Mg}]$ or $2[Mg^{\bullet\bullet}_{I}] = [e']$; (3) (J', O''_{I}) for either $2[V''_{Mg}] = [Fe^{\bullet}_{Mg}]$ or $2[Mg^{\bullet\bullet}_{I}] = [e']$. Oxygen interstitials are unfavorable defects because of their relatively large size. The recent findings of M. Poumellec and O. Jaoul (private communication, 1983) strongly support this assumption. These authors found that the self-diffusion coefficient of oxygen in iron-bearing olivine varies as $p_{O_2}^{1/6}$; this result rules out an interstitial mechanism. Therefore, the jog-interstitial defect pair is an unlikely candidate to explain our creep results. If creep is controlled by a kink mechanism, the steady-state strain rate should be independent of the activity of enstatite [Stocker and Smyth, 1978, Table III]; this conclusion contradicts our experimental results. Therefore, we favor positive jogs with oxygen vacancies as being the controlling defect combination for the creep of buffered olivine. For this defect pair, point defect calculations predict that the creep

TABLE 6. Dependence of Deformation Mechanisms on P_{O_2} in Buffered Iron-Bearing Olivine for Several Charge Neutrality Conditions

Jog/Kink	Diffusion Mechanism	Neutrality Conditions						
		$[V_{Mg}^{''}]=[Mg_I^{\bullet\bullet}]$	$2[V_{Mg}^{''}]=[Fe_{Mg}^{\bullet}]$	$[Fe_I^{\bullet\bullet}]=[V_{Mg}^{''}]$	$[Fe_{Mg}^{\bullet}]=[e']$	$2[Si_I^{\bullet\bullet\bullet\bullet}]=[V_{Mg}^{''}]$	$[V_O^{\bullet\bullet}]=[V_{Mg}^{''}]$	$2[Mg_I^{\bullet\bullet}]=[e']$
$K^{\bullet}$		1/4	1/6	1/4	0	1/4	1/4	1/6
$K^{\bullet}$		-1/4	-1/6	-1/4	0	-1/4	-1/4	-1/6
$J^{\bullet}$	$V_O^{\bullet\bullet}$	1/4	0	1/4	-1/2	1/4	1/4	0
$J^{\bullet}$	$V_O^{\bullet\bullet}$	-1/4	-1/3	-1/4	-1/2	-1/4	-1/4	-1/3
$J^{\bullet}$	$O_I^{''}$	1/4	1/3	1/4	1/2	1/4	1/4	1/3
$J^{\bullet}$	$O_I^{''}$	-1/4	0	-1/4	1/2	-1/4	-1/4	0
$J^{\bullet}$	$V_{Si}^{''''}$	1/4	1/2	1/4	1	1/4	1/4	1/2
$J^{\bullet}$	$V_{Si}^{''''}$	-1/4	1/6	-1/4	1	-1/4	-1/4	1/6
$J^{\bullet}$	$Si_I^{\bullet\bullet\bullet\bullet}$	1/4	-1/6	1/4	-1	1/4	1/4	-1/6
$J^{\bullet}$	$Si_I^{\bullet\bullet\bullet\bullet}$	-1/4	-1/2	-1/4	-1	-1/4	-1/4	-1/2

rate should increase as a_{en}^{+1}; this result is in good agreement with our experimental observation, $\dot{\varepsilon} \propto a^{+1.2}$. Of the two possible charge neutrality conditions, $2[V_{Mg}^{''}] = [Fe_{Mg}^{\bullet}]$ and $2[Mg_I^{\bullet\bullet}] = [e']$, the former is most likely to apply [Sockel, 1974; Nakamura and Schmalzried, 1983]. The defect reaction which controls charge neutrality, then, involves the oxidation of iron,

$$3Fe_{Mg}^{x} + FeSiO_3 + 1/2\ O_2 =$$

$$2Fe_{Mg}^{\bullet} + V_{Mg}^{''} + Fe_2SiO_4 \qquad (11)$$

and the law of mass action yields

$$\frac{[V_{Mg}^{''}]}{a_{en}} \frac{[Fe_{Mg}^{\bullet}]^2}{p_{O_2}^{1/2}} = K_{11} \qquad (12)$$

The creep rate of unbuffered olivine crystals is proportional to $p_{O_2}^{1/6}$. In this case, no excess oxide phase is present, and the majority defects are related through the oxidation reaction as follows:

$$2Mg_{Mg}^{x} + 8Fe_{Mg}^{x} + Si_I^{\bullet\bullet\bullet\bullet} + 2O_2 =$$

$$8Fe_{Mg}^{\bullet} + 2V_{Mg}^{''} + Mg_2SiO_4 \qquad (13)$$

To test the controlling defect pair ($J^{\bullet}$, $V_O^{\bullet\bullet}$) proposed for the buffered case, the law of mass action is applied to Eq. 13:

$$\frac{[V_{Mg}^{''}]^2 \ [Fe_{Mg}^{\bullet}]^8}{[Si_I^{\bullet\bullet\bullet\bullet}] \ p_{O_2}^{2}} = K_{13} \qquad (14)$$

K_{13} is a function of composition, namely the Fe/Mg ratio, which is assumed to be constant in the present case. If the charge neutrality condition involves all the charged defects in Eq. 13, then,

$$2[V_{Mg}^{''}] = 4[Si_I^{\bullet\bullet\bullet\bullet}] + [Fe_{Mg}^{\bullet}] \qquad (15)$$

Because the concentrations of these three defects are linearly related, they vary with p_{O_2} in the same way, and Eq. 14 reduces to

$$[Fe_{Mg}^{\bullet}] \propto p_{O_2}^{2/9} \qquad (16)$$

Because the concentration of iron is large, its oxidized state, $Fe_{Mg}^{\bullet}$, is approximated to an infinitely large source of holes. Thus, in the present case,

$$[Fe_{Mg}^{\bullet}] = [h^{\bullet}] \propto p_{O_2}^{2/9} \qquad (17)$$

This assumption must, of course, break down for compositions close to the pure magnesium end-member of the Mg-Fe olivine solid solution series. The concentration of positively charged jogs can now be expressed as

$$[J^{\bullet}] \propto [h^{\bullet}] \propto p_{O_2}^{2/9} \qquad (18)$$

The incorporation of oxygen into the olivine lattice may be written as

$$V_O^{\bullet\bullet} + 1/2\ O_2 = O_O^x + 2h^{\bullet} \qquad (19)$$

The law of mass action for Eq. 19 combined with Eq. 17 then yields

$$[V_O^{\bullet\bullet}] \propto p_{O_2}^{-1/18} \qquad (20)$$

From Eqs. 18 and 19, the dependence of the strain rate on the partial pressure of oxygen for the defect model proposed earlier, $(J^{\bullet}, V_O^{\bullet\bullet})$, is

$$\dot{\varepsilon} \propto p_{O_2}^{1/6} \qquad (21)$$

This dependence is the same as that measured in creep experiments performed with unbuffered olivine. Nakamura and Schmalzried [1984] proposed that the creep of unbuffered olivine is controlled by the diffusion of silicon vacancies since their concentration varies of $p_{O_2}^{1/5.5}$. However, a silicon vacancy mechanism cannot account for the observed dependence of the creep rate on the activity of enstatite ($\dot{\varepsilon} \propto a_{en}^{1.2}$); thus their model is inadequate. These authors [1983] suggest that Fe'_{Si} is a more likely candidate for the third dominant defect than is $Si_I^{\bullet\bullet\bullet\bullet}$. The Fe'_{Si} defect has not been considered in the present study.

4.2. Forsterite

The experimental creep results for forsterite are quite different from those for iron-bearing olivine. No p_{O_2} dependence was observed whether or not the samples were buffered, and no significant difference in strain rate was measured between buffered and unbuffered runs. It would be

TABLE 7. P_{O_2} and a_{MgO} Dependences of the Concentrations of Silicon and Oxygen Point Defects

	Charge Neutrality Conditions			
	$[V_{Mg}''] = [Mg_I^{\bullet\bullet}]$		$2[V_O^{\bullet\bullet}] = [e']$	
Defect	m	p	m	p
V_{Si}''''	0	+2	+1/3	+2
$Si_I^{\bullet\bullet\bullet\bullet}$	0	-2	-1/3	-2
$V_O^{\bullet\bullet}$	0	-1	-1/6	0
O_I''	0	+1	+1/6	0

TABLE 8. p_{O_2} and a_{MgO} Dependences of Deformation Mechanisms in Buffered Forsterite for Two Charge Neutrality Conditions

		Neutrality Conditions			
		$[V_{Mg}'']=[Mg_I^{\bullet\bullet}]$		$2[V_O^{\bullet\bullet}]=[e']$	
Jog/Kink	Diffusion	m	p	m	p
$K^{\bullet}$		+1/4	-1/2	+1/6	0
K'		-1/4	+1/2	-1/6	0
$J^{\bullet}$	$V_O^{\bullet\bullet}$	+1/4	+1/2	0	0
J'	$V_O^{\bullet\bullet}$	-1/4	-1/2	-1/3	0
$J^{\bullet}$	O_I''	+1/4	-1/2	+1/3	0
J'	O_I''	-1/4	+1/2	0	0
$J^{\bullet}$	V_{Si}''''	+1/4	-2	+1/2	+2
J'	V_{Si}''''	-1/4	+2	+1/6	+2
$J^{\bullet}$	$Si_I^{\bullet\bullet\bullet\bullet}$	+1/4	+2	-1/6	-2
J'	$Si_I^{\bullet\bullet\bullet\bullet}$	-1/4	-2	-1/2	-2

tempting to ascribe the different creep behavior exhibited by olivine and forsterite to a different operating deformation mechanism, for instance, to a change from climb to glide-controlled creep. However, the similarity of the dislocation microstructures observed in crystals of the two materials deformed under the same conditions makes this position difficult to defend.

Hobbs [1983] has suggested that aluminum could control the point defect chemistry of the forsterite crystals used by various research groups. We are aware that even small amounts of dopant drastically affect, for instance, the electrical properties of semiconductors and insulators. However, in our case, the chemical analyses suggest that our crystals may indeed be in the intrinsic regime. Therefore, we consider here two possible intrinsic point defect models. In the first one, magnesium vacancies and interstitials are the majority defects; based on electrical conductivity measurements, Pluschkell and Engell [1968] argue that these defects dominate when the crystal is stoichiometric or nearly stoichiometric. In the second model, oxygen vacancies and electrons comprise the disorder type. This condition appears to apply [Pluschkell and Engell; 1968, Shock and Duba, this volume] to the p_{O_2} range in which our experiments were run. In Table 7, the p_{O_2} and a_{MgO} dependences of various silicon and oxygen defect concentrations are listed. None of these defect concentrations varies as $p_{O_2}^0$ and a_{MgO}^0 simultaneously. In Table 8, the p_{O_2} and a_{MgO} dependences of various jog or kink-controlled deformation mechanisms for buffered forsterite are summarized for the two charge neutrality conditions referred to above. In only two cases is the creep rate independent of both

TABLE 9. Summary of the Defect Models Proposed Iron-Bearing Olivine and Forsterite

Sample	Buffer	Charge Neutrality	Defect Model
Iron-olivine	unbuffered	$[Mg_I^{\bullet\bullet}]=[V_{Mg}'']$	$J^\bullet,V^{\bullet\bullet}$
	(Mg,Fe)O	$2[Mg_I^{\bullet\bullet}]=[e']$	$J^\bullet,V_O^{\bullet\bullet}$
	$(Mg,Fe)SiO_3$	$2[V_{Mg}'']=[Fe_{Mg}^\bullet]$	$J^\bullet,V_O^{\bullet\bullet}$
Forsterite	unbuffered	$2[V_{Mg}'']=[e']$	$J^\bullet,V_O^{\bullet\bullet}$
	MgO	$2[V_O^{\bullet\bullet}]=[e']$	$J^\bullet,V_O^{\bullet\bullet}$
	$MgSiO_3$	$2[V_O^{\bullet\bullet}]=[e']$	$J^\bullet,V_O^{\bullet\bullet}$

the partial pressure of oxygen and of the activity of periclase: (J', O_i'') and ($J^\bullet$,$V_O^{\bullet\bullet}$) for the charge neutrality condition $2[V_O^{\bullet\bullet}] = [e']$. Oxygen interstitials being unlikely defects, the ($J^\bullet$, $V_O^{\bullet\bullet}$) defect pair is preferred. It should be emphasized that this conclusion is only a model and that it needs to be tested by carefully designed experiments.

5. Comparison With Other Works

The defect pair, ($J^\bullet$, $V_O^{\bullet\bullet}$), can account for the creep behavior observed for buffered and unbuffered forsterite and for buffered and unbuffered iron-bearing olivine, Table 9. In addition, the charge neutrality conditions used in our calculations are consistent with reported dependence of electrical conductivity on oxygen partial pressure for both forsterite and olivine [Schock and Duba, this volume]. However, while the charge neutrality condition for olivine yields the observed $p_{O_2}^{-1/6}$ dependence for the oxygen diffusity (assuming that the diffusion samples were unbuffered) (O. Jaoul and M. Poumellec, private communication, 1983), that for forsterite does not give the reported $p_{O_2}^0$ dependence for the oxygen diffusivity (whether or not the samples were buffered) [Jaoul et al., 1981].

The ($J^\bullet$, $V_O^{\bullet\bullet}$) model also suggests that oxygen is the slowest diffusing species. Indeed, Sockel et al. [1980] found that silicon diffuses more than a factor of ten faster than oxygen. However, more recent diffusion measurements by Jaoul et al. [1983] indicate that silicon diffuses four orders of magnitude more slowly than oxygen. Further deformation and diffusion experiments are necessary to resolve these discrepancies. Recent creep tests on Co_2SiO_4, Fe_2SiO_4 and vanadium-doped Mg_2SiO_4 should help constrain the mechanism of creep in olivine [Ricoult and Kohlstedt, 1984a,b]. Also, experimental and theoretical work is needed to completely define the point defect chemistry of the olivine family. For instance, defect complexes, such as (V_{Mg}'' $Si_I^{\bullet\bullet\bullet\bullet}$ V_{Mg}''), which were not considered here could prove to be important [Smyth and Stocker, 1975; Nakamura and Schmalzried, 1984]. Partly ionized defects should also be taken into consideration.

6. Defect Equilibration Time

The present experiments demonstrate that the creep rate adjusts quickly to a new steady state level not only with a change in p_{O_2} but also with a change in solid oxide buffer. In both cases, the defect relaxation time is less than one hour at 1400°C. The iron to oxygen ratio changes rapidly in response to a change in p_{O_2} (Eq. 11) because only the diffusion of magnesium vacancies, charged compensated by the diffusion of holes, is required. A change in solid oxide buffer produces a change in the Mg/Si ratio. The fact that a change in oxide buffer affects the creep strength suggests that the cationic ratio is also changing by the diffusion of point defects. The self-diffusion of ionic species in olivine [Jaoul et al., 1981; Sockel et al., 1980] is much too slow to account for the observed relaxation times.

Under the assumption that silicon diffuses by an interstitial mechanism, it is possible to calculate the concentration of silicon interstitials necessary for a 2mm-wide sample to equilibrate with a solid buffer in less than 40 minutes at 1675°K. The velocity, v_{Si}, of the silicon ions is related to the thermodynamic driving force, that is, the activity gradient existing between the surface and the center of the sample [Schmalzried, 1981, p. 60]:

$$v_{Si} = D_{Si}(\partial \ln a_{en}/\partial x) \tag{22}$$

where D_{Si} is the self-diffusion coefficient. The average velocity of the silicon ions is

$$v_{Si} = D_{Si} \ln(a_{en_1}/a_{en_2})/\Delta x \tag{23}$$

where Δx is half the width of the sample. The flux of silicon ions can be expressed as

$$j_{Si} = v_{Si}/\Omega \tag{24}$$

where Ω is the molecular volume. Since the flux of ions in a given direction must be equal to that of the interstitials, j_{SiI}, in the opposite direction,

$$j_{Si} = j_{SiI} = v_{SiI}\, c_{SiI} \tag{25}$$

with v_{SiI} being the velocity of the silicon interstitials and c_{SiI} their concentration. Finally, from Eqs. 24 and 25,

$$c_{SiI} = j_{Si}/v_{SiI} = D_{Si}\ln(a_{en_1}/a_{en_2})/\Omega\Delta x v_{SiI} \tag{26}$$

Using the silicon tracer diffusion data of Jaoul

et al. [1981] as an approximation, a concentration of silicon interstitial of less than 1.3 x 10^{-7} is required to explain the observed relaxation time of less than 40 minutes. Silicon defects are expected to be less numerous than the other cationic defects. In Co_2SiO_4, Greskovich and Schmalzried [1970] report 10^{-5} vacancies of cobalt per mole. From thermogravimetric measurements on Fe_2SiO_4, Nakamura and Schmalzried [1983] inferred a concentration of 10^{-3}-10^{-4} iron vacancies; their results are expected to hold for $(Mg,Fe)_2SiO_4$. If the concentration of magnesium vacancies in $(Mg,Fe)_2SiO_4$ is in the same order of magnitude, then the value computed here for the silicon interstitial concentration is not unreasonable.

7. Conclusions

1. The creep rate of olivine single crystals deformed in the absence of a solid oxide buffer varies as $p_{O_2}^{1/6}$.
2. The creep rate of olivine single crystals deformed in the presence of a solid oxide buffer, either enstatite or magnesiowustite, is independent of p_{O_2}.
3. The creep rate of olivine single crystals buffered against enstatite is one order of magnitude faster than that of crystals buffered against magnesiowustite.
4. The creep rate of forsterite single crystals is insensitive to both oxygen partial pressure and oxide activity.
5. For both olivine and forsterite, the observed dependences of creep rate on p_{O_2} and a_{ox} are consistent with a model for the creep process in which the rate of deformation is limited by climb of dislocations; dislocation climb, in turn, is rate-limited by the nucleation and displacement of jogs the mobility of which is determined by oxygen self-diffusion via a vacancy mechanism. In this model, charge neutrality in olivine is controlled by the point defect reaction describing the oxidation of iron (Eqs. 11 and 13), and charge neutrality in forsterite is governed by the point defect reaction governing the incorporation of oxygen into the olivine lattice (Eq. 19).
6. The steady state creep rate of olivine crystals changes to a new value in less than one hour at 1675°K not only after a change in p_{O_2}, but also after a change in a_{ox}. These rapid relaxation times indicate that the equilibration kinetics are controlled by defect diffusion.

Acknowledgments. The authors wish to thank R. N. Schock from Lawrence Livermore National Laboratory for kindly providing the forsterite crystals and for helpful comments. Valuable discussions with H. Schmalzried are gratefully acknowledged. The research reported in this paper was supported by the National Science Foundation through the Materials Science Center at Cornell University.

References

Abelard, P., and F. J. Baumard, A new graphical representation for a systematic study of the defect structure in ternary oxides, with a specific application to forsterite Mg_2SiO_4, _J. Phys. Chem. Solids_, _43_(7), 617-625, 1982.

Bilderback, D. H., A real-time back-reflection Laue camera, _J. Appl. Crystallogr._, _12_, 95-98, 1979.

Blacic, J. D., Effect of water on the experimental deformation of olivine, in _Flow and Fracture of Rocks_, _Geophys. Monogr. Ser._, vol. 16, edited by H. C. Heard, I. Y. Borg, N. L. Carter, and C. B. Raleigh, pp. 109-115, AGU, Washington, D.C., 1972.

Carter, N. L., and H. G. AveLallemant, High-temperature flow of dunite and peridotite, _Geol. Soc. Am. Bull._, _81_, 2181-2202, 1970.

Chopra, P. N., and M. S. Paterson, The experimental deformation of dunite, _Tectonophysics_, _78_, 453-473, 1981.

Darot, M., Deformation experimentale de l'olivine et de la forsterite, thèse d'état, Nantes Univ., Nantes, France, 1980.

Darot, M., and Y. Gueguen, High-temperature creep of forsterite single crystals, _J. Geophys. Res._, _86_(B7), 6219-6234, 1981.

Durham, W. B., and C. Goetze, Plastic flow of oriented single crystals of olivine, 1, Mechanical data, _J. Geophys. Res._, _82_(36), 5737-5753, 1977.

Durham, W. B., C. Froidevaux, and O. Jaoul, Transient and steady-state creep of pure forsterite at low stress, _Phys. Earth Planet. Inter._, _19_, 263-274, 1979.

Durham, W. B., D. Ricoult, and D. L. Kohlstedt, Interaction of slip systems in olivine, in _Point Defects in Minerals_, _Geophys. Monogr. Ser._, edited by R. Schock, AGU, Washington, D.C., this volume.

Greskovich, C., and H. Schmalzried, Nonstoichiometry and electronic defects in Co_2SiO_4 and in $CoAl_2O_4$-$MgAl_2O_4$ crystalline solutions, _J. Phys. Chem. Solids_, _31_, 639-646, 1970.

Gueguen, Y., High-temperature olivine creep: Evidence for control by edge dislocations, _Geophys. Res. Lett._, _6_(5), 357-360, 1979.

Hamilton, D. L., and C. M. B. Henderson, The preparation of silicate compositions by a gelling method, _Mineral. Mag._, _36_, 832-838, 1968.

Hobbs, B. E., The influence of metamorphic environment upon the deformation of minerals, _Tectonophysics_, 78, 335-383, 1981.

Hobbs, B. E., Constraints on the mechanism of deformation of olivine imposed by point defect chemistry, _Tectonophysics_, _92_, 35-69, 1983.

Hornack, P. G., The effect of oxygen fugacity on the creep of olivine, M.S. thesis, 99 pp., Cornell Univ., Ithaca, N.Y., 1978.

Jaoul, O., C. Froidevaux, W. B. Durham, and M. Michaut, Oxygen self-diffusion in forsterite: Implications for the high-temperature creep mechanism, Earth Planet. Sci. Lett., 47, 391, 1980.

Jaoul, O., M. Poumellec, C. Froidevaux, and A. Havette, Silicon diffusion in forsterite: A new constraint for understanding mantle deformation, in Anelasticity in the Earth, edited by F. D. Stacey, M. S. Paterson, and A. Nicolas, pp. 95-100, AGU, Washington, D.C., 1981.

Jaoul, O., B. Houlier, and F. Abel, Study of ^{18}O diffusion in magnesium orthosilicate by nuclear microanalysis, J. Geophys. Res., 88, 613-624, 1983.

Kohlstedt, D. L., and C. Goetze, Low-stress high-temperature creep in olivine single crystals, J. Geophys. Res., 79(14), 2045-2051, 1974.

Kohlstedt, D. L. and D. L. Ricoult, High-temperature creep of silicate olivines, in Plastic Deformation of Ceramic Materials, edited by R. C. Bradt and R. E. Tressler, in press, Plenum Press, New York, 1984.

Kohlstedt, D. L., and P. Hornack, Effect of oxygen partial pressure on the creep of olivine, in Anelasticity in the Earth, Geodynamics Ser., vol. 4, edited by F. D. Stacey, M. S. Paterson, and A. Nicolas, pp. 101-107, AGU, Washington, D.C., 1981.

Kohlstedt, D. L., C. Goetze, and W. B. Durham, Experimental deformation of single crystal olivine with application to flow in the mantle, in Petrophysics, The Physics and Chemistry of Minerals and Rocks, edited by R. G. J. Strens, pp. 35-49, John Wiley, New York, 1976.

Kroger, F. A., The Chemistry of Imperfect Crystals, North-Holland, Amsterdam, 1964.

Medaris, L. G., Jr., Partitioning of Fe and Mg between coexisting synthetic olivine and orthopyroxene, Am. J. Sci., 267, 945-968, 1969.

Nakamura, A. and H. Schmalzried, On the nonstoichiometry and point defects of olivine, Phys. Chem. Minerals, 10, 27-37, 1983.

Nakamura, A. and H. Schmalzried, On the Fe^{2+}-Mg^{2+} interdiffusion in olivine, Ber. Bunsenges. Phys. Chem., in press, 1984.

Nafziger, R. H., and A. Muan, Equilibrium phase compositions and thermodynamic properties of olivines and pyroxenes in the system MgO-"FeO"-SiO_2, Am. J. Sci., 52, 1364-1385, 1967.

Oberheuser, G., H. Kathrien, G. Demortier, H. Gonska, and F. Freund, Carbon in olivine single crystals analyzed by the $^{12}C(d,p)^{13}C$ method and by photoelectron spectroscopy, Geochim. Cosmochim. Acta, 47, 1117-1129, 1983.

Pluschkell, W., and H. J. Engell, Ionen- und Elektronen-leitung im Magnesiumorthosilikat, Ber. Dtsch. Keram. Ges., 45, 388, 1968.

Post, R. L., High-temperature creep of Mt. Burnett dunite, Tectonophysics, 42, 75-110, 1977.

Raleigh, C. B., Mechanisms of plastic deformation of olivine, J. Geophys. Res., 73(14), 5391-5406, 1968.

Relandeau, C., High temperature creep of forsterite polycrystalline aggregates, Geophys. Res. Lett., 8, 733-736, 1981.

Ricoult, D. L., and D. L. Kohlstedt, Creep behavior of single-crystal vanadium-doped forsterite, J. Am. Ceram. Soc., in press, 1984a.

Ricoult, D. L., and D. L. Kohlstedt, Compression creep of Fe_2SiO_4 and Co_2SiO_4 single-crystals in controlled thermodynamic environments, Phil. Mag., in press, 1984b.

Schmalzried, H., Reactivity and point defects of double oxides with emphasis on simple silicates, Phys. Chem. Miner., 2, 279-294, 1978.

Schmalzried, H., Solid state reactions, 2nd ed., 252 pp., Verlag Chemie, Weinheim, Germany, 1981.

Schock, R. N., and A. G. Duba, Point defects and the mechanisms of electrical conduction in olivine, in Point Defects in Minerals, Geophys. Monogr. Ser., edited by R. Schock, AGU, Washington, D.C., this volume

Smyth, D. M., Thermodynamic characterization of ternary compounds, I, The case of negligible defect association, J. Solid State Chem., 16, 73-81, 1976.

Smyth, D. M., and R. L. Stocker, Point defects and non-stoichiometry in forsterite, Phys. Earth Planet. Inter., 10, 183-192, 1975.

Sockel, H. G., Defect structure and electrical conductivity of crystalline ferrous silicate, in Defects and Transport in Oxides, edited by M. S. Smeltzer and R. J. Jaffee, pp. 341-354, Plenum, New York, 1974.

Sockel, H. G., D. Hallwig, and R. Schochtner, Investigations of slow exchange processes at metal and oxide surfaces and interfaces using secondary ion mass spectroscopy, Mater. Sci. Eng., 42, 59-64, 1980.

Stocker, R. L., Point defect formation parameters in olivine, Phys. Earth Planet. Inter., 17, 108-117, 1978a.

Stocker, R. L., Influence of oxygen pressure on defect concentrations in olivine with a fixed cationic ratio, Phys. Earth Planet. Inter., 17, 118-129, 1978b.

Stocker, R. L., and D. M. Smyth, Effect of enstatite activity and oxygen partial pressure on the point defect chemistry of olivine, Phys. Earth Planet. Inter., 16, 145-156, 1978.

Swalin, R. A., Thermodynamics of Solids, 387 pp., John Wiley, New York, 1972.

INTERACTION OF SLIP SYSTEMS IN OLIVINE

W. B. Durham

Lawrence Livermore National Laboratory, University of California
Livermore, California 94550

D. L. Ricoult and D. L. Kohlstedt

Department of Materials Science and Engineering, Cornell University
Ithaca, New York 14853

Abstract. Eleven deformation experiments have been performed on five selectively oriented single crystals of olivine (Fo_{92}) for the purpose of exploring interactions between slip systems. The crystals were deformed such that two slip systems operated either simultaneously or sequentially. No evidence, either in the mechanical behavior or in the dislocation structure, was found to indicate that the presence of dislocations of one slip system impedes the motion of other dislocations. This result, combined with previously published observations on the role of climb in the deformation of olivine crystals, suggests that single crystal creep results may be used to predict the behavior of polycrystalline olivine. Recently reported creep data on dry polycrystalline olivine are in good agreement with the data for single crystals.

Introduction

Deformation experiments on single crystals of olivine have generally been performed to study grain-matrix flow mechanisms and their application to the deformation of polycrystalline olivine. Single crystal experiments are attractive because they simplify both the experiment (principally by eliminating the need for confining pressures and sample jacketing) and the analysis of deformed samples [Kohlstedt and Goetze, 1974]. The disadvantage of using single crystals is that the extrapolation from single crystal flow to polycrystalline flow is not well understood for olivine. Another disadvantage is that single crystals provide little information regarding grain-boundary mechanisms of deformation. Under conditions where grain-boundary mechanisms dominate, grain-matrix flow laws provide only a lower bound to strain rate.

The purpose of the present work is to explore one of the central points which bears on the use of single crystal results to predict the flow behavior of polycrystalline olivine: interaction of slip systems. In earlier studies of high-temperature creep of single-crystal olivine [Durham and Goetze, 1977; Durham et al., 1977] the polycrystalline flow law was hypothesized to lie amongst the flow laws for the three known independent slip systems, based partially on the observation that the two independent slip systems which operate at the $[101]_c$ orientation of applied stress showed no mechanical or visible (in the dislocation structure) signs of impeding one another's operation. (The notation $[hkl]_c$ indicates the Miller index of a direction in an imaginary cubic lattice whose principal directions are aligned with those of the orthorhombic olivine lattice. Thus, for instance, $[101]_c$ is the direction in the olivine (010) plane 45° from [100] and [001]. [hkl] and $[hkl]_c$ are in general not parallel.) This observation implied that recovery processes, in particular dislocation climb and lattice diffusion, proceeded with relative ease and in turn implied that the observation could be extended to all potential slip system interactions. If slip systems were non-interactive, then the three flow laws for individual slip systems could be simply superposed (with appropriate geometric considerations) to predict the flow of polycrystalline olivine.

Experimental Details

Single crystals of San Carlos olivine (Fo_{92}) provided the starting material. One sample was cut in the standard rectangular parallelepiped form, roughly 3 mm x 3 mm x 5 mm. The long dimension, intended as the compression direction, was aligned along

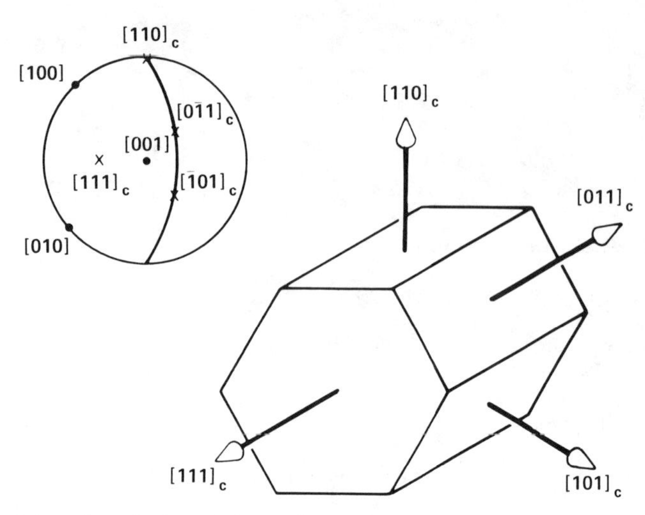

Fig. 1. Orientation of the hexagonal parallelepipeds used in the experiments.

$[101]_c$ and one set of vertical faces was aligned parallel to (010), the anticipated plane of plane strain. The remaining four samples were cut in the form of hexagonal parallelepipeds with faces oriented as shown in Figure 1. The distance between opposite rectangular faces, as well as between opposite hexagonal faces, was about 5 mm. Orientation was determined optically from thin sections cut near the sample faces. Including misorientation across tilt boundaries in the starting material, overall orientation accuracy was ± 5°.

Samples were deformed under dead weight loading in the apparatus described by Kohlstedt and Hornack [1981]. The tungsten resistance heating element was a free-standing mesh rather than a wire strand, so that an Al_2O_3 tube was not needed in the chamber. For some runs a dummy Al_2O_3 tube was introduced in order to better simulate earlier work. Oxygen fugacity was held at 1-3 x 10^{-3} Pa with an H_2/CO_2 gas mixture at one atmosphere.

The sample ends and mating platen faces were ground flat and polished to 0.03 μm using Al_2O_3 grit prior to each run. Multiple runs were usually performed on a given sample. At the start of the first run on a given sample, no concern was given to the possibility of annealing, so the temperature was raised gently and the load was applied only after run temperature had been reached. To avoid annealing dislocation structures at the end of a run, the samples were cooled at a rate of 1.5-2 K/s with the load applied. This stress-temperature history is believed not to alter significantly the dislocation structure from its high-temperature state [Durham et al., 1977]. The load was removed when the temperature fell below 1500 K, where the dislocation structure was no longer in danger of annealing [Goetze and Kohlstedt, 1973; Toriumi and Karato, 1978; Ricoult, 1979]. For the second and subsequent runs annealing was minimized during initial heating by loading the sample at T < 1500 K and proceeding to run temperature (usually near 1900 K) at a rate slightly above 2 K/s.

Sample shortening, ε_ℓ = 1 - length/initial length (sometimes referred to as engineering strain), during a run was monitored by displacement transducers. Total sample shortening was determined from micrometer measurements of sample length before and after a run. Accuracy of the micrometer measurement was better than ±5 μm. For the small strain runs, the shortening divided by the run time provided a more accurate determination of the average shortening rate than did the chart record.

For the hexagonal samples, where stress is non-uniform owing to the variable horizontal sectional area, it is useful to speak of an "equivalent stress." The equivalent stress, labeled σ_{eq}, is that stress which would produce the same shortening rate in a rectangular sample of identical height and depth, with width intermediate between the maximum and minimum width of the hexagonal sample, under an identical load. For a stress exponent of n = 3.5 [Durham and Goetze, 1977] and a perfectly hexagonal vertical section, σ_{eq} is calculated to be 0.778 times the maximum stress difference (which occurs at the ends of the sample).

The history of a typical hexagonal sample is illustrated in Figure 2: in the first run the sample was deformed through several percent strain to establish a steady-state dislocation microstructure [Durham et al., 1977] and to identify a "steady-state" hexagonal shortening rate, critical for later comparisons. In run 2 the sample was deformed briefly at a new orientation. Presumably, any effect caused by the preexisting microstructure would be most pronounced at lowest strains at the new orientation and thus would be most easily measured in run 2. In run 3 the sample was deformed through several percent at the new orientation to look for strain-dependent mechanical effects and ultimately to identify again a steady-state shortening rate for use in comparisons.

Following each run, a thin slice was taken from a vertical sample face in order to record and study the dislocation structure resulting from the run. The slices were ground to thin section thickness and decorated for optical microscopic observation [Kohlstedt et al., 1976] or ion-thinned for transmission electron microscopic (TEM) observation of dislocations.

Results

Mechanical data. Five samples were deformed in eleven runs. The temperature was 1873 ± 5 K for all runs except the one run on sample 2 and part of the first run on sample 3, both of which were conducted at 1823 K. Equivalent stresses in the eleven runs varied from 20.0 to 38.3 MPa. Resulting shortening rates, $\dot{\varepsilon}_{\ell}$, are given in Table 1. For comparison in Table 1, approximate steady-state values, $\dot{\varepsilon}_{ss}$, are calculated on the basis of several of the first run values of $\dot{\varepsilon}_{\ell}$. For the orientation changes $[101]_c \rightarrow [110]_c$ and $[101]_c \rightarrow [111]_c$, the first increment of strain at the new orientation proceeds at a rate which is somewhat faster than the steady-state strain rate. For $[110]_c \rightarrow [101]_c$, the first increment proceeds slightly more slowly than the steady-state rate. The magnitude and sign of $\dot{\varepsilon}_{ss} - \dot{\varepsilon}_{\ell}$ appears to correlate with the magnitude and sign of the stress change which accompanies the orientation change. The strain rate at the new orientation is independent of strain.

The correlation of anomalous hardness at $[101]_c$ with absence of the Al_2O_3 tube (runs 2(1) and 4(1)) is acknowledged without further comment. The authenticity and possible causes of this phenomenon are currently being investigated.

Optical microscopy. Decorated sections are labeled by their sample (run) number, as in Table 1. The sections of primary interest for direct observation of dislocations were the three "strain-increment" sections taken after the run 2 strain increment at a new orientation, sections 3(2), 4(2), and 5(2). No sign of extensive tangling or of increased densities of dislocations, the "coldwork" structures that might indicate dislocation interference during multiple slip, was observed in any decorated section. Coldwork structures were not observed for the one case where multiple slip actually took place, $[101]_c$, consistent with earlier observations [Durham et al., 1977; Darot and Gueguen, 1981].

The structures in the three strain-increment sections are not as simple and homogeneous as the steady-state "fingerprint" microstructures described by Durham et al. [1977]. For example, section 3(2), ($[110]_c \rightarrow [101]_c$), shows some $[101]_c$ fingerprint, some $[110]_c$ fingerprint, and some other structures including the "[100]-organization" fingerprint of the $[111]_c$ orientation. The various structures tend to be segregated into volumes a few tens of microns across. Rarely are two recognizable patterns, in particular the $[110]_c$ and $[101]_c$ fingerprints, superposed, so it is difficult to describe the process by which one fingerprint changes to another. It is certain, however, that the sample is in the process of converting to the $[101]_c$ fingerprint because section 3(3) (4.8% further strain at $[101]_c$) shows almost exclusively the $[101]_c$ fingerprint.

Similarly, section 4(2) ($[101]_c \rightarrow [110]_c$) shows a variety of structures, including

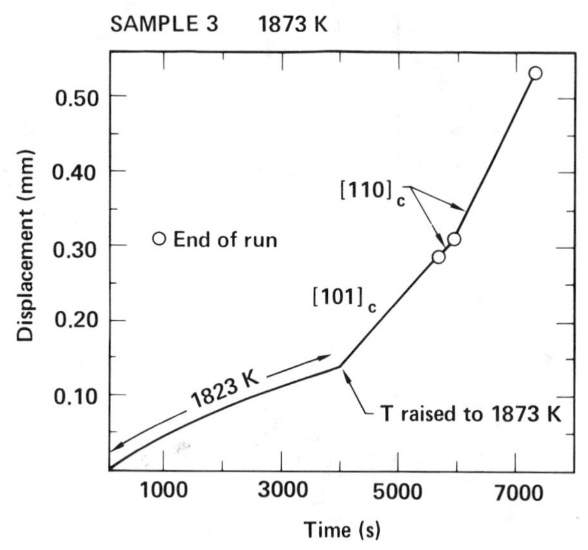

Fig. 2. Typical mechanical data from a sample. Shown are data from sample 3.

[100]-organization. It shows more homogeneity than section 3(2), but does not seem to show any of the $[110]_c$ fingerprint. Once again, however, section 4(3) (3.5% further strain at $[101]_c$) is dominated by the $[110]_c$ fingerprint.

Section 5(2) ($[101]_c \rightarrow [111]_c$) shows the most homogeneous structure of the three strain increment sections (Figure 3). It is interesting that there is some superposition of the two expected structures, the $[101]_c$ and $[111]_c$ fingerprints. The section also contains a high density of small dislocation loops, debris perhaps from the interacting slip systems, but the loops do not seem to interfere with the overall dislocation structures.

Electron microscopy. Section 5(2) was the only strain increment section examined by TEM. The most striking feature of the dislocation structure is the high density of low-angle ($<$ or $\ll 1°$) grain boundaries (Figure 4a), the majority of which lie in the (100) plane, forming a portion of what is viewed optically as the $[111]_c$ fingerprint, i.e., the [100]-organization. Numerous (001) tilt boundaries and (010) twist boundaries exist also.

The microstructure varies with position at the TEM scale. Long, somewhat curved $\underline{b}$ = [100] edge dislocations dominate in some regions. In other areas long, straight $\underline{b}$ = [001] screw dislocations are prominent. In still other areas both $\underline{b}$ = [100] and $\underline{b}$ = [001] dislocations are found (Figures 4b, 4c, 4d). These observations are consistent with the optical observation that remnants of the $[101]_c$ fingerprint still exist in the section. The [100] edge dislocations are often observed climbing into a low-angle tilt boundary,

TABLE 1. Mechanical Results

Vertical Section	Sample (Run)	Orientation	T, K	σ_{eq}, MPa	ε_ℓ, %	$\dot{\varepsilon}_\ell$, 10^{-5} s^{-1}	$\dot{\varepsilon}_{ss}$, 10^{-5} s^{-1}	Comments
Rectangular	1(1)*	$[101]_c$	1873	30.0	14.7	1.9		
Rectangular	1(2)*	$[101]_c$	1873	30.0	11.6	1.35		
Hexagonal	2(1)	$[101]_c$	1823	38.3	4.3	0.57		
Hexagonal	3(1)	$[110]_c$	1873	25.0	6.0	2.5		T = 1823 K for first 3% ε_ℓ
Hexagonal	3(2)*	$[101]_c$	1873	35.0	0.43	2.1	2.6	$\dot{\varepsilon}_{ss}$ calculated from runs 1(1), 1(2), 5(1)
Hexagonal	3(3)*	$[101]_c$	1873	35.0	4.8	3.3		
Hexagonal	4(1)	$[101]_c$	1873	35.0	2.4	0.8		
Hexagonal	4(2)	$[110]_c$	1873	25.0	0.59	3.5	2.5	$\dot{\varepsilon}_{ss}$ calculated from run 3(1)
Hexagonal	4(3)	$[110]_c$	1873	25.0	3.5	15.7		microcracks formed during run
Hexagonal	5(1)*	$[101]_c$	1873	35.0	3.8	2.1		
Rectangular	5(2)*	$[111]_c$	1873	20.0	0.45	2.3	0.81	$\dot{\varepsilon}_{ss}$ derived from run 3(1) and Durham and Goetze [1977] flow laws

*$A\ell_2O_3$ tube around column.

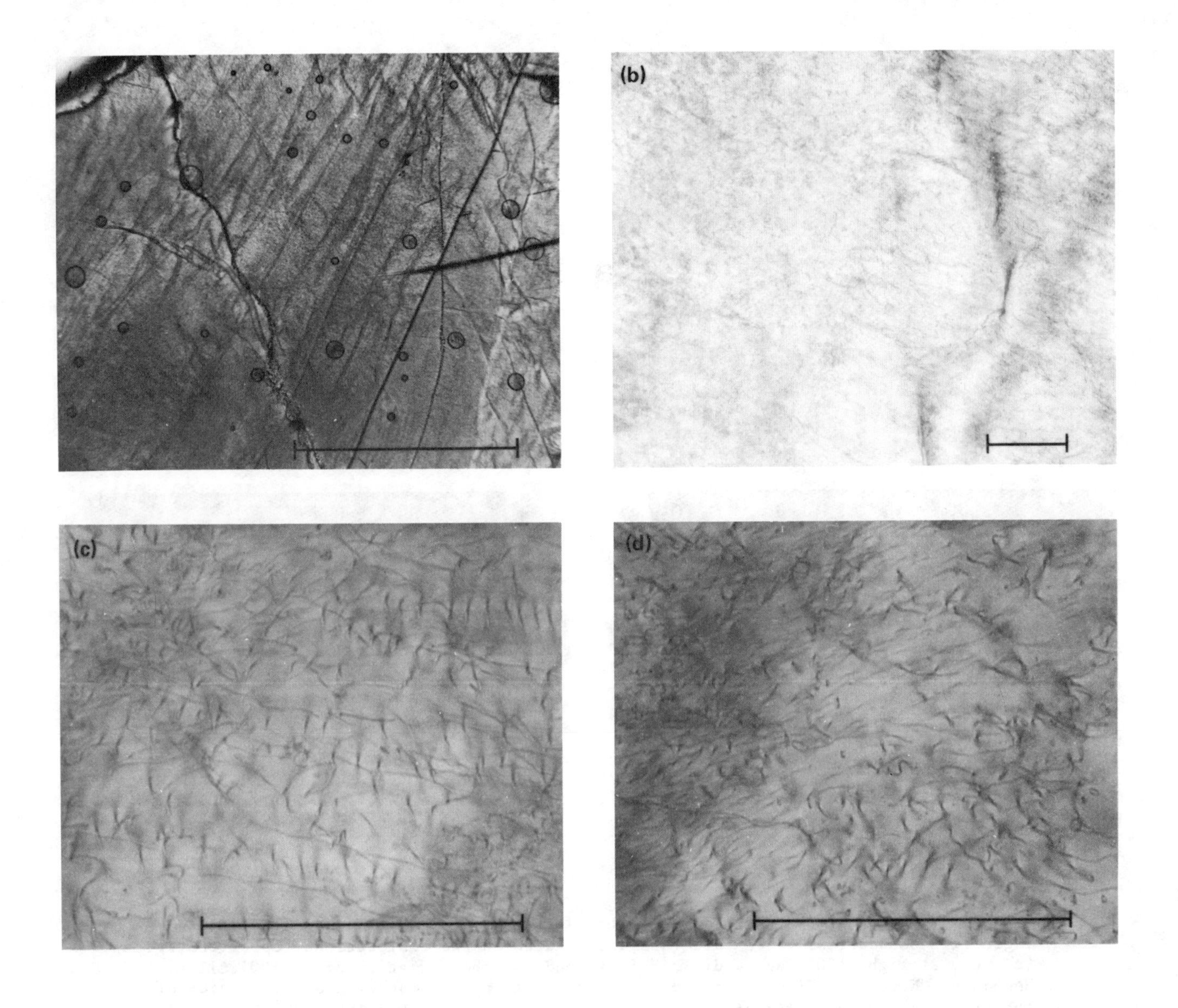

Fig. 3. Transmission optical micrographs of section 5(2). The viewing direction in all four photos is along $[111]_c$, with the (100) plane striking approximately NNE, the (001) plane approximately NNW, and the (010) plane approximately E-W. (a) Numerous subgrain boundaries seen as dark bands with individual dislocations not resolvable. Subgrain boundaries in the (100) plane are more numerous than those in the (001) plane. (b) Segregation of two structures, one dominated by developing [100]-organization (right), the other still showing signs of the $[101]_c$ finger print (left). The plane dividing the two structures lies near (100). (c) [100]-organization well underway. Edge dislocations in the (100) tilt boundaries appear as a swath of points in the lower right and upper left. The (100) boundaries in Figure 3a seem to be composed of groups of tilt boundaries of 4-10 dislocations each. A few [100] screws are visible. The distinct rows of dislocations passing between the tilt boundaries lie near in the (010) plane and bear a similarity to glide plane rows of (010) [100] prominent at the $[110]_c$ orientation of stress. (d) Remnants of the $[101]_c$ fingerprint with incipient organization of a (100) tilt boundary at the left. Most of the numerous dislocations running ENE are [001] screw dislocations. Note the many dislocation loops at the lower right. Scale bar in Figure 3a is 1 mm; in Figures 3b-3d, it is 50 μm.

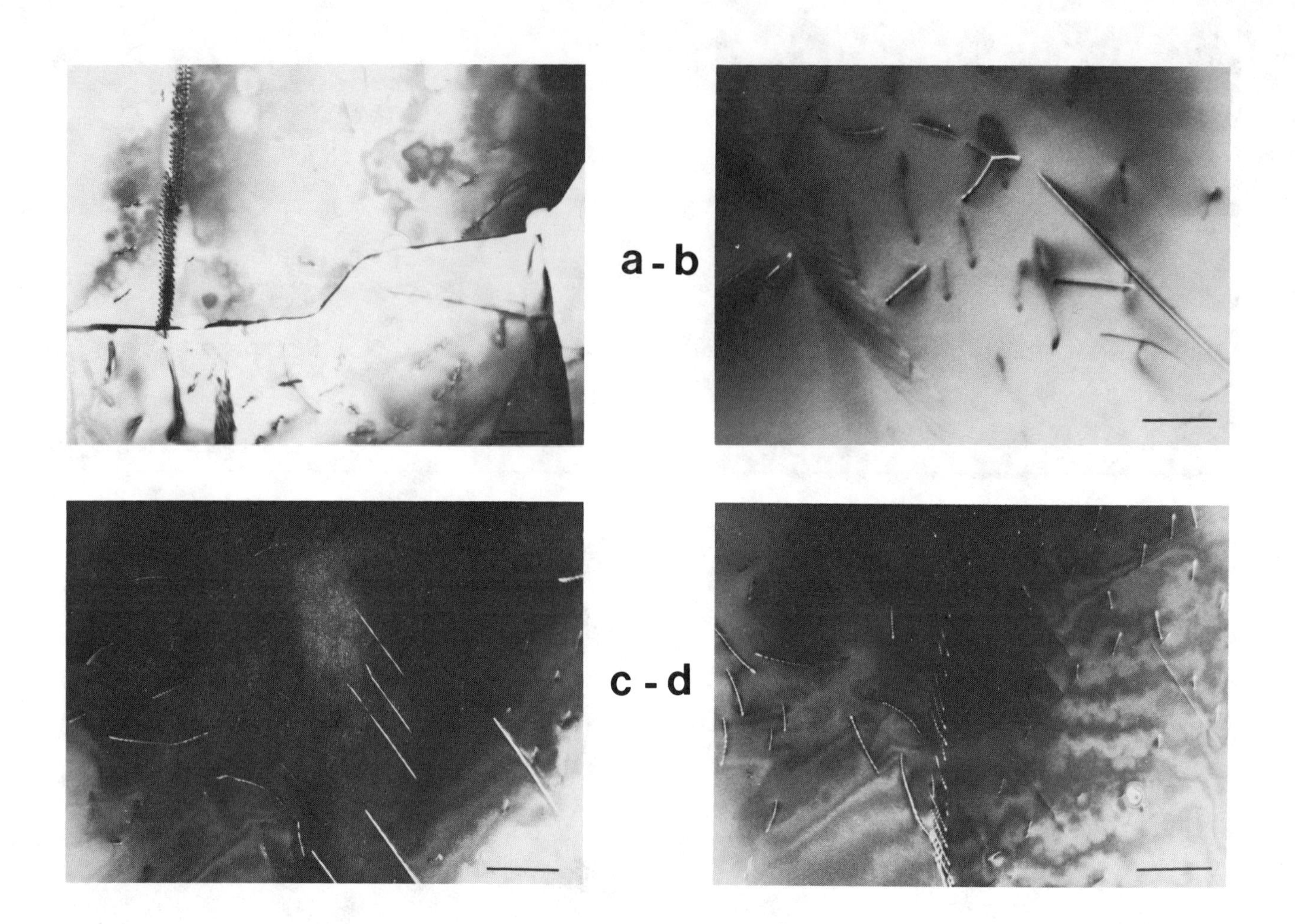

Fig. 4. Transmission electron micrographs of sample 5(2) illustrating (a) low-angle tilt boundaries, (b) a rare interaction between two dislocations, upper center, (c) very straight [001] screw dislocations, and (d) the formation of a (100) tilt boundary. Note that Figures 4c and 4d are taken in approximately the same region of the sample but under diffracting conditions which emphasize $\underline{b}$ = [001] and $\underline{b}$ = [100] dislocations, respectively. Scale bars represent 1 μm.

perhaps in the process of forming the [100]-organization (Figure 4d). Importantly, dislocations with different Burgers vectors are observed interacting with each other only on rare occasion (Figure 4b).

Discussion

Both (1) the deformation behavior and (2) the dislocation structures indicate that, for the slip systems tested here, the presence of dislocations of one system does not interfere with the action of a second system.

1. For a crystal first compressed along one direction and then along a second direction to activate a new slip system, the creep rate at the second orientation goes immediately to a steady-state value. If the dislocations introduced at the first orientation were obstacles to the new set of dislocations, the creep rate might be expected to accelerate as the obstacles were removed by recovery (climb and cross slip) processes.

2. Dislocations are not bowed out between forest dislocations and no pile-ups are observed. There is no evidence of dislocation tangling, even on the smallest scale (Figures 3, 4). Some sign of local interference might be anticipated where $\underline{b}$ = [100] and $\underline{b}$ = [001] dislocations intersect (such as at the $[101]_c$ orientation and under the $[101]_c \rightarrow [110]_c$ rotation). Recovery of the two-dislocation tangles which must form after such intersections [Friedel, 1967, section 5.3] appears to be nearly instantaneous. Furthermore, the high density of low-angle tilt and twist

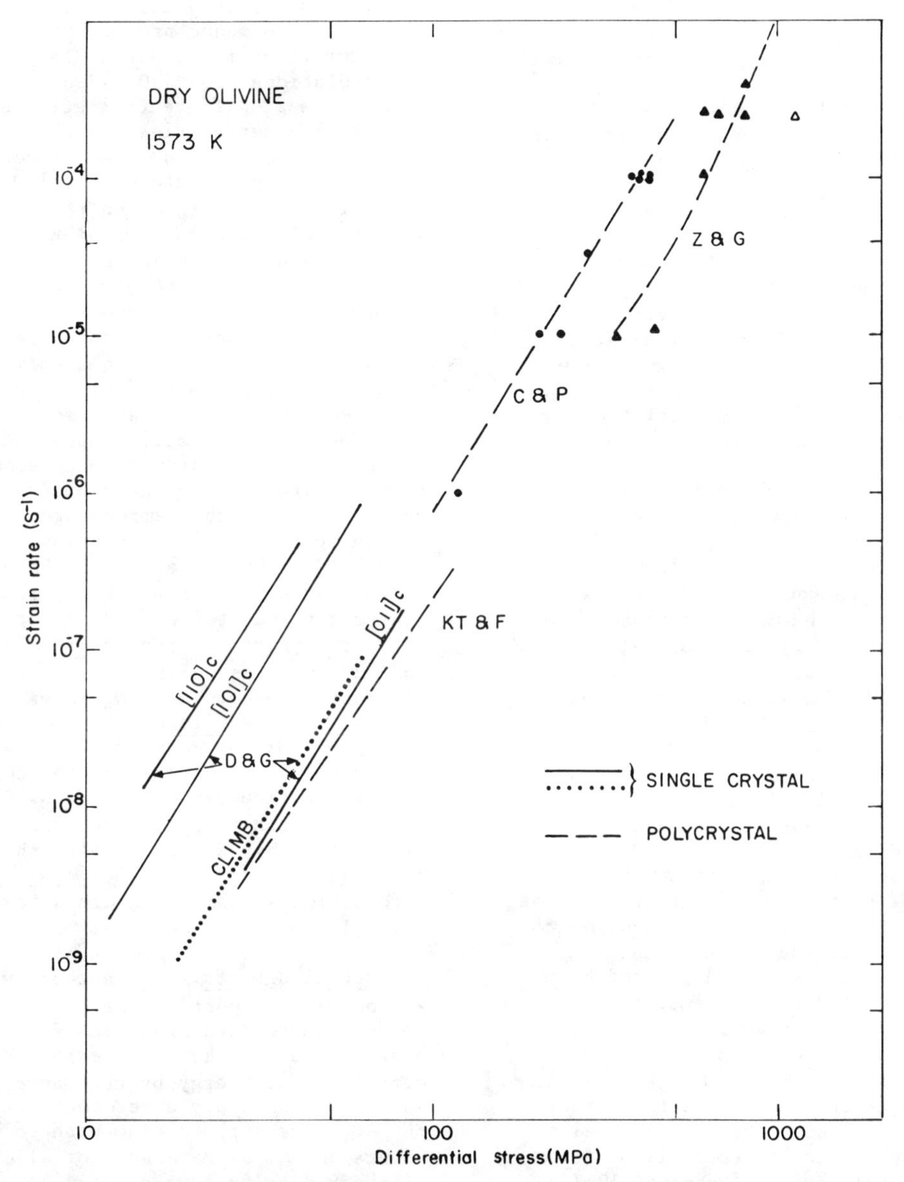

Fig. 5. Comparison of experimental flow laws for single crystal and polycrystalline olivine. Studies are coded as follows: D & G, Durham and Goetze [1977]; KT&F, Karato et al. [1982]; C & P, P. N. Chopra and M. S. Paterson (unpublished manuscript, 1983); Z & G, Zeuch and Green [1979]. Data from Durham and Goetze [1977] are extrapolated from 1873 K using an activation energy of 0.523 MJ/mole. Data from Karato et al. [1982] are extrapolated from 1923 K using 0.544 MJ/mole. Neglible extrapolation was necessary for the remaining data, so actual published points are shown. Careful infrared absorption measurements by Chopra and Paterson established their samples to be dry (solid circles). The runs by Zeuch and Green [1979] were conducted in an anhydrous environment (triangles). The same represented by the open triangle was distinguished by being "carefully dried" before the run. The dotted line, from Durham and Goetze [1977] represents deformation due to climb, measured at the $[101_c]$ orientation to be 10% of the total deformation.

boundaries (Figures 3a, 4a) demonstrates that climb and cross slip must be sufficiently rapid in olivine to allow dislocations to move past one another at velocities on the order of the glide velocity.

This study has not been exhaustive in that all possible slip system interactions have not been explored. For example, the slip system activated at the $[011]_c$ orientation did not enter the study. However, the interactions studied here involve the basic mechanisms of recovery (climb and cross slip motion of dislocations), so the results likely apply to all dislocation interactions in olivine.

Durham and Goetze [1977] and Durham et al. [1977] hypothesized that the polycrystalline flow law for olivine must be softer than the hardest of the several single crystal flow laws they had measured. Their argument was based on the following points: (1) flow laws for four independent deformation systems in olivine can be measured in the laboratory, (2) the existence of four independent deformation systems is a sufficient condition for real (inhomogeneous) flow, and (3) simultaneous action of one slip system does not impede the action of a second slip system. (A "deformation system" in the above statements is any real mechanism of plastic flow which contributes to one or more components of the small strain tensor. Deformation systems include slip systems, whose shape change contribution is simple shear, and climb systems, whose contribution is pure shear. It should be emphasized that climb systems referred to here are real mechanisms, meaning they comprise the strain due to the climbing dislocation and that due to bulk (or pipe) diffusion as required to conserve the volume of the system. In contrast, the "climb systems" discussed by Groves and Kelly [1969] are mathematical formulations which are not volume-conservative and which cannot exist unaccompanied in real plastic flow.)

Point 3 has been the subject of this study and is clearly supported. Point 1 is based on the observation that simultaneous climb of $\underline{b}$ = [100] and [001] dislocations [Nabarro, 1967] contributes at least 10% of the total strain in single crystal flow at the $[101]_c$ orientation [Durham et al., 1977]. Point 2 is the assertion that the von Mises criterion (5 independent deformation systems are sufficient for homogeneous flow [von Mises, 1928]) is over restrictive in a real situation where inhomogeneous flow can accommodate a small population of unfavorably oriented grains. This assertion is supported theoretically in hexagonal materials [Hutchinson, 1977] and by computer simulations of deformation in quartz [Lister et al., 1978].

Existing data on flow of dry polycrystalline olivine are compared to single crystal flow laws in Figure 5, which shows good support for the relationship advanced here between single crystal and polycrystalline flow. The relationship is by no means proven, however. To make the comparison in Figure 5, large temperature extrapolations (300-350 K) were required of data taken at differential stresses <100 MPa; resultant uncertainties are at least half an order of magnitude in strain rate. Also, the polycrystals of Karato et al. [1982] were in fact single crystals which recrystallized to a polycrystalline structure at high strains. Two sets of data pertain to samples which began as polycrystals and which were necessarily deformed in the presence of high confining pressure. The measurements of Zeuch and Green [1979] were carried out in a solid-medium apparatus while those of P. N. Chopra and M. S. Paterson (unpublished manuscript, 1983) were done in a gas-medium apparatus. The latter must be considered more accurate since they are free of the large uncertainties associated with the temperature gradients and confining media in solid-medium deformation rigs. P. N. Chopra and M. S. Paterson (unpublished manuscript, 1983) found, in contrast to dunites crept with water available from hydrous phases [Chopra and Paterson, 1981], that dry dunites deform with little or no grain boundary sliding; that is, they deform primarily by grain-matrix (dislocation) mechanism. The flow laws for the recrystallized single crystals of Karato et al. [1982] and for the dry polycrystalline samples of P. N. Chopra and M. S. Paterson (unpublished manuscript, 1983) lie comfortably within or near the field defined by the flow laws for the four deformation systems determined for single crystals.

Acknowledgments. The authors wish to acknowledge support of the National Science Foundation through grant EAR-7919725. Work performed under the auspices of the U.S. Department of Energy by the Lawrence Livermore National Laboratory under contract W-7405-ENG-48. The authors also wish to thank P. N. Chopra and M. S. Paterson for allowing the use of their data in advance of publication.

References

Chopra, P. N., and M. S. Paterson, The experimental deformation of dunite, Tectonophysics, 78, 453-473, 1981.

Darot, M., and Y. Gueguen, High-temperature creep of forsterite single crystals, J. Geophys. Res., 86, 6219-6234, 1981.

Durham, W. B., and C. Goetze, Plastic flow of oriented single crystals of olivine, 1, Mechanical data, J. Geophys. Res., 82, 5737-5753, 1977.

Durham, W. B., C. Goetze, and B. Blake, Plastic flow in oriented single crystals of olivine, 2, Observations and interpretations of dislocation structures, J. Geophys. Res., 82, 5755-5770, 1977.

Friedel, J., Dislocations, Addison-Wesley, Reading, Mass., 1967.

Goetze, C., and D. L. Kohlstedt, Laboratory study of dislocation climb and diffusion in olivine, J. Geophys. Res., 78, 5961-5971, 1973.

Groves, G. W., and A. Kelly, Change of shape due to dislocation climb, Philos. Mag., 19, 977-986, 1969.

Hutchinson, J. W., Creep and plasticity of hexagonal polycrystals as related to single crystal slip, Trans. Metall. Soc. AIME, 8A, 1465-1469, 1977.

Karato, S., M. Toriumi, and T. Fujii, Dynamic recrystallization and high temperature rheology of olivine, in High-Pressure Research in Geophysics, Adv. in Earth and Planet. Sci., vol. 12, edited by S. Akimoto and M. H. Manghnani, pp. 171-189, Center for Academic Publication Japan, Tokyo, 1982.

Kohlstedt, D. L., and C. Goetze, Low-stress high-temperature creep in olivine single crystals, J. Geophys. Res., 79, 2045-2051, 1974.

Kohlstedt, D. L., and P. Hornack, Effect of oxygen partial pressure on the creep of olivine, in Anelasticity in the Earth, Geodynamics Ser., vol. 4, edited by F. D. Stacey, M. S. Paterson, and A. Nicolas, pp. 101-107. AGU, Washington, D. C., 1981.

Kohlstedt, D. L., C. Goetze, W. B. Durham, and J. B. Vander Sande, A new technique for decorating dislocations in olivine, Science, 191, 1045-1046, 1976.

Lister, G. S., M. S. Paterson, and B. E. Hobbs, The simulation of fabric development in plastic deformation and its application to quartzite--The model, Tectonophysics, 45, 107-158, 1978.

Nabarro, F. R. N., Steady-state diffusional creep, Philos. Mag., 16, 231-237, 1967.

Ricoult, D., Experimental annealing of a natural dunite, Bull. Mineral., 102, 86-91, 1979.

Toriumi, M., and S. Karato, Experimental studies on the recovery process of deformed olivines and the mechanical state of the upper mantle, Tectonophysics, 49, 79-95, 1978.

von Mises, W., Mechanik der plastischen Formanderung von Kristallen, Z. Angew. Math. Mech., 8, 161, 1928.

Zeuch, D. H., and H. W. Green, Experimental deformation of an "anhydrous" synthetic dunite, Bull. Mineral., 102, 185-187, 1979.

EXPERIMENTAL DIFFUSIONAL CRACK HEALING IN OLIVINE

B. J. Wanamaker and Brian Evans[1]

Department of Geological and Geophysical Sciences, Princeton University
Princeton, New Jersey 08544

Abstract. Both natural and laboratory produced cracks in San Carlos peridot heal by a two stage process involving the initial formation of cylindrical voids and the subsequent formation of spherical pores when subjected to heat treatments at 1250°C - 1400°C; laboratory produced cracks also heal rapidly at 1000°C. In addition to being thermally activated, the kinetics of the healing process depend on the crack dimensions and apparently either the chemical speciation or the pressure of the fluid filling the crack. Interpretation of the results of these experiments using theories of diffusive crack healing developed for metals and ceramics indicates that surface diffusion is probably the dominant mechanism of crack healing. The effective activation energy of the break-up of the cylindrical voids is approximately 53 ± 22 kcal/mole and the time necessary to produce a spherical void from a cylinder is proportional to the fourth power of the radius of the cylinder. The region of healed crystal between the spherical pores contains dislocations which are apparently formed during the healing process as accommodation of strain between the free crack surfaces.

Introduction

The study of fluid inclusions in mantle xenoliths has provided information on the chemical speciation, fugacity, and migration of fluids at depth in the earth [Roedder, 1965; Murck et al., 1978; Green and Radcliffe, 1975]. Fluid inclusions may be dispersed throughout the xenolith grains or may occur in associations along curved planes of healed fractures [Roedder, 1965, 1981; Simmons and Richter, 1976]. Healed fractures may be either intra- or intergranular, and may be filled with the same material as the host grains (healed cracks) or infilled with different minerals (sealed cracks) (Batzle and Simmons [1976], Padovani et al. [1982], and Figure 1 in this paper). In the case of intragranular healed microcracks in olivine, many cracks appear to have been initiated at pre-existing fluid inclusions, apparently due to decrepitation of these inclusions during uplift of the xenolith (e.g., Figure 19 of Roedder [1965] and Figure 2 in this paper). Because such properties as seismic velocity, seismic attenuation, and electrical conductivity depend on crack shape as well as total crack porosity [Shankland et al., 1981], any process which is capable of modifying crack shape or inter-connectivity is an important ingredient in our understanding of the physical properties of the mantle.

Crack healing has been observed experimentally in fluid filled fractures in sodium nitrate [Lemmlein and Kliya, 1952; Lemmlein, 1956] and quartz [Shelton and Orville, 1980; Pecher, 1981; Smith and Evans, 1984]. However, there is abundant evidence to indicate that the presence of a pore fluid is not necessary for crack healing if the temperature allows atomic or ionic mobility through lattice diffusion, surface diffusion, or vapor transport [Nichols and Mullins, 1965a, b; Yen and Coble, 1972; Gupta, 1975, 1976, 1978, 1980; Nichols, 1976]. In this paper we present the results of some experiments on crack healing in single crystals of olivine at temperatures from 1000°C to 1400°C. In general, the results conform to the theories developed for diffusive crack healing and, hence, will help to increase our understanding of the thermo-mechanical history of the peridotite nodules and megacrysts in which they are found.

Experimental Method

Polished samples of single crystal San Carlos peridot (1 cm x 1 cm x 1 mm) were heat treated in an H_2 - CO_2 buffered atmosphere at several temperatures. During each experiment, the oxygen fugacity inside the furnace was maintained between the iron-wustite and QFM buffers by mixing appropriate volumes of hydrogen and carbon dioxide gases at a total flow rate of 100 cc/min [Deines et al., 1974]. We judged this method to

[1]Now at Earth, Atmospheric, and Planetary Sciences, Massachusetts Institute of Technology, Cambridge, Massachusetts 02139.

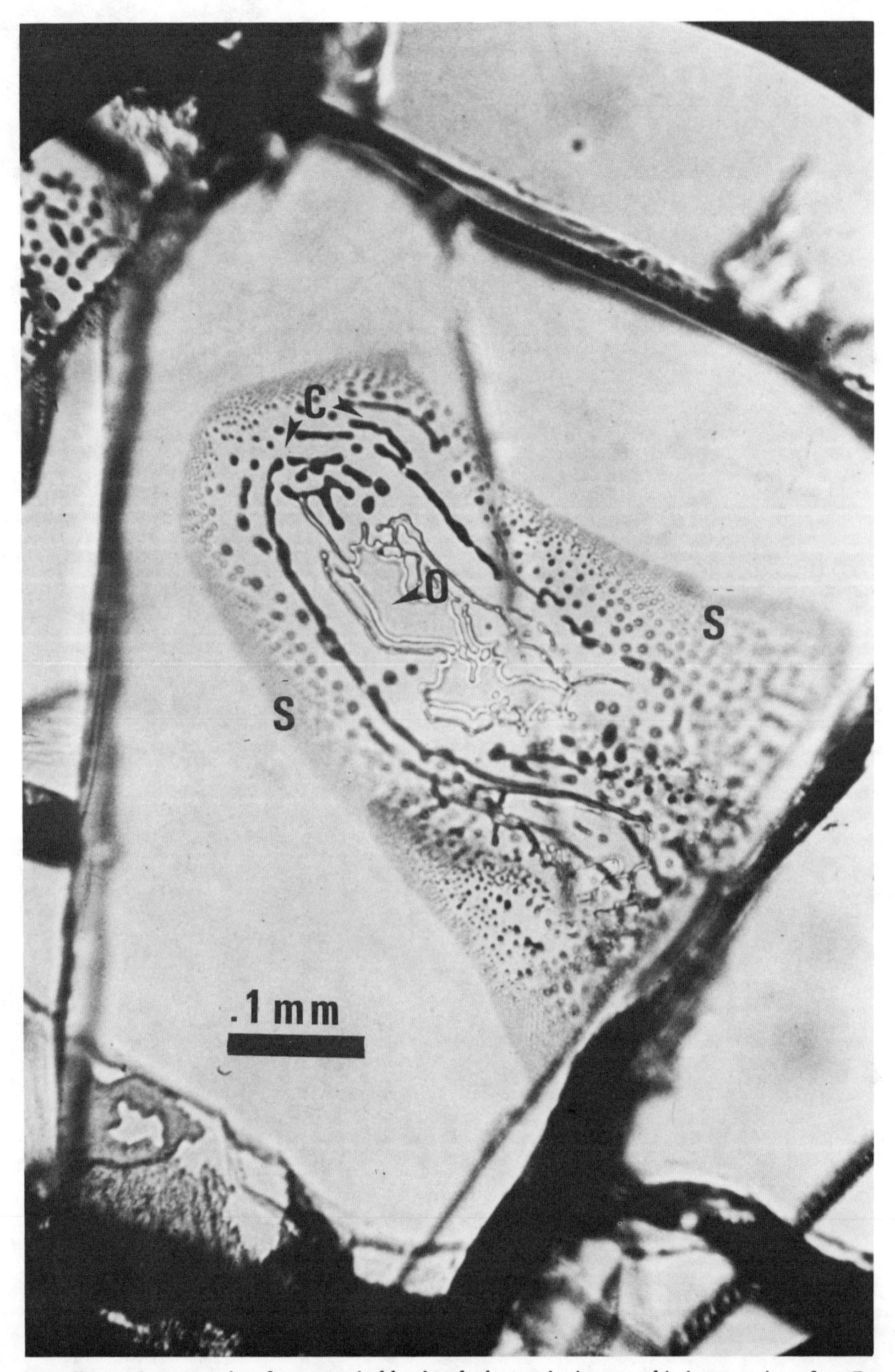

Fig. 1. Photomicrograph of a partially healed crack in an olivine grain of a Type 1 nodule from Lunar Crater, Nevada. The central unhealed portion of the crack (O) is surrounded by cylindrical pores (C) and then by isolated spherical pores (S) near the tip of the original crack. Olivine, CO_2 liquid plus vapor and silicate glass now fill the crack [Bergman, 1982].

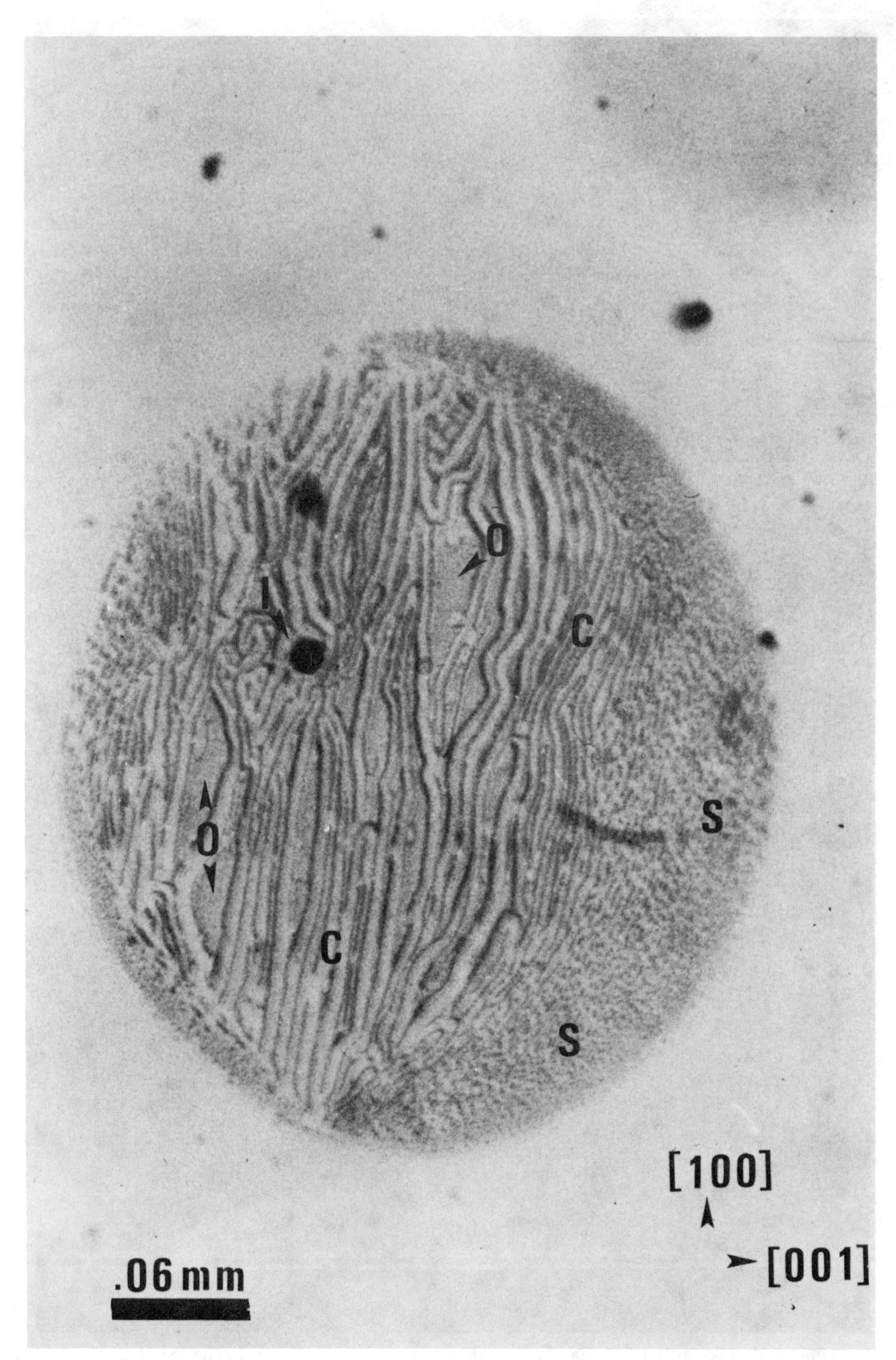

Fig. 2. Photograph (in plan view) of a partially healed natural crack in San Carlos peridot formed by the decrepitation of a pre-existing fluid inclusion. Unhealed areas (O), cylindrical pores (C), isolated spherical pores (S), the orginal pre-existing fluid inclusion (I), and crystallographic directions are labelled.

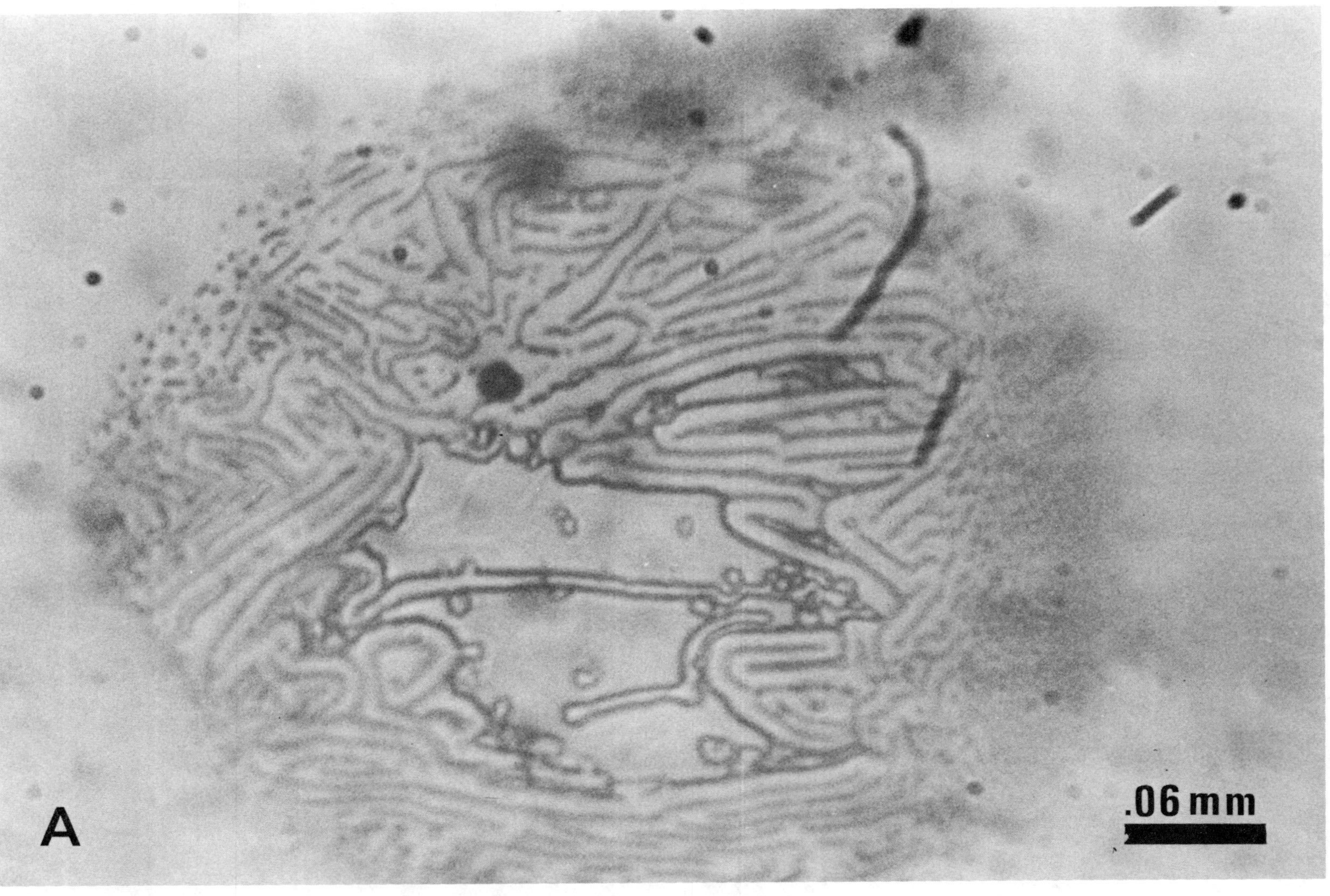

Fig. 3. (a) Partially healed decrepitated fluid inclusion crack in San Carlos peridot before heat treatment. (b) Same feature as in Figure 3a after heat treatment for 64 hrs. at 1400°C. (c) A second partially healed natural crack from the same sample as Figure 3a before heat treatment. (d) Same feature as in Figure 3c after heat treatment for 64 hrs. at 1400°C. Spheroidization (SP) and ovulation (OV) events are labelled.

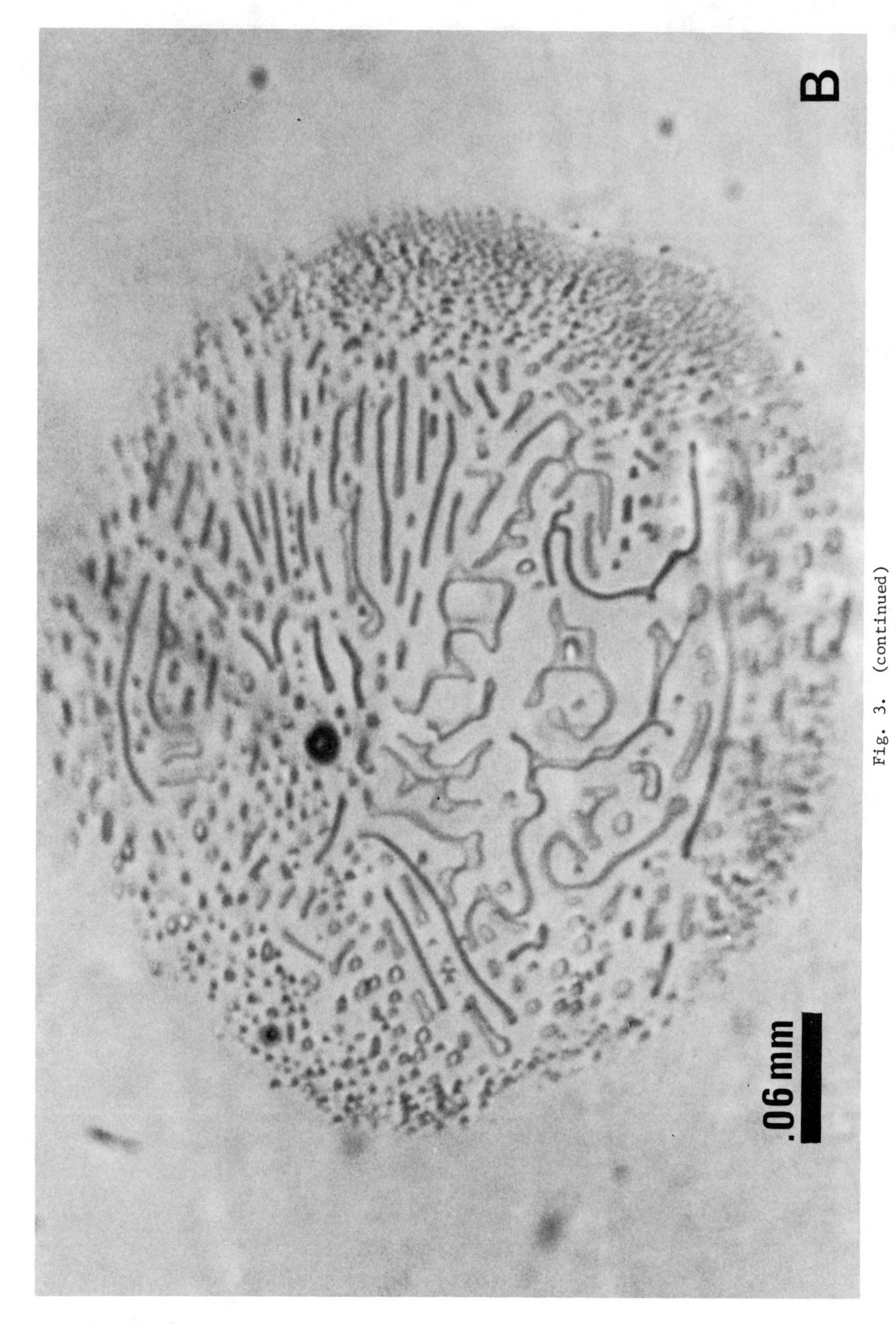

Fig. 3. (continued)

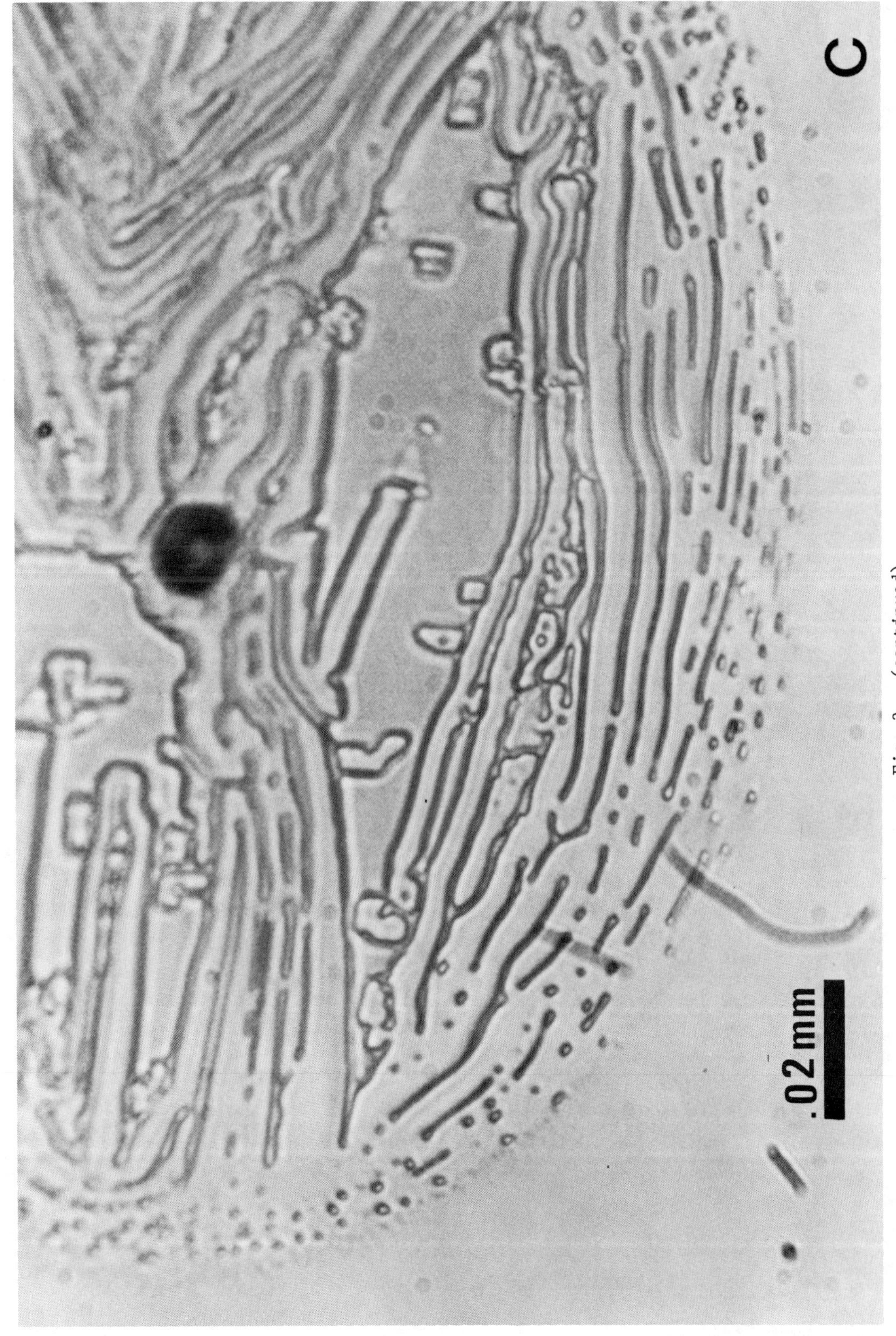

Fig. 3. (continued)

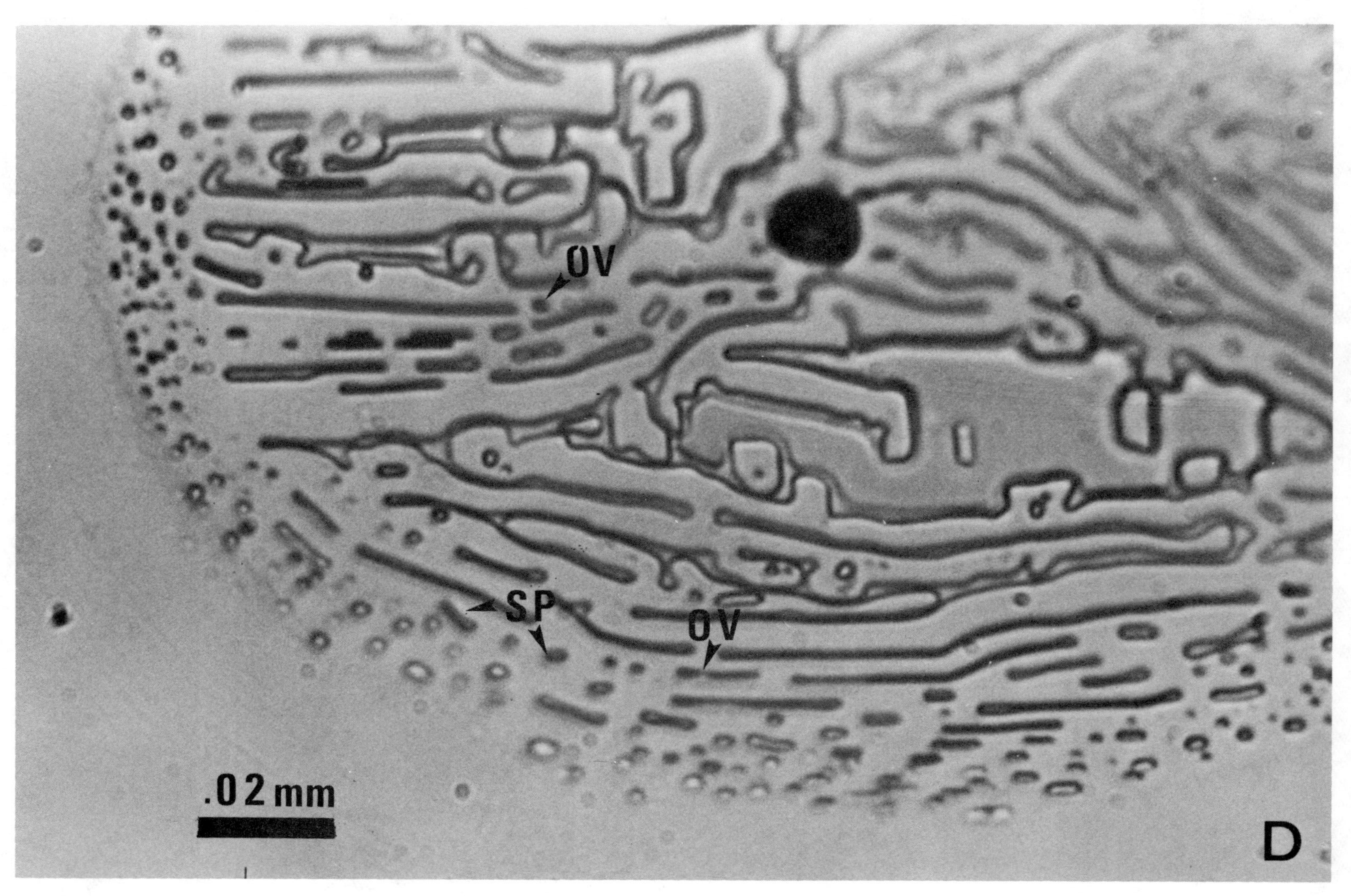

Fig. 3. (continued)

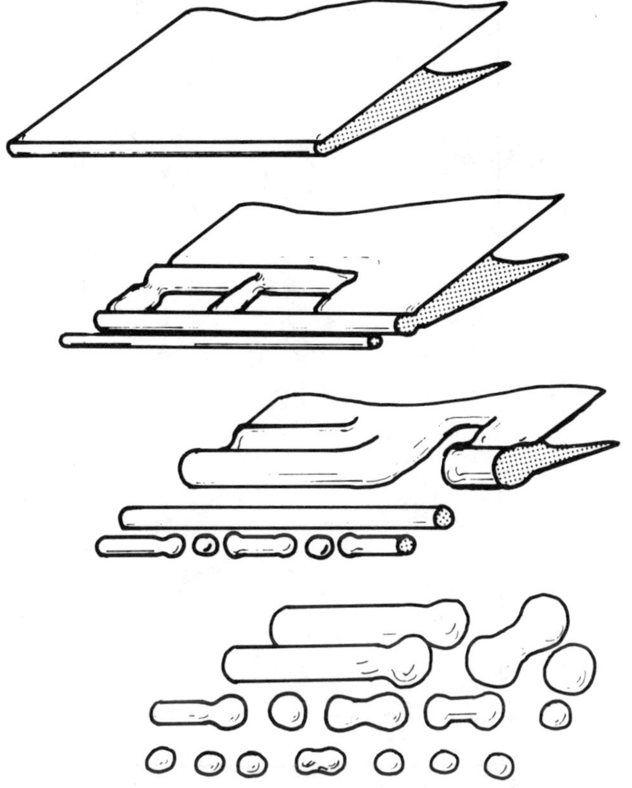

Fig. 4. Schematic illustration of typical changes in pore morphology during crack healing. The open crack tip recedes by forming cylindrical pores which then undergo ovulation or spheroidization to form isolated spherical pores.

be successful since the samples lacked oxidation or reduction products when subsequently studied under the petrographic microscope. Samples were repolished after each heat treatment to remove thermal etching and surface tension effects. Samples heat treated at 1000°C and 1250°C were suspended in a shallow platinum cup and those annealed at higher temperatures were supported by a second piece of olivine in the bottom of an alumina cup.

We observed crack healing with an optical microscope by comparing changes in pore morphology before and after the heat treatment (Figure 3). Pore shape changes were observed in both natural, partially healed cracks apparently formed by the decrepitation of fluid inclusions [see Roedder, 1965; Kirby and Green, 1980] and in cracks produced in the laboratory by thermal shock. Cracks produced in the laboratory intersected at least one surface of the sample and were thus exposed to the buffering atmosphere, whereas the natural cracks contained the fluids which were present at depth. Both intact and decrepitated natural fluid inclusions in these samples contain dominantly CO_2 (S. C. Bergman (personal communication, 1982) and work done in the course of this study) as has been found in previous studies of other xenoliths [Roedder, 1965; Murck et al., 1978; Bergman, 1982].

Theory and Results

Morphological changes during crack healing are shown in Figure 3. The tip of an open crack recedes at first by forming cylindrical pores at

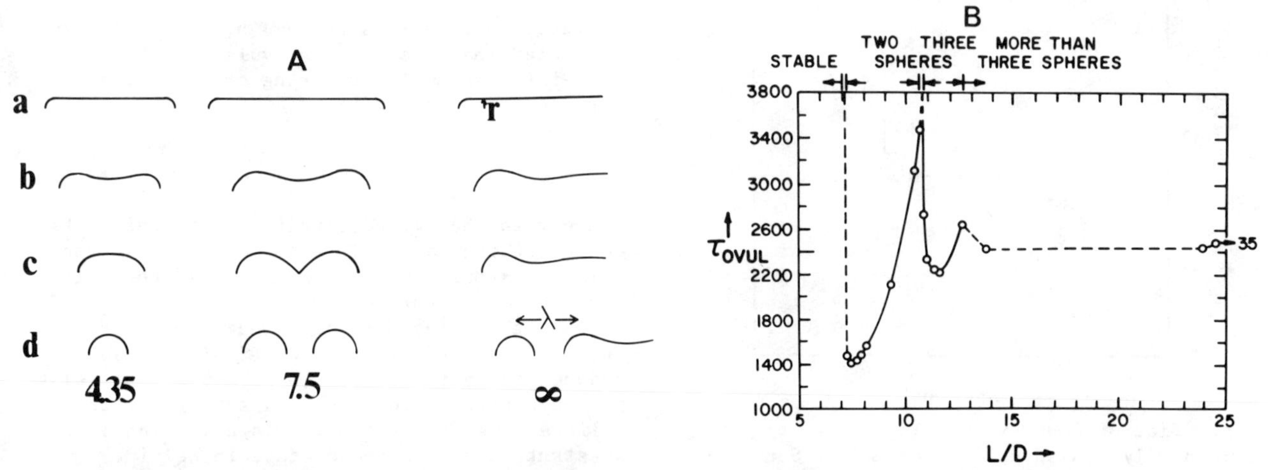

Fig. 5. (a) Cross-section of sequential morphological changes a-d in cylindrical pores of various aspect ratios (length divided by diameter) of 4.35, 7.5, and ∞. Cylinders with an aspect ratio greater than 7.2 will form a single, large spherical aspect ratio less than 7.2 will form a single, large spherical pore (spheroidization) while those with an aspect ratio greater than 7.2 will form two or more spherical pores (ovulation) [after Nichols, 1976]. (b) The effect of the aspect ratio (L/D) of a cylindrical pore on the dimensionless ovulation time ($\tau_{ovul} = (18/\pi r)^4 \tau \Omega^2 \nu D \gamma$) [after Nichols, 1976]. Symbols are defined in the text.

its edge. Because the cylindrical pores are thermodynamically unstable, they proceed to form isolated spherical pores during continued heat treatment. This process conserves total pore volume and is shown schematically in Figure 4.

Nichols and Mullins [1965a, b] have theoretically modelled the changes in shape of a semi-infinite cylinder with a hemispherical tip. These changes are driven by gradients in chemical potential caused by perturbations in surface curvature. Any longitudinal surface perturbation with a wavelength greater than the cylinder circumference is unstable and results in the formation of an isolated sphere at the end of the cylinder. This process is termed "ovulation." Cylinders below a critical aspect ratio will "spheroidize" into a single large sphere instead of ovulating [Nichols, 1976]. The dominant diffusional path during ovulation can be identified by the ratio of the spacing between spheres (λ) and the radius of the parent cylinder (r): λ/r is 8.89 for surface diffusion, 9.02 for diffusion through the pore fluid, and 12.96 for lattice diffusion [Nichols and Mullins, 1965a]. Figure 5a illustrates the relationships discussed above and Figure 5b details the effect of aspect ratio on the time for ovulation or spheroidization. For both laboratory produced and natural cracks, the ratio λ/r was measured to be 7.8 ± 1.9.

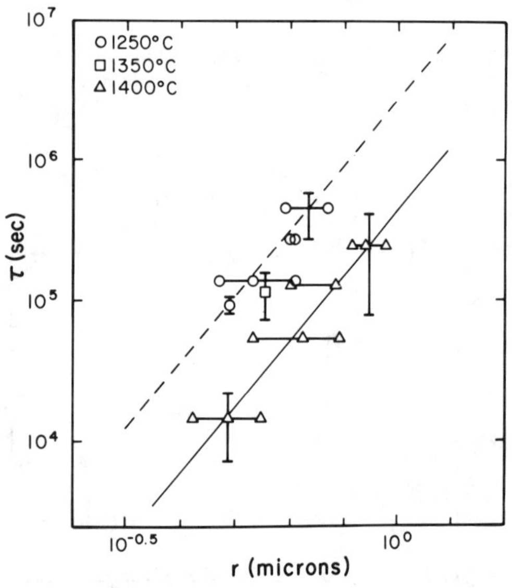

Fig. 6. Plot of log τ versus log r for experimentally healed natural cracks in San Carlos peridot. Typical error bars are shown. For a given τ, data points which overlap because of similar radii have not been shown as separate points. See Table 1 for the complete data set. The solid line is the least squares fit to all the data at 1400°C. The dashed line has been drawn parallel to the solid line but through the data at 1250°C.

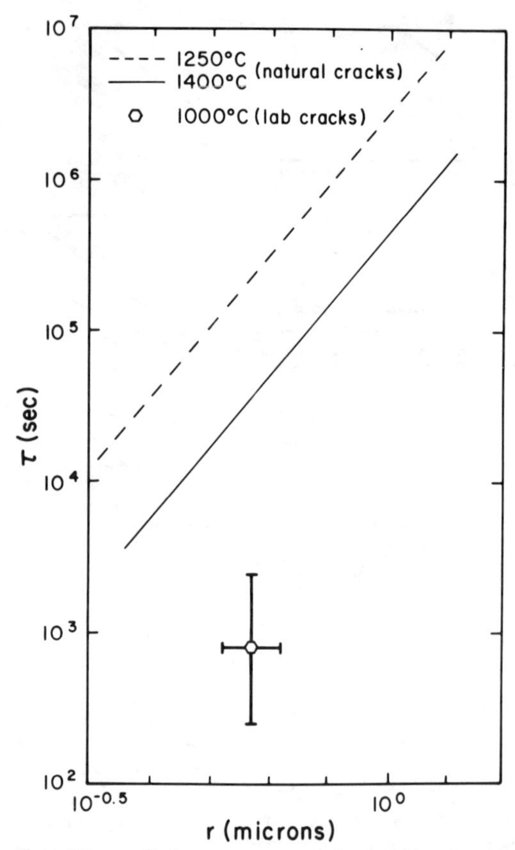

Fig. 7. Plot of log τ versus log r for experimentally healed laboratory produced cracks in San Carlos peridot. Shown for comparison are the lines from Figure 6 for natural cracks.

Nichols and Mullins [1965a, b] theoretically predicted the time for the ovulation of a semi-infinite cylinder during crack healing by surface diffusion to be

$$\tau = \frac{\beta r^4 k T}{D \gamma \Omega^2 \nu} \qquad (1)$$

where r is the cylinder radius, γ is the surface energy of the material (assumed to be isotropic), Ω is the atomic volume (we have used the unit cell volume of forsterite), ν is the number of diffusing species per unit area ($\nu \equiv \Omega^{-2/3}$), D is the coefficient of surface diffusion (also assumed to be isotropic), β is a constant related to the aspect ratio (length/diameter) of the cylinder (see Figure 5b), k is Boltzmann's constant, and T is temperature in Kelvin.

We measured the ovulation times of cylinders of various radii by observing the time (τ) necessary for complete separation of a sphere from the end of a cylinder at a particular temperature. The data are graphed as log τ versus log r for natural cracks in Figure 6 and for natural and laboratory cracks in Figure 7 and are tabulated along with the experimental conditions in

TABLE 1. Summary of Experimental Data

T, °C	r, microns	τ_{ovul}	τ, x10^5 sec	τ^*, x10^5 sec	Comments
1000	0.66	2400	0.00252	0.00191	data from sequential heat treatments on one laboratory produced crack in sample SC-105b
	0.52	2400	0.024	0.023	
1250	0.49	2400	0.936 ± 0.072	1.02	data from sequential heat treatments of 4 hrs., 6 hrs., 12 hrs., 4 hrs., 25 hrs., and 50 hrs. on one natural decrepitation crack in sample SC-201
	0.64	2650	1.39 ± 0.45	0.47	
	0.63	2400	2.74 ± 0.90	1.09	
	0.64	2400	2.74 ± 0.90	0.89	
	0.47	2300	1.39 ± 0.45	1.86	
	0.62	2400	3.64 ± 0.90	1.54	
	0.54	2400	1.39 ± 0.45	1.02	
1350	0.57	2400	1.15 ± 0.43	0.68	data from sequential heat treatments of 8 hrs. and 24 hrs. on one natural decrepitation crack in sample SC-302
1400	0.55	2400	0.144 ± 0.07	0.098	data from sequential heat treatments of 4 hrs. and 64 hrs. on four natural decrepitation cracks in sample SC-204c
	0.54	2300	0.144 ± 0.07	0.110	
	0.50	1900	0.144 ± 0.07	0.182	
	0.49	2200	0.144 ± 0.07	0.170	
	0.49	2700	0.144 ± 0.07	0.139	
	0.56	2300	0.144 ± 0.07	0.095	
	0.42	2400	0.144 ± 0.07	0.298	
	0.51	2200	0.144 ± 0.07	0.145	
	0.49	2400	0.144 ± 0.07	0.156	
	0.49	2400	0.144 ± 0.07	0.156	
	0.67	2400	0.54 ± 0.27	0.167	
	0.78	2400	0.54 ± 0.27	0.091	
	0.54	2400	0.54 ± 0.27	0.381	
	0.83	2400	2.44 ± 1.65	0.32	
	0.95	2400	2.44 ± 1.65	0.19	
	0.87	2400	2.44 ± 1.65	0.27	
	0.86	2400	2.44 ± 1.65	0.28	
	0.85	2400	2.44 ± 1.65	0.29	
	0.82	2400	2.44 ± 1.65	0.34	
	0.64	2400	1.29 ± 0.58	0.48	
	0.77	2400	1.29 ± 0.58	0.20	
	0.65	2650	1.29 ± 0.58	0.41	

$\tau^* = (2400/\tau_{ovul})\tau(r^*/r)^4$

Table 1. A linear least squares fit to the data at 1400°C (solid line) yields a line with a slope of 4.7 ± 0.9. The scatter in the data of Figures 6 and 7 is adequately accounted for by an experimental uncertainty in the exact time of ovulation of one half the duration of the last heat treatment, an uncertainty of 0.15 microns in the cylinder radius due to the optical resolution of the microscope, and the effect of varying aspect ratio. In Table 1, the measured τ for each data point has been corrected for the effect of aspect ratio using Figure 5b and recalculated for a standard radius (r^*) of 0.5 microns assuming an r^4 dependence. The mean value of the corrected ovulation times (τ^*) for each temperature is used in the Arrhenius plot of Figure 8.

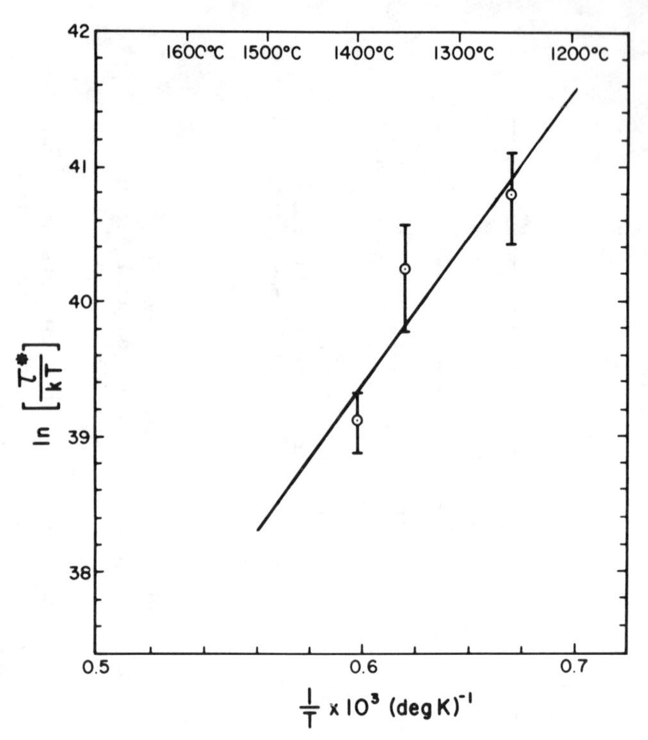

Fig. 8. Arrhenius plot using the mean for each temperature of the measured ovulation times after correcting for the effect of aspect ratio and recalculated for a standard radius of 0.5 microns (see Table 1). The effective activation energy for the ovulation process can be determined from the slope of the best fit line assuming that the activation energy is a constant.

Discussion

The data for subsequent healing in the laboratory of the natural, partially healed cracks in San Carlos peridot are consistent with the kinetic theory of Nichols and Mullins [1965a, b] in that ovulation is a thermally activated process and is proportional to r^4 within experimental error (Figure 6). Both ovulation and spheroidization events can be seen in Figure 3. The range of the measured ratio of λ/r encompasses the predicted value for both surface diffusion and diffusion through a pore fluid [see Nichols and Mullins, 1965b]. However, silicates exhibit low solubility in CO_2--on the order of 0.2 wt. % even at 1300°C and 20 kb (2 GPa) [Eggler and Rosenhauer, 1978]. Since CO_2 would always be saturated by a small amount of olivine, this suggests that surface diffusion, rather than pore fluid diffusion, is the more likely mechanism of crack healing in these experiments with CO_2 (vapor)-filled natural microcracks. In any case, one may determine from Figure 8 an effective activation energy of 53 ± 22 kcal/mole ((2 ± 1) x 10^5 J/mole) for the ovulation process in olivine. Using a surface energy of 1600 ergs/cm^2 (1.6 J/m^2) [Cooper and Kohlstedt, 1982], an atomic volume for forsterite of 0.29 (nm)^3, $\beta = 2400(\pi/18)^4$, and τ^* for 1400°C from Figure 8 and from (1), we calculate $\nu\Omega D$ for surface diffusion to be on the order of 10^{-16} (cm^2/sec) x cm at 1400°C. The assumption of isotropic surface energy may not be strictly true, however, since these natural cracks are crystallographically oriented parallel to (010) with the long dimension of the cylinders commonly aligned in the [100] direction.

Healing rates were vastly accelerated in the laboratory produced cracks as compared to natural cracks which still retained their original filling contents (Figure 7). In several experiments where the sample surface intersected the central portion of a natural crack, accelerated kinetics were also observed in that portion of the crack which had been exposed to the gas flow at 1 bar. It is possible that adsorbed molecular water or some other hydroxyl-bearing species introduced by the buffering gas plays an important role in aiding surface diffusion or promoting pore fluid diffusion, but at present the effect of pressure cannot be discounted. In addition, we have observed different healing rates in natural cracks in the same sample. Thus we believe that the crack healing kinetics are sensitive to either the chemical speciation or the internal pressure exerted on the crack walls by the filling contents rather than some intrinsic difference in the solid.

The natural, partially healed cracks in San Carlos peridot show several interesting relationships with the dislocation structure as revealed by decoration [Kohlstedt et al., 1976]. Fluid inclusions are sometimes located on and distorted from spherical symmetry by subgrain boundaries. In addition, as noted by Kirby and Green [1980] the fluid inclusions are often surrounded by dislocation nets (Figure 9a) indicating that they underwent stretching prior to decrepitation and crack formation. Finally, dislocations can be seen along the healed fracture surface bridging open cylinders (Figure

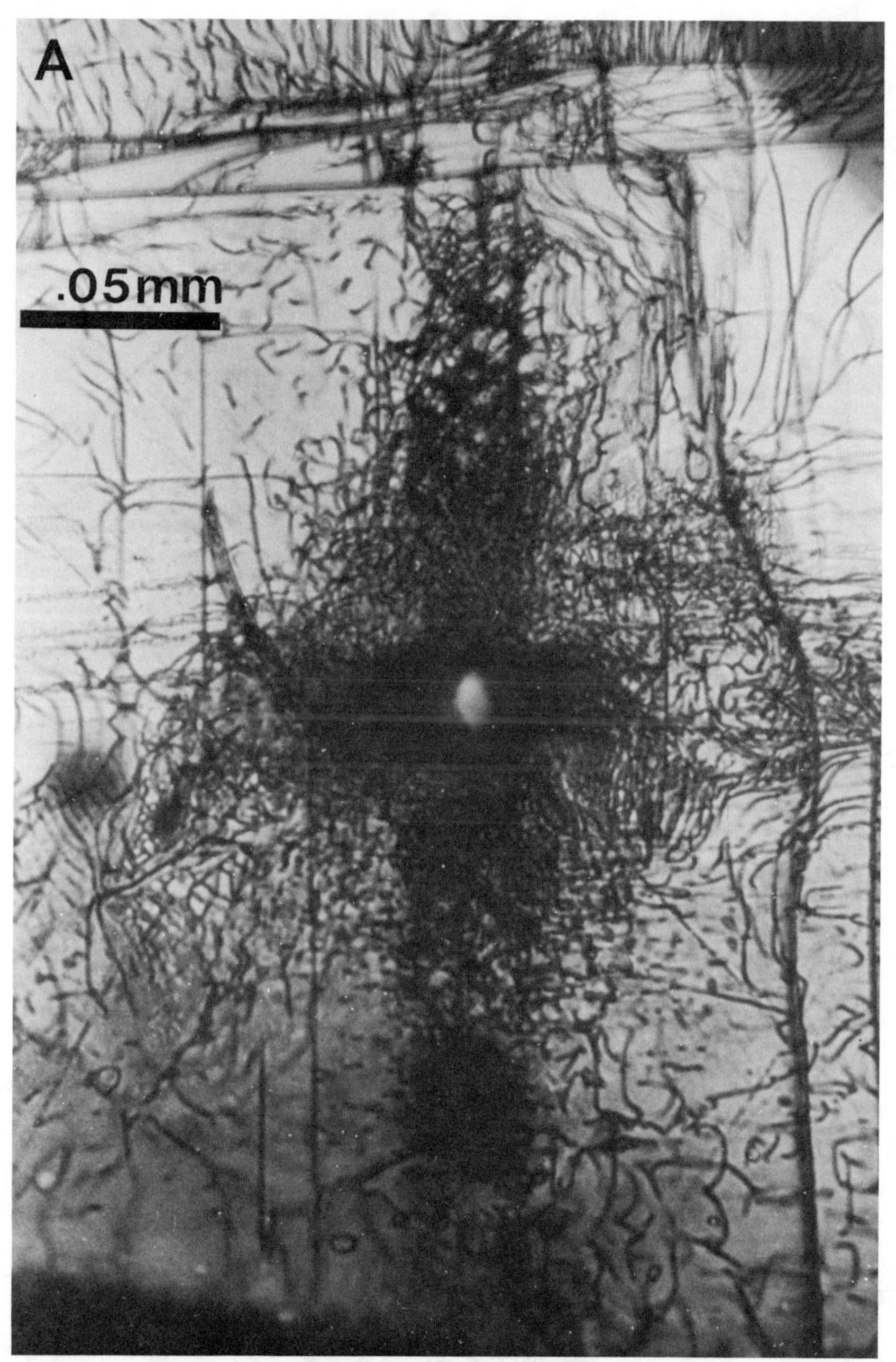

Fig. 9. (a) Optical micrograph in transmitted light of a dislocation net surrounding the original fluid inclusion of a natural crack in San Carlos peridot. (b) Optical micrograph in transmitted light of dislocations bridging open cylindrical pores on a natural crack in San Carlos peridot. (c) Optical micrograph in transmitted light of spherical pores connected by dislocations on a natural crack in San Carlos peridot. (d) Optical micrograph in transmitted light of dislocations which are not decorated with bubbles on a natural crack in San Carlos peridot.

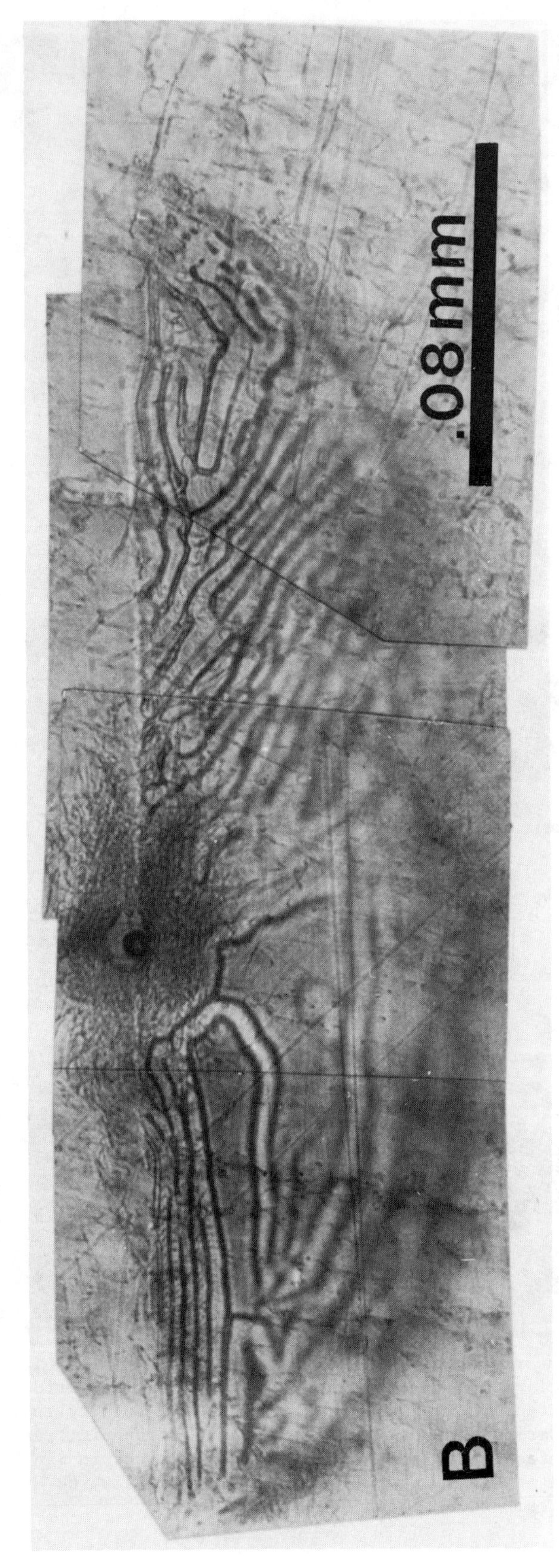

Fig. 9. (continued)

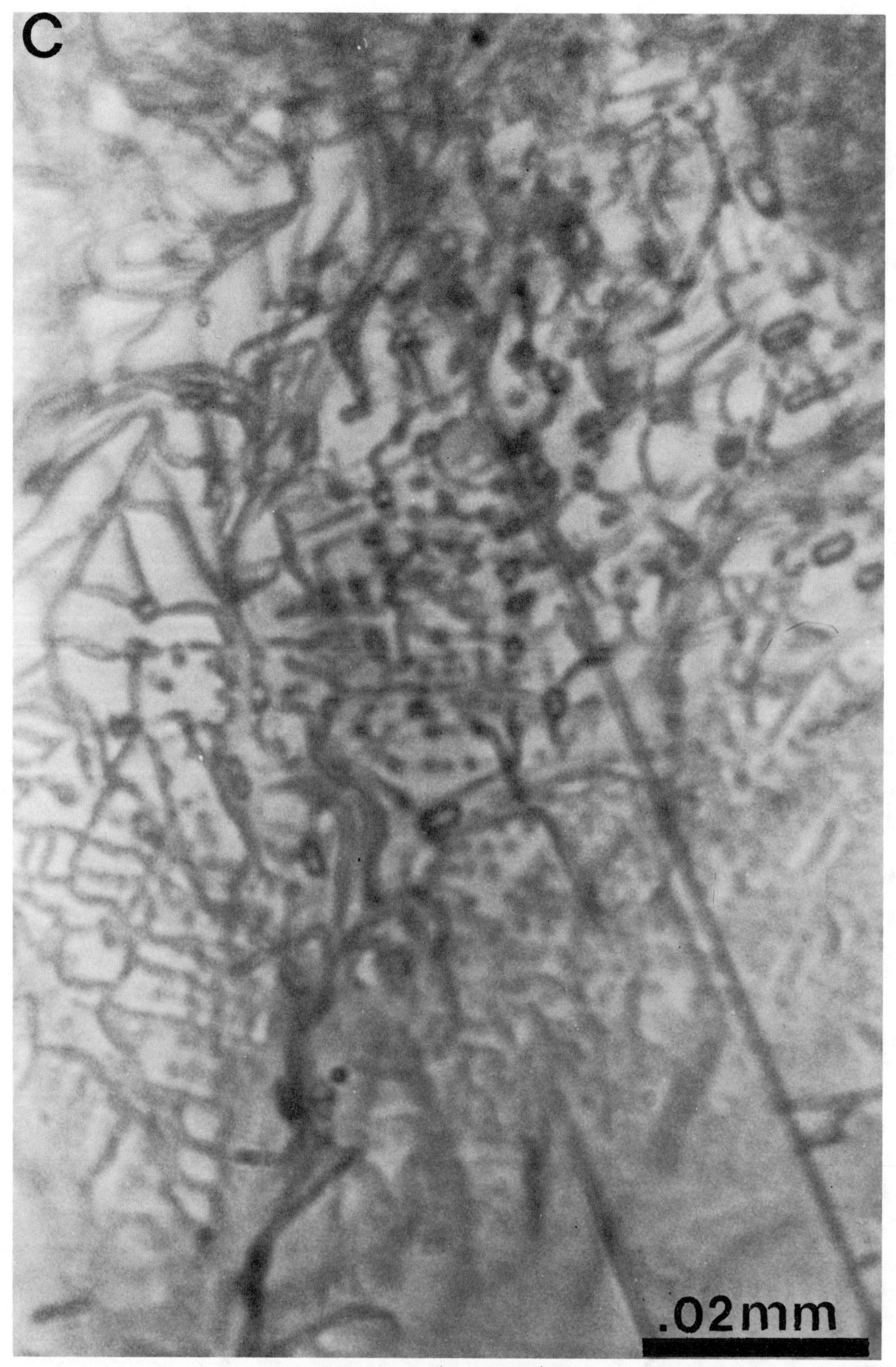

Fig. 9. (continued)

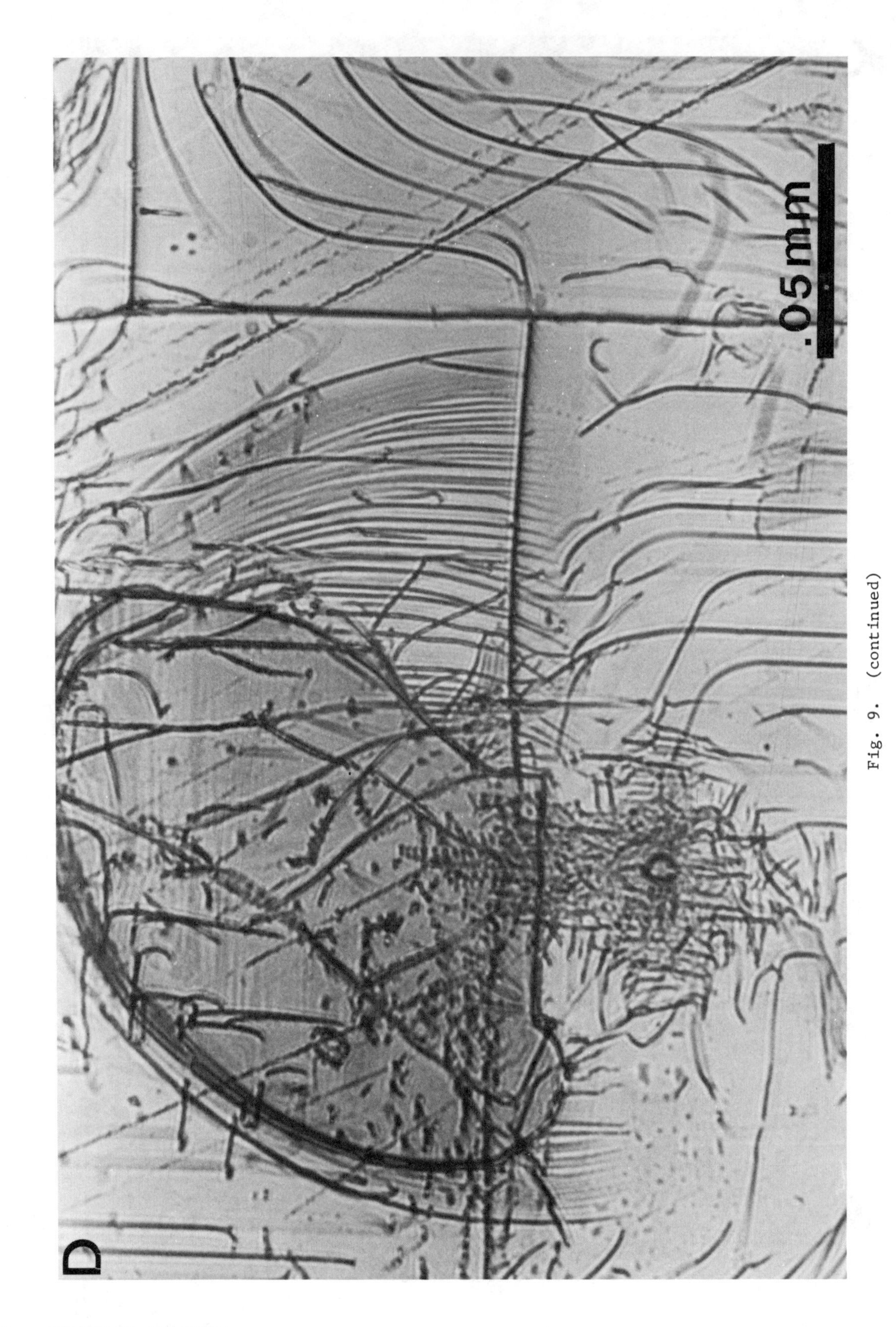

Fig. 9. (continued)

9b) and also stringing together small, isolated fluid inclusions (Figure 9c). In other cases, dislocation nets have been observed which are not decorated with bubbles but which appear to be associated with a decrepitated fluid inclusion (Figure 9d). Both the bubble decorated and undecorated nets probably formed as an accommodation of lattice mismatch developed during relaxation of the crystal during the production of the crack. Similar nets have been observed in association with cracks healed in high vacuum in alumina [Wiederhorn et al., 1973]. Green and Radcliffe [1975] and Kirby and Green [1980] have suggested that CO_2 bubbles distributed throughout a suite of volcanic olivine nodules may have been formed by precipitation from solid state solution in the mantle and that the dislocation nets may either collect bubbles during the dislocation movement or provide preferential bubble nucleation sites. Crack healing is a second possible method of producing bubble decorated dislocations. Both processes may occur in the mantle.

Conclusions

Experimental crack healing in olivine is consistent with the diffusive models of Nichols and Mullins [1965a, b]. As predicted, crack healing is a thermally activated process during which the tip of an open crack recedes by forming cylindrical pores which subsequently undergo ovulation or spheroidization to form isolated spherical pores. We have observed the time for ovulation (τ) to follow an r^4 dependence (where r is cylinder radius) and have measured the ratio of sphere spacing to cylinder radius (λ/r) to be consistent with that predicted for either surface or internal pore fluid diffusion. Laboratory produced cracks exposed to atmospheres at 1 bar heal faster than natural decrepitated fluid inclusion cracks which contain their original filling contents indicating that either the pressure or the chemical speciation of the pore fluid in a crack dramatically effects the kinetics of crack healing. We have calculated an effective activation energy of 53 ± 22 kcal/mole for diffusive crack healing in olivine and have determined $\nu\Omega D$ at 1400°C to be on the order of 10^{-16} (cm^2/sec) x cm.

Acknowledgments. We would like to thank L. Hollister and S. Bergman for their constructive criticism, thought-provoking discussions, and expert help on many aspects of this paper. We have also benefited from discussions with B. Moskowitz, V. Sisson, D. Smith, and F. Spera. This work was supported by grant EAR-8108560 from the National Science Foundation.

References

Batzle, M. L., and G. Simmons, Microfracture in rocks from two geothermal areas, Earth Planet. Sci. Lett., 30, 71-93, 1976.

Bergman, S. C., Petrogenetic aspects of the alkali basaltic lavas and included megacrysts and nodules from the lunar crater volcanic field, Nevada, USA, Ph.D. thesis, Princeton Univ., Princeton, N.J., 1982.

Cooper, R. F., and D. L. Kohlstedt, Interfacial energies in the olivine-basalt system, in Advances in Earth and Planetary Science, High Pressure Res. in Geophys., vol. 12, edited by S. Akimoto and M. H. Manghnani, pp. 217-228, Center for Academic Publications, Tokyo, 1982.

Deines, P., R. H. Nafziger, G. C. Ulmer, and E. Woermann, Temperature-oxygen fugacity tables for selected gas mixtures in the system C-H-O at one atmosphere total pressure, Bull. Earth Miner. Sci. Exp. Stn. Penn. State Univ., 88, 1974.

Eggler, D. H., and M. Rosenhauer, Carbon dioxide in silicate melts, II, Solubilities of CO_2 and H_2O in $CaMgSi_2O_6$ (diopside) liquids and vapors at pressures to 40 kb, Am. J. Sci., 278, 64-94, 1978.

Green, H. W., and S. V. Radcliffe, Fluid precipitates in rocks from the earth's mantle, Geol. Soc. Am. Bull., 86, 390-398, 1975.

Gupta, T. K., Crack healing in thermally shocked MgO, J. Am. Ceram. Soc., 58, 143, 1975.

Gupta, T. K., Kinetics of strengthening of thermally shocked MgO and Al_2O_3, J. Am. Ceram. Soc., 59, 448-449, 1976.

Gupta, T. K., Instability of cylindrical voids in alumina, J. Am. Ceram. Soc., 61, 191-195, 1978.

Gupta, T. K., Effect of crack healing on thermal stress fracture, in Thermal Stresses in Severe Environments, edited by D. P. H. Hasselman and R. A. Heller, pp. 365-380, Plenum, New York, 1980.

Kirby, S. H., and H. W. Green, The dunite xenoliths of Hualalai volcano: Evidence for mantle diapiric flow beneath the island of Hawaii, Am. J. Sci., 280-A, 550-575, 1980.

Kohlstedt, D. L., C. Goetze, W. B. Durham, and J. B. Vander Sande, New technique for decorating dislocations in olivine, Science, 191, 1045-1046, 1976.

Lemmlein, G., Formation of fluid inclusions and their use in geological thermometry, Geochemistry, 6, 630-642, 1956.

Lemmlein, G. G. and M. O. Kliya, Distinctive features of the healing of a crack in a crystal under conditions of declining temperature, (English trans.) Akad. Nauk SSSR Dokl., 87, 957-960, 1952.

Murck, B., R. Buruss, and L. Hollister, Phase equilibria in fluid inclusions in ultramafic xenoliths, Am. Mineral., 63, 40-46, 1978.

Nichols, F. A., Spheroidization of rod-shaped particles of finite length, J. Mater. Sci., 11, 1077-1082, 1976.

Nichols, F. A., and W. W. Mullins, Morphological changes in a surface of revolution due to a capillarity-induced surface diffusion, J. Appl. Phys., 36, 1826-1835, 1965a.

Nichols, F. A., and W. W. Mullins, Surface (interface) and volume diffusion contributions to morphological changes driven by capillarity, Trans. Am. Inst. Min. Metall. Pet. Eng., 233, 1940-1948, 1965b.

Padovani, E. R., S. B. Shirey, and G. Simmons, Characteristics of microcracks in amphibolite and granulite facies grade rocks from southeastern Pennsylvania, J. Geophys. Res., 87, 8605-8630, 1982.

Pecher, A., Experimental decrepitation and re-equilibration of fluid inclusions in synthetic quartz, Tectonophysics, 78, 567-583, 1981.

Roedder, E., Liquid CO_2 inclusions in olivine-bearing nodules and phenocrysts from basalts, Am. Mineral., 50, 1746-1782, 1965.

Roedder, E., Origin of fluid inclusions and changes that occur after trapping, in Fluid Inclusions: Applications to Petrology Short Course Handbook, 6, edited by L. S. Hollister and M. L. Crawford, pp. 101-137, MAC, Calgary, Alta., 1981.

Shankland, T. J., R. J. O'Connell, and H. S. Waff, Geophysical constraints on partial melt in the upper mantle, Rev. Geophys. Space Phys., 19, 394-406, 1981.

Shelton, K. L., and P. Orville, Formation of synthetic fluid inclusions in natural quartz, Am. Mineral., 65, 1233-1236, 1980.

Simmons, G., and D. Richter, Microcracks in rocks, in The Physics and Chemistry of Minerals and Rocks, edited by R. G. J. Strens, pp. 105-137, Interscience, New York, 1976.

Smith, D., and B. Evans, Diffusional crack healing in quartz, J. Geophys. Res., in press, 1984.

Wiederhorn, S. M., B. J. Hockey, and D. E. Roberts, Effect of temperature on the fracture of sapphire, Philos. Mag., 28, 783-796, 1973.

Yen, C. F., and R. L. Coble, Spheroidization of tubular voids in Al_2O_3 crystals at high temperatures, J. Am. Ceram. Soc., 55, 507-509, 1972.

DEFECT MECHANISMS FOR THE SOLID STATE REDUCTION OF OLIVINE

J. N. Boland

Institute of Earth Sciences, State University of Utrecht
3508 TA Utrecht, The Netherlands

A. Duba

Lawrence Livermore National Laboratory, University of California

Abstract. The development of the microstructures in experimentally reduced single crystals of San Carlos olivine (Fo92) has been studied by optical and electron microscopy. At 1400°C in a reducing atmosphere of nearly pure CO, reduction proceeds by propagation of a reaction zone from the surface to the center of the crystals. This zone consists of metallic precipitates of Fe-Ni in a matrix of a more forsterite-rich olivine. No silica-rich phase has been detected. The reduction process can be divided into a number of stages. There is an incubational pre-precipitation period followed by two micro-structurally-distinct stages. In the first stage, the reaction zone is clearly delineated by its optically dense rim. Based on a mass transfer model, the kinetics of this stage is controlled by oxygen vacancy diffusivity. The next stage, with its increased reduction rate, has a more diffuse reaction-zone front accompanied by coarsening of the metallic precipitates. Dislocations, whether free or bound to subboundaries, act as high diffusivity pathways for oxygen vacancies. The ratio of dislocation pipe diffusion to lattice diffusion (D_p/D_l) is 2.5 x 10^5 at 1400°C. Additionally, dislocations are preferred nucleation sites for the metallic precipitates. Nickel is selectively reduced ahead of the advancing reaction zone, producing metallic precipitates with up to 80 weight percent nickel. This reaction, which is unrelated to pre-existing dislocations, is at least an order of magnitude faster than the oxygen vacancy mechanism and indicates that electronic defects are responsible. Pores are observed close to the gas-mineral interface. Their formation represents a precipitation reaction involving vacancy clustering in olivine that has become supersaturated with vacancies.

Introduction

One important class of exsolution reactions in olivine involves oxidation and reduction processes. The products of oxidation and their microstructures have been reviewed by Haggerty [1976]. Kohlstedt and Vander Sande [1975], appreciating the role of electronic and point defects in oxidation, postulated the reaction: olivine → magnetite plus defective olivine; the defects envisaged were oxygen vacancies each of which trapped two electrons for charge balance. The rate of oxidation was too rapid to be explained by oxygen diffusion either through the lattice or along dislocations. They implied a fast electron transport process from the surface to the interior but did not elaborate on the details of the mechanism.

There have been few studies of reduction reactions in olivine although meteoritic olivines commonly contain metallic particles [see Boland and Duba, 1981; 1983; Kracher et al., 1983]. This present work is a micro-structural study of the precipitation of iron-nickel particles in reduced San Carlos olivine. We show that pre-existing dislocations are both high diffusivity pathways and preferred nucleation sites for the metallic precipitates. The development of submicron-sized pores close to the gas-mineral interface infers a super-saturation of vacancies in the reduced olivine. Finally, the reduction process is modelled in terms of the formation and migration of both point and electronic defects.

Experimental Conditions

Oriented slices in the form of parallelopipeds about 2 mm on a side and 0.5 mm thick were cut from a large single crystal of San Carlos olivine with a nominal composition of Fo92 and a NiO content of 0.365 ± .008 weight percent. One face of each slice was polished to a 1 μm-diamond finish with the other surfaces being in the as-cut condition. This distinction between the surface finishes proved essential to our understanding of the resultant microstructures. The samples were placed in iron capsules such that only four of the corners (or two of the

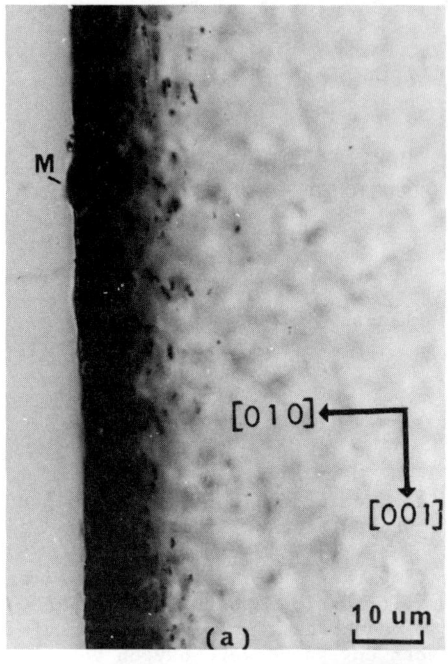

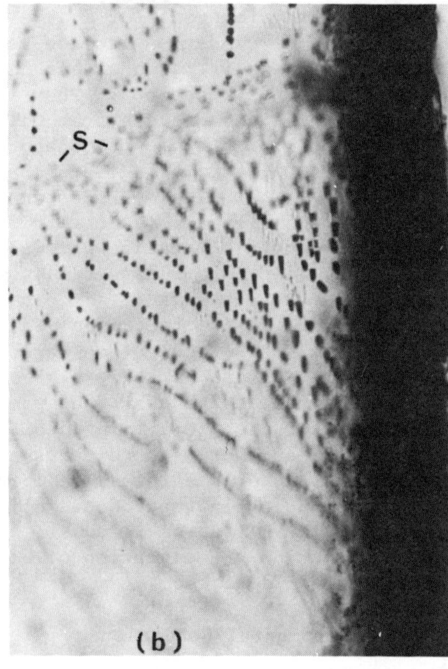

Figure 1. Optical micrographs in transmitted, plane-polarized light. (a) Depth of penetration of the reduction reaction is evident from the metallic phase distribution inward from the polished surface on the left; M is an Fe condensate. (b) Heterogeneous nucleation of Fe-Ni precipitates on pre-existing dislocations. S defines a near-vertical subboundary into which the decorated dislocations plunge; the rough surface is on the right (BSC-3).

edges) were in direct contact with the capsule during the reduction run. The capsules were suspended from an alumina tube in a vertical furnace operating at 1400°C for periods of 0.25, 1.0, 3.0, 10.7, and 30 hours with flowing CO gas as the reductant. Electrical conductivity was not monitored during these runs. Frequently the samples recovered from the reduction runs had a carbon film adhering to their surface. This indicates that the oxygen fugacity (fO_2) of the gas is most likely buffered by the reaction $2C + O_2 = 2CO$, in the vicinity of the surface. This condition would produce an fO_2 of 10^{-11} Pa at 1400°C ($\sim 10^{-16}$ atm).

The pre-polished surface of each sample was examined in a scanning electron microscope (SEM). The optical and transmission electron microscopy (TEM) sections were cut so that the microstructure could be studied in profile from the pre-polished surface into the bulk of the single crystal. Chemical analyses were performed in both SEM and TEM but the advantage of the TEM-mode of operation is that particles less than 1000 Å could be readily analyzed.

Results

Optical Microscopy

Representative microstructures of a short-term reduction reaction are shown in Figure 1. There is an optically dense reaction rim - hereinafter referred to as the reaction zone - that extends from the gas-mineral interface into the sample. The zone itself consists of iron-nickel metallic precipitates and for short reaction times ($\leq$ 3 hours), the boundary between the zone and the unreacted crystal is clearly defined by the sharp decrease in number density, ρ_p, of precipitates. The depth of penetration of the zone is a function of surface finish, crystallographic direction and time; see Table 1.

For reaction times greater than 3 hours, two microstructural developments are observed: (1) the metallic precipitates coarsen and (2) the zone boundary becomes more diffuse. The coarsening effect results in a more optically transparent rim and it is no longer possible to measure the width of the reaction zone unambiguously. This effect is shown in Figure 2 for BSC-8, in which the optically dense rim is about 26 μm but there is still a high density of precipitates (2×10^{11} cm^{-3}) 100 microns from the surface.

In addition to this development, it should be emphasized that precipitation is not strictly limited to the reaction zone itself. For example, in BSC-2 the depth to which optically observable precipitation occurs in the bulk of the olivine is 140 μm while the dislocations that form the (100) subboundaries [Boland et al., 1971] are decorated with metallic precipitates across the total width of the samples. After 3 hours (BSC-3) precipitates are

TABLE I. Reduction Profiles for San Carlos Olivine in CO at 1400°C

Sample	Time, Hours	Penetration Along [010], X, μm — In Reaction Zone: Pre-Polished	Rough	Isolated Precipitates[c]
BSC-1	0.25	∿0.5	3	70
BSC-2	1.0	5	8	140
BSC-3	3.0	10	20	160 (middle)
BSC-8	10.75	∿65[a]	∿65[b]	202 (middle)
BSC-5	30	see below[a]	.. [b]	165 (middle)

a. In BSC-8 and BSC-5, the precipitates are coarsening.
b. No significant difference between the surface finishes.
c. In all samples, precipitation had occurred in the middle along pre-existing dislocations.

observable in the center of the samples. Consequently, there must be extremely rapid processes occurring both along the dislocations and through the structure, although such processes represent only a small fraction of the changes in the samples. The influence of pre-existing dislocations on the process of precipitation is illustrated in Figure 1b. These dislocations were revealed as a "string-of-beads" pattern along their length. The dislocations appeared to lie in a subboundary array that undulated about (100), the plane of the section - the "out-of focus" images of the precipitates indicated that strict parallelism to (100) was not adhered to. The dislocation directions can be determined by through-focussing and those shown in Figure 1b curve sharply out of (100) into a near-vertical subboundary S. Another important influence of the pre-existing dislocations on microstructural development is shown in Figure 2. This was the existence of a precipitation-free zone on either side of the vertically aligned subboundary, SB. Furthermore, the precipitates decorated the dislocations in the subboundary across the whole section indicating complete penetration along

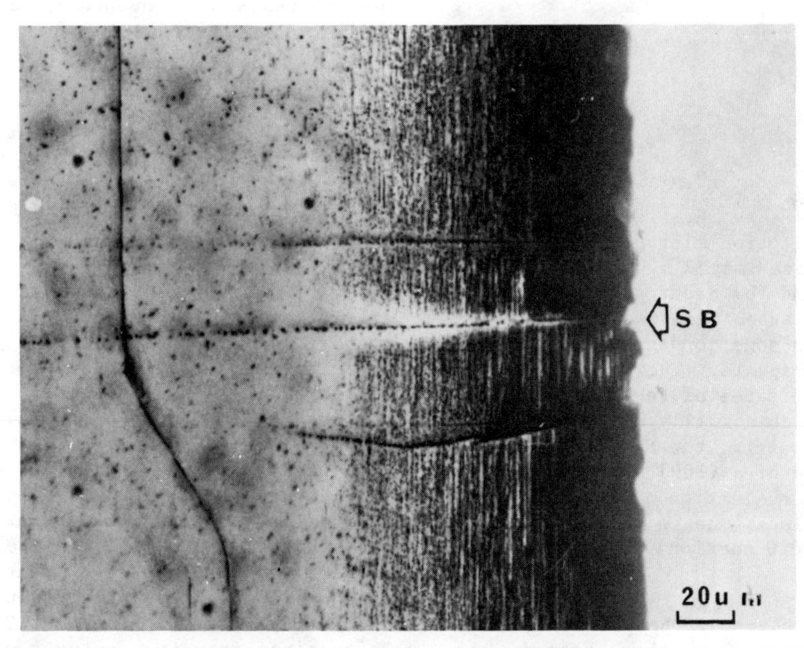

Figure 2. Optical micrograph showing the preferential precipitation of Fe-Ni particles along a (100) subboundary-SB. There is a precipitation free zone on either side of SB; rough surface is on the right. BSC-8 (001 section).

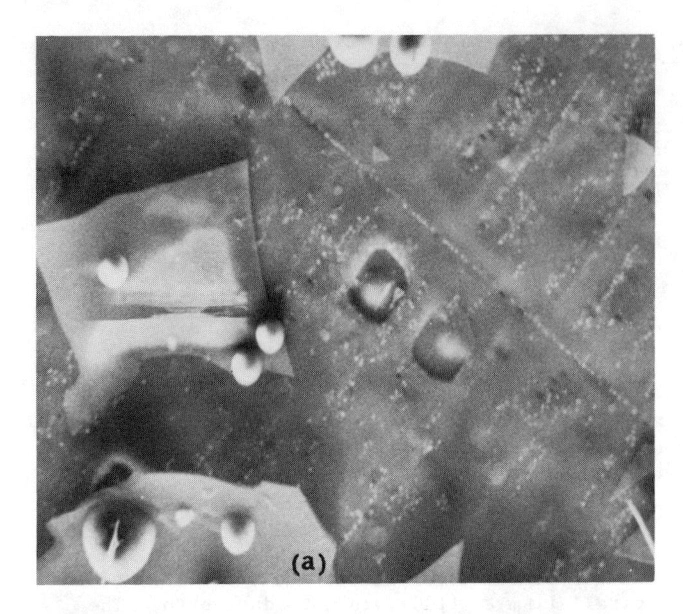

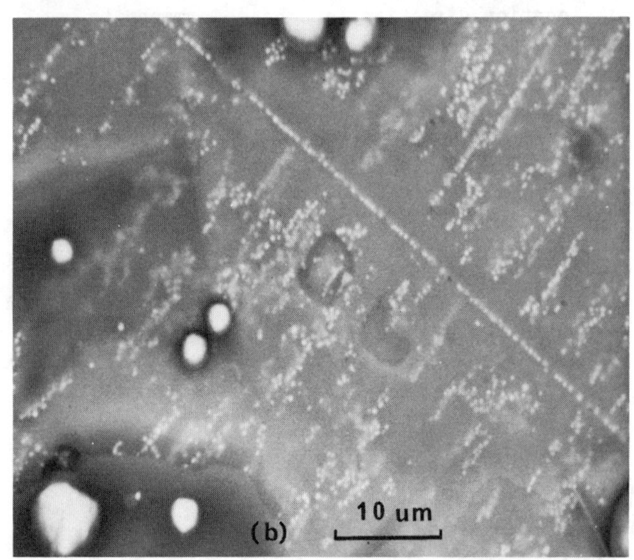

Figure 3. Scanning electron micrographs of the pre-polished surface imaged by (a) secondary electrons and (b) backscattered electrons. A patchy carbon film was formed on the surface along with some condensates of Fe originating from the capsule. The submicron-sized precipitates of Fe-Ni are reduction products which normally nucleate homogeneously in the matrix; the NW-SE lineation is the trace of a (100) subboundary showing heterogeneous nucleation of the metallic phase on dislocations in the subboundary. BSC-3 (010 section).

the subboundary. The sizes of the precipitates varied, but in all the samples, those formed along dislocations were larger than those in the matrix (whether in the reaction-zone or elsewhere). For example, in Figure 3, the precipitates at the dislocation were about 1.0 μm or larger while those in the matrix were generally less than 0.5 μm - care had to be exercised in these optical measurements since numerical aperture effects may influence the apparent size of the precipitates [McLaren et al., 1970]. The depth of penetration was not uniform along the gas-mineral interface. For example, in BSC-2, the penetration varied from 3 to 8 microns from the polished surface. This variation, although not correlated with any existing microstructural feature, may have been caused by pre-existing defects (such as cracks) that have subsequently been annealed out (or healed) during the reaction run.

Scanning Electron Microscopy

The pre-polished surfaces were carbon-coated and examined in the SEM. The microstructures are shown in Figures 3a and 3b, imaged with secondary electrons (SEI) and back-scattered electrons (BEI) respectively. There are two scales of precipitation, the larger, micron-sized, particles that were analyzed as Fe-only and the smaller, sub-micron precipitates (more clearly imaged in BEI) that were both Fe- and Ni-bearing. The presence of the larger precipitates may not be inferred from the surface rumpling at M in Figure 1a. Associated with each of these larger precipitates was a surface depression or pit, but notably, not all these pits had an Fe-particle attached to them. The matrix composition in the regions apparently devoid of surface precipitates was Fo97. From the TEM results given below, this value over-estimated the Fe content of the reduced olivine - the additional Fe results from electrons scattered by fine-scaled Fe-Ni precipitates beneath the surface. No other phases were detected in the surface. The metallic precipitates are aligned along [100] and [001] with the (100) subboundary clearly delineated as the NW-SE trending feature in Figure 3.

Precipitate Size and Distribution

As is apparent from the optical microscopy, there is a characteristic precipitate distribution across the samples with the highest number-density of precipitates in the surface regions, tailing off to a small value at the center. Transmission electron microscopy is the ideal technique for resolving details of microstructure in optically dense rims. Furthermore, for shorter reduction runs up to three hours, the best technique for determining precipitate distribution within the first 20 microns from the gas-olivine interface is the scanning transmission mode at 200 kV (STEM). This arises from two features of the STEM mode: (1) the signal can be processed electronically to give an interpretable image of both larger areas and thicker sections than the normal TEM; (2) the dynamical contrast effects are very much reduced leaving the precipitates clearly identifiable although the image is partially blurred [Humphreys, 1979]. Ion-thinned samples from BSC-3 were prepared and examined at 100 and 200 kV.

The metallic precipitates have a crystallographic shape with approximate dimensions of 0.2 μm x 0.15 μm x 0.05 μm. Pores or negative crystals are observed close to the gas-olivine interface - Figure 4a. Similar metallic precipitate morphologies and size are found in the sample center but the number density has decreased by several orders of magnitude. No pores are detectable in the central region - Figure 4b.

Figure 5 shows a precipitate number-density distribution constructed for BSC-3 using measurements from the STEM. The density at the surface is $\sim 10^{13}/cm^3$ while in the center it is $\sim 10^9/cm^3$. The sharp reaction zone, as observed by optical microscopy, is a more diffuse boundary region when studied using STEM - see Discussion below.

Chemical Compositions of the Metallic Phase and Matrix

Chemical analyses of the precipitate and matrix have been made by using the small beam diameter obtainable in the STEM mode - the average beam-size was about 800 Å. The contamination rate was so high that only 15 second counts were used. Because of the small size of the precipitates it was practically impossible to analyze them without the interfering effects of the matrix. Consequently, all analyses have been corrected for such interference but it is worth noting that on the rare occasion when the precipitate was located on the very edge of the foil, it was found to contain only Fe and Ni and no Si.

The composition of the metallic precipitates in BSC 3 followed a pattern. Those precipitated in the surface regions had nickel contents averaging 25 weight percent (ranging from 8 to 30 weight percent), while those in the center averaged 62 percent (the latter had a range of 52 to 80 weight percent nickel and the relative error in any one analysis was +10 percent). One of the precipitates in Figure 4b had a composition approximated by Ni_3Fe. The high-nickel precipitates produced diffraction patterns consistent with a face-centered-cubic structure (fcc, a_o = 3.54 Å); some of the precipitates were twinned. Several of the low-nickel particles were indexed on a body-centered cubic structure (bcc, a_o = 2.87 Å) but the composition of the fcc to bcc transition was not determined. Over 100 precipitates were analyzed and all of them were found to be Fe-Ni bearing only; despite an intensive search for a more silica-rich phase such as a pyroxene, no

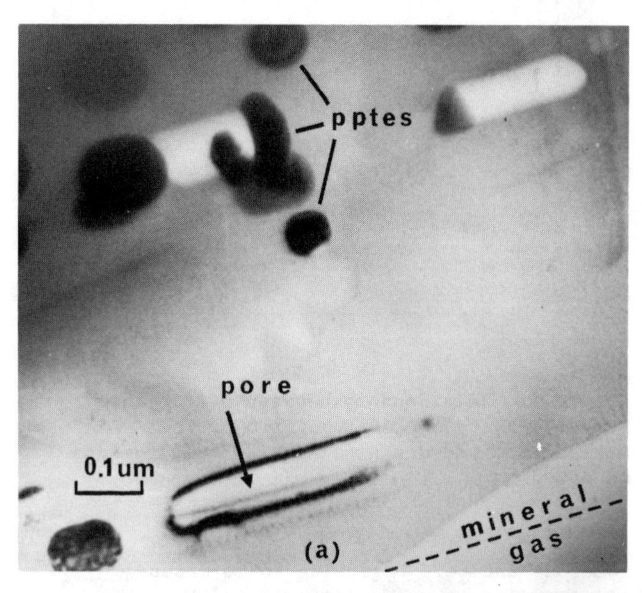

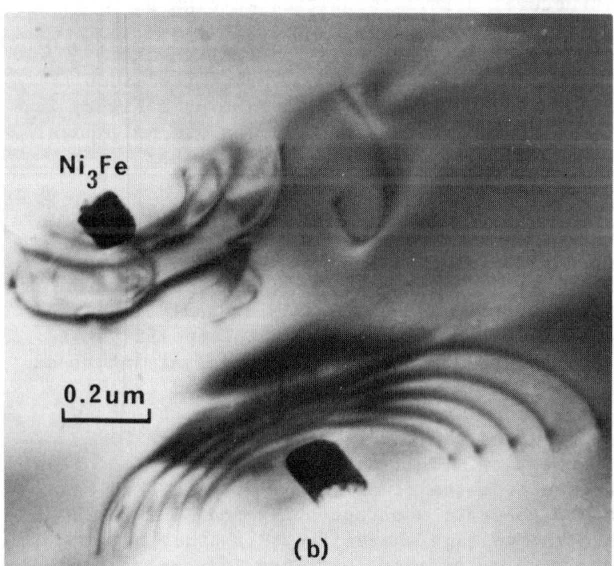

Figure 4. Transmission electron micrographs showing precipitates (a) near the gas-mineral interface and (b) in the central region of the crystal. The latter are Ni-rich, while those near the interface are Fe-rich. Within the first few microns of the gas interface, negative crystals or pores are observed. BSC-3 (001 section).

phases other than olivine and Fe-Ni precipitates were observed in any of the samples in this series.

Dislocations were generated during the precipitation process. In the central area, they were localized about the precipitates and appeared to have been punched out into the matrix during precipitate growth - Figure 4b. In the surface region, dislocations were more

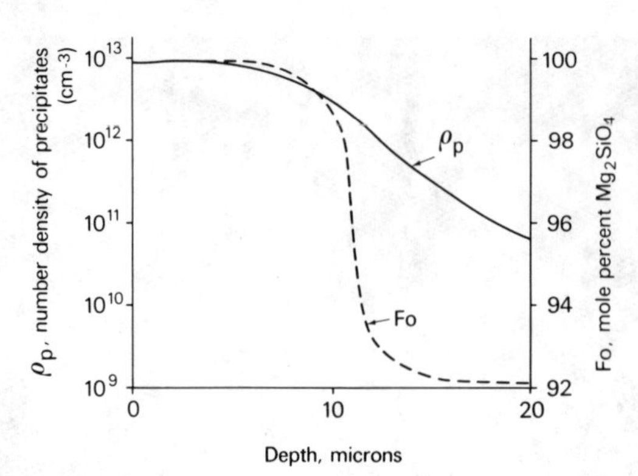

Figure 5. Relationship between precipitate number density, ρ_p, composition of the matrix, Fo, and depth using STEM measurements on BSC-3.

numerous and were beginning to form substructural arrays (densities as high as $10^9/cm^2$ were observed from a starting density of $10^6/cm^2$). In contrast to the heterogeneous nucleation of precipitates on pre-existing dislocations as shown in Figure 2b, these newly created dislocations did not appear to act as nucleating sites for further precipitation.

Olivine matrix composition may be determined by two methods: (1) a difference method based on the analysis of the precipitate number-density plus the average size and composition of the precipitates: (2) STEM microanalysis of the matrix. Using method (1), the composition of the reduced matrix at the gas-mineral interface was estimated as Fo_{100}; using method (2), a small amount of Fe was recorded in the x-ray spectrum of the matrix representing an Fo97 composition. This discrepancy in the Fe-content arises from the fluorescence effects that are known to occur when operating the STEM at 200 kV; the voltage chosen for this study because of the need to examine sufficiently representative volumes of the samples. This anomalous effect may also account for the lower accuracy of the reported analyses compared with other researchers (see Nord [1982] for a review of the analytical techniques using TEM). When the analyses were made at 100 kV, the matrix was found to be approximately Fo100 within 0.5 μm of the gas interface, a value more in keeping with the calculation by method (1).

Discussion

Thermodynamic State of the Reduced Samples

According to the heterogeneous oxidation-reduction equilibria in the system $MgO-SiO_2-FeO$ [Nitsan, 1974; Morse, 1980], the following equation would be applicable to the 1400°C reduction:

$$Fe_2SiO_4 \rightleftarrows Fe + FeSiO_3 + \frac{1}{2} O_2 \text{ (gas) (OPI)}$$

O, olivine; P, pyroxene; I, iron. Neither pyroxene nor silica were detected in the run products and there was no evidence of gas-bubble formation at the advancing reaction zone boundary. One must conclude that the resulting olivine is highly non-stoichiometric. The actual structural state of the olivine is the subject of further investigation but it should be recorded that this defective olivine is highly sensitive to electron irradiation damage; such damage is not normally observed in olivines.

The system can be considered partially open to the exchange of oxygen between the solid and gas. At 1400°C the volatility of the other components is assumed negligible. Surface contamination is observed, mainly from iron condensation from the capsule, but because of the extremely reducing conditions this iron segregates into nickel-free balls on the surface; see Figure 3.

Modeling the Reduction Process

Unlike reduction of iron oxides in which a compact product layer of iron is formed at the gas-oxide interface [Schmalzried, 1981], iron particles in reduced olivine are nucleated in a reaction zone which propagates into the sample from the surface. This is shown schematically in Figure 7, where the reaction zone is defined by optically dense rims (see Figure 1) consisting of a high number-density, ρ_p, of iron-nickel precipitates, but across the boundary ρ_p decreases rapidly. With increasing reaction times, the zone widens and the boundary region becomes more diffuse. This type of microstructural development is akin to that observed during the internal oxidation of metal alloys (Kofstad [1966]; see Chapter 8), but instead of selective oxidation of the less "noble" metal, there is a selective reduction of the more "noble" cation.

The kinetics of the process has been determined from the rate of propagation of the reaction zone as indicated in Figures 5 and 6 (STEM study of BSC-3 showed that the reaction zone boundary was slightly more diffuse than the optical data had indicated but only by a few microns). The limited extent of the data did not allow unique identification of the type of rate law: logarithmic, linear, parabolic or cubic. However, all four possibilities indicated a distinct break in the rate around three hours. For simplicity, a parabolic law of the following form was applied:

$$X^2 = 2\bar{k}t \tag{1}$$

where X is the reaction zone thickness; t, the time; and $\bar{k}$, the reaction rate constant.

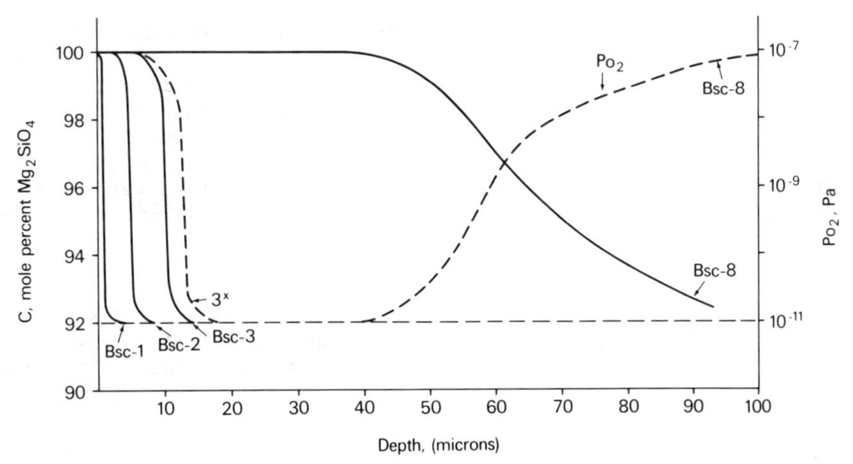

Figure 6. Composition-depth profiles for three of the shorter-term runs; 3* refers to the STEM profile. The composition and fO_2 profile for BSC-8 is based on optical measurements and assuming heterogeneous equilibrium at OPI.

Furthermore, provided diffusion processes are rate controlling, advance of the boundary may be described by:

$$X = \sqrt{D_{eff} t} \qquad (2)$$

where D_{eff} is effective diffusivity of the rate controlling mechanism [Shewmon, 1969].

According to our observations, the overall reaction can be divided into the following sequence:

(1) an incubation period, less than 0.25 hours, prior to precipitation of iron-nickel particles; (2) an initial stage of propagation of the reaction zone in which D_{eff} is $10^{-10} cm^2/s$ and pores are developed close to the gas interface; (3) a second stage represented by an accelerated reduction rate in which the reaction zone boundary becomes more diffuse and precipitates coarsen - D_{eff} is $1.5 \times 10^{-9} cm^2/s$.

Each of these subdivisions represents the development of a particular set of defect structures. These will be analyzed separately.

Incubation Period

The incubation period is that time prior to onset of precipitation. It represents an ill-defined situation in which olivine is attempting to equilibrate with imposed fO_2 and T conditions. It represents the state of the olivine immediately ahead of the advancing reaction zone boundary - see Figures 5, 6 and 7. The types of defects generated during this stage have been postulated by Smyth and Stocker [1975], Stocker and Smyth [1978], Stocker [1978a, b, c] and Hobbs [1981]. However, a summary of the most important defects (atomic and electronic) expected in olivine under reducing conditions is given in Appendix 1.

As oxygen fugacity decreases in the single phase field - regimes A and B, Figure A1 - the concentration of $Si_I^{\bullet\bullet\bullet\bullet}$ and e' increases while the majority defects V_{Mg}'', $Mg_I^{\bullet\bullet}$ (plus $Fe_I^{\bullet\bullet}$ and $Ni_I^{\bullet\bullet\bullet}$) are invariant. As an example of this development, the gradient in fO_2 for sample BSC-8 is depicted in Figure 6. This type of fO_2 gradient induces a gradient in concentration of defects which in turn induces a flux of defects. The surface fO_2 conditions of about 10^{-11} Pa are controlled by the gas phase while the internal fO_2 conditions will be determined by rate of propagation of the reaction zone boundary.

Precipitation of a nickel-rich metallic phase is observed under these fO_2 conditions which indicates a selective reduction involving the removal of nickel although some iron is also present in the precipitate. This selective reduction would be opposed by the expected higher crystal field stabilization energy of Ni compared with Fe in olivine [Burns, 1970; Wood, 1981]. Nor does the effect reflect the relative diffusivities of Fe and Ni since D_{Fe}/D_{Ni} at 1400°C is 12 [Morioka, 1981]. Using the Ni-NiO and Fe-FeO data in Darken and Gurry [1953] and assuming an fO_2 condition just below the iron-wüstite stability, then the difference in relative partial molar free energy of oxygen for the two systems, ΔG is 30 kcal/mole, where

$$\Delta G = (RT\ln PO_2)_{Ni-NiO} - (RT\ln PO_2)_{Fe-FeO}$$

It is this free energy change which provides the driving force for selective nickel reduction. However, when fO_2 conditions become extremely reducing, this small difference becomes relatively unimportant and the composition of the metallic precipitates reflects the initial

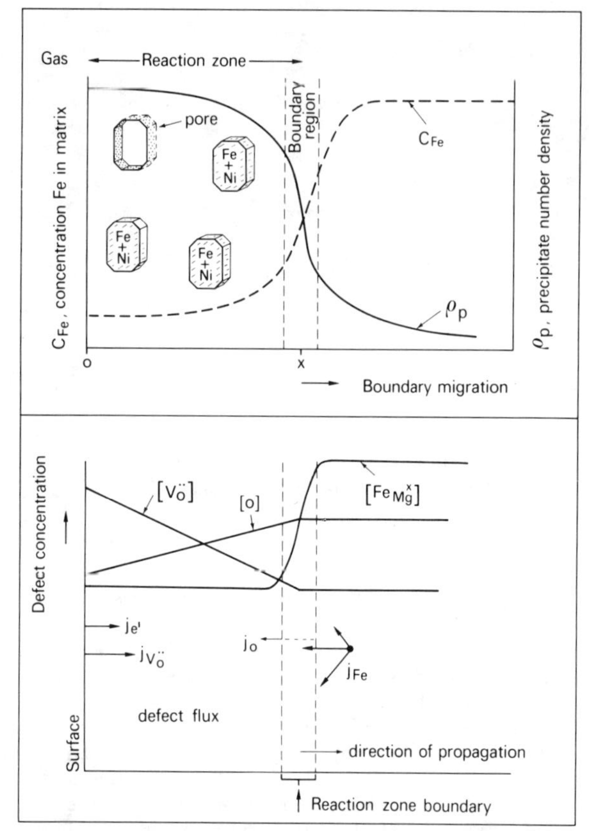

Figure 7. (a) Schematic representation of the reduction process defining the reaction zone and its boundary, precipitate distribution (ρ_p) and concentration of iron (C_{Fe}). (b) Various fluxes and concentration gradients for dominant defects (see text).

concentration of iron and nickel in the starting material [see Boland and Duba 1981].

This selective reduction of nickel precedes the reduction reaction associated with propagation of the reaction zone. Possible defect reactions are (see also Appendix 1)

$$Ni_{Mg}^{x} + 2e' \rightleftarrows Ni + V_{Mg}'' \quad (3)$$

$$Ni_{I}^{\cdot\cdot} + 2e' \rightleftarrows Ni \quad (4)$$

Reaction (3) expresses the reduction of a nickel atom on an M-site in olivine, followed by its subsequent removal from the site to form a vacancy - i.e., a Frenkel defect. The Frenkel defect formation is presumed to precede reaction (4), thus the vacancy precedes (4). These neutral nickel atoms reach supersaturation and the metallic phase nucleates and grows. Dislocations are not the sole nucleation sites and homogeneous nucleation is commonly observed (Figure 4b).

There is a volume change associated with the precipitation and this in turn leads to internal stresses. These stresses are sufficient to induce plastic deformation of olivine by generation of dislocations (Figure 4b). The effect is very local for these first-formed nickel-rich precipitates but it can lead to a complete change in dislocation substructures during reaction zone formation; see below.

Equations (3) and (4) show that electrons are required to reduce the cations. Their source is most likely at the surface where either of the following reactions occurs:

$$O_O^x \rightleftarrows \frac{1}{2}O_2 + V_O^{\cdot\cdot} + 2e' \quad (5)$$

$$2Fe_{Mg}^{x} + Si_{Si}^{x} + 4O_O^{x} \rightleftarrows 2Fe + Si_I^{\cdot\cdot\cdot\cdot} + 2O_2 + 4e' \quad (6)$$

although (5) is favored in this incubation period before the formation of the reaction zone. It is possible to estimate effective diffusivity of migrating charged species from the Nernst-Einstein relation:

$$\frac{\sigma}{D_{eff}} = \frac{Ne^2}{kT} Z^2 \quad (7)$$

σ, electrical conductivity; D_{eff}, effective diffusivity, N, concentration of conducting species of charge Z. From the diffusivity data of Misener [1974] and Buening and Buseck [1973], σ_{cal} can be estimated assuming that conductivity is occurring solely by Fe-cationic defects - N, being the Fe concentration in the starting olivine, Fo92. This value can be compared with the observations, σ_{obs}, of Duba and Nicholls [1973] giving $\sigma_{cal}/\sigma_{obs}$ of 4×10^{-2} at 1400°C. This result is slightly at variance with the value of 3.5×10^{-1} obtained from Shankland [1981]. However, the situation is more complex because of the varying fO_2 conditions. What emerges from these tentative calculations is that olivine should be seen as a mixed conductor [see Tuller, 1981] and that, provided all diffusive migration are not ambipolar, there is the possibility of fast electronic transport from the surface to the center of the sample where electrons participate in the reduction reactions of types (3) and (4) above.

Formation and Initial Propagation of Reaction Zone

After the incubation period, the microstructure develops by widening of the reaction zone - Figure 7. In order to understand this process, one has to determine diffusive fluxes operating. Two limiting cases are shown in Figure 7b.

1. Since iron precipitates locally at the reaction zone boundary, flux lines may be approximated as radiating lines. This indicates that transport distance for Fe-species is given by the mean separation of precipitates. For a precipitate density of $10^{13}/cm^3$, the average separation is about 0.5 µm. Hence the average diffusion length for Fe is 0.25 µm. Using D_{Fe} values from Buening and Buseck [1973] (extrapolated to 1400°C), the time required to establish such an Fe-precipitate distribution is about 1 minute. Parameter j_{Fe} is the Fe-flux from the unreacted olivine to the reaction zone for this limiting case.

2. The flux lines connect the surface with the advancing reaction zone boundary. In this case, the following fluxes are considered: j_O, flux of oxygen from the reaction front to the surface; j_{VO}, flux of oxygen vacancies from the surface to the interior; $j_{e'}$, flux of electrons. Other fluxes may be expected because of the inhomogeneous distribution of point defects as shown in Figure A1. However, since no other phases such as pyroxene of periclase are observed, the diffusive transport of Mg and Si is not directly related to development of microstructures.

The flux equations for the migrating species have the form:

$$j_i = L_{ii} \frac{d\eta_i}{dx} \qquad (8)$$

where L_{ii} are the transport coefficients (off-diagonal elements are assumed zero) and η_i the electrochemical potential of species i; further

$$L_{ii} = \frac{D_i C_i}{RT} \qquad (9)$$

D_i, the diffusivity of species i with a concentration C_i [Schmalzried, 1981, chapter 5; Anderson, 1981].

The fundamental partial reaction for the reduction is generalized by the reaction:

$$M_{Mg}^{x} + 2e' \rightleftarrows M + V_{Mg}'' \qquad (10)$$

where M may be Fe or Ni. More importantly, the necessary condition for internal reduction is that $D_{e'}C_{e'} >> D_{Fe}C_{Fe}$; otherwise one would expect a continuous iron film to form on the surface of the olivine. If the migration rate of electrons is coupled with flow of oxygen vacancies from the surface according to the charge neutrality condition $[e'] = 2[V_O\ \ ]$, then the effective diffusivity for this coupled diffusion can be shown to be

$$D_{eff} = 3D_{V_O} \qquad (11)$$

for the condition $D_e >> D_{V_O}$ [Kröger, 1964, chapter 20]. From our kinetic studies D_{eff} is 10^{-10}cm /s, so D_V is $3 \times 10^{-11}cm^2$ /s.

The relationship between oxygen diffusivity D_O and oxygen vacancy diffusivity D_{V_O} is $D_O = D_{V_O} C_{V_O}$. Jaoul et al., [1983] found that the ^{18}O diffusivity in forsterite (Fo100) was PO_2-independent over the range 10^{-4} to 10Pa. At 1400°C, D_O is $10^{-14}cm^2/s$. This would imply a C_{V_O} of 3×10^{-4}, which would be a lower limit of C_{V_O} since one may expect D_O to show some PO_2-independence over the range 10^{-7} to 10^{-11} Pa. The value of D_{V_O} is comparable with the Fe-Mg interdiffusion coefficients obtained by Misener [1974] for Fo93 ($3 \times 10^{-11}cm^2/s$) but it is slightly larger than the extrapolated value for Fo100; $3 \times 10^{-12}cm^2/s$.

Continued Propagation of Reaction Zone

For times longer than three hours, propagation of the reaction zone leads to a more diffuse reaction zone boundary (Figure 2). Effective diffusivity increases which may be directly related to the rapid increase in dislocation density. From a starting density of $10^6/cm^2$, the number of dislocations increases to $10^9/cm^2$ after 3 hours; in Appendix 2 it is shown that such a dislocation density can lead to an order-of-magnitude increase in the effective diffusivity. Dislocation-assisted diffusivity is also evident in Figure 1b where the width of the reaction zone at the rough surface is about twice that at the polished surface. The difference between rough and polished surfaces is density of defects such as microcracks and dislocations. From the results of Boland and Hobbs [1973] and Gaboriaud et al., [1981] on the semi-brittle behavior of olivine, dislocations densities as high as $10^9/cm^2$ are expected in the first few microns from the rough surface.

Pipe-Diffusion Along Dislocations

Dislocations have been shown to be high diffusivity pathways [Shewmon, 1963]. It is possible to estimate pipe diffusivity along dislocations by comparing extent of reaction at the polished surface with maximum depth along which dislocations are decorated by metallic precipitates. This has been done on BSC-1 where the "bulk" effects from the polished surface indicate a depth of 0.50 µm while precipitates occur at dislocations in the sample center, i.e., a depth of 0.25 mm. From Equation (2), the ratio of dislocation (pipe) diffusivity, D_p, to bulk diffusivity, D_1, is 2.5×10^5 at 1400°C.

Pore Formation and Surface Rumpling

Formation of pores as negative crystals or voids adjacent to the gas interface indicates a precipitation reaction involving any of the following: (1) vacancies, (2) gaseous species or (3) a mixture of both. Reactions (5) and (A22) generate $V_O^{\bullet\bullet}$ while reaction (10) produces V_{Mg}''; reaction (6) implies the removal of one formular unit - Fe_2SiO_4 - producing; (1) iron precipitates, (2) $Si_I^{\bullet\bullet\bullet}$, (3) molecular oxygen that escapes into the gas phase, (4) electrons and (5) vacancies at all atomic sites i.e., V_{Mg}'', V_{Si}'''', $V_O^{\bullet\bullet}$. Initially, these reactions are specifically located at the surface, so any supersaturation of vacanicies leads to clustering and condensation at the surface; this will account for surface rumpling reported in Figures 1a and 3.

As the reaction zone widens, the location for each of the reactions changes. Oxygen transfer reactions (5) and (A22) are likely to be fixed at the surface while (6) and (10) migrate with the reaction zone boundary. Fluxes and depth-dependent concentration of active atomic species shown in Figure 7b indicate that vacancy concentration is expected to be highest near the surface with a flux of oxygen across the reaction zone. A supersaturation of vacancies develops adjacent to the surface leading to pore formation shown in Figure 4a but oxygen is likely to follow a short-circuit diffusion pathway via the pores. The latter process is known to occur in CoO and it leads to the migration of pores from the low PO_2 region (the surface) to the higher PO_2 region (the sample center) [see Schmalzried et al., 1979].

Conclusions

1. Solid-state reduction of San Carlos olivine (Fo92) occurs by internal reduction reactions involving propagation of a reaction zone that consists of Fe-Ni metallic particles precipitated in a more forsterite- rich matrix.
2. By modeling the reduction in terms of point defect mechanisms, the rate controlling process appears to be the diffusion of oxygen vacancies coupled with electron migration.
3. An anomalously rapid selective reduction of nickel cations precedes the propagation of the reaction zone and is explained by highly mobile electronic defects.
4. Rumpling of prepolished surfaces and pores, in the form of negative crystals, close to the gas-mineral interface results from the precipitation of excess vacancies.
5. Dislocations act as high diffusivity pathways. The ratio of dislocation pipe diffusion, D_p, to bulk diffusion, D_l, is 2.5×10^5 at 1400°C.

Appendix 1

Defect Relationships

Following the work of Stocker [1978b] and using notation recommended by Kröger [1974], some important defect formation reactions in olivine are derived. It is convenient to consider four regimes related to degree of departure of oxygen partial pressure, PO_2, from the stoichiometric condition given by given by $P\mathring{O}_2$ ($P\mathring{O}_2 = fO_2$ for an ideal gas at standard conditions).

Regime A:

$PO_2 < P\mathring{O}_2$. Under the charge neutrality condition,

$$[V_{Mg}''] = [Mg_I^{\bullet\bullet}] \quad \text{(A1)}$$

the only defects sensitive to the PO_2 conditions are silicon vacancies and interstitials plus oxygen vacancies and interstitials. Silicon interstitials may form according to the reaction:

$$2Mg_{Mg}^{x} + Si_{Si}^{x} + 4O_O^{x} \rightleftarrows 2Mg_I^{\bullet\bullet} + Si_I^{\bullet\bullet\bullet\bullet} + 2O_2 + 8e' \quad \text{(A2)}$$

The mass-action equation is:

$$K_{A2} = [Mg_I^{\bullet\bullet}]^2 [Si_I^{\bullet\bullet\bullet\bullet}] [PO_2]^2 [e']^8 / a_{Fo}^2 \quad \text{(A3)}$$

where a_{Fo} is the activity of Mg_2SiO_4 in olivine. Similar reactions for Fe and Ni cations can be written, although Frenkel defects - implied in (A1) - may be more important for these cations, e.g.,

$$Fe_{Mg}^{x} \rightleftarrows Fe_I^{\bullet\bullet} + V_{Mg}'' \quad \text{(A4)}$$

$$Ni_{Mg}^{x} \rightleftarrows Ni_I^{\bullet\bullet} + V_{Mg}'' \quad \text{(A5)}$$

Both $[Fe_I^{\bullet\bullet}]$ and $[Ni_I^{\bullet\bullet}]$ would be PO_2-independent.

Equation (A2) represents the removal of lattice sites with release of oxygen as a gaseous phase and with cations forming interstitial defects. Another reaction representing removal of oxygen is the following:

$$O_O^{x} \rightleftarrows \tfrac{1}{2}O_2 + V_O^{\bullet\bullet} + 2e' \quad \text{(A6)}$$

whence

$$K_{A6} = [PO_2]^{1/2} [V_O^{\cdot\cdot}] [e']^{2} \qquad (A7)$$

If [e'] is independent of PO_2, then

$$[V_O^{\cdot\cdot}] \propto PO_2^{-1/2} \qquad (A8)$$

Assuming both $[Mg_I^{\cdot\cdot}]$ and [e'] are independent of PO_2, we have from (A3)

$$[Si_I^{\cdot\cdot\cdot\cdot}] \propto PO_2^{-2}$$

<u>Regime B:</u>

$PO_2 << PO_2$ - initiation of metallic precipitates. As PO_2 is decreased further, the concentration of $Si_I^{\cdot\cdot\cdot\cdot}$ increases until the condition

$$[e'] = 4 [Si_I^{\cdot\cdot\cdot\cdot}] \qquad (A9)$$

We can rewrite (A3) such that

$$[e'] \propto PO_2^{-2/9} \qquad (A10)$$

If the electron concentration increases further, the charge neutrality condition of (A1) is replaced by

$$[e'] = 2([Mg_I^{\cdot\cdot}]+[Fe_I^{\cdot\cdot}]+[Ni_I^{\cdot\cdot}])+4[Si_I^{\cdot\cdot\cdot\cdot}] \qquad (A11)$$

Provided there is no removal of the divalent cations by reduction, and the ratio of divalent to tetravalent silicon is a constant, then from (A11) and (A3):

$$[e'] \propto PO_2^{-2/11} \qquad (A12)$$

A critical situation must soon arise in which the PO_2 conditions lead to reduction of iron and especially nickel. The exact form of the reaction is not certain. If reduction occurs while the cation is in the normal site then:

$$Fe_{Mg}^{x} + 2e' \rightleftarrows Fe + V_{Mg}'' \qquad (A13)$$

If the cation is in the interstitial site

$$Fe_I^{\cdot\cdot} + 2e' \rightleftarrows Fe \qquad (A14)$$

In these reactions, atomic sites are initially conserved. However, (A2) may also be modified to give metallic iron:

$$2Fe_{Mg}^{x} + Si_{Si}^{x} + 4O_O^{x} \rightleftarrows 2Fe + Si_I^{\cdot\cdot\cdot\cdot} + 2O_2 + 4e' \qquad (A15)$$

This reaction, like (A2), removes atomic sites but new sites must be found for precipitating iron (and nickel). If PO_2 conditions are such that all iron and nickel are removed, then the resulting olivine is highly non-stoichiometric with an Mg/Si ratio of 1.84. For reactions (A13) and (A15), we can write the mass-action equation:

$$K_{A13} = a_{Fe} [V_{Mg}''] a_{Fa}^{-1} [e']^{-2} \qquad (A16)$$

$$K_{A15} = a_{Fe}^{2} [Si_I^{\cdot\cdot\cdot\cdot}] PO_2^{2} [e']^{4} a_{Fa}^{-2} \qquad (A17)$$

where a_{Fe} is activity of iron in the metal (assumed to be unity) and a_{Fa} is activity of Fe_2SiO_4 in olivine; assuming ideal behavior, this latter activity is proportional to the fayalite content of the olivine. Before simplifying these equations it will be necessary to determine the charge neutrality conditions.

The removal of iron without the accompanying precipitation of a silica-rich phase (see equation OPI) implies the incorporation of excess silica in the olivine. Stocker [1978b] has covered this situation in his equations (12) and (13) both of which create atomic sites. However, the only reduction reaction hypothesized was the exsolution of periclase. If $Si_I^{\cdot\cdot\cdot\cdot}$ and e' are the dominant defects, with $[e'] = 4 [Si_I^{\cdot\cdot\cdot\cdot}]$ as the neutrality condition, then

$$[e'] \propto PO_2^{-2/5} \qquad (A18)$$

which is different from that expected in the single phase regime in Equation (A12). This situation is also applicable in non-stoichiometric forsterite [see Figure 2 from Stocker 1978b] under extreme conditions.

<u>Regime C:</u>

<u>Two-phase regime with olivine composition changing to Fo_{100}.</u> The precipitation reactions mentioned above proceed until all iron

(and nickel) is removed from olivine. The end product is a highly defective olivine. The process is represented microstructurally at the advancing reaction zone boundary.

Regime D:

Coarsening of metallic precipitates and equilibration of forsterite. After removal of iron and nickel from olivine, a new process of precipitate coarsening begins [Boland, 1980; Boland and Duba, 1981]. This involves transfer of iron through olivine from smaller precipitates to larger, faster growing particles. Diffusive mass transfer is most likely via a vacancy mechanism.

Oxygen Vacancies

Normally, oxygen vacancies are expected to be of very minor importance. However, under non-stoichiometric conditions in regime D above, a new charge neutrality condition can readily develop. Excess silica may be incorporated by creating oxygen vacancies and magnesium vacancies according to

$$SiO_2 \rightleftarrows 2V_{Mg}'' + Si_{Si}^x + 2V_O^{\bullet\bullet} + 2O_O^x \quad (A19)$$

where mass-balance is:

$$K_{A19} = [V_{Mg}'']^2 [V_O^{\bullet\bullet}]^2 \quad (A20)$$

If charge neutrality changes from $[V_{Mg}''] = [V_O^{\bullet\bullet}]$, then from (A20) and (A7)

$$[e'] = 2[V_O^{\bullet\bullet}] \;\; \alpha \;\; PO_2^{-1/6} \quad (A21)$$

[see also Table I, Stocker 1978b].

Iron precipitation reactions are given in reactions (A13), (A14) and (A15). As with precipitation of Ni-rich particles, reactions (A13) and (A14) only require electron transport to reaction sites at the boundary. However, (A15) implies generation of oxygen gas. This gas may be dissolved in the olivine structure as a neutral species and transported to the surface via a vacancy mechanism, readily supplied in reaction (A6).

From the information available, the sign of the migrating species is not yet known. Equation (A6) implies generation of charged vacancies which may diffuse by an ambipolar diffusion process. Also, one may postulate the following reaction:

$$O_O^x \rightleftarrows \frac{1}{2} O_2 + V_O^x \quad (A22)$$

which generates uncharged vacancies. The dependence of such a reaction on oxygen fugacity is:

$$[V_O^x] \; \alpha \; PO_2^{-1/2} \quad (A23)$$

which is the same for charged vacancies according to (A8) but contrasts with the PO_2 dependence expected for the highly reducing conditions in the forsterite matrix - see Equation (A21).

Electronic Defects

So far, the only electronic defect considered is the electron. It is possible to rewrite the equations with electron holes, $h^{\bullet}$, in place of the electrons. However, hole concentrations are expected to be very small under the reducing conditions in these experiments. Under more oxidizing conditions, the dominant defect is $Fe_{Mg}^{\bullet}$ which is an effective electron hole. Contrary to the oxidizing model of Kohlstedt and Vander Sande [1975], it appears that electrons are more important under reducing conditions while holes or, in this case, small polarons [see Adler, 1973] are more numerous under oxidizing condition [see also Schock and Duba, this volume].

Brouwer Diagram

The PO_2 dependence of defects has been plotted in Figure A1 - a Brouwer-type diagram. Although actual concentrations are not known, their relative values have been estimated from chemical composition and plots given in Stocker [1978b]. Once the two-phase field is entered, Fe concentration is plotted as if the heterogeneous equilibrium curve - OPI - is followed. The non-stoichiometric forsterite, while having Mg/Si less than 2, may also have an oxygen deficiency which is considered much smaller than the cationic deficiency, simply because of the slower rate of diffusive transfer of oxygen from solid to gas phase [Jaoul et al., 1981; Reddy et al., 1980]. The discontinuities at the OPI and "reducing gas" boundaries are contrived to fit the two better-defined regimes A and D.

Appendix 2

Dislocation Pipe Diffusion

It is known that dislocations are high diffusivity pathways [Shewmon, 1963]. The extent of enhancement in diffusivity depends on the property being measured. It is possible to measure pipe diffusion along an isolated dislocation using the diffusive flux model shown

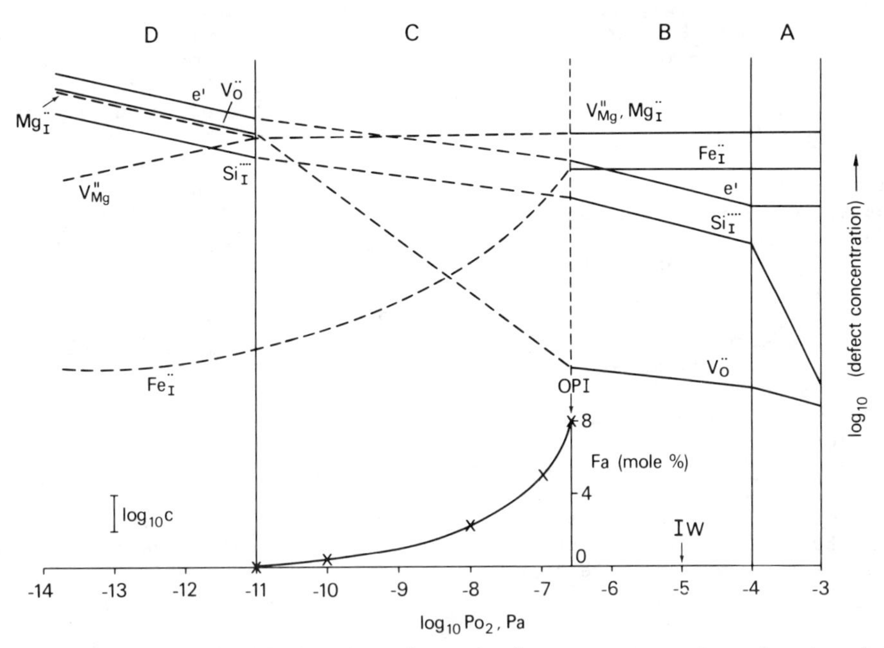

Figure A1. Brouwer diagram showing in schematic form concentration of point defects in four regimes A, B, C and D (see Appendix 1 for details).

in Figure 7. Following the same arguments with respect to the rate-controlling step for the selective internal reduction reactions, the ratio of dislocation pipe diffusion, D_p, to bulk diffusion, D_l, can be calculated from Equation (2), i.e.,

$$\frac{D_p}{D_l} = \left(\frac{X_p}{X_l}\right)^2 \qquad \text{(B1)}$$

where X_p is the depth at which precipitates are observed along the dislocation and X_l is the depth from the polished surface. For BSC-1, this ratio was shown to be 2.5×10^5.

Dislocation-Assisted Diffusion

The above process is specific to the dislocation itself and accounts for the location of precipitates at great depths in the crystal. There is another effect associated with dislocations referred to as dislocation-assisted diffusivity, D_a; it relates to the bulk effects of a high density of dislocation, ρ_d. Using the idea of pipe diffusion given by Shewmon [1963], we may write

$$D_a = D_p g + D_l \qquad \text{(B2)}$$

where g is a measure of enhancement due to dislocations. From Equations (B1) and (B2) we have

$$\frac{D_a}{D_l} = \left(\frac{X_a}{X_l}\right)^2 = 1 + \left(\frac{D_p}{D_l}\right) g \qquad \text{(B3)}$$

where X_a is the penetration depth of the reaction zone from the rough surface that has a high dislocation density in the matrix and X_l is the depth from the polished surface. For BSC-3 (see Figure 1), X_l = 10 μm and X_a = 20 μm, hence D_a/D_l is 4.

One can estimate g from the relationship:

$$g = \frac{\rho_d}{a}$$

where a is the fraction of the atoms per cm^2 that lie "inside" the cores of dislocations. If we allow about 10 atoms per dislocation, a is approximately $10^{14}/cm^2$. For ρ_d equal to $10^9/cm^2$, we can calculate that D_p/D_l is 3×10^5. This is in good agreement with the value obtained from Equation (2) in the text.

Acknowledgments. The helpful suggestions and critical comments on earlier versions of this research by R. N. Schock and R. L. Stocker are gratefully acknowledged. Discussions with T. J. Shankland and J. P. Poirier helped clarify certain points. The electron microscopy was funded by the Netherlands Organization for the Advancement of Pure Research (ZWO - WACOM). Financial support was provided by a North Atlantic Treaty Organization Grant (NATO RG 112-81). A portion of this work was supported by a grant from the Office of Basic Energy Science of the Department of Energy, and conducted under the auspices of Contract No. W-7405-ENG-48 at Lawrence Livermore National Laboratory.

References

Adler, D., Electronic configuration and electrical conductivity in ceramics, Am. Ceram. Soc. Bull. 52, 154-159, 1973.

Anderson, D. E., Diffusion in electrolyte mixtures, in Kinetics of Geochemical Processes, Rev. in Mineral., vol. 8, edited by A. C. Lasaga and R. J. Kirkpatrick, pp. 211-260, Mineralogical Society of America, Washington D. C., 1981.

Boland, J. N., Electron microscopy of mineral phase transformations in metamorphic reactions, in: Electron Microscopy, Vol. 1, Physics, edited by P. Brederoo and G. Boom, pp. 444-451, North-Holland, Amsterdam, 1980.

Boland, J. N., A. Duba, Solid-state reduction of iron in olivine - Planetary and meteoritic evolution, Nature 294, 142-144, 1981.

Boland, J. N. and A. Duba, The role of crystalline defects in the distribution of metallic particles in olivine: Meteoritic implications (abstract), Lunar Planet Science XIV, 57-58, 1983.

Boland, J. N., B. E. Hobbs, Microfracturing processes in experimentally deformed peridotite, Int. J. Rock Mech. Min. Sci. and Geomech. Abstr. 10, 623-626, 1973.

Boland, J. N., A. C. McLaren, and B. E. Hobbs, Dislocations associated with optical features in naturally-deformed olivine, Contrib. Mineral. Petrol., 29, 104-115, 1971.

Buening, D. K.and P. R. Buseck, Fe-Mg lattice diffusion in olivine, J. Geophys. Res., 78, 6852-6862, 1973.

Burns, R. G., Mineralogical Applications of Crystal Field Theory, Cambridge University Press, New York, 1970.

Darken, L. S., R. W. Gurry, Physical Chemistry of Metals, McGraw-Hill, New York, 1953.

Duba, A. G., and I. A. Nicholls, The influence of oxidation state on the electrical conductivity of olivine, Earth Planet. Sci. Lett. 18, 59-64, 1973.

Gaboriaud, R. J., M. Darrot, Y. Gueguen, and J. Woirgard, Dislocations in olivine indented at low temperatures, Phys. Chem. Minerals, 7, 100-104, 1981.

Haggerty, S. E., Oxidation of opaque mineral oxides in basalts. in Oxide Minerals, Rev. in Mineral., vol. 3, edited by D. Rumble III, pp. Hg1-Hg100 Minerlogical Scoiety of America, Washington, D. C., 1976.

Hobbs, B. E., The influence of metamorphic environment upon the deformation of minerals, Tectonophysics, 78, 355-383, 1981.

Humphreys, C. J., STEM imaging of crystals and defects, in Introduction to Analytical Electron Microscopy, edited by J. J. Hren, J. I. Goldstein, and D. C. Joy, pp. 305-322, Plenum, New York, 1979.

Jaoul, O., M. Poumellec, C. Froidevaux, and A. Havette, Silicon diffusion in forsterite: A new constraint for understanding mantle deformation, in Anelasticity in the Earth, Geodynamics Ser., vol. 4, edited by F. D. Stacey, M. S. Paterson, and A. Nicolas, pp. 95-100, AGU, Washington D. C., 1981.

Jaoul, O., B. Houlier, and F. Abel, Study of ^{18}O diffusion in magnesium orthosilicate by nuclear microanalysis, J. Geophys. Res. 88, 613-624, 1983.

Kofstad, D., High-Temperature Oxidation of Metals, John Wiley, New York, 1966.

Kohlstedt, D. L., and J. B. Vander Sande, An electron microscopy study of naturally occurring oxidation produced precipitates in iron-bearing olivines, Contrib. Mineral. Petrol., 53, 13-24, 1975.

Kracher, A., E.R.D. Scott, K. Keil, Dusty olivines in the Vigarano (CV3) chondrite: Evidence for an ubiquitous reduction process (abstract), Lunar Planet. Sci. XIV, 407-408, 1983.

Kröger, F. A., The Chemistry of Imperfect Crystals, North-Holland, Amsterdam, 1964.

Kröger, F. A., The Chemistry of Imperfect Crystals, vol. 2. Imperfection Chemistry of Crystalline Solids, 2nd rev. ed., North-Holland, Amsterdam, 1974.

McLaren, A. C., R. G. Turner, J. N. Boland, and B. E. Hobbs, Dislocation structure of the deformation lamellae in synthetic quartz; a study by electron and optical microscopy, Contrib. Mineral. Petrol., 29, 104-115, 1970.

Misener, D. J., Cationic diffusion in olivine to 1400°C and 35 kbar, Carnegie Inst. Washington Publ 634, 117-129, 1974.

Morioka, M., Cation diffusion in olivine-II, Ni-Mg, Mn-Mg, Mg and Ca, Geochim. Cosmochim. Acta, 45, 1573-1580, 1981.

Morse, S. A., Basalts and Phase Diagrams, Springer-Verlag, New York, 1980.

Nitsan, U., Stability field of olivine with respect to oxidation and reduction, J. Geophys. Res., 79, 706-722, 1974.

Nord, G. L., Analytical electron microscopy in mineralogy, exsolved phases of pyroxenes, Ultramicroscopy, 8, 109-120, 1982.

Reddy, K. P. R., S. M. Oh, L. D. Major, and A. R. Cooper, Oxygen diffusion in forsterite, J. Geophys. Res., 85, 322-326, 1980.

Schmalzried, H., Solid State Reactions, 2nd rev. ed., Verlag Chemie, Weinheim, Federal Republic of Germany, 1981.

Schmalzried, H. W., W. Laqua, and P. L. Lin, Crystalline oxide solid solutions in oxygen potential gradients, Z. Naturforsch. A, 34, 192-199, 1979.

Schock, R.N., and Duba, A. G., Point defects and the mechanisms of electrical conduction in olivine, in Point Defects in Minerals, Geophys. Monogr. Ser., edited by R. N. Schock, this volume.

Shankland, T. J., Electrical conduction in mantle materials, in: Evolution of the Earth, Geodynamics Ser., vol. 5, edited by R. J. O'Connell and W. S. Fyfe, pp. 256-263, AGU, Washington, D. C., 1981.

Shewmon, P. G., Diffusion in Solids, McGraw-Hill, New York, 1963.

Shewmon, P. G., Transformation in Metals, McGraw- Hill, New York, 1969.

Smyth, D. M., and R. L. Stocker, Point defects and non-stoichiometry in forsterite, Phys. Earth Planet. Inter., 10, 183-192, 1975.

Stocker, R. L., Point-defect formation parameters in olivine, Phys. Earth Planet. Inter., 17, 108-117, 1978a.

Stocker, R. L., Influence of oxygen pressure on defect concentrations in olivine with a fixed cationic ratio, Phys. Earth Planet. Inter., 17, 118-129, 1978b.

Stocker, R. L., Variation of electrical conductivity in enstatite with oxygen partial pressure: Comparison of observed and predicted behavior, Phys. Earth. Planet. Inter., 17, 34-40, 1978c.

Stocker, R. L., and D. M. Smyth, Effect of enstatite activity and oxygen partial pressure on the point-defect chemistry of olivine, Phys. Earth Planet.Inter., 16, 145-156, 1978.

Tuller, H. L., Mixed conduction in non-stoichiometric oxides, in Nonstoichiometric Oxides, edited by O.T. Sorensen, pp. 271-337, Academic, New York, 1981.

Wood, B. J., Crystal field electronic effects on the thermodynamic properties of Fe^{2+} minerals, in Thermodynamics of Minerals and Melts, edited by R. C. Newton, A. Navrotsky, and B. J. Wood, pp. 63-84, Springer-Verlag, New York, 1981.

COUPLED EXSOLUTION OF FLUID AND SPINEL FROM OLIVINE: EVIDENCE FOR O^- IN THE MANTLE?

H. W. Green II

Department of Geology, University of California, Davis
Davis, California 95616

Abstract. At least some of the carbon in the earth's upper mantle is dissolved in the silicates. Recent experimental studies on olivine by Freund and coworkers have indicated carbon solubility of several hundred ppm at atmospheric pressure and very rapid diffusivities. These workers have proposed that carbon and hydrogen dissolve in olivine via a process involving a charge transfer (CT) reaction in which O^- ions are formed. O^- is predicted to be much smaller than O^{2-}, hence pressure should enhance such reactions. To explore this hypothesis, I here consider the chemistry of composite fluid and solid precipitates in olivine from peridotite xenoliths in kimberlite pipes. The precipitates consist of hemispheres of CO_2-rich fluid and platelets of spinel. The two phases exsolved simultaneously, suggesting a link between the dissolution of carbon and the trivalent cations of the spinel. Qualitative analysis of the spinel shows that Al is subordinate to both Cr and Fe, suggesting that the exsolution process is essentially represented by

$$(Fe^{\bullet}_{Mg},Cr^{\bullet}_{Mg})_2C^{o}_{Si}O_4 + Fe^{o}_{Mg}V''_{Mg}C^{o}_{Si}O_4 \rightarrow Fe(Fe,Cr)_2O_4 + 2CO_2$$

if oxygen is exclusively O^{2-}, or

$$3(Fe^{o}_{Mg},Cr^{o}_{Mg})_2C''''_{Si}O^{\bullet}_4 \rightarrow 2Fe(Fe,Cr)_2O_4 + 2CO_2 + C$$

if O^- is involved. These two reactions differ in the fluid-solid ratio of the products and in the development of elemental carbon (or CO) in the second equation. The volumetric fluid-solid ratio of the precipitates is about 3:1, and the bubbles appear to have an amorphous film lining their surfaces, suggesting that the second equation is the better choice at this time. Quantitative analysis of the two phases should offer a clear choice. Previous studies involving pressure-induced reduction of aliovalent cations have invoked other CT mechanisms which are macroscopically equivalent to the O^- hypothesis. Other decompression-induced reactions and the controversy about the fO_2 of the mantle also may reflect operation of such CT mechanisms.

Introduction

Small amounts of carbon apparently are present throughout the earth's upper mantle. Evidence includes carbonatitic volcanism, CO_2 emanations accompanying volcanic eruptions of all types, fluid inclusions trapped in silicates within xenoliths [Roedder, 1965]; and phenocrysts [Delaney et al., 1978], graphite and diamond in xenoliths and kimberlite, and C-rich fluids precipitated in silicates of peridotite xenoliths [Green, 1972; Green and Radcliffe, 1975; Green and Gueguen, 1983]. The last of these observations on natural rocks indicates that at least some mantle carbon is dissolved in the silicates at high pressure. Two sets of recent experimental data bear on this problem. Ernst et al. [1983] have done high pressure equilibration studies on peridotite nodules with abundant CO_2 inclusions. They found that above 9.0 GPa, all fluid and carbonate dissolved into olivine. In addition, in an extensive study of carbon dissolution in MgO and olivine, F. Freund and co-workers [cf. Freund et al., 1980; Freund, 1981a, b; Oberheuser et al., 1983] report carbon solubilities in excess of several hundred ppm at atmospheric pressure and extraordinarily rapid diffusivities. Of particular importance in the latter studies is the prediction that charge transfer (CT) reactions involving the production of O^- are involved in dissolution of both hydrogen and carbon [Freund, 1981a, b]. Such CT reactions should be strongly enhanced by increased pressure because of the much smaller size of O^- compared to O^{2-}. If this hypothesis is correct, it could have major consequences for our understanding of the entire mantle (and core?).

In this paper, I shall examine a number of previously perplexing observations in the context of the possible existence of O^- in mantle silicates and show that these observations are consistent with the existence of O^- under mantle pressures. Indeed, some of these earlier studies have proposed essentially the same interpretation. The discussion will point to measurements and experiments which should be able to answer the question unambiguously.

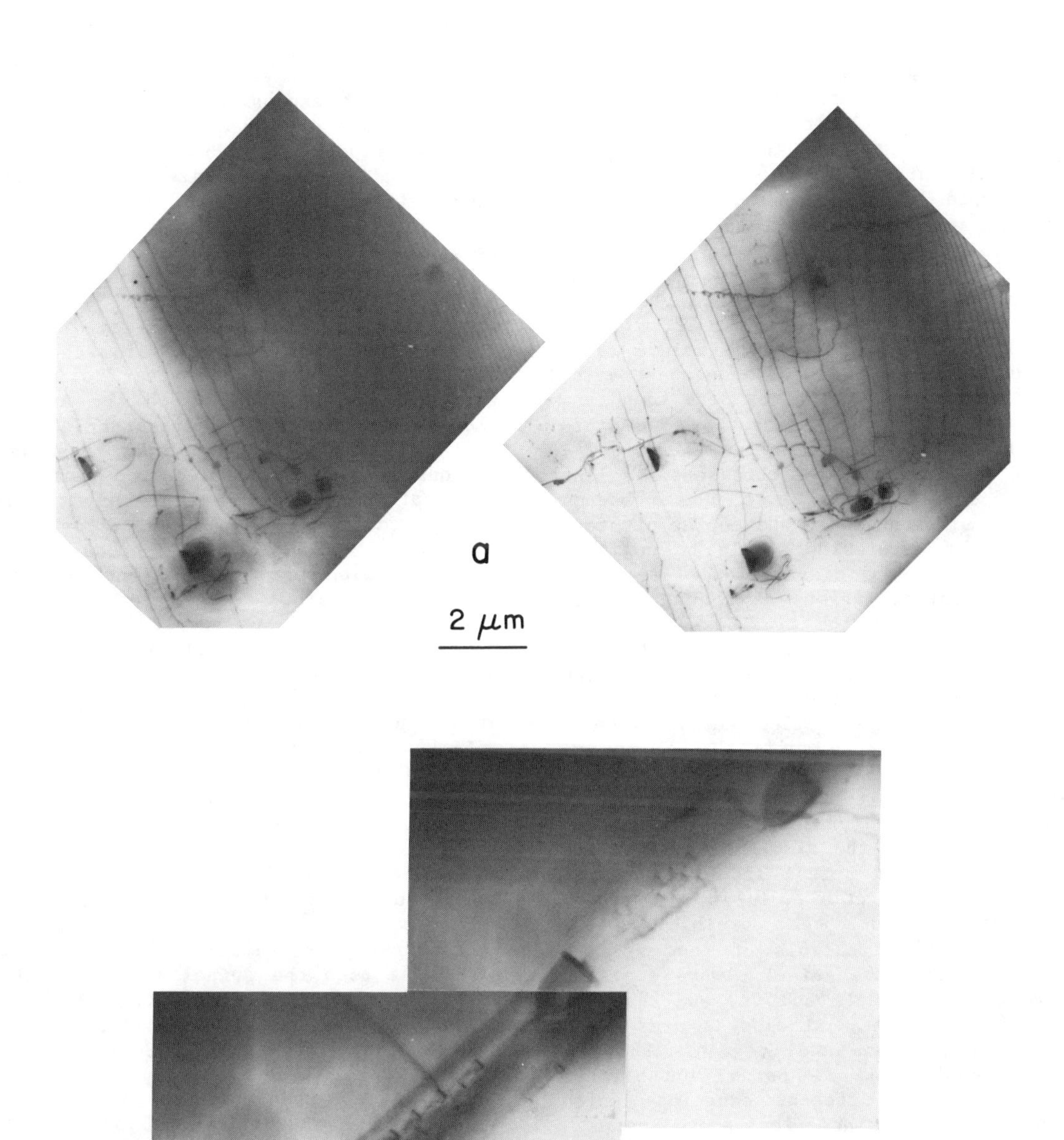

Fig. 1. Electron micrographs of precipitates on low-angle boundaries in olivine (1000 kV). (a) Stereo pair of (010) twist boundary in olivine; dislocations in strong contrast are [001] screws and those in weak contrast are [100] screws. Black triangles are {111} platelets of spinel lying in (100) of olivine. Note that all precipitates lie on dislocations; most lie at intersections. (b) Low angle boundary close to (001). All bubbles lie on dislocation intersections. Bubbles of a given size are equally spaced.

Precipitation Reactions

Green and Gueguen [1983] reported composite fluid and solid precipitates which form in olivine crystals of peridotite xenoliths during their journey to the surface in kimberlite magmas. The precipitates consist of hemispherical- to lozenge-shaped bubbles of CO_2-rich fluid and semicoherent platelets of spinel (Figure 1); nucleation and growth proceed continually throughout a significant portion of the path to the surface. In one specimen studied in great detail, the precipitates range in size from about 0.6μm to less than 0.01μm, yet the volume ratio remains about three parts fluid to one part spinel. No single phase precipitates were found and no other phases were found (except for an amorphous film lining the bubble surface; Figure 2). The simplest interpretation of these observations is that the constituents of both phases became supersaturated simultaneously and coprecipitated. Such simultaneous saturation should be unlikely unless the mechanisms of dissolution of the carbon and the trivalent cations of the spinel are related. (I have suggested previously that the details of spinel morphology might indicate bubble initiation preceded spinel nucleation [Green and Gueguen, 1983]. However, such consecutive precipitation is not reconciled easily with a constant ratio of fluid to solid through an extended precipitation history, and the observed morphology is equally consistent with simultaneous growth of the two phases.) This would be precisely analogous to the simultaneous precipitation of pyroxene and spinel from olivine reported previously [Bell et al., 1975].

Although it has not been possible yet to obtain a quantitative analysis of the exsolved phases, we know that the spinel always contains major concentrations of Fe and Cr, and that the fluid is rich in carbon. Al also is present in some precipitates [Green and Gueguen, 1983; H. Green, P. Vaughan, and R. Borch, unpublished data, 1983]. Four different equations which satisfy the observations described above are:

$$\begin{aligned} &Fe^{\circ}_{Mg}(Fe^{\bullet}_{Mg},Cr^{\bullet}_{Mg},Al^{\bullet}_{Mg})Al'_{Si}O_4 \\ &+ (Fe^{\bullet}_{Mg},Cr^{\bullet}_{Mg},Al^{\bullet}_{Mg})_2C^{\circ}_{Si}O_4 + (Fe^{\circ}_{Mg},Cr^{\circ}_{Mg})V''_{Mg}C^{\circ}_{Si}O_4 \\ &\rightarrow 2Fe(Fe,Cr,Al)_2O_4 + 2CO_2 \end{aligned} \qquad (1)$$

$$\begin{aligned} &Fe^{\circ}_{Mg}(Fe^{\bullet}_{Mg},Cr^{\bullet}_{Mg}Al^{\bullet}_{Mg})Al'_{Si}O_4 + (Fe^{\circ}_{Mg},Cr^{\circ}_{Mg})C^{\bullet\bullet}_{Mg}Al'_{Si}O_4 \\ &+ C^{\bullet\bullet}_{Mg}V''_{Mg}Al'_{Si}O_4 \rightarrow 2Fe(Fe,Cr,Al)_2O_4 + 2CO_2 \end{aligned} \qquad (2)$$

$$\begin{aligned} &2Fe^{\circ}_{Mg}(Fe^{\bullet}_{Mg},Cr^{\bullet}_{Mg},Al^{\bullet}_{Mg})Al'_{Si}O_4 \\ &+ 3(Fe^{\circ}_{Mg},Cr^{\circ}_{Mg})_2C''''_{Si}O^{\bullet}_4 \\ &\rightarrow 4Fe(Fe,Cr,Al)_2O_4 + CO_2 + 2CO \end{aligned} \qquad (3)$$

$$\begin{aligned} &3Fe^{\circ}_{Mg}C''_{Mg}Al'_{Si}O^{\bullet}_4 + 3(Fe^{\circ}_{Mg},Cr^{\circ}_{Mg})_2Al'_{Si}O_4 \\ &\rightarrow 5Fe(Fe,Cr,Al)_2O_4 + CO_2 + 2CO \end{aligned} \qquad (4)$$

I here utilize the effective charge notation of Kroger [1974], where a superscript dot represents a defect electron (hole) compared to the normal occupancy of a given site, a superscript slash represents an excess electron, and a superscript zero signifies no excess charge. Equations (1) and (2) represent carbon solubility in tetrahedral and octahedral sites, respectively, with oxygen in all cases O^{2-}. Carbon and trivalent cations in octahedral sites are stabilized via vacancies on other octahedral sites and by Al substitution for Si. Equations (3) and (4) are analogous to equations (1) and (2) except that on the left side, carbon is effectively neutral; it has accepted four electrons from the surrounding oxygens, converting the latter to O^-. The oxygens are reduced to O^{2-} on the right side by electron transfer. The transfer is hypothesized to take place in part from four divalent Fe or Cr cations, producing the required number of trivalent cations for the spinel and leaving a partially reduced fluid. A comparison of equations (1) and (3) with (2) and (4) shows that the spinel composition is a strong function of the site assumed for carbon. The absence of Si in the precipitated products requires extensive substitution of Al in the tetrahedral site if carbon is assumed to occupy octahedral sites. As a consequence, equations (2) and (4) require a minimum of 75 and 60% Al respectively in the trivalent spinel site, whereas equations (1) and (3) easily can be written with no Al at all (as in the Abstract); Al was included because it is present in some precipitates. We see, therefore, that the presence of O^- at high pressure predicts a partially reduced fluid and that octahedral carbon requires highly aluminous spinels (unless significant Fe^{3+} could be accommodated in the tetrahedral site of olivine). Quantitative analysis of both the spinel and the fluid present severe analytical difficulties but are being pursued actively. At this point, the qualitative evidence is : (1) both the Cr and Fe contents of the spinel are greater than the Al content, suggesting that carbon occupies the tetrahedral site and pointing to equations (1) or (3) as the best fit (2) the amorphous film visible on bubble surfaces (Figure 2) suggests the reaction

$$2CO \rightarrow C + CO_2 \qquad (5)$$

may have proceeded during the trip to the surface. If this is correct, then equation (3) is indicated and O^- survives the first test. Evidence for $C+CO_2$ in xenoliths also has been reported previously by Mathez and Delaney (1981) and by Watanabe et al., (1982). The latter authors report carbon isotope measurements which imply equilibrium between the two phases.

However, the reaction

$$CO_2 \rightarrow C + O_2 \qquad (6)$$

during upwelling is also a possible origin for an amorphous carbon film (Mathez and Delaney, 1981). Equation (5) has a significantly negative ΔV,

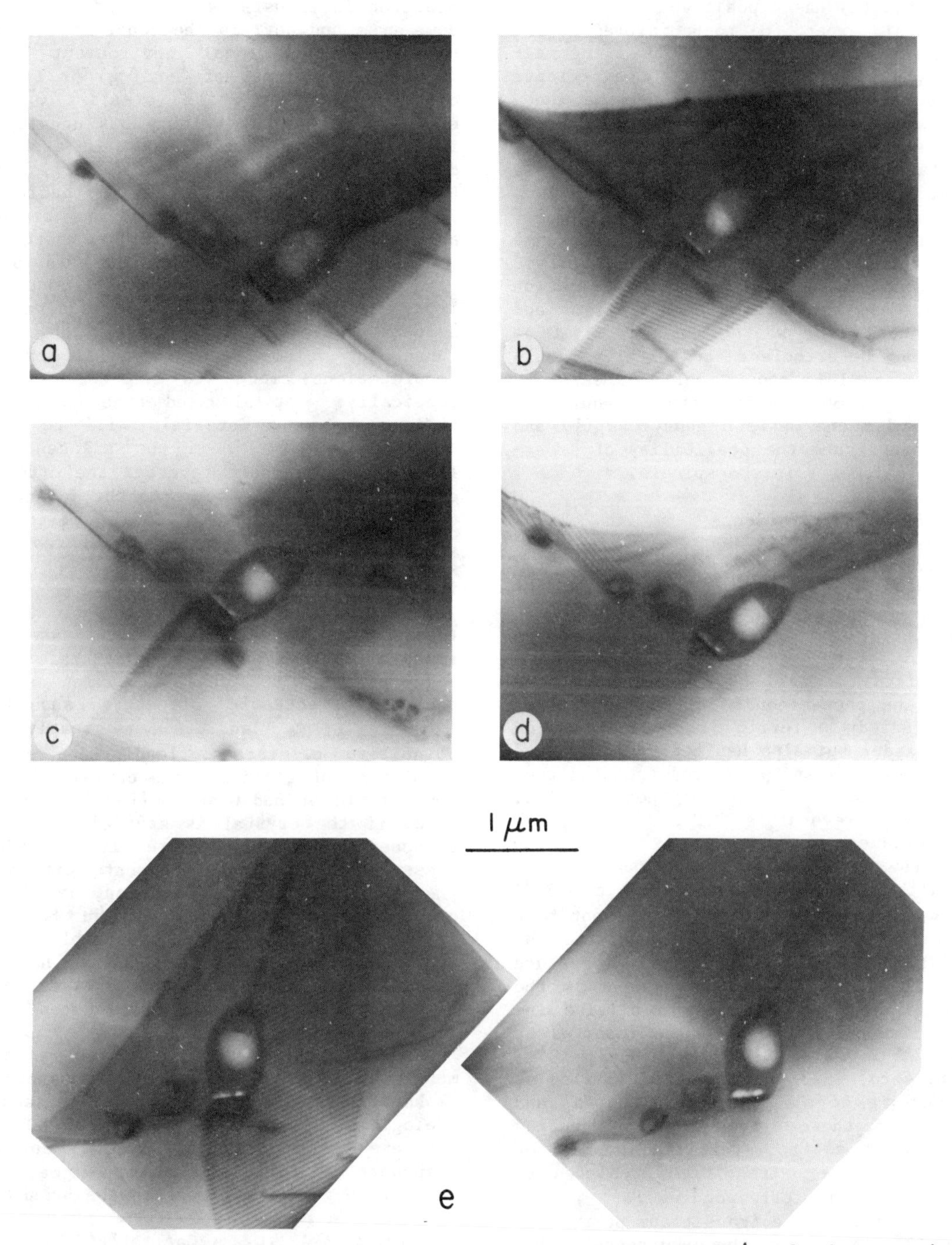

Fig. 2. Series of electron micrographs of a large bubble located at the intersection of two subgrain boundaries, detailing the nature of the surface film and its rearrangement during irradiation in the electron beam (1000kV). Approximately 5 minutes between exposures. (a) Bubble surface evenly mottled. (b) Surface somewhat less mottled; white areas appearing along spinel lamella. (c) White areas along spinel enlarged. (d) White spot appears near center of bubble. (e) Stereo pair showing three-dimensional relationships. Note that mottling is on bubble surface, not within it. White spot of Figure 2d is enlarged in left member of Figure 2e and still more enlarged in right member of Figure 2e. Varying contrast on dislocations and bubble wall is a result of specimen tilting between exposures.

whereas equation (6) has a positive ΔV. At first glance, one would expect (6) to be favored during decompression of a xenolith in an upwelling magma. However, fluid inclusions do not experience directly the reduction of pressure on the xenolith. The excess pressure in the fluid produces stress (and strain) concentrations at the fluid-solid interface which can exceed the yield stress of the host crystal, punching out dislocations and/or producing microcracks [Green and Gueguen, 1983]. Operation of equation (5) would decrease the stress concentrations and should be favored under these conditions, but operation of equation (6) would increase the stress concentrations and hence it is not clear whether or not it is a likely reaction. Nevertheless, quantitative analysis of both the fluid and spinel are necessary to distinguish unambiguously between equations (5) and (6) and hence between equations (1) and (3) and to investigate the possibility of departure from stoichiometry in the spinels.

Other Evidence

A number of other surprising observations reported of recent years may bear on the O^- problem. Paramount among these is the pressure-induced reduction of aliovalent cations at high pressures [e.g. Burns et al., 1972a]. Burns [1975] summarized the evidence for many of these transitions. The phenomenon has been particulary thoroughly studied in ferric iron-bearing halogenides and oxides but also has been demonstrated in five other systems and predicted from indirect evidence for six more. In most of these studies, the phenomenon is reversible. For example, high pressure Mössbauer spectroscopy shows disappearance of the Fe^{3+} peak with increasing pressure, but with reduction of pressure, the Fe^{3+} peak returns. Although such studies have not been done on olivine, Burns (1975) has predicted the $Cr^{2+} \rightarrow Cr^{3+}$ transition in olivine during pressure reduction. The basis of Burns' argument is the measurement of high Cr contents in olivine inclusions in diamonds. The data are reported as up to 0.16 wt % Cr_2O_3 [Meyer and Boyd, 1972; Meyer and Svisero, 1973; Sobolev, 1972; Prinz et al., 1973; Hervig et al., 1980] but the Na^+ and Al^{3+} contents of these olivines are low,ruling out interpretation as $Cr^{\bullet}_{Mg}$ stabilized by Na'_{Mg} or Al'_{Si}. The $Cr^{2+} \rightarrow Cr^{3+}$ transition predicted by Burns (1975), and the corresponding $Fe^{2+} \rightarrow Fe^{3+}$ transition are exactly the transitions postulated in equation (3). Moreover, the occurrence as inclusions in diamond should guarantee carbon saturation of the olivine. The conventional interpretation of the results is that Cr^{2+} implies a reduced oxygen fugacity (fO_2) [Hervig et al., 1980], but Burns (1975) argued that Cr^{2+} could be stabilized in olivine by pressure-enhanced distortion of the M1 site and might not imply reduced oxygen fugacity in the mantle.

The Mössbauer spectra of iron-bearing compounds show only half of what is happening at high pressure. As the pressure is raised, an electron is transferred onto Fe^{3+}. Where does it come from and why? Is the donor the general negatively charged local environment (reflecting increased covalency of bonding) or is it a specific neighboring anion (creating O^-, F°, Cl°, etc.)? Increased covalency has usually been the implicit interpretation, but in a particularly detailed study on a synthetic ferric amphibole, Burns et al., [1972b] suggested the formation of OH° as a new defect species and proposed that such anion oxidation could be responsible for published alkali-deficient analyses of natural glaucophanes. The arguments of Freund [1981a, b] imply that it is a simple consequence of pressure-induced "oxidation" of O^{2-} to O^- and consequent volume reduction. All of these interpretations amount to the same thing macroscopically: At elevated pressures, aliovalent cations in many materials will be reduced by electron transfer from the adjacent anion(s); lowering of pressure reverses the effect and the electron transfers back onto the anion(s). Whatever the specific nature of the CT reaction, the high pressure state also should favor electron transfer onto carbon and significant increase in its solubility.

If the pressure is reduced while the temperature remains elevated, as is the case for unconfined olivine crystals in xenoliths, equation (3) suggests that the destabilization of the high pressure electronic structure should produce oxidation of Fe^{2+} and Cr^{2+} and greatly reduce the solubility of carbon, leading to simultaneous exsolution of fluid and the trivalent cations (as found by Green and Gueguen [1983]). On the other hand, if the crystal is stuffed with carbon and quenched to low temperature without reduction of pressure (due to entrapment within diamond [Leung, 1974; Leung and Manson, 1974], the high pressure state conceivably might be frozen in. If the CT reaction is not quenchable, the olivine inclusions in diamond also should have precipitates similar to those found by Green and Gueguen [1983] or lattice distortions induced by enlargement of oxygen ions.

Another paradoxical observation concerning mantle rocks involves strongly nonstoichiometric omphacitic pyroxenes from kyanite and coesite ecologites reported by Smyth [1980]. These pyroxenes break down very rapidly during upward transport in the kimberlite magma, exsolving quartz. Smyth [1980] modeled the breakdown as:

$$CaV''Al^{\bullet}_2Si_4O_{12} \rightarrow CaAl^{\bullet}Al'SiO_6 + 3SiO_2 \quad (7)$$

The essence of this reaction is that the solubility of silica is greatly enhanced at high pressure and that the solubility is accommodated by shifting additional Al to six-fold coordination in M1, and stabilizing a vacancy in M_2 for each such transition of Al. A vacancy is inherently a dilational phenomenon. Creation of a vacancy requires removal of an ion to a grain boundary, where its volume will be little different from that in the crystal interior. Thus, the volume of the vacancy itself, however small, amounts to an expansion of the structure. As a consequence, although (7) is a balanced reaction and apparently provides an adequate chemical description of the breakdown process, it is difficult to understand why the high pressure pyroxene containing the vacancy-bearing molecule on the left is 4% more dense than the vacancy-free pyroxene on the right [Smyth, 1980]. Formally, one can say that the partial molar volume of SiO_2 must be sufficiently small to more than offset the dilatant effect of the vacancies, but why should silica be so dense? Coesite, which coexists with the high pressure pyroxene has a specific gravity of 2.9. One possibility is that omphacite, like olivine, has a certain amount of O^- stabilized at high pressure, and that the dissolution of significant silica is coupled to O^- formation, analogous to the enhanced carbon solubility represented by equations (3) and (4). In fact, the data do not rule out carbon involvement in these pyroxenes; Smyth [1980] identified the breakdown products by x-ray precession photography, for which fluids are invisible. Optical examination of similar pyroxenes in my collections shows abundant secondary fluid inclusions.

Discussion

Burns' [1975] argument that Fe^{3+} is not to be expected in mantle olivine because of the pressure-induced reduction phenomenon makes it very difficult to propose equation (1) or (2) to explain the spinel precipitates. On the other hand, the O^- (or equivalent high pressure electronic structure) postulated in equations (3) and (4) predicts reduced cations and provides a ready explanation for their decompression-induced oxidation. Taken by itself, the reduction of trivalent cations produces an expansion, hardly what one would expect to occur spontaneously with increasing pressure. Coupled with the large volume decrease of O^- formation, however, the other half of the CT reaction is identified, and the pressure dependence makes sense. The identification of Cr^{2+} in mantle olivine provides further support to the argument.

Equations (1) and (2) for olivine and equation (7) for clinopyroxene are analogous in that their left-hand members are hypothetical nonstoichiometric high pressure phases which have been proposed to explain observed low-pressure exsolution products. The CT hypothesis is an alternative explanation which requires no (or lesser) departure from stoichiometry, while simultaneously providing an easily understandable explanation for the pressure dependence of the reactions.

The CT reactions discussed here also may bear on the current controversy concerning the oxygen fugacity of the mantle reviewed recently by Haggerty and Hopkins [1983]. Briefly, the major pieces of the puzzle are: (1) Fe^{3+}/Fe^{2+} ratios in mantle-derived magmas yield high fO_2's (near the quartz-fayalite-magnetite buffer); (2) Spinel peridotite xenoliths of the suboceanic mantle show very low intrinsic oxygen fugacities (near iron-wüstite), and (3) Ilmenite-bearing assemblages from the subcontinental mantle give intermediate fO_2's, both measured directly and calculated from phase equilibria (approximately wüstite-magnetite). Haggerty and Hopkins [1983] postulate a layered structure to the mantle, in which the shallow mantle source of basalts is at QFM, the residual depleted mantle of the oceanic lithosphere is at IW and the more fertile upper mantle at greater depths is at MW. Some authors (e.g. Sato, 1978; Arculus and Delano, 1981) have questioned the reliability of the high fO_2's of magmas. The apparent incompatibility between reduced suboceanic mantle and oxidized magmas has been rationalized by Sato (1978) as due to the reaction

$$2FeO + H_2O \rightarrow Fe_2O_3 + H_2\uparrow \qquad (8)$$

where the loss of H_2 yields oxidized magmas from reduced source material. As an alternative, the discussion presented here suggests the reactions

$$2Fe^{2+}_{Mg} + [C^{\prime\prime}_{Si}{}^{\prime\prime}O^{\bullet}_4]^{4-} \rightarrow Fe_2O_3 + CO \qquad (9)$$

and (equation (5))

$$2CO \rightarrow C + CO_2\uparrow$$

whereby reduced iron is oxidized, graphite is produced and CO_2 is lost from the system. A formulation equivalent to (9) is easily written with H_2O replacing CO_2. The intermediate fO_2's involve ilmenite -- exactly the crystal structure predicted by Burns et al. [1972a] as the structure most likely to stabilize Fe^{3+} in the mantle. If CT reactions involving oxygen are important in the upper mantle, then the whole concept of oxygen fugacity becomes confused [Freund, 1981b]. At best, it appears that the problem of the fO_2 of the mantle remains open.

Conclusions

The foregoing observations and arguments are not definitive, but they present a simple scenario coupling diverse phenomena: (1) the O^- hypothesis (or its equivalent increase in covalency) provides a convenient explanation of why and how carbon might dissolve in olivine at high pressure; (2) decompression-induced instability of the O^- (or covalent bond) would naturally lead to the observed simultaneous exsolution of fluid and spinel in xenoliths, (3) the predicted partially reduced fluid is consistent with the amorphous film deposited on bubble surfaces in these xenoliths, and (4) the systematics of the observed and predicted CT reactions seem to have parallels with the paradoxical fO_2 data for the mantle.

If further work confirms O^- as a significant species in the mantle, then many trace elements currently considered "incompatible" with mantle silicates may be soluble at high pressure. As a consequence, interpretations of melting and metasomatic phenomena based on the low pressure behavior of these elements may need revision.

The presentation here suggests a series of tests to the hypothesis of pressure-induced stabilization of O^- in the mantle:

(1) Quantitative determination of the composition of fluid and spinel precipitates in mantle olivine.

(2) Determination of the high pressure solubility and diffusivity of carbon in olivine.

(3) Analysis of the carbon content of Cr^{2+}-bearing olivines and testing for O^-. If O^- is metastably preserved in the lattice, heating under vacuum or wet chemical analysis should convert Cr^{2+} to Cr^{3+} and release CO or produce C.

(4) Transmission electron microscopical examination of omphacitic pyroxenes like those studied by Smyth [1980] should be able to verify the presence or absence of a fluid involved in the exsolution reaction (7).

The first two of these tests currently are being initiated in my laboratory.

References

Arculus, R. J., and J. W. Delano, Intrinsic oxygen fugacity measurements: techniques and results for spinels from upper mantle peridotites and megacryst assemblages, Geochim. Cosmochim. Acta, 45, 899-913, 1981.

Bell, P. M., H. K. Mao, E. Roedder, and P. W. Weiblen, The problem of the origin of symplectites in olivine-bearing lunar rocks, Proc. Lunar Planet. Sci. Conf. 6th, 1, 231-236, 1975.

Burns, R. G., On the occurrence and stability of divalent chromium in olivines included in diamonds, Contrib. Mineral. Petrol., 51, 213-221, 1975.

Burns, R. G., F. E. Huggins, and H. G. Drickamer, Applications of high-pressure Mössbauer spectroscopy of mantle mineralogy, Proc. Int. Geol. Congr. 24th, Sect. 14, 113-123, 1972a.

Burns, R. G., J. A. Tossell, and D. J. Vaughan, Pressure-induced reduction of a ferric amphibole. Nature, 240, 33-35, 1972b.

Delaney, J. R., D. W. Muenow, and D. G. Graham, Abundance and distribution of water, carbon and sulfur in the glassy rims of submarine pillow basalts, Geochim. Cosmochim. Acta., 42, 581-594, 1978.

Ernst, T., R. B. Schwab, B. Scheubel, and P. Brosch, High pressure experiments on the equilibrium olivine-CO_2-enstatite-carbonate in the natural system, in High Pressure in the Geosciences, edited by W. Schreyer, Nagele Verlag, Stuttgart, Federal Republic of Germany, in press, 1983.

Freund, F., Charge transfer and O^- formation in high and ultrahigh pressure phase transitions, Bull. Mineral., 104, 177-185, 1981a.

Freund, F., Mechanism of the water and carbon dioxide solubility in oxides and silicates and the role of O^-, Contrib. Mineral. Petrol., 76, 474-482, 1981b.

Freund, F., H. Kathrein, H. Wengeler, R. Knoble, and G. Demortier, Atomic carbon in magnesium oxide, I, Carbon analysis by the $^{12}C(d,p)^{13}C$ method, Mater. Res. Bull., 15, 1011-1018, 1980.

Green, H. W., A CO_2-charged asthenosphere. Nature Phys. Sci., 238, 2-5, 1972.

Green, H. W., and Y. Gueguen, Deformation of peridotite in the mantle and extraction by kimberlite: A case history documented by fluid and solid precipitates in olivine, Tectonophysics, 92, 71-92, 1983.

Green, H. W., and S. V. Radcliffe, Fluid precipitates in rocks from the earth's mantle, Geol. Soc. Am. Bull., 86, 846-852, 1975.

Haggerty, S. E., and L. A. Hopkins, Redox state of earth's upper mantle from kimberlitic ilmenites, Nature, 303, 295-300, 1983.

Hervig, R. L., J. V. Smith, I. M. Steele, J. J. Gurney, J. O. A. Meyer, and J. W. Harris, Diamonds: Minor elements in silicate inclusions: Pressure-temperature implications. J. Geophys. Res., 85, 6919-6929, 1980.

Kroger, F. A., The Chemistry of Imperfect Crystals, vol. 2, 988 pp., Elsevier, New York, 1974.

Leung, I. S., An x-ray study of diopside inclusions in natural diamonds, Geol. Soc. Am. Abstr. Programs, 6, 842-843, 1974.

Leung, I. S., and D. V. Manson, Orientations of olivine inclusions in natural diamond (abstract), Eos Trans. AGU, 55, 481-482, 1974.

Mathez, E. A., and J. R. Delaney, The nature and distribution of carbon in submarine basalts and peridotite nodules, Earth Planet. Sci. Letts., 56, 217-232, 1981.

Meyer, H. O. A. and F. R. Boyd, Composition and origin of crystalline inclusions in natural diamonds, Geochim. Cosmochim. Acta., 36, 1255-1273, 1972.

Meyer, H. O. A., and D. P. Svisero, Mineral inclusions in Brazilian diamonds, Phys. Chem. Earth, 9, 785-796, 1975.

Oberheuser, G., H. Kathrein, G. Demortier, H. Gonska, and F. Freund, Carbon in olivine single crystals analyzed by the $^{12}C(d,p)^{13}C$ method and by photoelectron spectroscopy, Geochim. et Cosmochim. Acta., 47, 1117-1130, 1983.

Prinz, M., D. V. Manson, P. F. Hlava, and K. Keil, Inclusions in diamonds: garnet lherzolite and eclogite assemblages, Phys. Chem. Earth, 9, 797-816, 1975.

Roedder, E., Liquid CO_2 inclusions in olivine-bearing nodules and phenocrysts from basalts. Am. Mineral., 50, 1746-1782, 1965.

Sato, M., Oxygen fugacity of basaltic magmas and the role of gas-forming elements, Geophys. Res. Letts., 5, 447-449, 1978.

Smyth, J. R., Cation vacancies and the crystal chemistry of breakdown reactions in kimberlitic omphacites. Am. Mineral., 65, 1185-1191, 1980.

Sobolev, N. V., Petrology of xenoliths in kimberlite pipes and indications of their abyssal origin, Proc. Int. Geol. Congr. 24th, Sect. 2, 297-302, 1972.

Watanabe, S., K. Mishima, and S. Matsuo, Chemical form, amount and isotope ratio of carbon bearing compounds in olivine crystal, paper presented at the Fifth Int. Conf. on Geochronology, Cosmochronology, and Isotope Geol., Nikko Natl. Park, Japan, 1982.